Recent Advances in Boundary Element Methods

G.D. Manolis · D. Polyzos
Editors

Recent Advances in Boundary Element Methods

A Volume to Honor Professor Dimitri Beskos

Springer

Editors
George D. Manolis, Professor
Aristotle University
Department of Civil Engineering
54124 Thessaloniki
Greece
gdm@civil.auth.gr

Demosthenes Polyzos, Professor
University of Patras
Department of Mechanical Engineering
and Institute of Chemical
Engineering & High Temperature Processes
265 00 Patras
Greece
polyzos@mech.upatras.gr

ISBN 978-90-481-8190-2 e-ISBN 978-1-4020-9710-2

DOI 10.1007/978-1-4020-9710-2

Printed on acid-free paper

9 8 7 6 5 4 3 2 1

springer.com

This book is dedicated to Dimitri Beskos, his family and his friends.

Editorial

This task of preparing a volume to honor Dimitri Beskos on occasion of his 65th birthday was something I undertook with zest. It indirectly brought me back to the days when I was a graduate student in the Civil Engineering department at the University of Minnesota in Minneapolis, when Dimitri looked after my welfare and I, in return, ploughed through the boundary element fields trying to produce something interesting and worthwhile in elastodynamics. I was not sure at first if anything would come out, but Dimitri had indeed chosen a gem for a doctoral subject, something that started to dawn on me as time progressed. The endeavor had a happy end, I finished quickly and almost instantly I got a job as an assistant professor with the Civil Engineering department at the State University of New York in Buffalo. There I joined a group of BEM people, namely Dick Shaw, Prashanta Banerjee (the senior people) and Trevor Davies (my contemporary) and continued work among similarly-minded people. As time went on, Dimitri made a name for himself and eventually returned to Greece. I followed a number of years later, and to this day we have cooperated, off and on, on a number of BEM-related research subjects. In sum, this volume is a small tribute to the man who guided me into academics and was more of a friend than supervisor to me during the last 30 years.

George D. Manolis

I had already been appointed lecturer at the Mechanical Engineering department in Patras, having complete my doctoral degree on elastic wave scattering in particulate composite materials and in search of new research topics to focus on when I ran into Dimitri Beskos. He had been appointed chaired professor in the neighboring Civil Engineering department, and the news was that here is a young professor from the US with many research ideas springing from his head and a formidable research record. His pleasant, kind and mild-mannered personality, as well as his ability to be extremely helpful to students and colleagues immediately convinced me that Dimitri is an ideal person to work with. Since then and for the last 15 years we have had an intimate friendship and a strong collaboration on BEM-related subjects. Dimitri is an excellent friend, both at the professional and personal level. I consider this friendship a great privilege for me and as my colleague George Manolis says, this volume is a small tribute to a man who gave so much to his friends, colleagues and students.

Demosthenes Polyzos

Contents

Contributors

E. L. Albuquerque Faculty of Mechanical Engineering, State University of Campinas, Campinas, Brazil, ederlima@fem.unicamp.br

M. H. Aliabadi Department of Aeronautics, Imperial College London, London, UK, m.h.aliabadi@imperial.ac.uk

Nick K. Anifantis Mechanical and Aeronautics Engineering Department, University of Patras, GR-26500, Greece, nanif@mech.upatras.gr

S. N. Atluri Center for Aerospace Research & Education, University of California, Irvine, 5251 California Avenue, Suite 140, Irvine, CA, 92612, USA, satya.atluri@gmail.com

P. M. Baiz Department of Aeronautics, Imperial College London, London, UK, p.m.baiz@imperial.ac.uk

Milind Bapat Department of Mechanical Engineering, University of Cincinnati, Cincinnati, Ohio 45221-0072, USA, bapatms@email.uc.edu

Gernot Beer Institute for Structural Analysis, Graz University of Technology, Lessingstrasse 25/II, 8010 Graz, Austria, gernot.beer@tugraz.at

Marc Bonnet LMS, Ecole Polytechnique, France, bonnet@lms.polytechnique.fr

Dominik Brunner Institute of Applied and Experimental Mechanics, Pfaffenwaldring 9, 70550 Stuttgart, Germany, brunner@iam.uni-stuttgart.de

Alexis Rodrguez Carranza Faculty for Physics and Mathematics, Universidad Nacional de Trujillo, Trujillo, Peru, alexlar20@yahoo.es

Stéphanie Chaillat LMS, Ecole Polytechnique, France, chaillat@lms.polytechnique.fr

Jeng-Tzong Chen Department of Mechanical and Mechatronic Engineering, Department of Harbor and River Engineering, National Taiwan Ocean University, Keelung, Taiwan, jtchen@mail.ntou.edu.tw

Evgenia Christodoulou Department of Mathematics and Statistics, University of Cyprus, PO Box 20537, 1678 Nicosia, Cyprus, map4ec3@yahoo.gr

Adrian P. Cisilino Welding and Fracture Division - INTEMA, Faculty of Engineering, University of Mar del Plata, Mar del Plata, Argentina, cisilino@fi.mdp.edu.ar

Marco Dondero Welding and Fracture Division - INTEMA, Faculty of Engineering, University of Mar del Plata, Mar del Plata, Argentina, mdondero@fi.mdp.edu.ar

C.Y. Dong Department of Mechanics, School of Science, Beijing Institute of Technology, Beijing 100081, China, cydong218@yahoo.com.cn

Marijan Dravinski USC Viterbi School of Engineering, University of Southern California, Los Angeles, CA 90089-1453, USA, mdravins@usc.edu

Attilio Frangi Department of Structural Engineering, Politecnico of Milano, Milano, Italy, e-mail: attilio.frangi@polimi.it

F. García-Sánchez Department of Civil Engineering, of Materials and Manufacturing, University of Málaga, 29013 Málaga, Spain, fgsanchez@uma.es

Lothar Gaul Institute of Applied and Experimental Mechanics, Pfaffenwaldring 9, 70550 Stuttgart, Germany, gaul@iam.uni-stuttgart.de

Georgios Georgiou Department of Mathematics and Statistics, University of Cyprus, PO Box 20537, 1678 Nicosia, Cyprus, georgios@ucy.ac.cy

Georgios I. Giannopoulos Mechanical and Aeronautics Engineering Department University of Patras, GR-26500, Greece, g-giann@mech.upatras.gr

S. Guimarães Programa de Engenharia Civil, COPPE/UFRJ, Caixa Postal 68506, CEP 21945-970, Rio de Janeiro, RJ, Brazil, sol@coc.ufrj.br

Z. D. Han Center for Aerospace Research & Education, University of California, Irvine, 5251 California Avenue, Suite 140, Irvine, CA, 92612, USA, han.zhidong@gmail.com

George D. Hatzigeorgiou Department of Environmental Engineering, Democritus University of Thrace, GR-67100, Xanthi, Greece , gchatzig@env.duth.gr

Nikolaos I. Ioakimidis Division of Applied Mathematics and Mechanics, Department of Engineering Sciences, School of Engineering, University of Patras, GR-265.04, Rion, Patras, Greece, n.ioakimidis@upatras.gr; n.ioakimidis@otenet.gr

Michael Junge Institute of Applied and Experimental Mechanics, Pfaffenwaldring 9, 70550 Stuttgart, Germany, junge@iam.uni-stuttgart.de

Andreas Karageorghis Department of Mathematics and Statistics, University of Cyprus, P.O.Box 20537, 1678 Nicosia, Cyprus, andreask@ucy.ac.cy

G.F. Karlis Department of Mechanical and Aeronautical Engineering, University of Patras, GR-26500 Patras, Greece, gerasimos.karlis@gmail.com

John T. Katsikadelis Office of Theoretical and Applied Mechanics, Academy of Athens & School of Civil Engineering, National Technical University, Athens, GR-15773, Greece, jkats@central.ntua.gr

Loukas K. Keppas Mechanical and Aeronautics Engineering Department, University of Patras, GR-26500, Greece, lkeppas@mech.upatras.gr

Lars Kielhorn Institute of Applied Mechanics, Graz University of Technology, Austria, l.kielhorn@tugraz.at

Vladimr Kompiš Department of Mechanical Engineering, Academy of Armed Forces of General M. R. Štefánik, Demänovská 393, Liptovský Mikuláš, 03119, Slovakia, kompis@aoslm.sk

Ying-Te Lee Department of Harbor and River Engineering, National Taiwan Ocean University, Keelung, Taiwan, ytlee@mail.ntou.edu.tw

Wei-Ming Lee Department of Mechanical Engineering, China Institute of Technology, Taipei, Taiwan, wmlee@mail.chit.edu.tw

Yijun Liu Department of Mechanical Engineering, University of Cincinnati, Cincinnati, Ohio 45221-0072, USA, Yijun.Liu@uc.edu

George D. Manolis Department of Civil Engineering, Aristotle University, Thessaloniki, GR-54124, Greece, gdm@civil.auth.gr

Webe João Mansur Department of Civil Engineering, COPPE - Federal University of Rio de Janeiro, CP 68506, CEP 21945-970, Rio de Janeiro, RJ, Brazil, webe@coc.ufrj.br

E. Pan Department of Civil Engineering, The University of Akron, Akron, OH 44325-3905, USA, pan2@uakron.edu

Andre Pereira Institute for Structural Analysis, Graz University of Technology, Lessingstrasse 25/II, 8010 Graz, Austria, andre.pereira@tugraz.at

D. Polyzos Department of Mechanical and Aeronautical Engineering, University of Patras, GR-26500 Patras, Greece; Institute of Chemical Engineering and High Temperature Chemical Process, GR-26504 Patras, Greece, polyzos@mech.upatras.gr

C. P. Providakis Department of Applied Sciences, Technical University of Crete, GR-73100 Chania, Greece, cpprov@mred.tuc.gr

Qing-Hua Qin Department of Engineering, Australian National University, Canberra, ACT, Australia, 0200, Qinghua.qin@anu.edu.au

Dimitris C. Rizos Department of Civil and Environmental Engineering, University of South Carolina, Columbia, SC 29208, USA, rizos@engr.sc.edu

Thomas Ruberg Institute for Structural Analysis, Graz University of Technology, Austria, rueberg@tugraz.at

Evangelos J. Sapountzakis School of Civil Engineering, National Technical University of Athens, Zografou Campus, GR-157 80, Greece, cvsapoun@central.ntua.gr

Martin Schanz Institute of Applied Mechanics, Graz University of Technology, Austria, m.schanz@tugraz.at

Jean-François Semblat LCPC, Paris, France, semblat@lcpc.fr

Liang Shen Department of Mechanical Engineering, University of Cincinnati, Cincinnati, Ohio 45221-0072, USA, Liang.Shen2@boeing.com

J. Sladek Institute of Construction and Architecture, Slovak Academy of Sciences, 84503 Bratislava, Slovakia, jan.sladek@savba.sk

V. Sladek Institute of Construction and Architecture, Slovak Academy of Sciences, 84503 Bratislava, Slovakia, vladimir.sladek@savba.sk

Yiorgos-Sokratis Smyrlis Department of Mathematics and Statistics, University of Cyprus, P.O.Box 20537, 1678 Nicosia, Cyprus, smyrlis@ucy.ac.cy

Delfim Soares Jr. Structural Engineering Department, Federal University of Juiz de Fora, Cidade Universitária, CEP 36036-330, Juiz de Fora, MG, Brazil, delfim@coc.ufrj.com

Georgios Stavroulakis Department of Production Engineering and Management, Technical University of Crete, Chania, Greece, gestavr@dpem.tuc.gr

Mário Štiavnický Department of Mechanical Engineering, Academy of Armed Forces of General M. R. Štefánik, Demänovská 393, Liptovský Mikuláš, 03119, Slovakia, stiavnicky@aoslm.sk

J. C. F. Telles Programa de Engenharia Civil, COPPE/UFRJ, Caixa Postal 68506, CEP 21945-970, Rio de Janeiro, RJ, Brazil, telles@coc.ufrj.br

S.V. Tsinopoulos Department of Mechanical Engineering, Technological and Educational Institute of Patras, GR-26334 Patras, Greece, stsinop@teipat.gr

Vasileios J. Tsipiras School of Civil Engineering, National Technical University of Athens, Zografou Campus, GR-157 80, Greece, tsipiras@gmail.com

Hui Wang College of Civil Engineering and Architecture, Henan University of Technology, Zhengzhou, 450052, China, huiwang.china@yahoo.com.cn

Arnaldo Warszawski National Petroleum Agency – ANP CEP 20090-004, Rio de Janeiro, RJ, Brazil, warszawski@coc.ufrj.br

Christos Xenophontos Department of Mathematics and Statistics, University of Cyprus, PO Box 20537, 1678 Nicosia, Cyprus, xenophontos@ucy.ac.cy

F. Xie Department of Mechanics, School of Science, Beijing Institute of Technology, Beijing 100081, China, f.xie@bit.edu.cn

Zhenhan Yao Department of Engineering Mechanics, Tsinghua University, Beijing, 100084, China, demyzh@tsinghua.edu.cn

Ch. Zhang Department of Civil Engineering, University of Siegen, D-57068 Siegen, Germany, c.zhang@uni-siegen.de

Biography of Dimitri Beskos

Dimitri Beskos was born in Athens, Greece on January 26, 1946 and attended the elite Varvakion high school. He entered the National Technical University of Athens in 1964 following nation-wide entrance examinations, ranking first among all the candidates in Civil Engineering for that year! Upon graduation with a Diploma in Civil Engineering in 1969, he was accepted at Cornell University in Ithaca, New York, for graduate studies under the supervision of Constantine Dafermos. Following his Master of Science degree in 1971, he continued with doctoral work under the supervision of Bruno Boley and received his PhD in 1974. He was immediately hired as Assistant Professor with the Civil Engineering department at the University of Minnesota and embarked on an academic career that has been extremely productive to this date, with many contributions to the fields of Computational Mechanics (Boundary Element and Finite Element Methods), Structural Engineering (Structural Dynamics, Earthquake Engineering and Steel Structures) and Applied Mechanics (Mechanics of Materials, Wave Propagation, Porous Media Flow). Given below is a detailed account of his academic achievements and of his work.

Personal Information

Married to Sophia Papargyri-Beskou. Three daughters (Niki, Denise and Thelma Beskou) from a previous marriage, a son (Ioannis Pegios) and a daughter (Smaro Pegiou) from a previous marriage of his wife.

Address

Department of Civil Engineering, University of Patras, GR-26500, Patras, Greece
Tel: +30-2610-996559, Fax: +30-2610-996579, E-mail: d.e.beskos@upatras.gr

Distinctions and Honours

National Foundation Scholarship during all five undergraduate years (1964–1969) in Greece.

John McMullen Graduate Fellowship, 1969–1970, Cornell University.

Reviewer for many Research Funding Agencies in the USA, Europe and Asia.

Member of the Electronic Computation Committee (Subcommittee on Automated Analysis and Design) of the Structural Division of ASCE (1980–1988).

Member of the Administrative Council of the European Center on the Prevention and Forecasting of Earthquakes, Athens, Greece (1989–1992).

Member of the Congress Committee of the International Union of Theoretical and Applied Mechanics (IUTAM) (2002–2010).

Fellow of the American Society of Civil Engineers (ASCE), Fellow of the Wessex Institute of Technology of Great Britain, Fellow of the International Association for Computational Mechanics, Fellow of the New York Academy of Sciences.

Member of the Academia Europaea, Member of the European Academy of Sciences and Arts.

E. Reissner Medal during ICCES 2003 for sustained and outstanding contributions to the development of BEMs.

P.S. Theocaris award of 2007 for the best paper in Applied Mechanics during 2004–2007 by the Academy of Athens.

Co-Editor of "Computational Mechanics" (Springer–Verlag) and Associate Editor of "Soil Dynamics and Earthquake Engineering" (Elsevier) and "Applied Mechanics Reviews" (ASME).

Editorial Board Member of "International Journal for Numerical Methods in Engineering (John Wiley & Sons), "Engineering Analysis with Boundary Elements (Elsevier), "European Journal of Mechanics, A/Solids" (Elsevier), "International Journal of Computers and Structures" (Elsevier), "Computer Modelling in Engineering and Sciences" (Tech Science Press), "International Journal of Computational Engineering Science (Taylor & Francis Press), "Acta Mechanica Solida Sinica" (Huazhong University, China), "Facta Universitatis–Architecture and Civil Engineering" (University of Nis, Serbia), "Journal of the Serbian Society of Computational Mechanics" (University of Kragujevac, Serbia).

Co-Editor of the International Book Series on "Advances in Earthquake Engineering" of Computational Mechanics Publications and Editorial Board Member of "International Series on Computational Engineering Books" of Computational Mechanics Publications and Elsevier Applied Science.

Member, Bureau of Theoretical and Applied Mechanics of the Academy of Athens.

Administrative Positions

Director, Structural Engineering Division, Department of Civil Engineering, University of Patras, Patras, Greece (1982–1983, 1986–1988, 1989–1991, 1992–1999).

Member of the Executive Council of the Committee of Research, University of Patras, Greece (1995–2004).

Associate Chairman, Department of Civil Engineering, University of Patras, Patras, Greece (2003–2007).

Member, Administrative Council of the European Center on the Prevention and Forecasting of Earthquakes, Athens, Greece (1989–1992).

Member, Governing Board of the "Patras Science Park" (1991–1994).

National Representative of Greece for BRITE/EURAM Program in Aeronautics of the EC (1990–1992).

Member of the Greek National Council of Research (1990–1992).

Substitute Member of the Technical Council of the Academy of Athens (2000–2009).

Academic Coordinator of the Earthquake Engineering Graduate Program of the Greek Open University, Patras (2003-present).

Substitute Member of the Executive Council of the Hellenic Organization for Aseismic Design and Protection (OASP), Athens, Greece (2005–2009).

Societies

Member of the Technical Chamber of Greece.

Member of the Greek Society of Civil Engineers.

Fellow of the American Society of Civil Engineers (ASCE).

Member of the American Academy of Mechanics (1980–1990).
Member of the Hellenic Society of Theoretical and Applied Mechanics (Secretary General, 1990–2003).
Member of the Metal Structures Research Society of Greece (President, 2000–2003).
Member of the International Association of Computational Mechanics – IACM (Member of General Council, 1991–1998).
Member of the International Association for Boundary Element Methods – IABEM (Vice President, 1993–1998).
Member of Scientific Steering Committee of the International Society for Boundary Elements (ISBE).
Member of the Greek Association of Computational Mechanics (Founder and First President: 1991–1996).
Member of the Greek Association for Earthquake Engineering.

Academic Positions

Teaching Assistant (1970–1973), Department of Theoretical and Applied Mechanics, Cornell University, Ithaca, New York.
Instructor (1973–1974), Department of Theoretical and Applied Mechanics, Cornell University, Ithaca, New York.
Assistant Professor (1974–1979), Department of Civil and Mineral Engineering, University of Minnesota, Minneapolis.
Associate Professor (1979–1981), Department of Civil and Mineral Engineering, University of Minnesota, Minneapolis.
Professor (1981-present), Chair of Structural Synthesis II, Department of Civil Engineering, University of Patras, Greece (the Chaired Professor system abolished in Greece in 1983).
Visiting Professor (1981–1983), Department of Civil and Mineral Engineering, University of Minnesota, Minneapolis.
Visiting Professor (1985), Department of Civil Engineering, Colorado State University, Fort Collins.
Visiting Scientist (1987), Applied Mechanics Division, Ispra Establishment, Joint Research Center, 21020 Varese, Italy.
Visiting Professor (1989), Department of Civil and Mineral Engineering, University of Minnesota, Minneapolis.

Publication Record

His publication record in English includes 19 books, 4 guest edited journal issues, 16 chapters in books, 148 articles in refereed journals and about 200 articles in conference proceedings. In particular, he has co-authored the first text on BEM as applied to elastodynamics (1988), edited the first handbook on BEM (1987) and edited or co-edited three more books on BEM (1989, 1991, 2003) and a handbook on Earthquake Engineering. In addition, he has published 8 books and 19 journal and conference proceedings articles in Greek. His published work has received more than 1500 citations.

List of Publications – Books

1. D.E. BESKOS, T. KRAUTHAMMER and I.G. VARDOULAKIS, Editors, "Dynamic Soil-Structure Interaction", A.A. Balkema Publishers, Rotterdam, The Netherlands, 1984.
2. D.E. BESKOS, T. KRAUTHAMMER and I.G. VARDOULAKIS, Editors, Proceedings of the International Workshop on Dynamic Soil-Structure Interaction, University of Minnesota, Minneapolis, Minnesota, December 1985.
3. D.E. BESKOS, Editor, "Boundary Element Methods in Mechanics", North-Holland Publishing, Amsterdam, The Netherlands, 1987.
4. D.E. BESKOS, "Numerical Methods in Dynamic Fracture Mechanics", EUR 11300 EN, Joint research Centre of E.C., Ispra Establishment, Ispra, Italy, 1987.
5. G.D. MANOLIS and D.E. BESKOS, "Boundary Element Methods in Elastodynamics", Unwin Hyman Publishing, London, 1988 (translated into Chinese and published by Tianjin Science Press, Tianjin, 1991).
6. D.E. BESKOS, Editor, "Boundary Element Methods in Structural Analysis", American Society of Civil Engineers, New York, 1989.
7. S.N. ATLURI, D.E. BESKOS, R. JONES and G. YAGAWA, Editors, "Computational Mechanics 91", ICES Publications, Atlanta, 1991.
8. D.E. BESKOS, Editor, "Boundary Element Analysis of Plates and Shells", Springer-Verlag, Berlin, 1991.
9. D.E. BESKOS and F. ZIEGLER, Editors, "Advances in Dynamic Systems and Stability", Springer-Verlag, Wien, 1992.
10. D.E. BESKOS et al., Editors, Proceedings of the 1st National Congress on Computational Mechanics, Patras University Press, Patras 1992.
11. D.A. SOTIROPOULOS and D.E. BESKOS, Editors, Proceedings of the 2nd National Congress on Computational Mechanics, Technical University of Crete, Chania, 1996.
12. G.D. MANOLIS, D.E. BESKOS and C.A. BREBBIA, Editors, Proceedings of 1st International Conference on Earthquake Resistant Engineering Structures, ERES 1996, Thessaloniki, Greece, Computational Mechanics Publications, Southampton, 1996.
13. D.E. BESKOS and S.A. ANAGNOSTOPOULOS, Editors, Computer Analysis and Design of Earthquake Resistant Structures: A Handbook, Computational Mechanics Publications, Southampton, 1997.
14. J.T. KATSIKADELIS, D.E. BESKOS and E.E. GDOUTOS, Editors, "Recent Advances in Applied Mechanics: Honorary Volume for Professor A.N. Kounadis, National Technical University of Athens, Greece.

15. D.E. BESKOS, D.L. KARABALIS and A.N. KOUNADIS, Editors, Proceedings of the 4th National Conference on Steel Structures, Typorama, Patras, 2002.
16. D.E. BESKOS and G. MAIER, Editors, Boundary Element Advances in Solid Mechanics, Springer-Verlag, Wien, 2003.
17. S.N. ATLURI, D.E. BESKOS and D. POLYZOS, Editors, Proceedings (in CD-ROM) of International Conference on Computational and Experimental Engineering and Sciences, ICCES 03, Corfu, Greece, 2003.
18. C.A. BREBBIA, D.E. BESKOS, G.D. MANOLIS and C.C. SPYRAKOS, Editors, Proceedings of 5th International Conference on Earthquake Resistant Engineering Structures, ERES 2005, Skiathos, Greece, WIT Press, 2005.
19. N. BAZEOS, D.L. KARABALIS, D. POLYZOS, D.E. BESKOS and J.T. KATSIKADELIS, Editors, Proceedings of 8th HSTAM International Congress on Mechanics, Patras, Greece, 2007.

List of Publications – Guest Editorials

1. D.E. BESKOS, "Soil Dynamics and Dynamic Soil Structure Interaction", Special Issue Part 1 in "Engineering Analysis with Boundary Elements", V. 8, No. 3, 1991.
2. D.E. BESKOS, "Soil Dynamics and Dynamic Soil Structure Interaction", Special Issue Part 2 in "Engineering Analysis with Boundary Elements", V. 8, No. 4, 1991.
3. D.E. BESKOS, "Plasticity", Special Issue in "Engineering Analysis with Boundary Elements", V.14, No. 1, 1994.
4. D.E. BESKOS and D. SOTIROPOULOS, "Greek National Congress on Computational Mechanics", Special Issue in "Computational Mechanics", V. 21, No. 4/5, 1998.

List of Publications – Book Chapters

1. D.E. BESKOS, Introduction to Boundary Element Methods, Chapter 1 (pp. 1–21), in: Boundary Element Methods in Mechanics, D.E. Beskos, Editor, North-Holland Publishing, Amsterdam, The Netherlands, 1987.
2. D.E. BESKOS, Potential Theory, Chapter 2 (pp. 23–106), in: Boundary Element Methods in Mechanics, D.E. Beskos, Editor, North-Holland Publishing, Amsterdam, The Netherlands, 1987.
3. D.L. KARABALIS and D.E. BESKOS, Dynamic Soil-Structure Interaction, Chapter 11 (pp. 499–562), in: Boundary Element Methods in Mechanics, D.E. Beskos, Editor, North-Holland Publishing, Amsterdam, The Netherlands, 1987.
4. I.G. VARDOULAKIS and D.E. BESKOS, Dynamic Consolidation of Nearly Saturated Granular Media, Chapter 5 (pp. 167–208), in: Developments in Soil

Mechanics and Foundation Engineering, Vol.3, Dynamic Behaviour of Foundations and Buried Structures, P.K. Banerjee, and R. Butterfield, Editors, Elsevier Applied Science, London, 1987.

5. D.L. KARABALIS and D.E. BESKOS, Three-Dimensional Soil-Structure Interaction by Boundary Element Methods, Chapter 1 (pp. 1–26), in: Topics in Boundary Element Research, Vol. 4, Applications in Geomechanics, C.A. Brebbia, Editor, Springer-Verlag, Berlin,1987.
6. D.E. BESKOS, Dynamic Analysis of Beams, Plates and Shells, Chapter 6 (pp. 139–161), in: Boundary Element Methods in Structural Analysis, D.E. Beskos, Editor, ASCE, New York, 1989.
7. D.E. BESKOS, Present Status and Future Developments, Chapter 12 (pp. 337–341), in: Boundary Element Methods in Structural Analysis, D.E. Beskos, Editor, ASCE, New York, 1989.
8. D.E. BESKOS, Dynamic Analysis of Plates and Shallow Shells by the D/BEM, Chapter 8 (pp. 177–196), in: Advances in the Theory of Plates and Shells, G.Z. Voyiadjis and D. Karamanlidis, Editors, Elsevier, Amsterdam, 1990.
9. D.E. BESKOS, Dynamic Analysis of Plates, Chapter 2 (pp. 35–92), in: Boundary Element Analysis of Plates and Shells, D.E. Beskos, Editor, Springer-Verlag, Berlin, 1991.
10. D.E. BESKOS, Static and Dynamic Analysis of Shells, Chapter 3 (pp. 93–140), in: Boundary Element Analysis of Plates and Shells, D.E. Beskos, Editor, Springer-Verlag, Berlin, 1991.
11. D.E. BESKOS, Applications of the Boundary Element Method in Dynamic Soil-Structure Interaction, Chapter 4 (pp. 61–90), in: Developments in Dynamic Soil-Structure Interaction, P. Gulkan and R.W. Clough, Editors, Kluwer Academic Publishers, Dordrecht, 1993.
12. G.D. MANOLIS, T.G. DAVIES and D.E. BESKOS, Overview of Boundary Element Techniques in Geomechanics, Chapter 1 (pp. 1–35), in: Boundary Element Techniques in Geomechanics, G.D. Manolis and T.G. Davies, Editors, Computational Mechanics Publications, Southampton, 1993.
13. D.E. BESKOS, Wave Propagation Through Ground, Chapter 11 (pp.359–406), in: Boundary Element Techniques in Geomechanics, G.D. Manolis and T.G. Davies, Editors, Computational Mechanics Publications, Southampton, 1993.
14. D.L. KARABALIS and D.E. BESKOS, Numerical Methods in Earthquake Engineering, Chapter 1 (pp. 1–102) in: Computer Analysis and Design of Earthquake Resistant Structures, D.E. Beskos and S.A. Anagnostopoulos, Editors, Computational Mechanics Publications, Southampton, 1997.
15. G.D. MANOLIS and D.E. BESKOS, Underground and Lifeline Structures, Chapter 16 (pp. 775–837) in: Computer Analysis and Design of Earthquake Resistant Structures, D.E. Beskos and S.A. Anagnostopoulos, Editors, Computational Mechanics Publications, Southampton, 1997.
16. D.E. BESKOS, Dynamic Analysis of Structures and Structural Systems, Chapter 1 (pp. 1–50) in: Boundary Element Advances in Solid Mechanics, D.E. Beskos and G. Maier, Editors, Springer-Verlag, Wien, 2003.

List of Publications – Journal Papers

1. D.E. BESKOS, "Universal solutions for fiber-reinforced compressible isotropic elastic materials", Journal of Elasticity, V. 2, pp. 153–168 (1972).
2. D.E. BESKOS, "Universal solutions for fiber-reinforced incompressible isotropic elastic materials", International Journal of Solids and Structures, V. 9, pp. 553–567 (1973).
3. D.E. BESKOS, "Fracture of plain concrete under biaxial stresses", Cement and Concrete Research, V. 4, pp. 979–985 (1974).
4. Y.C. WAUNG, D.E. BESKOS and W.H. SACHSE, "Ultrasonic velocity measurement of elastic constants of Al-CuAl2 eutectic composite", Journal of Materials Science, V. 10, pp. 109–112 (1975).
5. D.E. BESKOS and J.T. JENKINS, "A mechanical model for mammalian tendon", Journal of Applied Mechanics, V. 42, pp. 755–758 (1975).
6. D.E. BESKOS and B.A. BOLEY, "Use of dynamic influence coefficients in forced vibration problems with the aid of Laplace transform", Computers and Structures, V. 5, pp. 263–269 (1975).
7. D.E. BESKOS, A reply to K. Rajagopalan's discussion on "Fracture of plain concrete under biaxial stresses", Cement and Concrete Research, V. 5, p. 527 (1975).
8. E.C. AIFANTIS and D.E. BESKOS, "Dynamic universal solutions for fiber-reinforced incompressible isotropic elastic materials," Journal of Elasticity, V. 6, pp. 1–15 (1976).
9. D.E. BESKOS, "The lumping mass effect on frequencies of beam-columns", Journal of Sound and Vibration, V. 47, pp. 139–142 (1976).
10. A. AVAKIAN and D.E. BESKOS, "Use of dynamic stiffness influence coefficients in vibrations of non-uniform beams", Journal of Sound and Vibration, V. 47, pp. 292–295 (1976).
11. D.E. BESKOS, "The effect of damping on the vibration-stability relation of linear systems", Mechanics Research Communications, V. 3, pp. 373–377 (1976).
12. D.E. BESKOS and B.A. BOLEY, "The Gauss-Seidel convergence criterion for elastic structural stability", Mechanics Research Communications, V. 3, pp. 457–462 (1976).
13. E.C. AIFANTIS and D.E. BESKOS, "Inflation, bending, extension and azimuthal shearing of a fiber-reinforced elastic sector of a circular tube", Acta Mechanica, V. 26, pp. 159–170 (1977).
14. E.C. AIFANTIS, W.W. GERBERICH and D.E. BESKOS, "Diffusion equations for a mixture of an elastic solid and an elastic fluid", Archives of Mechanics, V. 29, pp. 339–353 (1977).
15. E.C. AIFANTIS, D.E. BESKOS and W.W. GERBERICH, "Diffusion in dislocation fields", Archives of Mechanics, V. 29, pp. 723–740 (1977).
16. D.E. BESKOS, "Framework stability by finite element method", Journal of the Structural Division of ASCE, V. 103, ST 11, pp. 2273–2276 (1977).

17. G.V. NARAYANAN and D.E. BESKOS, "Use of dynamic influence coefficients in forced vibration problems with the aid of fast Fourier transform", Computers and Structures, V. 9, pp. 145–150 (1978).
18. G.D. MANOLIS and D.E. BESKOS, "Plastic design aids for pinned-base gabled frames", Engineering Journal of AISC, V. 16, pp. 1–10 (1979).
19. D.S. KAMDAR and D.E. BESKOS, "Numerical methods for structural stability analysis", Computer Methods in Applied Mechanics and Engineering, V. 19, pp. 205–222 (1979).
20. D.E. BESKOS, "Dynamics and stability of plane trusses with gusset plates", Computers and Structures, V. 10, pp. 758–795 (1979).
21. G.D. MANOLIS and D.E. BESKOS, "Thermally induced vibrations of beam structures", Computer Methods in Applied Mechanics and Engineering, V. 21, pp. 337–355 (1980).
22. E.C. AIFANTIS and D.E. BESKOS, "Heat extraction from hot dry rocks", Mechanics Research Communications, V. 7, pp. 165–170 (1980).
23. D.E. BESKOS and B.A. BOLEY, "Critical damping in linear discrete dynamic systems", Journal of Applied Mechanics, V. 47, pp. 627–630 (1980).
24. G.D. MANOLIS and D.E. BESKOS, "Dynamic stress concentration studies by boundary integrals and Laplace transform," International Journal for Numerical Methods in Engineering, V.17, pp. 573–599 (1981).
25. D.E. BESKOS and J.B. OATES, "Dynamic analysis of ring-stiffened circular cylindrical shells", Journal of Sound and Vibration, V. 75, pp. 1–15 (1981).
26. D.E. BESKOS and B.A. BOLEY, "Critical damping in certain linear continuous dynamic systems", International Journal of Solids and Structures, V. 17, pp. 575–588 (1981).
27. W.D. RUTLEDGE and D.E. BESKOS, "Dynamic analysis of linearly tapered beams", Journal of Sound and Vibration, V. 79, pp. 457–462 (1981).
28. G.V. NARAYANAN and D.E. BESKOS, "Dynamic soil-structure interaction by numerical Laplace transform", Engineering Structures, V. 4, pp. 53–62 (1982).
29. G.D. MANOLIS and D.E. BESKOS, "Dynamic response of framed underground structures, Computers and Structures", V. 15, pp. 521–531 (1982).
30. C.C. SPYRAKOS and D.E. BESKOS, "Dynamic response of frameworks by fast Fourier transform", Computers and Structures, V. 15, pp. 495–505 (1982).
31. G.V. NARAYANAN and D.E. BESKOS, "Numerical operational methods for time-dependent linear problems", International Journal for Numerical Methods in Engineering, V. 18, 1829–1854 (1982).
32. D.L. KARABALIS and D.E. BESKOS, "Static, dynamic and stability analysis of structures composed of tapered beams", Computers and Structures, V. 16, pp. 731–748 (1983).
33. D.E. BESKOS and G.V. NARAYANAN, "Dynamic response of frameworks by numerical Laplace transform", Computer Methods in Applied Mechanics and Engineering, V. 37, pp. 289–307 (1983).

34. G.D. MANOLIS and D.E. BESKOS, "Internal force distribution effect on framework stability", Journal of the Structural Division of ASCE, V. 109, ST 1, pp. 250–257 (1983).
35. G.D. MANOLIS and D.E. BESKOS, "Dynamic response of lined tunnels by an isoparametric boundary element method", Computer Methods in Applied Mechanics and Engineering, V. 36, pp. 291–307 (1983).
36. M.Y. KHALED, D.E. BESKOS and E.C. AIFANTIS, "On the theory of consolidation with double porosity – III: A finite element formulation", International Journal for Numerical and Analytical Methods in Geomechanics, V. 8, pp. 101–123 (1984).
37. D.E. BESKOS, I. OKUTANI and P.G. MICHALOPOULOS, "Testing of dynamic models for signal controlled intersections," Trasportation Research – B, V. 18B, pp. 397–408 (1984).
38. D.L. KARABALIS and D.E. BESKOS, "Dynamic response of 3-D rigid surface foundations by time domain boundary element method," Earthquake Engineering and Structural Dynamics, V. 12, pp. 73–93 (1984).
39. D.E. BESKOS and A.Y. MICHAEL, "Solution of plane transient elastodynamic problems by finite elements and Laplace transform," Computers and Structures, V. 18, pp. 695–701 (1984).
40. P.G. MICHALOPOULOS, D.E. BESKOS and J. LIN, "Analysis of interrupted traffic flow by finite difference methods," Transportation Research – B, V. 18B, pp. 409–421 (1984).
41. P.G. MICHALOPOULOS, D.E. BESKOS and Y. YAMAUCHI, "Multilane traffic flow dynamics: Some macroscopic considerations," Transportation Research – B, V. 18B, pp. 377–395 (1984).
42. D.E. BESKOS and K.L. LEUNG, "Dynamic analysis of plate systems by combining finite differences, finite elements and Laplace transform", Computers and Structures, V. 19, pp. 763–775 (1984).
43. D.E. BESKOS and P.G. MICHALOPOULOS, "An application of the finite element method in traffic signal analysis", Mechanics Research Communications, V. 11, pp. 185–189 (1984).
44. D.E. BESKOS, P.G. MICHALOPOULOS and J.K. LIN, "Analysis of traffic flow by the finite element method", Applied Mathematical Modelling, V. 9, pp. 358–364 (1985).
45. D.L. KARABALIS and D.E. BESKOS, "Dynamic response of 3-D flexible foundations by time domain BEM and FEM", Soil Dynamics and Earthquake Engineering, V. 4, pp. 91–101 (1985).
46. G.D. MANOLIS, D.E. BESKOS and B.J. BRAND, "Elastoplastic analysis and design of gabled frames", Computers and Structures, V. 22, pp. 693–697 (1986).
47. C.C. SPYRAKOS and D.E. BESKOS, "Dynamic response of rigid strip foundations by time domain boundary element method," International Journal for Numerical Methods in Engineering, V.23, pp. 1547–1565 (1986).
48. I.G. VARDOULAKIS and D.E. BESKOS, "Dynamic behavior of nearly saturated porous media", Mechanics of Materials, V. 5, pp. 87–108 (1986).

49. D.E. BESKOS, B. DASGUPTA and I.G. VARDOULAKIS, "Vibration isolation using open or filled trenches. Part I: 2-D Homogeneous soil", Computational Mechanics, V. 1, pp. 43–63 (1986).
50. G.D. MANOLIS, D.E. BESKOS and M.F. PINEROS, "Beam and plate stability by boundary elements", Computers and Structures, V. 22, pp. 917–923 (1986).
51. D.L. KARABALIS and D.E. BESKOS, "Dynamic response of 3-D embedded foundations by the boundary element method", Computer Methods in Applied Mechanics and Engineering, V. 56, pp. 91–120 (1986).
52. C.C. SPYRAKOS and D.E. BESKOS, "Dynamic response of flexible strip foundations by boundary and finite elements", Soil Dynamics and Earthquake Engineering, V. 5, pp. 84–96 (1986).
53. C.P. PROVIDAKIS and D.E. BESKOS, "Dynamic analysis of beams by the boundary element method", Computers and Structures, V. 22, pp. 957–964 (1986).
54. I. OKUTANI, D.E. BESKOS and P.G. MICHALOPOULOS, "Finite element analysis of freeway dynamics", Engineering Analysis, V. 3, pp. 85–92 (1986).
55. D.E. BESKOS and E.C. AIFANTIS, "On the theory of consolidation with double porosity – II", International Journal of Engineering Science, V. 24, pp. 1697–1716 (1986).
56. D.E. BESKOS, "Boundary Element Methods in Dynamic Analysis", Applied Mechanics Reviews, V. 40, pp. 1–23 (1987).
57. D.P.N. KONTONI, D.E. BESKOS and G.D. MANOLIS, "Uniform half-plane elastodynamic problems by an approximate boundary element method", Soil Dynamics and Earthquake Engineering, V. 6, pp. 227–238 (1987).
58. I.G. VARDOULAKIS, D.E. BESKOS, K.L. LEUNG, B. DASGUPTA and R.L.STERLING, "Computation of vibration levels in underground space," International Journal of Rock Mechanics and Mining Sciences & Geomechanics Abstracts, V. 24, pp. 291–298 (1987).
59. P.G. MICHALOPOULOS, J. LIN and D.E. BESKOS, "Integrated modelling and numerical treatment of freeway flow", Applied Mathematical Modelling, V. 11, pp. 447–457 (1987).
60. D.P.N. KONTONI and D.E. BESKOS, "Boundary element formulation for dynamic analysis of nonlinear systems", Engineering Analysis, Vol. 5, pp. 114–125, (1988).
61. G.D. MANOLIS and D.E. BESKOS, "Integral equation formulation and fundamental solutions of dynamic poroelasticity and thermoelasticity", Acta Mechanica, V. 76, pp. 89–104 (1989).
62. D.E. BESKOS, "Dynamics of saturated rocks-I: equations of motion", Journal of Engineering Mechanics of ASCE, V. 115, pp. 982–995 (1989).
63. D.E. BESKOS, I. VGENOPOULOU and C.P. PROVIDAKIS, "Dynamics of saturated rocks II: body waves", Journal of Engineering Mechanics of ASCE, V. 115, pp. 996–1016 (1989).
64. D.E. BESKOS, C.N. PAPADAKIS and H.S. WOO, "Dynamics of saturated rocks-III: Rayleigh waves", Journal of Engineering Mechanics of ASCE, V. 115, pp. 1017–1034 (1989).

65. C.P. PROVIDAKIS and D.E. BESKOS, "Free and forced vibrations of plates by boundary and interior elements", International Journal for Numerical Methods in Engineering, V. 28, pp. 1977–1994 (1989).
66. C.P. PROVIDAKIS and D.E. BESKOS, "Free and forced vibrations of plates by boundary elements", Computer Methods in Applied Mechanics and Engineering, V. 74, pp. 231–250 (1989).
67. D.L. KARABALIS and D.E. BESKOS, "Discussion on the paper – Time domain transient elastodynamic analysis of 3-D solids by BEM, "International Journal for Numerical Methods in Engineering, V.29, pp. 211–215 (1990).
68. G.D. MANOLIS and D.E. BESKOS, Errata in "Integral formulation and fundamental solutions of dynamic poroelasticity and thermoelasticity", Acta Mechanica, V. 83, pp. 223–226 (1990).
69. I. VGENOPOULOU, D.E. BESKOS and I.G. VARDOULAKIS, "High frequency wave propagation in nearly saturated porous media", Acta Mechanica, V. 85, pp. 115–123 (1990).
70. B. DASGUPTA, D.E. BESKOS and I.G. VARDOULAKIS, "Vibration isolation using open or filled trenches. Part 2: 3-D Homogeneous soil", Computational Mechanics, V. 6, pp. 129–142 (1990).
71. K.L. LEUNG, D.E. BESKOS and I.G. VARDOULAKIS, "Vibration isolation using open or filled trenches. Part 3: 2-D Non-homogeneous soil", Computational Mechanics, V. 7, pp. 137–148 (1990).
72. K.L. LEUNG, I.G. VARDOULAKIS, D.E. BESKOS and J.L. TASSOULAS, "Vibration isolation by trenches in continuously nonhomogeneous soil by the BEM", Soil Dynamics and Earthquake Engineering, V. 10, pp. 172–179 (1991).
73. A.H.D. CHENG, T. BADMUS and D.E. BESKOS, "Integral equation for dynamic poroelasticity in frequency domain with boundary element solution", Journal of Engineering Mechanics of ASCE, V. 117, pp. 1136–1157 (1991).
74. O.VON ESTORFF, A.A. STAMOS, H. ANTES and D.E. BESKOS, "Dynamic interaction effects in underground traffic systems", Engineering Analysis with Boundary Elements, V. 8, pp. 167–175 (1991).
75. C.P. PROVIDAKIS and D.E. BESKOS, "Free and forced vibrations of shallow shells by boundary and interior elements", Computer Methods in Applied Mechanics and Engineering, V. 92, pp. 55–74 (1991).
76. I. VGENOPOULOU and D.E. BESKOS, "Dynamic poroelastic soil column and borehole problem analysis", Soil Dynamics and Earthquake Engineering, V.11, pp. 335–345 (1992).
77. I. VGENOPOULOU and D.E. BESKOS, "Dynamics of saturated rocks. IV: column and borehole problems", Journal of Engineering Mechanics of ASCE, V. 118, pp. 1795–1813 (1992).
78. I. VGENOPOULOU and D.E. BESKOS, "Dynamic behavior of saturated poroviscoelastic media", Acta Mechanica, V. 95, pp. 185–195 (1992).
79. D.P.N. KONTONI and D.E. BESKOS, "Transient dynamic elastoplastic analysis by the dual reciprocity BEM", Engineering Analysis with Boundary Elements, V. 12, p.p. 1–16 (1993).

80. C.P. PROVIDAKIS, D.A. SOTIROPOULOS and D.E. BESKOS, "BEM analysis of reduced dynamic stress concentration by multiple holes", Communications in Numerical Methods in Engineering, V. 9, pp. 917–924 (1993).
81. N. BAZEOS and D.E. BESKOS, "A computer program for determining wind pressures on 2-D structures", Advances in Engineering Software, V. 17, pp. 173–187 (1993).
82. D.D. THEODORAKOPOULOS and D.E. BESKOS, "Flexural vibrations of fissured poroelastic plates", Archives of Mechanics, V. 63, pp. 413–423 (1993).
83. D.D. THEODORAKOPOULOS and D.E. BESKOS, Flexural vibrations of poroelastic plates, Acta Mechanica, V. 103, pp. 191–203 (1994).
84. D. POLYZOS, A.A. STAMOS and D.E. BESKOS, "BEM computation of DSIF's in cracked viscoelastic plates", Communications in Numerical Methods in Engineering, V. 10, pp. 81–87 (1994).
85. C.P. PROVIDAKIS, D.E. BESKOS and D.A. SOTIROPOULOS, "Dynamic analysis of inelastic plates by the D/BEM", Computational Mechanics, V. 13, pp. 276–284 (1994).
86. N. BAZEOS and D.E. BESKOS, "Numerical determination of time averaged winds loads on complex structures in 2-D flows", Engineering Structures, V. 16, pp. 190–200 (1994).
87. A.A. STAMOS, O. VON ESTORFF, H. ANTES and D.E. BESKOS, "Vibration isolation in road-tunnel traffic systems", International Journal for Engineering Analysis and Design, V. 1, pp. 111–121 (1994).
88. A.S. LYRINTZIS, P. YI, P.G. MICHALOPOULOS and D.E. BESKOS, "Advanced continuum traffic flow models for congested freeways", Journal of Transportation Engineering of ASCE, V. 120, pp. 461–477 (1994).
89. G.D. PAVLATOS and D.E. BESKOS, "Dynamic elastoplastic analysis by BEM/FEM", Engineering Analysis with Boundary Elements, V. 14, pp. 51–63 (1994).
90. C.P. PROVIDAKIS and D.E. BESKOS, "Dynamic analysis of elastoplastic flexural plates by the D/BEM", Engineering Analysis with Boundary Elements, V. 14, pp. 75–80 (1994).
91. D. POLYZOS, G. DASSIOS and D.E. BESKOS, "On the equivalence of dual reciprocity and particular integrals approaches in the BEM", Boundary Elements Communications, V. 5, pp. 285–288 (1994).
92. K.L. LEUNG, P.B. ZAVAREH and D.E. BESKOS, "2-D elastostatic analysis by a symmetric BEM/FEM scheme", Engineering Analysis with Boundary Elements, V. 15, pp. 67–78 (1995).
93. J. QIAN and D.E. BESKOS, Dynamic interaction between 3-D rigid surface foundations: comparison with ATC-3 provisions, Earthquake Engineering and Structural Dynamics, V. 24, pp. 419–437 (1995).
94 A.A. STAMOS and D.E. BESKOS, "Dynamic analysis of large 3-D underground structures by the BEM", Earthquake Engineering and Structural Dynamics, V. 24, pp. 917–934 (1995).

95. D.E. BESKOS. "Dynamic inelastic structural analysis by boundary element methods", Archives of Computational Methods in Engineering, V. 2, pp. 1–33 (1995).
96. J.P. AGNANTIARIS, D. POLYZOS and D.E. BESKOS, "Some studies on dual reciprocity BEM for elastodynamic analysis", Computational Mechanics, V. 17, pp.270–277 (1996).
97. J. QIAN and D.E. BESKOS, Harmonic wave response of two 3-D rigid surface foundations, Soil Dynamics and Earthquake Engineering, V. 15, pp. 95–110 (1996).
98. A.A. STAMOS and D.E. BESKOS, "3-D seismic response analysis of long lined tunnels in half-space", Soil Dynamics and Earthquake Engineering, V. 15, pp. 111–118 (1996).
99. N. BAZEOS and D.E. BESKOS, "Torsional moments on buildings subjected to wind loads", Engineering Analysis with Boundary Elements, V. 18, pp. 305–310 (1996).
100. D.E. BESKOS, "Boundary element methods in dynamic analysis: Part II (1986–1996), Applied Mechanics Reviews, V. 50, pp. 149–197 (1997).
101. D. POLYZOS, S.V. TSINOPOULOS and D.E. BESKOS, "Static and dynamic boundary element analysis of incompressible linear elasticity", European Journal of Mechanics: A/Solids, V. 17, pp. 515–536 (1998).
102. J.P. AGNANTIARIS, D. POLYZOS and D.E. BESKOS, "Three – dimensional structural vibration analysis by the dual reciprocity BEM", Computational Mechanics, V. 21, pp. 372–381 (1998).
103. D. POLYZOS and D.E. BESKOS, "A new time domain BEM formulation and solution procedure for generalized dynamic coupled thermoelasticity", Engineering Analysis with Boundary Elements, V. 22, pp. 111–116 (1998).
104. S.V. TSINOPOULOS, S.E. KATTIS, D. POLYZOS and D.E. BESKOS, "An advanced boundary element method for axisymmetric elastodynamic analysis", Computer Methods in Applied Mechanics and Engineering, V. 175, pp. 53–70 (1999).
105. S.E. KATTIS, D. POLYZOS and D.E. BESKOS, "Modelling of pile wave barriers by effective trenches and their screening effectiveness", Soil Dynamics and Earthquake Engineering, V. 18, pp. 1–10 (1999).
106. S. E. KATTIS, D. POLYZOS and D.E. BESKOS, "Vibration isolation by a row of piles using a 3-D frequency domain BEM", International Journal for Numerical Methods in Engineering, V. 46, pp. 713–728 (1999).
107. C.P. PROVIDAKIS and D.E. BESKOS, "Dynamic analysis of plates by boundary elements", Applied Mechanics Reviews, V. 52, pp. 213–236 (1999).
108. G.D. HATZIGEORGIOU, D.E. BESKOS, D.D. THEODORAKOPOULOS and M. SFAKIANAKIS, "Static and dynamic analysis of the Arta Bridge by finite elements", Facta Universitatis-Architecture and Civil Engineering Series, V. 2, pp. 41–51, (1999).
109. C.P. PROVIDAKIS and D.E. BESKOS, "Inelastic transient dynamic analysis of Reissner–Mindlin plates by the D/BEM", International Journal for Numerical Methods in Engineering, V. 49, pp. 383–397 (2000).

110. D.D. THEODORAKOPOULOS, A.P. CHASIAKOS and D.E. BESKOS, "Dynamic pressures on rigid cantilever walls retaining poroelastic soil media. Part I: exact solution", Soil Dynamics and Earthquake Engineering, V. 21, pp.315–338(2001).
111. D.D. THEODORAKOPOULOS, A.P. CHASIAKOS and D.E. BESKOS, "Dynamic pressures on rigid cantilever walls retaining poroelastic soil media. Part II: approximate solution", Soil dynamics and Earthquake Engineering, V. 21, pp. 339–364 (2001).
112. J.P. AGNANTIARIS, D. POLYZOS and D.E. BESKOS, "Free vibration analysis of non-axisymmetric and axisymmetric structures by the dual reciprocity BEM", Engineering Analysis with Boundary Elements, V. 25 pp.713–723 (2001).
113. G. HATZIGEORGIOU, D. BESKOS, D. THEODORAKOPOULOS and M. SFAKIANAKIS, "A simple concrete damage model for FEM applications", International Journal of Computational Engineering Science, V. 2, pp.267–286 (2001).
114. S. PAPARGYRI-BESKOU and D.E. BESKOS, "On critical viscous damping determination in linear discrete dynamic systems", Acta Mechanica, V. 153, pp.33–45 (2002).
115. N. BAZEOS, G.D. HATZIGEORGIOU, I.D. HONDROS, H. KARAMANEAS, D.L. KARABALIS and D.E. BESKOS, "Static, seismic and stability analyses of a prototype wind turbine steel tower", Engineering Structures, V. 24, pp. 1015–1025 (2002).
116. G.D. HATZIGEORGIOU and D.E. BESKOS, "Static analysis of 3-D damaged solids and structures by BEM", Engineering Analysis with Boundary Elements, V. 26, pp. 521–526 (2002).
117. G.D. HATZIGEORGIOU and D.E. BESKOS, "Dynamic elastoplastic analysis of 3-D structures by the D/BEM", Computers and Structures, V. 80, pp. 339–347 (2002).
118. G.D. HATZIGEORGIOU and D.E. BESKOS, "Dynamic response of 3-D damaged solids and structures by BEM", Computer Modeling in Engineering and Sciences, V. 3, pp. 791–801 (2002).
119. K.G. TSEPOURA, S. PAPARGYRI-BESKOU, D. POLYZOS and D.E. BESKOS, "Static and dynamic analysis of gradient elastic bars in tension", Archive of Applied Mechanics, V. 72, pp. 483–497 (2002).
120. G.D. HATZIGEORGIOU and D.E. BESKOS, "Dynamic analysis of 2-D and 3-D quasi-brittle solids and structures by D/BEM". Theoretical and Applied Mechanics, V. 40, pp. 39–48 (2002).
121. S.E. KATTIS, D.E. BESKOS and A.H.D. CHENG, "2-D Dynamic response of unlined and lined tunnels in poroelastic soil to harmonic body waves", Earthquake Engineering and Structural Dynamics, V. 32, pp. 97–110 (2003).
122. S. PAPARGYRI-BESKOU, K.G. TSEPOURA, D. POLYZOS and D.E. BESKOS, "Bending and stability analysis of gradient elastic beams", International Journal of Solids and Structures, V. 40, pp. 385–400 (2003).

123. D.D. THEODORAKOPOULOS and D.E. BESKOS, "Dynamic pressures on a pair of rigid walls experiencing base rotation and retaining poroelastic soil", Engineering Structures, V. 25, pp. 359–370 (2003).
124. D. POLYZOS, K.G. TSEPOURA, S.V. TSINOPOULOS and D.E. BESKOS, "A boundary element method for solving 2D and 3D static gradient elastic problems, Part I: Integral Formulation", Computer Methods in Applied Mechanics and Engineering, V. 192, pp. 2845–2873 (2003).
125. K.G. TSEPOURA, S.V. TSINOPOULOS, D. POLYZOS and D.E. BESKOS, "A boundary element method for solving 2D and 3D static gradient elastic problems, Part II: Numerical implementation", Computer Methods in Applied Mechanics and Engineering, V. 192, pp. 2875–2907 (2003).
126. D. POLYZOS, K.G. TSEPOURA and D.E. BESKOS, "BEM solutions of frequency domain gradient elastodynamic 3-D problems", Electronic Journal of Boundary Elements, V. 1, pp. 174–200 (2003).
127. S. PAPARGYRI-BESKOU, D. POLYZOS and D.E. BESKOS, "Dynamic analysis of gradient elastic flexural beams", Structural Engineering and Mechanics, V. 15, pp. 705–716 (2003).
128. D.D. THEODORAKOPOULOS, A.P. CHASIAKOS and D.E. BESKOS, "Dynamic effects of moving load on a poroelastic soil medium by an approximate method", International Journal of Solids and Structures, V. 41, pp. 1801–1822 (2004).
129. S. PAPARGYRI-BESKOU and D.E. BESKOS, "Response of gradient-viscoelastic bar to static and dynamic axial load", Acta Mechanica, V. 170, pp. 199–212 (2004).
130. G.D. HATZIGEORGIOU and D.E. BESKOS, "Minimum cost design of fiber-reinforced concrete-filled steel tubular columns", Journal of Constructional Steel Research, V. 61, pp. 167–182 (2004).
131. D. POLYZOS, K.G. TSEPOURA and D.E. BESKOS, "Transient dynamic analysis of 3-D gradient elastic solids by BEM", Computers and Structures, V. 83, pp. 783–792 (2005).
132. T.L. KARAVASILIS, N. BAZEOS and D.E. BESKOS, "Maximum displacement profiles for the performance based seismic design of plane steel moment resisting frames", Engineering Structures, V. 28, pp. 9–22 (2006).
133. A.A. VASILOPOULOS and D.E.BESKOS, "Seismic design of plane steel frames using advanced methods of analysis", Soil Dynamics and Earthquake Engineering, V. 26, pp. 1077–1100 (2006).
134. C. CHRYSANTHAKOPOULOS, N. BAZEOS and D.E. BESKOS, "Approximate formulae for natural periods of plane steel frames", Journal of Constructional Steel Research, V. 62, pp. 592–604 (2006).
135. D.D. THEODORAKOPOULOS and D.E. BESKOS, "Application of Biot's poroelasticity to some soil dynamics problems in civil engineering", Soil Dynamics and Earthquake Engineering, V. 26, pp. 666–679 (2006).
136. G.A. PAPAGIANNOPOULOS and D.E.BESKOS, "On a modal damping identification model of building structures", Archive of Applied Mechanics, V. 76, pp. 443–463 (2006).

137. G.D. HATZIGEORGIOU and D.E. BESKOS, "Direct damage controlled design of concrete structures", Journal of Structural Engineering of ASCE, V. 133, pp. 205–215 (2007).
138. A.V. ASIMAKOPOULOS, D.L. KARABALIS and D.E. BESKOS, "Inclusion of $P - \Delta$ effect in displacement-based seismic design of plane steel frames", Earthquake Engineering and Structural Dynamics, V. 36, pp. 2171–2188 (2007).
139. T.L. KARAVASILIS, N. BAZEOS and D.E. BESKOS, "Estimation of seismic drift and ductility demands in planar regular X-braced steel frames", Earthquake Engineering and Structural Dynamics, V. 36, pp. 2273–2289 (2007).
140. G.F. KARLIS, S.V. TSINOPOULOS, D. POLYZOS and D.E. BESKOS, "Boundary element analysis of mode I and mixed mode (I and II) crack problems of 2-D gradient elasticity", Computer Methods in Applied Mechanics and Engineering, V. 196, pp. 5092–5103 (2007).
141. T.L. KARAVASILIS, N. BAZEOS and D.E. BESKOS, "Behaviour factor for performance-based seismic design of plane steel moment resisting frames", Journal of Earthquake Engineering, V. 11, pp. 531–559 (2007).
142. T.L.KARAVASILIS, N.BAZEOS and D.E.BESKOS, "Seismic response of plane steel MRF with setbacks: Estimation of inelastic deformation demands", Journal of Constructional Steel Research, V. 64, pp. 644–654 (2008).
143. A.A. VASILOPOULOS, N. BAZEOS and D.E. BESKOS, "Seismic design of irregular space steel frames using advanced methods of analysis", Steel & Composite Structures, V.8, pp. 53–83 (2008).
144. G.F. KARLIS, S.V. TSINOPOULOS, D. POLYZOS and D.E. BESKOS, "2D and 3D boundary element analysis of mode I cracks in gradient elasticity", Computer Modelling in Engineering and Sciences, V. 26, pp. 189–207 (2008).
145. S. PAPARGYRI-BESKOU and D.E. BESKOS, "Static, stability and dynamic analysis of gradient elastic flexural Kirchhoff plates", Archive of Applied Mechanics, V. 98, pp. 625–635 (2008).
146. T.L. KARAVASILIS, N. BAZEOS and D.E. BESKOS, "Drift and ductility estimates in regular steel MRF subjected to ordinary ground motions: a design oriented approach", Earthquake Spectra, V. 24, pp. 431–451 (2008).
147. T.L. KARAVASILIS, N. BAZEOS and D.E. BESKOS, "Estimation of seismic inelastic deformation demands in plane steel MRF with vertical mass irregularities", Engineering Structures, V. 30, pp. 3265–3275 (2008).
148. A.A. VASILOPOULOS and D.E. BESKOS, "Seismic design of space steel frames using advanced methods of analysis", Soil Dynamics and Earthquake Engineering, V. 29, pp. 194–218 (2009).

Doctoral Students – University of Minnesota

1. G.V. Narayanan, "Numerical Operational Methods in Structural Dynamics", 1979; Present Position: Assistant Professor, Department of Engineering Technology, The University of Toledo, Toledo, Ohio 43606, USA.

2. G.D. Manolis, "Dynamic Response of Underground Structures", 1980; Present Position: Professor, Department of Civil Engineering, Aristotle University of Thessaloniki, GR-54124, Thessaloniki, Greece.
3. D.L. Karabalis, "Dynamic Response of 3-D Foundations by the Time Domain Boundary Element Method", 1984; Present position: Professor, Department of Civil Engineering, University of Patras, Patras 26500, Greece.
4. C.C. Spyrakos, "Dynamic Response of Strip Foundations by the Time Domain BEM-FEM Method", 1984; Present Position: Professor and Director of the Earthquake Engineering Laboratory, School of Civil Engineering, National Technical University of Athens, Athens 15773, Greece.
5. B. Dasgupta, "Vibration Isolation of Structures in a Homogeneous Elastic Soil Medium", 1987; Present Position: Staff Engineer, Southwest Research Institute, 6220 Culebra Road, San Antonio, Texas 78023, USA.
6. K.L. Leung, "Vibration Isolation of Structures from Ground-Transmitted Waves in Non-homogeneous Elastic Soil", 1989; Present Position: Lead Analyst, General Motors CAE Application Development, 7000 Chicago Road, Warren, Michigan 48092, USA.

Doctoral Students – University of Patras

1. C. P. Providakis, "Dynamic Analysis of Plates and Shells by the Boundary Element Method", 1988; Present Position: Professor, Department of Engineering Sciences, Technical University of Crete, 73100 Chania, Greece.
2. I. Vgenopoulou, "Analysis of the Dynamic Behavior of Poroelastic Media", 1988; Present Position: Professor, School of Applied Technology, Technical and Educational Institute of Patras, 26334 Patras, Greece .
3. N. Bazeos, "Numerical Determination of Wind Loads on Civil Engineering Structures", 1993; Present Position: Associate Professor, Department of Civil Engineering, University of Patras, 26500 Patras, Greece.
4. D.P.N. Kontoni, "Dynamic Elastoplastic Analysis by the Boundary Element Method", 1993; Present Position: Associate Professor, School of Applied Technology, Technical and Educational Institute of Patras, 26334 Patras, Greece.
5. A.A. Stamos, "Dynamic Response of Underground Structures", 1994; Present Position: A.C. Stamos Civil Engineering Consultants, 53 Sofocleous Str., 10553 Athens, Greece.
6. S. Kattis, "Numerical Solution of Problems of Isolation of Structures from Elastic Surface Waves", (Co-advisor Professor D. Polyzos), 1997; Present Position: Research Engineer, Envirocoustics ABEE, E. Venizelou 7 & Delfon, 14452 Athens, Greece.
7. G.D. Hatzigeorgiou, "Seismic Inelastic Analysis of Underground Structures with Boundary and Finite Elements", 2001; Present Position: Lecturer, Department of Environmental Engineering, Democritus University of Thrace, GR-67100, Xanthi, Greece.

8. A.V. Asimakopoulos, "Displacement Based Seismic Design of Steel Structures", 2003; Present Position: Technical Consultant, 13A Gounari Str., 26221, Patras, Greece.
9. A. Vasilopoulos, "Seismic Design of Steel Structures Using Advanced Methods of Analysis", 2005; Present Position: Civil Engineer, Region of Western Greece, 28 Patron-Athinon Str., 26223 Patras, Greece.
10. T.L. Karavasilis (Co-Advisor Prof. N. Bazeos), "A New Seismic Design Method for Steel Structures", 2007; Present Position: Visiting Research Scientist, ATLSS Engineering Center, Lehigh University, Bethlehem, PA 18015, USA.
11. G. A. Papagiannopoulos, "Seismic Design of Steel Structures using Equivalent Modal Damping Ratios or Modal Strength Reduction Factors", 2008; Present Position: Research Associate, Department of Civil Engineering, University of Patras, 26500 Patras, Greece.

Stability Analysis of Plates

M.H. Aliabadi, P.M. Baiz and E.L. Albuquerque

Abstract Boundary element only formulations for the stability analysis of isotropic and anisotropic plates are presented. Domain integrals which arise in the formulation are transformed into boundary integrals by the radial integration method and the dual reciprocity method. Plate buckling equations are formulated as standard eigenvalue problem. The accuracy of the proposed formulations including buckling coefficients and buckling modes is assessed by comparison with results from literature.

1 Introduction

The increasing use of slender constructions in aeronautical, civil, maritime and offshore structures requires accurate and efficient analysis of their stability under compressive loads. The relatively recent emergence of carbon-fibre composites, with their superior weight and stiffness, has further contributed to the development of light structures. Future generation of aircrafts will be utilizing greater level of composite in their primary structures with possibility of developing post buckling.

Analytical descriptions of plate stability have been presented by Timoshenko and Gere (1961). Numerical methods, particularly the finite element, have been widely used to investigate the problem (Kapania and Raciti, 1989). Recent developments on Experimental, Analytical and Numerical Studies can be found in Falzon and Aliabadi (2008).

Applications of the Boundary Element Method (BEM) to linear stability problems for plate structures have been investigated since the early 1980's. Manolis et al. (1986) developed one of the first direct boundary element formulation dealing with linear elastic stability analysis of bars and plates. Costa and Brebbia (1985) presented a general direct formulation of the problem while Syngellakis and Kang (1987) eliminated the curvatures from the domain and presented a solution which required the modelling in the domain of the deflections only. Reviews of

M.H. Aliabadi (✉)
Department of Aeronautics, Imperial College London, London, UK
e-mail: m.h.aliabadi@imperial.ac.uk

G.D. Manolis, D. Polyzos (eds.), *Recent Advances in Boundary Element Methods*,
DOI 10.1007/978-1-4020-9710-2_1, © Springer Science+Business Media B.V. 2009

the application of BEM to the stability analysis of thin plates can be found in Liu (1987) and Elzein (1991) providing a comprehensive work on plate stability by the boundary element method. Later, Syngellakis and Elzein (1994) presented a boundary element formulation for plate buckling based on Kirchhoff theory under any combination of loadings and support conditions. Nerantzaki and Katsikadelis (1996) developed a boundary element method for buckling analysis of plates with variable thickness. Elastic buckling analysis of plates using boundary elements can also be found in Lin et al. (1999). Buckling analysis of shear deformable isotropic plates was presented by Purbolaksono and Aliabadi (2005). In the case of buckling boundary element formulations applied to non-isotropic plates, the only works available are due to Shi (1990) who presented an orthotropic formulation with a domain discretization and Albuquerque et al. (2008) who presented a formulation without domain discretization.

Other contributions on plate and shell analysis by BEM are due to Providakis and Beskos (2000) for dynamic analysis of elastoplastic Reissner-Mindlin plates. Also Providakis and Beskos (1991) developed a direct BEM approach for the dynamic analysis of thin elastic shallow shells, free and forced vibrations and the effect of the external viscous or internal viscoelastic damping on the response were also taken into account. Baiz and Aliabadi (2006) introduced the first boundary element approach to linear shallow shell buckling, using a shear deformable formulation, buckling modes and buckling coefficients are obtained from the standard eigenvalue problem. Beskos (2003) presents and discusses recent advances in BEM and their solid mechanics applications. More works on plates and shells by BEM can be found in Beskos's book (1992).

In this chapter, boundary element formulation for the linear stability analysis of general plates with no domain discretization are presented. First, classical plate bending and plane elasticity anisotropic formulations are used and the domain integrals due to body forces are transformed into boundary integrals using the radial integration method. Later, a shear deformable plate formulation is presented, in which again domain integrals are transformed into boundary integrals, this time by using the dual reciprocity method. Numerical results at every stage are presented in order to assess the accuracy of the formulations. Buckling coefficients computed using the proposed method are in good agreement with results available in literature and finite element results of a commercial software.

2 Basic Concepts

There are two basic theories of plates. The first theory is known as the classical theory and it was first proposed by Kirchhoff in 1850. The other theory is known as the shear deformable theory and it was first proposed by Reissner in 1947. The difference between both theories rely on the shear deformation through the plate thickness, which is neglected in the former one.

2.1 Reissner Theory

Governing equations of shear deformable plate buckling can be written as:

$$M_{\alpha\beta,\beta} - Q_\alpha = 0 \tag{1}$$

$$Q_{\alpha,\alpha} + \left[\left(N_{\alpha\beta} w_{3,\beta}\right)_{,\alpha}\right] + q_3 = 0 \tag{2}$$

$$N_{\alpha\beta,\beta} + q_\alpha = 0 \tag{3}$$

where $M_{\alpha\beta}$ represent bending moments, Q_α is the shear forces for plate bending and $N_{\alpha\beta}$ denote membrane stresses. Five independent displacements are defined in shear deformable theory: w_i and u_α, where w_α denotes the change of the slope of the normal to the middle surface, w_3 denotes the out-of-plane displacement normal to the middle surface and u_α denotes in-plane displacements of the middle surface. The term in square brackets: $\left(N_{\alpha\beta} w_{3,\beta}\right)_{,\alpha}$ is a body term due to the large deflection of w_3, this term is the usual extra term that appears in the geometrically nonlinear equilibrium equations of plates and shells. Indicial notation is used throughout this chapter, Greek indices will vary from 1 to 2 and Roman indices from 1 to 3.

Strain tensors in shear deformable linear elastic plate theory can be derived from the deformation pattern of a differential plate element,

$$\varepsilon_{\alpha\beta} = \frac{1}{2}\left(u_{\alpha,\beta} + u_{\beta,\alpha}\right) \tag{4}$$

$$\gamma_{\alpha 3} = w_\alpha + w_{3,\alpha} \tag{5}$$

$$\kappa_{\alpha\beta} = 2\chi_{\alpha\beta} = w_{\alpha,\beta} + w_{\beta,\alpha} \tag{6}$$

where $\gamma_{\alpha 3}$ denotes transverse shear strains, $\kappa_{\alpha\beta}$ denote flexural strains and $\varepsilon_{\alpha\beta}$ represent membrane strains.

Relationships between stress resultants and strains were derived by Reissner (1947), as follows:

$$M_{\alpha\beta} = D\frac{1-\nu}{2}\left(2\chi_{\alpha\beta} + \frac{2\nu}{1-\nu}\chi_{\gamma\gamma}\delta_{\alpha\beta}\right) \tag{7}$$

$$Q_\alpha = C\gamma_{\alpha 3} \tag{8}$$

$$N_{\alpha\beta} = B\frac{1-\nu}{2}\left(2\varepsilon_{\alpha\beta} + \frac{2\nu}{1-\nu}\varepsilon_{\gamma\gamma}\delta_{\alpha\beta}\right) \tag{9}$$

where $\lambda_s = \sqrt{10}/h$ is the shear factor and $\delta_{\alpha\beta}$ is the Kronecker delta, $B(= Eh/(1-\nu^2))$ is the tension stiffness, $D(= Eh^3/[12(1-\nu^2)])$ is the bending stiffness of the plate, and $C(=\left[D(1-\nu)\lambda^2\right]/2)$ is the shear stiffness.

2.2 Kirchhoff Theory

In the case of classical plate theory, equations are formed in terms of only the out of plane deflection w_3 and its derivative. Assuming that transverse shear deformation can be neglected,

$$\gamma_{\alpha 3} = 0, \tag{10}$$

equation (6) can be written as:

$$\kappa_{\alpha\beta} = -\frac{1}{2}\left(w_{3,\alpha\beta} + w_{3,\beta\alpha}\right) \tag{11}$$

In the same way, equation (7) can be written as:

$$M_{\alpha\beta} = D\,(1-\nu)\left(\kappa_{\alpha\beta} + \frac{\nu}{1-\nu}\kappa_{\gamma\gamma}\delta_{\alpha\beta}\right) \tag{12}$$

The shear force Q_α, given by equation (8) for Reissner plate, is obtained from equilibrium equation (1) for Kirchhoff plates and is given by:

$$Q_\alpha = M_{\alpha\beta,\beta}, \tag{13}$$

where

$$Q_\alpha = -Dw_{3,\gamma\gamma\alpha}. \tag{14}$$

Equilibrium equation in terms of displacement, obtained using equations (2) and (14), is given by:

$$D\nabla^4 w_3 + \left[\left(N_{\alpha\beta} w_{3,\beta}\right)_{,\alpha}\right] + q_3 = 0, \tag{15}$$

where $\nabla^4 = \Delta^2$ and Δ is the laplacian.

Bending moments are given by:

$$M_n = M_\alpha n_\alpha n_\beta, \tag{16}$$

The equivalent Kirchhoff shear is the sum of the tangential derivative of the twisting moment M_{ns} and the shear force Q_n and is given by:

$$V_n = \frac{\partial M_{ns}}{ds} + Q_n, \tag{17}$$

where s denotes the tangential direction,

$$M_{ns} = M_{\alpha\beta} n_\alpha t_\beta, \tag{18}$$

and

$$Q_n = Q_\alpha n_\alpha. \tag{19}$$

For anisotropic plate behaviour, only constitutive equations have to be modified. In the case of symmetric laminate composites, equation (12) is given by:

$$\begin{Bmatrix} M_{11} \\ M_{22} \\ M_{12} \end{Bmatrix} = \begin{bmatrix} D_{11} & D_{12} & D_{16} \\ D_{12} & D_{22} & D_{26} \\ D_{16} & D_{26} & D_{66} \end{bmatrix} \begin{Bmatrix} \kappa_{11} \\ \kappa_{22} \\ \kappa_{12} \end{Bmatrix} \tag{20}$$

where D_{11}, D_{22}, D_{66}, D_{12}, D_{16}, and D_{26} are the anisotropic thin plate stiffness constants, as given in Lekhnitskii (1968).

Using equations (13) and (20), shear forces are given by:

$$\begin{Bmatrix} Q_1 \\ Q2 \end{Bmatrix} = \begin{bmatrix} D_{11} & D_{12} + D_{16} & D_{16} + D_{66} & 0 \\ 0 & D_{12} + D_{66} & 2D_{26} & D_{22} \end{bmatrix} \begin{Bmatrix} \kappa_{11,1} \\ \kappa_{12,1} \\ \kappa_{12,2} \\ \kappa_{22,2} \end{Bmatrix} \tag{21}$$

In the same way, equation (9) for membrane stress are given by:

$$\begin{Bmatrix} N_{11} \\ N_{22} \\ N_{12} \end{Bmatrix} = \begin{bmatrix} A_{11} & A_{12} & A_{16} \\ A_{12} & A_{22} & A_{26} \\ A_{16} & A_{26} & A_{66} \end{bmatrix} \begin{Bmatrix} \varepsilon_{11} \\ \varepsilon_{22} \\ \varepsilon_{12} \end{Bmatrix} \tag{22}$$

where A_{11}, A_{22}, A_{66}, A_{12}, A_{16}, and A_{26} are the anisotropic extensional stiffness constants, as given in Lekhnitskii (1968).

The equilibrium equation (15) is given by:

$$D_{11}w_{3,1111} + 4D_{16}w_{3,1112} + 2(D_{12} + D_{66})w_{3,1122} + 4D_{26}w_{3,1222} + D_{22}w_{3,2222} = (N_{\alpha\beta}w_{3,\beta})_{,\alpha} \tag{23}$$

Membrane equations for equilibrium (3), strain-displacement (4) and stress-strain (9) are the same for classical and shear deformable theories.

3 Shear Deformable Plate Boundary Integral Equations

Eigenvalue solutions are generally used to estimate the critical buckling loads of stiff structures which carry their design loads primarily by axial or membrane action, rather than by bending action. Their response usually involves very little deformation prior to buckling. Also a linear eigenvalue analysis usually represents the first step in the stability analysis, because the eigenmodes obtained from this procedure can be used in the investigation of sensitivity of the structure to imperfections.

To reduce the linear buckling problem of plates to an eigenvalue formulation, the distribution of membrane stresses $N^{(l)}$ should be known when the structure is subjected to the applied compressive load. Later, it is assumed that membrane stresses in the buckling state are given by the product between the stress at the pre-buckling state and their critical buckling coefficients.

Boundary integral equations for shear deformable plate problems can be derived by considering the integral representations of the governing equations as given by (Aliabadi, 2002),

$$c_{ij}(\mathbf{x}')w_j(\mathbf{x}') + \fint_\Gamma P^*_{ij}(\mathbf{x}', \mathbf{x})w_j(\mathbf{x})\,d\Gamma(\mathbf{x}) = \int_\Gamma W^*_{ij}(\mathbf{x}', \mathbf{x})p_j(\mathbf{x})\,d\Gamma(\mathbf{x}) + \int_\Omega W^*_{i3}(\mathbf{x}', \mathbf{X})q_3(\mathbf{X})d\Omega(\mathbf{X}) \tag{24}$$

$$c_{\theta\alpha}\left(\mathbf{x}'\right)u_\alpha(\mathbf{x}') + \fint_\Gamma T^{*(i)}_{\theta\alpha}(\mathbf{x}', \mathbf{x})u_\alpha(\mathbf{x})d\Gamma(\mathbf{x}) = \int_\Gamma U^*_{\theta\alpha}(\mathbf{x}', \mathbf{x})t_\alpha(\mathbf{x})d\Gamma(\mathbf{x}) + \int_\Omega U^*_{\theta\alpha}(\mathbf{x}', \mathbf{X})q_\alpha(\mathbf{X})d\Omega(\mathbf{X}) \tag{25}$$

where $\mathbf{x}', \mathbf{x} \in \Gamma$ are source and field points respectively. $c_{ij}(\mathbf{x}')$ represent the jump terms, whose value is equal to $0.5\delta_{ij}$ when $\mathbf{x}'$ is located on a smooth boundary. $U^*_{\theta\alpha}(\mathbf{X}', \mathbf{x})$ and $T^{(i)*}_{\theta\alpha}(\mathbf{X}', \mathbf{x})$ are the fundamental solutions for in-plane displacements and membrane tractions respectively. $W^*_{ij}(\mathbf{X}', \mathbf{x})$ and $P^*_{ij}(\mathbf{X}', \mathbf{x})$ are the fundamental solutions for rotations and out-of-plane displacements and bending and shear tractions respectively. All these fundamental solutions can be found in (Aliabadi, 2002).

Initially determination of membrane stress resultants in the domain is the first step solution for the analysis. Next, the plate buckling equations are arrange as an eigenvalue problem, resulting in an expression in terms of buckling modes and the critical loading factor (λ).

3.1 Membrane Integral Equations

After solving equations (24) and (25), boundary values $t_\gamma(\mathbf{x})$ and $u_\gamma(\mathbf{x})$. In the absence of membrane body forces ($q_\alpha = 0$), the membrane stress resultant integral equation can be written as follows (Aliabadi, 2002):

$$N_{\alpha\beta}\left(\mathbf{X}'\right) = \int_\Gamma U^*_{\alpha\beta\gamma}(\mathbf{X}', \mathbf{x})t_\gamma(\mathbf{x})d\Gamma(\mathbf{x}) - \int_\Gamma T^{(i)*}_{\alpha\beta\gamma}(\mathbf{X}', \mathbf{x})u_\gamma(\mathbf{x})d\Gamma(\mathbf{x}) \tag{26}$$

The kernels $U^*_{\alpha\beta\gamma}$ and $T^{(i)*}_{\alpha\beta\gamma}$ can be found in Aliabadi (2002).

3.2 Integral Formulations for Linear Buckling Problem

In a similar way as equation (24) was obtained, plate buckling integral equations can be written as follows (Purbolaksono and Aliabadi, 2005),

$$
\begin{aligned}
&c_{ij}(\mathbf{x}')w_j(\mathbf{x}') + \fint_\Gamma P^*_{ij}(\mathbf{x}',\mathbf{x})w_j(\mathbf{x})\,d\Gamma(\mathbf{x}) \\
&= \int_\Gamma W^*_{ij}(\mathbf{x}',\mathbf{x})p_j(\mathbf{x})\,d\Gamma(\mathbf{x}) + \lambda\int_\Omega W^*_{i3}(\mathbf{x}',\mathbf{X})q_3(\mathbf{X})d\Omega(\mathbf{X}) \\
&\quad +\lambda\int_\Omega W^*_{i3}(\mathbf{x}',\mathbf{X})(N_{\alpha\beta}w_{3,\beta})_{,\alpha}(\mathbf{X})d\Omega(\mathbf{X})
\end{aligned}
\tag{27}
$$

Equation (27) can be rewritten in a more compact form as shown in (Purbolaksono and Aliabadi, 2005),

$$
\begin{aligned}
&c_{ij}(\mathbf{x}')w_j(\mathbf{x}') + \fint_\Gamma P^*_{ij}(\mathbf{x}',\mathbf{x})w_j(\mathbf{x})\,d\Gamma(\mathbf{x}) \\
&= \int_\Gamma W^*_{ij}(\mathbf{x}',\mathbf{x})p_j(\mathbf{x})\,d\Gamma(\mathbf{x}) + \lambda\int_\Omega W^*_{i3}(\mathbf{x}',\mathbf{X})f_b(\mathbf{X})d\Omega(\mathbf{X})
\end{aligned}
\tag{28}
$$

Out-of-plane displacement $w_3(\mathbf{X}')$ is also required to arrange the eigenvalue equation:

$$
\begin{aligned}
w_3(\mathbf{X}') &= \int_\Gamma W^*_{3j}(\mathbf{X}',\mathbf{x})p_j(\mathbf{x})\,d\Gamma(\mathbf{x}) - \int_\Gamma P^*_{3j}(\mathbf{X}',\mathbf{x})w_j(\mathbf{x})\,d\Gamma(\mathbf{x}) \\
&\quad +\lambda\int_\Omega W^*_{33}(\mathbf{X}',\mathbf{X})f_b(\mathbf{X})d\Omega(\mathbf{X})
\end{aligned}
\tag{29}
$$

where

$$
f_b(\mathbf{X}) = (N_{\alpha\beta}w_{3,\beta})_{,\alpha}(\mathbf{X}). \tag{30}
$$

4 Classical Plate Boundary Integral Equations

Boundary integral equations for classical plate problems can be derived by considering the integral representations of the governing equations as given by (Aliabadi, 2002),

$$
\begin{aligned}
&c_{\alpha\beta}(\mathbf{x}')v_\beta(\mathbf{x}') + \fint_\Gamma P^*_{\alpha\beta}(\mathbf{x}',\mathbf{x})v_\beta(\mathbf{x})\,d\Gamma(\mathbf{x}) = \int_\Gamma W^*_{\alpha\beta}(\mathbf{x}',\mathbf{x})p_\beta(\mathbf{x})\,d\Gamma(\mathbf{x}) \\
&+ \int_\Omega W^*_{\alpha 3}(\mathbf{x}',\mathbf{X})q_3(\mathbf{X})d\Omega(\mathbf{X}) + \sum_{i=1}^{N_c} R^*_{\alpha i}(\mathbf{X})w_{3i} - \sum_{i=1}^{N_c} R_i v^*_{\alpha i}(\mathbf{X})
\end{aligned}
\tag{31}
$$

where $v_1 = w_3$, $v_2 = \partial w_3/\partial n$, $p_1 = V_n$, $p_2 = M_n$, R_i is the thin-plate reaction of corner i, $R^*_{2i} = \partial R^*_{1i}/\partial n$; w_{3i} is the transverse displacement of corner i, N_c is the number of corners;

Terms with asterisk represent the following thin plate fundamental solutions:

$$P^*_{\alpha\beta}(\mathbf{x}', \mathbf{x}) = \begin{bmatrix} V^*_n(\mathbf{x}', \mathbf{x}) & -M^*_n(\mathbf{x}', \mathbf{x}) \\ \dfrac{\partial V^*_n(\mathbf{x}', \mathbf{x})}{\partial m} & -\dfrac{\partial M^*_n(\mathbf{x}', \mathbf{x})}{\partial m} \end{bmatrix}, \tag{32}$$

and

$$W^*_{\alpha\beta}(\mathbf{x}', \mathbf{x}) = \begin{bmatrix} w^*_3(\mathbf{x}', \mathbf{x}) & -\frac{\partial w^*_3(\mathbf{x}', \mathbf{x})}{\partial n} \\ \dfrac{\partial w^*_3(\mathbf{x}', \mathbf{x})}{\partial m} & -\dfrac{\partial^2 w^*_3(\mathbf{x}', \mathbf{x})}{\partial n \partial m} \end{bmatrix}, \tag{33}$$

where m stands for the normal vector at the source point.

Fundamental solutions for isotropic thin plates can be found in Aliabadi (2002) and for anisotropic thin plates can be found in Albuquerque et al. (2006).

Membrane stress for thin plates are given, as for thick plates, by the integral equations (26).

5 Approximation of Domain Integrals

As can be seen in equations of this section, domain integrals arise in the plate formulation owing to the contribution of in-plane stresses to the out of plane direction. In order to transform these integrals into boundary integrals, consider that a body force b is approximated over the domain Ω as a sum of M products between approximation functions f_m and unknown coefficients γ_m, that is:

$$b(P) \cong \sum_{m=1}^{M} \gamma_m f_m. \tag{34}$$

The approximation function used in this work is:

$$f_m = 1 + R, \tag{35}$$

Equation (34) can be written in a matrix form, considering all boundary and domain source points, as:

$$\mathbf{b} = \mathbf{F}\gamma \tag{36}$$

Thus, γ can be computed as:

$$\gamma = \mathbf{F}^{-1}\mathbf{b} \tag{37}$$

Using equation (37) and following the procedure presented by Albuquerque (2007), domain integrals that come from these body forces can be transformed into boundary integrals.

6 Matrix Equations

Quadratic isoparametric boundary elements are used to describe the geometry and the function along Γ, while for the implementation of the dual reciprocity method or the radial integration method and the evaluation of the derivative terms several points are uniformly distributed all over the domain Ω. After the discretization of equations into boundary elements and collocation of the source points in all boundary nodes, a linear system is generated. It is worth notice that the only loads considered in the linear buckling equations are that related to the in-plane stress $N_{\alpha\beta}$ that is multiplied by the critical load factor λ. This means that all the known values (homogeneous boundary conditions) are set to zero.

Dividing the boundary into Γ_1 and Γ_2, this linear system can be written as:

$$\begin{bmatrix} \mathbf{H}_{11} & \mathbf{H}_{12} \\ \mathbf{H}_{21} & \mathbf{H}_{22} \end{bmatrix} \begin{Bmatrix} \mathbf{w}_1 \\ \mathbf{w}_2 \end{Bmatrix} - \begin{bmatrix} \mathbf{G}_{11} & \mathbf{G}_{12} \\ \mathbf{G}_{21} & \mathbf{G}_{22} \end{bmatrix} \begin{Bmatrix} \mathbf{V}_1 \\ \mathbf{V}_2 \end{Bmatrix} = \lambda \begin{bmatrix} \mathbf{M}_{11} & \mathbf{M}_{12} \\ \mathbf{M}_{21} & \mathbf{M}_{22} \end{bmatrix} \begin{Bmatrix} \mathbf{w}_1 \\ \mathbf{w}_2 \end{Bmatrix}, \tag{38}$$

where Γ_1 stands for stands for the part of the boundary where displacements and rotations are zero and Γ_2 stands for the part of the boundary where bending moment and tractions are zero. Indices 1 and 2 stand for boundaries Γ_1 and Γ_2, respectively. Matrices **H, G**, and **M** are influence matrices of the boundary element method due to integral terms of equations (28) and (29).

As $\mathbf{w}_1 = \mathbf{0}$ and $\mathbf{V}_2 = \mathbf{0}$, equation (38) can be written as:

$$\begin{aligned} \mathbf{H}_{12}\mathbf{w}_2 - \mathbf{G}_{11}\mathbf{V}_1 &= \lambda \mathbf{M}_{12}\mathbf{w}_2, \\ \mathbf{H}_{22}\mathbf{w}_2 - \mathbf{G}_{21}\mathbf{V}_1 &= \lambda \mathbf{M}_{22}\mathbf{w}_2 \end{aligned} \tag{39}$$

or

$$\hat{\mathbf{H}}\mathbf{w}_2 = \lambda \hat{\mathbf{M}}\mathbf{w}_2, \tag{40}$$

where $\hat{\mathbf{H}}$ and $\hat{\mathbf{M}}$ are given by:

$$\begin{aligned} \hat{\mathbf{H}} &= \mathbf{H}_{22} - \mathbf{G}_{21}\mathbf{G}_{11}^{-1}\mathbf{H}_{12}, \\ \hat{\mathbf{M}} &= \mathbf{M}_{22} - \mathbf{G}_{21}\mathbf{G}_{11}^{-1}\mathbf{M}_{12}. \end{aligned} \tag{41}$$

The matrix equation (40) can be rewritten as an eigen vector problem

$$\mathbf{A}\mathbf{w}_2 = \frac{1}{\lambda}\mathbf{w}_2, \tag{42}$$

where

$$\mathbf{A} = \hat{\mathbf{H}}^{-1}\hat{\mathbf{M}}. \tag{43}$$

Provided that **A** is non-symmetric, eigenvalues and eigenvectors of equation (42) can be found using standard numerical procedures for non symmetric matrices.

7 Numerical Results

The proposed techniques are applied to several benchmark problems to assess their accuracy and efficiency. The numerical results are presented in terms of the dimensionless parameter K_{cr} which, for isotropic materials, is given by:

$$K_{cr} = \frac{N_{cr} b^2}{D} \tag{44}$$

and, for anisotropic materials, is given by

$$K_{cr} = \frac{N_{cr} b^2}{D_{22}} \tag{45}$$

where N_{cr} is the critical load ($N_{cr} = \lambda \times$ the applied load) and b is the edge width of the plate (see Fig. 1).

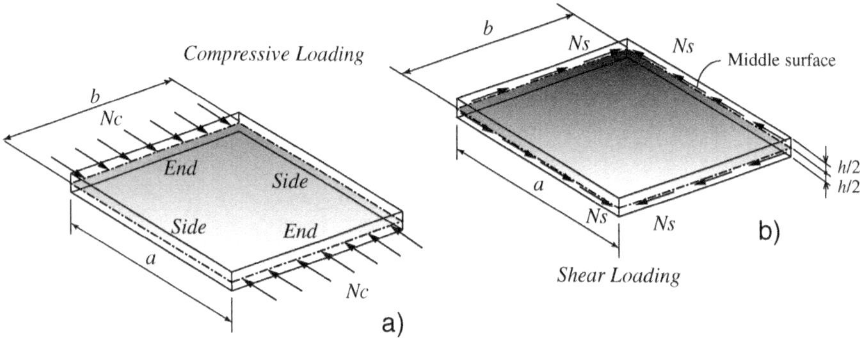

Fig. 1 Rectangular plates subjected to compressive and shear loads

7.1 *Rectangular Anisotropic Plate Subjected to Uniform Compression*

Consider a square graphite/epoxy plate under different boundary conditions. The thickness of the plate is $h = 0.01$ m. The material properties are: elastic moduli $E_1 = 181$ GPa and $E_2 = 10.3$ GPa, Poisson ratio $\nu_{12} = 0.28$, and shear modulus $G_{12} = 7.17$ GPa.

The mesh used has 28 quadratic discontinuous boundary elements of the same length (7 per edge) and 49 (7×7) uniformly distributed internal points.

Table 1 Critical load parameter K_{cr} for a graphite/epoxy plate with different boundary conditions

Case	Borders	Loadings	This work	Ref. Shi (1990)	Ref. Lekhnitskii (1968)
1	SSSS	$N_1 \neq 0$	130.82	–	129.78
2	SSSS	$N_2 \neq 0$	71.53	71.36	69.46
3	CCCC	$N_1 \neq 0$	493.70	481.21	–
4	CCCC	$N_2 \neq 0$	168.27	168.16	–
5	CSCS	$N_1 \neq 0$	161.47	163.24	162.03
6	CSCS	$N_2 \neq 0$	146.47	143.89	141.33

The plate is under uniformly uniaxial compression and the critical load parameter K_{cr} is computed considering a series of boundary conditions (sides and ends are depicted in Fig. 1):

CCCC : sides and ends clamped
SSSS : sides and ends simply supported
CSCS : ends clamped, sides simply supported
SFSC : one side free, one side clamped, end simply supported
SSSF : one side free, the other side and ends simply supported

The results are shown in Table 1 together with results obtained by Shi (1990) using a boundary element formulation with domain discretization and the analytical solution presented by Lekhnitskii (1968).

As it can be seen, there is a good agreement between the results obtained in this work and those presented in literature. Critical buckling modes for each case are shown in Fig. 2.

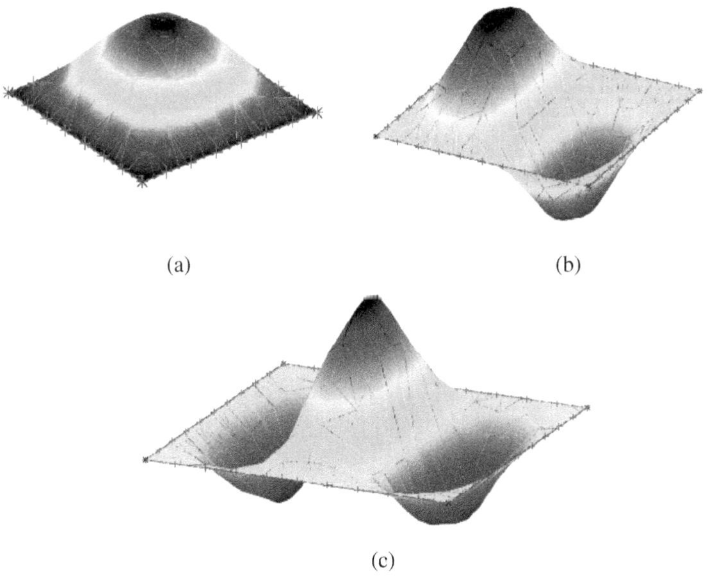

Fig. 2 Critical buckling modes: **(a)** cases 1, 3, and 5; **(b)** case 2; and **(c)** cases 4 and 6

7.2 Rectangular Plate Subjected to Uniform Compression

In this example shear deformable rectangular plates with different aspect ratios are subjected to compression loads, as shown in Fig. 1a. The buckling coefficients K for different aspect ratio a/b are presented in Fig. 3. The legend for the boundary conditions considered were given in the previous example.

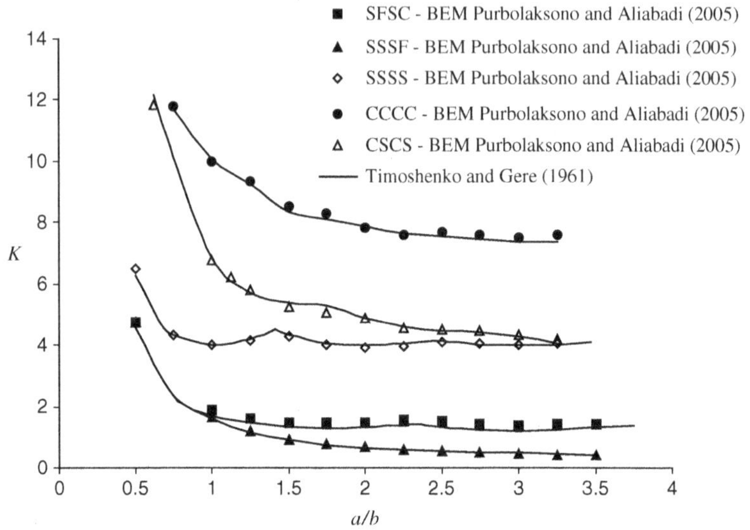

Fig. 3 Buckling coefficients for rectangular plates with different boundary conditions

7.3 Rectangular Plate Subjected to Shear Loads

In this final example, shear buckling (see Fig. 1b) of isotropic shear deformable rectangular plates with different aspect ratio a/b and boundary conditions are presented. Simply supported and clamped boundary conditions are considered. Buckling coefficients are compared with analytical and FEM solutions as shown in Fig. 4. Good agreement (less than 1.5% difference) is achieved in both cases.

8 Conclusions

This chapter presents boundary element formulations for the stability analysis of anisotropic Kirchhoff and Reissner plates where domain integrals are transformed into boundary integrals by the radial integration method and the dual reciprocity method. As the radial integration method does not demand particular solutions, it is easier to implement than the dual reciprocity boundary element method, which

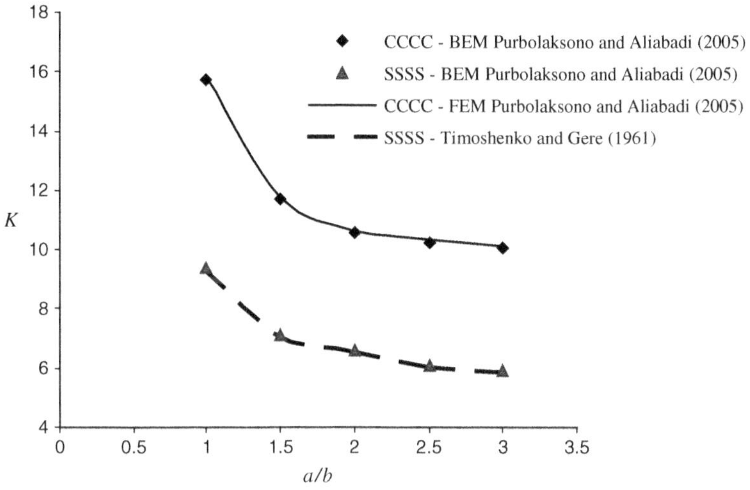

Fig. 4 Shear buckling coefficients for simply supported and clamped rectangular plates

requires particular solutions. Results obtained with the proposed formulations are in good agreement with results presented in the literature.

References

E. L. Albuquerque, P. M. Baiz, and M. H. Aliabadi. Stability analysis of composite plates by the boundary element method. *in International Conference on Boundary Element Techniques – 2008*, R. Abascal and M. H. Aliabadi (Editors), pp. 455–460, EC Ltd, Seville, Spain, 2008.

E. L. Albuquerque, P. Sollero, and W. P. Paiva. The radial integration method applied to dynamic problems of anisotropic plates. *Communications in Numerical Methods in Engineering*, 23:805–818, 2007.

E. L. Albuquerque, P. Sollero, W. Venturini, and M. H. Aliabadi. Boundary element analysis of anisotropic Kirchhoff plates. *International Journal of Solids and Structures*, 43: 4029–4046, 2006.

M. H. Aliabadi. *Boundary Element Method, the Application in Solids and Structures*. John Wiley and Sons Ltd, New York, 2002.

P. M. Baiz and M. H. Aliabadi. Linear buckling analysis of shallow shells by the boundary domain element method. *Computer Modelling in Engineering and Sciences*, 13:19–34, 2006.

D. E. Beskos (Editor). *Boundary element analysis of plates and shells*. Springer Verlag, New York, 1992.

D. E. Beskos. Dynamic analysis of structures and structural systems. *in Boundary Element Advances in Solid Mechanics*, D. E. Beskos and G. Maier (Editors). pp. 1–53, Springer, New York, 2003.

J. A. Costa and C. A. Brebbia. Elastic buckling of plates using the boundary element method. *in Boundary Element VII*, C. A. Brebbia (Editor). pp. 429–442, Springer-Verlag, Berlin, 1985.

A. Elzein. *Lecture notes in engineering 64*, C.A. Brebbia and S.A. Orszag (Editors). Springer-Verlag, New York, 1991.

B. G. Falzon and M. H. Aliabadi (Editors). *Buckling and Post buckling structures*. Imperial College Press, Singapore, 2008.

R. K. Kapania and S. Raciti. Recent advances in analysis of laminated beams and plates. Part I. Shear effects and buckling. *AIAA Journal*, 27:923–934, 1989.

S. G. Lekhnitskii. *Anisotropic Plates*. Gordon and Breach, New York, 1968.

J. Lin, R. C. Duffield, and H. Shi. Buckling analysis of elastic plates by boundary element method. *Engineering Analysis with Boundary Element*, 23:131–137, 1999.

Y. Liu. Elastic stability analysis of thin plate by the boundary element method – new formulation. *Engineering Analysis*, 4:160–164, 1987.

G. D. Manolis, D. E. Beskos and M. F. Pineros. Beam and plate stability by boundary elements. *Computers and Structures*, 22:917–923, 1986.

M. S. Nerantzaki and J. T. Katsikadelis. Buckling of plates with variable thickness, an analog equation solution. *Engineering Analysis with Boundary Element*, 18:149–154, 1996.

C. P. Providakis and D. E. Beskos. Free and forced vibrations of shallow shells by boundary and interior elements. *Computer Methods in Applied Mechanics and Engineering*, 92:55–74, 1991.

C. P. Providakis and D. E. Beskos. Inelastic transient dynamic analysis of Reissner-Mindlin plates by the D/BEM. *International Journal for Numerical Methods in Engineering*, 49: 383–397, 2000.

J. Purbolaksono and M. H. Aliabadi. Buckling analysis of shear deformable plates by boundary element method. *International Journal for Numerical Methods in Engineering*, 62: 537–563, 2005.

E. Reissner. On bending of elastic plates. *Quarterly of Applied Mathematics*, 5:55–68, 1947.

G. Shi. Flexural vibration and buckling analysis of orthotropic plates by the boundary element method. *Journal of Solids and Structures*, 26:1351–1370, 1990.

S. Syngellakis and E. Elzein. Plate buckling loads by the boundary element method. *International Journal for Numerical Methods in Engineering*, 37:1763–1778, 1994.

S. Syngellakis and M. Kang. A boundary element solution of the plate buckling problem. *Engineering Analysis*, 4:75–81, 1987.

S. Timoshenko and J. M. Gere. *Theory of elastic stability (2nd edn)*. McGraw-Hill, New York, 1961.

A. C. Walker. A brief review of plate buckling research. *in Behaviour of thin-walled structures*, J. Rhodes and J. Spence (Editors). Elsevier, London, 1984.

Multi-Level Fast Multipole BEM for 3-D Elastodynamics

Marc Bonnet, Stéphanie Chaillat and Jean-François Semblat

Abstract To reduce computational complexity and memory requirement for 3-D elastodynamics using the boundary element method (BEM), a multi-level fast multipole BEM (FM-BEM) based on the diagonal form for the expansion of the elastodynamic fundamental solution is proposed and demonstrated on numerical examples involving single-region and multi-region configurations where the scattering of seismic waves by a topographical irregularity or a sediment-filled basin is examined.

1 Introduction

The boundary element method (BEM) is a mesh reduction method, subject to restrictive constitutive assumptions but yielding highly accurate solutions. It is in particular well suited to unbounded-domain idealizations commonly used e.g. in seismic wave modelling (Dangla et al., 2005 and Guzina et al., 2001). Many early references on BEMs and their application to elastodynamics can be found in the review articles (Beskos, 1987, 1997) and in e.g. (Bonnet, 1999).

To reduce computational complexity and memory requirement for 3-D elastodynamics using the boundary element method (BEM), a multi-level fast multipole BEM (FM-BEM) treatment is proposed (see e.g. (Gumerov and Duraiswami, 2005, and Nishimura, 2002) for expositions of FM-BEM for Helmholtz-type problems and related references). By adapting to this context recent implementations of the FMM for the Maxwell equations (Darve, 2000 and Sylvand, 2002), it brings significant improvement over previously published elastodynamics FM-BEM (Fujiwara, 2000). The diagonal form (Rokhlin, 1993) for the expansion of the elastodynamic fundamental solution is used, with a truncation parameter adjusted to the subdivision level, a feature necessary for achieving optimal computational efficiency. The formulation is extended to problems featuring piecewise-homogeneous media via a multi-region FM-BEM whose unknowns feature displacements and tractions

M. Bonnet (✉)
LMS, Ecole Polytechnique, France
e-mail: bonnet@lms.polytechnique.fr

G.D. Manolis, D. Polyzos (eds.), *Recent Advances in Boundary Element Methods*,
DOI 10.1007/978-1-4020-9710-2_2, © Springer Science+Business Media B.V. 2009

on interfacial boundary elements. The correctness and computational performances of the proposed single- and multi-region versions of the elastodynamic FMM are demonstrated here on numerical examples featuring up to $O(2\times 10^5)$ DOFs run on a single-processor PC, including a 3-D site effect benchmark (semi-spherical empty canyon or sediment-filled basin, with previously published results (Mossessian and Dravinski, 1990; Reinoso et al., 1997 and Sánchez-Sesma, 1983) for low-frequency cases allowing comparisons).

2 Elastodynamic Boundary Element Method

Let $\Omega \subset \mathbb{R}^3$ denote the region of space occupied by a three-dimensional elastic solid with isotropic constitutive properties defined by μ (shear modulus), ν (Poisson's ratio) and ρ (mass density). Time-harmonic motions, with circular frequency ω, induced by a prescribed traction distribution $\boldsymbol{t}^D$ on the boundary $\partial\Omega$ and in the absence of body forces are considered for ease of exposition, other boundary conditions needing only simple modifications to the treatment presented therein. The displacement $\boldsymbol{u}$ on the boundary is governed by the well-known integral equation:

$$(\mathcal{K}u)(\boldsymbol{x}) = f(\boldsymbol{x}) \qquad (\boldsymbol{x} \in \partial\Omega), \tag{1}$$

with the linear integral operator $\mathcal{K}$ and the right-hand side f defined by

$$(\mathcal{K}u)(\boldsymbol{x}) = c_{ik}(\boldsymbol{x})u_i(\boldsymbol{x}) + (\text{P.V.})\int_{\partial\Omega} u_i(\boldsymbol{y})T_i^k(\boldsymbol{x},\boldsymbol{y})dS_y \tag{2}$$

$$f(\boldsymbol{x}) = \int_{\partial\Omega} t_i^D(\boldsymbol{y})U_i^k(\boldsymbol{x},\boldsymbol{y})dS_y, \tag{3}$$

where (P.V.) indicates a Cauchy principal value (CPV) singular integral and the *free-term* $c_{ik}(\boldsymbol{x})$ is equal to $0.5\delta_{ik}$ in the usual case where $\partial\Omega$ is smooth at $\boldsymbol{x}$. Moreover, $U_i^k(\boldsymbol{x},\boldsymbol{y})$ and $T_i^k(\boldsymbol{x},\boldsymbol{y})$ denote the i-th components of the elastodynamic fundamental solution (Eringen, 1975), i.e. of the displacement and traction, respectively, generated at $\boldsymbol{y} \in \mathbb{R}^3$ by a unit point force applied at $\boldsymbol{x} \in \mathbb{R}^3$ along the direction k:

$$U_i^k(\boldsymbol{x},\boldsymbol{y}) = \frac{1}{k_S^2\mu}\Big((\delta_{qs}\delta_{ik} - \delta_{qk}\delta_{is})\frac{\partial}{\partial x_q}\frac{\partial}{\partial y_s}G(r;k_S) + \frac{\partial}{\partial x_i}\frac{\partial}{\partial y_k}G(r;k_P)\Big), \tag{4a}$$

$$T_i^k(\boldsymbol{x},\boldsymbol{y}) = \mu\Big[\frac{2\nu}{1-2\nu}\delta_{ij}\delta_{h\ell} + \delta_{ih}\delta_{j\ell} + \delta_{jh}\delta_{i\ell}\Big]\frac{\partial}{\partial y_\ell}U_h^k(\boldsymbol{x},\boldsymbol{y})n_j(\boldsymbol{y}), \tag{4b}$$

in which $r = |\boldsymbol{y} = \boldsymbol{x}|$ and k_S, k_P are the respective S and P wavenumbers, so that

$$k_S^2 = \frac{\rho\omega^2}{\mu}, \qquad k_P = \gamma k_S, \qquad \gamma^2 = \frac{1-2\nu}{2(1-\nu)}, \tag{5}$$

$G(\cdot;k)$ is the free-space Green's function for the Helmholtz equation with wavenumber k, given by

$$G(r;k) = \frac{\exp(\mathrm{i}kr)}{4\pi r}, \tag{6}$$

$\boldsymbol{n}(\boldsymbol{y})$ is the unit normal to $\partial\Omega$ directed outwards of Ω.

The numerical solution of boundary integral equation (1) is based on a discretization of the surface $\partial\Omega$ into isoparametric boundary elements, with piecewise-linear interpolation of displacements and piecewise-constant interpolation of tractions, based on three-noded triangular boundary elements, being used here. On collocating (1) at displacement nodes, a square complex-valued matrix equation of the form

$$[K]\{u\} = \{f\}, \tag{7}$$

is obtained, where the N-vector $\{u\}$ collects the sought degrees of freedom (DOFs), here the nodal displacement components, while the $N \times N$ matrix of influence coefficients $[K]$ and the N-vector $\{f\}$ arise from (2) and (3), respectively. Setting up the matrix $[K]$ classically requires the computation of all element integrals for each collocation point, thus needing a computational time of order $O(N^2)$.

The influence matrix $[K]$ is fully-populated. Storing $[K]$ is thus limited, on ordinary computers, to BEM models of size not exceeding $N = O(10^4)$. Direct solvers, such as the LU factorization, have a $O(N^3)$ complexity and are thus also limited to moderately-sized BEM models. Both limitations are overcome by (i) resorting to an iterative solver, here GMRES (Saad and Schultz, 1986), and (ii) accelerating the matrix-vector products $[K]\{u\}$ requested by each iteration of GMRES using the fast multipole method (FMM) so as the complexity of this operation becomes lower than the $O(N^2)$ operations entailed by standard BEM methods.

3 Elastodynamic Fast Multipole Method

The FMM is based on a reformulation of the fundamental solutions in terms of products of functions of $\boldsymbol{x}$ and of $\boldsymbol{y}$. This allows to re-use integrations with respect to $\boldsymbol{y}$ when the collocation point $\boldsymbol{x}$ is changed, thereby lowering the $O(N^2)$ complexity per iteration entailed by standard BEMs. The elastodynamic fundamental solutions (4a) and (4b) are linear combinations of derivatives of the Green's function (6) for the Helmholtz equation. On recasting the position vector $\boldsymbol{r} = \boldsymbol{y} - \boldsymbol{x}$ in the form $\boldsymbol{r} = \boldsymbol{r}_0 + (\boldsymbol{y} - \boldsymbol{y}_0) - (\boldsymbol{x} - \boldsymbol{x}_0)$, where $\boldsymbol{x}_0$ and $\boldsymbol{y}_0$ are two poles and $\boldsymbol{r}_0 = \boldsymbol{y}_0 - \boldsymbol{x}_0$ (Fig. 1), the Helmholtz Green's function is shown (Epton and Dembart, 1995) to admit the decomposition

$$G(|\boldsymbol{r}|;k) = \lim_{L\to+\infty} \int_{\hat{\boldsymbol{s}}\in S} e^{\mathrm{i}k\hat{\boldsymbol{s}}.(\boldsymbol{y}-\boldsymbol{y}_0)} \mathcal{G}_L(\hat{\boldsymbol{s}};\boldsymbol{r}_0;k) e^{-\mathrm{i}k\hat{\boldsymbol{s}}.(\boldsymbol{x}-\boldsymbol{x}_0)} \mathrm{d}\hat{\boldsymbol{s}}, \tag{8}$$

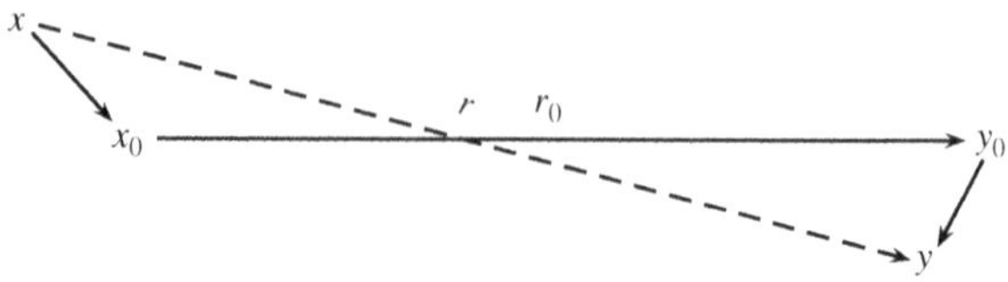

Fig. 1 Decomposition of the position vector: notation

where S is the unit sphere of $\mathbb{R}^3$ and the *transfer function* $\mathcal{G}_L(\hat{\boldsymbol{s}};\boldsymbol{r}_0;k)$ is defined in terms of the Legendre polynomials P_p and the spherical Hankel functions of the first kind $h_p^{(1)}$ by:

$$\mathcal{G}_L(\hat{\boldsymbol{s}};\boldsymbol{r}_0;k) = \frac{\mathrm{i}k}{16\pi^2}\sum_{0\le p\le L}(2p+1)\mathrm{i}^p h_p^{(1)}(k|\boldsymbol{r}_0|)P_p\big(\cos(\hat{\boldsymbol{s}},\boldsymbol{r}_0)\big). \tag{9}$$

The decomposition (8) and (9) is seen to achieve the desired separation of variables $\boldsymbol{x}$ and $\boldsymbol{y}$. Then, to recast the elastodynamic fundamental solutions in a form similar to (8) and (9), one simply substitutes decomposition (8) and (9) into (4a) and (4b), to obtain the following multipole decomposition of the elastodynamic fundamental solutions:

$$\begin{aligned} U_i^k(\boldsymbol{x},\boldsymbol{y}) = &\lim_{L\to+\infty}\int_{\hat{\boldsymbol{s}}\in S} e^{\mathrm{i}k_\mathrm{P}\hat{\boldsymbol{s}}.(\boldsymbol{y}-\boldsymbol{y}_0)}\,\mathcal{U}_{i,L}^{k,\mathrm{P}}(\hat{\boldsymbol{s}};\boldsymbol{r}_0)\,e^{-\mathrm{i}k_\mathrm{P}\hat{\boldsymbol{s}}.(\boldsymbol{x}-\boldsymbol{x}_0)}\mathrm{d}\hat{\boldsymbol{s}} \\ &+\lim_{L\to+\infty}\int_{\hat{\boldsymbol{s}}\in S} e^{\mathrm{i}k_\mathrm{S}\hat{\boldsymbol{s}}.(\boldsymbol{y}-\boldsymbol{y}_0)}\,\mathcal{U}_{i,L}^{k,\mathrm{S}}(\hat{\boldsymbol{s}};\boldsymbol{r}_0)\,e^{-\mathrm{i}k_\mathrm{S}\hat{\boldsymbol{s}}.(\boldsymbol{x}-\boldsymbol{x}_0)}\mathrm{d}\hat{\boldsymbol{s}}, \end{aligned} \tag{10}$$

$$\begin{aligned} T_i^k(\boldsymbol{x},\boldsymbol{y}) = &\lim_{L\to+\infty}\int_{\hat{\boldsymbol{s}}\in S} e^{\mathrm{i}k_\mathrm{P}\hat{\boldsymbol{s}}.(\boldsymbol{y}-\boldsymbol{y}_0)}\,\mathcal{T}_{i,L}^{k,\mathrm{P}}(\hat{\boldsymbol{s}};\boldsymbol{r}_0)\,e^{-\mathrm{i}k_\mathrm{P}\hat{\boldsymbol{s}}.(\boldsymbol{x}-\boldsymbol{x}_0)}\mathrm{d}\hat{\boldsymbol{s}} \\ &+\lim_{L\to+\infty}\int_{\hat{\boldsymbol{s}}\in S} e^{\mathrm{i}k_\mathrm{S}\hat{\boldsymbol{s}}.(\boldsymbol{y}-\boldsymbol{y}_0)}\,\mathcal{T}_{i,L}^{k,\mathrm{S}}(\hat{\boldsymbol{s}};\boldsymbol{r}_0)\,e^{-\mathrm{i}k_\mathrm{S}\hat{\boldsymbol{s}}.(\boldsymbol{x}-\boldsymbol{x}_0)}\mathrm{d}\hat{\boldsymbol{s}}, \end{aligned} \tag{11}$$

with the elastodynamic transfer functions given in terms of the acoustic transfer function $\mathcal{G}_L$ by

$$\mathcal{U}_{i,L}^{k,\mathrm{P}}(\hat{\boldsymbol{s}};\boldsymbol{r}_0) = \frac{\gamma^2}{\mu}\hat{s}_i\hat{s}_k\mathcal{G}_L(\hat{\boldsymbol{s}};\boldsymbol{r}_0;k_\mathrm{P}), \tag{12a}$$

$$\mathcal{T}_{i,L}^{k,\mathrm{P}}(\hat{\boldsymbol{s}};\boldsymbol{r}_0) = \frac{\mathrm{i}k_\mathrm{S}\gamma^3}{\mu}C_{ijh\ell}\hat{s}_\ell\hat{s}_h\hat{s}_k\mathcal{G}_L(\hat{\boldsymbol{s}};\boldsymbol{r}_0;k_\mathrm{P})n_j(\boldsymbol{y}), \tag{12b}$$

$$\mathcal{U}_{i,L}^{k,\mathrm{S}}(\hat{\boldsymbol{s}};\boldsymbol{r}_0) = \frac{1}{\mu}(\delta_{ik}-\hat{s}_k\hat{s}_i)\mathcal{G}_L(\hat{\boldsymbol{s}};\boldsymbol{r}_0;k_\mathrm{S}), \tag{12c}$$

$$\mathcal{T}_{i,L}^{k,\mathrm{S}}(\hat{\boldsymbol{s}};\boldsymbol{r}_0) = \frac{\mathrm{i}k_\mathrm{S}}{\mu}(\delta_{hk}-\hat{s}_k\hat{s}_h)C_{ijh\ell}\hat{s}_\ell\mathcal{G}_L(\hat{\boldsymbol{s}};\boldsymbol{r}_0;k_\mathrm{S})n_j(\boldsymbol{y}). \tag{12d}$$

In practice, the limiting process $L\to+\infty$ in (8) or (10), (11) cannot be performed exactly and is replaced with an evaluation for a suitably chosen finite value of L, empirically established from numerical experiments. One such formula, known from previous studies on FMMs for Maxwell equations (Darve, 2000), reads:

$$L(d) = \sqrt{3}kd + C_\epsilon\log_{10}(\sqrt{3}kd+\pi), \tag{13}$$

where d denotes the linear cell size. In this work, distinct truncation levels L_{P} and L_{S} are defined according to (13) with $k = k_{\text{P}}$ and $k = k_{\text{S}}$, respectively.

3.1 Single-Region FM-BEM

To have maximal efficiency, FM-BEM algorithms must confine non-FM calculations to the smallest possible portion of the boundary while clustering whenever possible the computation of influence terms into the largest possible non-adjacent groups. This is achieved by the multi-level FMM (Darve, 2000; Nishimura, 2002 and Sylvand, 2002), based on using large cells and hierarchically subdividing each cell into $2\times2\times2 = 8$ children cubic cells. This cell-subdivision approach is systematized by means of an oct-tree structure of cells. The level $\ell = 0$, composed of only one cubic cell containing the whole surface $\partial\Omega$, is the tree root. The level-0 cell is divided into $2\times2\times2 = 8$ children cubic cells, which constitute the level $\ell = 1$. All level-1 cells being adjacent, the FMM cannot be applied to them. The level $\ell = 2$ is then defined by dividing each level-1 cell into 8 children cells, and so contains 64 cells. The subdivision process is further repeated until the finest level $\ell = \bar{\ell}$, implicitly defined by a preset subdivision-stopping criterion, is reached. Level-$\bar{\ell}$ cells are usually termed *leaf cells*. The FMM is applied from level $\ell = 2$ to level $\ell = \bar{\ell}$, i.e. features $\bar{\ell} - 1$ "active" levels.

The multi-level approach basically consists in first applying the FMM to all influence computations between disjoint level-2 cells (so as to use the largest clusters whenever possible), and then recursively tracing the tree downwards, applying the FMM to all interaction between disjoint level-ℓ cells that are children of adjacent level-($\ell-1$) cells (Fig. 2). Finally, interactions between adjacent leaf cells are treated using traditional (i.e. non FM-based) BE techniques. This approach thus minimizes the overall proportion of influence computations requiring the traditional treatment.

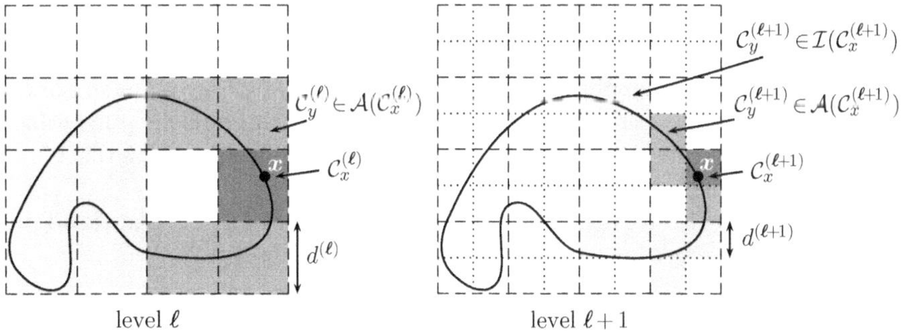

Fig. 2 Multi-level fast multipole algorithm. Only multipole moments from non-adjacent (*light-grey*) cells $C_y^{(\ell)} \notin \mathcal{A}(C_x^{(\ell)})$ may provide (through transfer) FM-computed contributions to $(\mathcal{K}u)^{\text{FM}}(\boldsymbol{x})$ at collocation points $\boldsymbol{x}$ lying in cell $C_x^{(\ell)}$. Upon cell subdivision (*right*), new FM-computed contributions to collocation points in cell $C_x^{(\ell+1)}$ originate from cells $C_y^{(\ell+1)}$ in the *interaction list* $\mathcal{I}(C_x^{(\ell+1)})$ of $C_x^{(\ell+1)}$, while the adjacent region $\mathcal{A}(C_x^{(\ell+1)})$ reduces in size

The computation of the discretized linear operator (2), i.e. of the matrix-vector product $[K]\{u\}$, by the multi-level elastodynamic FMM hence consists of the following main steps:

1. *Initialization*: compute multipole moments for all lowest-level cells $\mathcal{C}_y = \mathcal{C}_y^{\bar{\ell}}$:

$$\mathcal{R}_k^{\mathrm{S},u}(\hat{\boldsymbol{s}};\mathcal{C}_y^{(\bar{\ell})}) = -\mathrm{i}k_\mathrm{S} A_{ijk}^{\mathrm{S},u} \int_{\partial\Omega\cap\mathcal{C}_y^{(\bar{\ell})}} u_i(\boldsymbol{y}) n_j(\boldsymbol{y}) e^{\mathrm{i}k_\mathrm{S}\hat{\boldsymbol{s}}.(\boldsymbol{y}-\boldsymbol{y}_0^{(\bar{\ell})})} dS_y \tag{14a}$$

$$\mathcal{R}^{\mathrm{P},u}(\hat{\boldsymbol{s}};\mathcal{C}_y^{(\bar{\ell})}) = -2\mathrm{i}k_\mathrm{S}\gamma^3 A_{ij}^{\mathrm{P},u} \int_{\partial\Omega\cap\mathcal{C}_y^{(\bar{\ell})}} u_i(\boldsymbol{y}) n_j(\boldsymbol{y}) e^{\mathrm{i}k_\mathrm{P}\hat{\boldsymbol{s}}.(\boldsymbol{y}-\boldsymbol{y}_0^{(\bar{\ell})})} dS_y \tag{14b}$$

$$\mathcal{R}_k^{\mathrm{S},t}(\hat{\boldsymbol{s}};\mathcal{C}_y^{(\bar{\ell})}) = \frac{1}{\mu} A_{ki}^{\mathrm{S},t} \int_{\partial\Omega\cap\mathcal{C}_y^{(\bar{\ell})}} t_i(\boldsymbol{y}) e^{\mathrm{i}k_\mathrm{S}\hat{\boldsymbol{s}}.(\boldsymbol{y}-\boldsymbol{y}_0^{(\bar{\ell})})} dS_y \tag{14c}$$

$$\mathcal{R}^{\mathrm{P},t}(\hat{\boldsymbol{s}};\mathcal{C}_y^{(\bar{\ell})}) = \frac{\gamma^2}{\mu} \int_{\partial\Omega\cap\mathcal{C}_y^{(\bar{\ell})}} \hat{s}_a t_a(\boldsymbol{y}) e^{\mathrm{i}k_\mathrm{P}\hat{\boldsymbol{s}}.(\boldsymbol{y}-\boldsymbol{y}_0^{(\bar{\ell})})} dS_y \tag{14d}$$

   having set $A_{ijk}^{\mathrm{S},u} = \delta_{ik}\hat{s}_j + \delta_{jk}\hat{s}_i - 2\hat{s}_i\hat{s}_j\hat{s}_k$, $A_{ij}^{\mathrm{P},u} = \nu(1-2\nu)^{-1}\delta_{ij} + \hat{s}_i\hat{s}_j$ and $A_{ki}^{\mathrm{S},t} = \delta_{ki} - \hat{s}_k\hat{s}_i$.
2. *Upward pass*: recursively aggregate multipole moments by moving upward in the tree until level 2 is reached. Denoting by $\mathcal{S}(\mathcal{C})$ the set of children of a given cell $\mathcal{C}$, the transition from a level-$(\ell+1)$ cell to its parent level-ℓ cell is based on identities

$$\mathcal{R}_k^{\mathrm{S},u}(\hat{\boldsymbol{s}};\mathcal{C}_y^{(\ell)}) = \sum_{\mathcal{C}_y^{(\ell+1)}\in\mathcal{S}(\mathcal{C}_y^{(\ell)})} \exp\big[-\mathrm{i}k_\mathrm{S}\hat{\boldsymbol{s}}.\Delta\boldsymbol{y}_0^{(\ell)}\big]\mathcal{R}_k^{\mathrm{S},u}(\hat{\boldsymbol{s}};\mathcal{C}_y^{(\ell+1)}) \tag{15a}$$

$$\mathcal{R}^{\mathrm{P},u}(\hat{\boldsymbol{s}};\mathcal{C}_y^{(\ell)}) = \sum_{\mathcal{C}_y^{(\ell+1)}\in\mathcal{S}(\mathcal{C}_y^{(\ell)})} \exp\big[-\mathrm{i}k_\mathrm{P}\hat{\boldsymbol{s}}.\Delta\boldsymbol{y}_0^{(\ell)}\big]\mathcal{R}^{\mathrm{P},u}(\hat{\boldsymbol{s}};\mathcal{C}_y^{(\ell+1)}). \tag{15b}$$

   (with $\Delta\boldsymbol{y}_0^{(\ell)} = \boldsymbol{y}_0^{(\ell+1)} - \boldsymbol{y}_0^{(\ell)}$). A crucial feature of the elastodynamic multi-level FMM is that the number and location of the quadrature points on S are level-dependent, a consequence of the previously-mentioned dependence (13) of L on the cell size. Hence, application of identities (15a) and (15b) requires an extrapolation procedure (Darve, 2000 and Chaillat et al., 2008) furnishing the values of $\mathcal{R}_k^{\mathrm{S},u}$ and $\mathcal{R}^{\mathrm{P},u}$ at the level-ℓ quadrature points from those at the level-$(\ell+1)$ quadrature points.
3. *Transfer*: initialize local expansions for each level-ℓ cell $\mathcal{C}_x^{(\ell)}$ and at each level $2 \le \ell \le \bar{\ell}$ using

$$\mathcal{L}_k^{\mathrm{S},u}(\hat{\boldsymbol{s}}^{(\ell)};\mathcal{C}_x^{(\ell)}) = \sum_{\mathcal{C}_y^{(\ell)}\in\mathcal{I}(\mathcal{C}_x^{(\ell)})} \mathcal{G}_L(\hat{\boldsymbol{s}}^{(\ell)};\boldsymbol{r}_0;k_\mathrm{S})\mathcal{R}_k^{\mathrm{S},u}(\hat{\boldsymbol{s}}^{(\ell)};\mathcal{C}_y^{(\ell)}) \tag{16a}$$

$$\mathcal{L}^{\mathrm{P},u}(\hat{\boldsymbol{s}}^{(\ell)};\mathcal{C}_x^{(\ell)}) = \sum_{\mathcal{C}_y^{(\ell)}\in\mathcal{I}(\mathcal{C}_x^{(\ell)})} \mathcal{G}_L(\hat{\boldsymbol{s}}^{(\ell)};\boldsymbol{r}_0;k_\mathrm{P})\mathcal{R}^{\mathrm{P},u}(\hat{\boldsymbol{s}}^{(\ell)};\mathcal{C}_y^{(\ell)}) \tag{16b}$$

where $\mathcal{I}(\mathcal{C})$, the *interaction list* of a given cell $\mathcal{C}$ (Fig. 2), is the set of *same-level* cells which are not adjacent to $\mathcal{C}$ while having a parent cell adjacent to that of $\mathcal{C}$. For a level-2 cell, $\mathcal{I}(\mathcal{C}^2)$ collects all level-2 cells not adjacent to $\mathcal{C}^2$.

4. *Downward pass*: for all levels $3 \leq \ell \leq \bar{\ell}$, the local expansion for each level-ℓ cell $\mathcal{C}_x^{(\ell)}$ is updated with the contribution from the parent level-$(\ell-1)$ cell, by means of the identity

$$\mathcal{L}_k^{\mathrm{S},u}(\hat{s};\mathcal{C}_x^{(\ell)}) = \mathcal{L}_k^{\mathrm{S},u}(\hat{s};\mathcal{C}_x^{(\ell)}) + \exp\big[-ik_{\mathrm{S}}(\hat{s}.\Delta\boldsymbol{x}_0^{(\ell)})\big]\mathcal{L}_k^{\mathrm{S},u}(\hat{s};\mathcal{C}_x^{(\ell-1)}) \tag{17a}$$

$$\mathcal{L}^{\mathrm{P},u}(\hat{s};\mathcal{C}_x^{(\ell)}) = \mathcal{L}^{\mathrm{P},u}(\hat{s};\mathcal{C}_x^{(\ell)}) + \exp\big[-ik_{\mathrm{P}}(\hat{s}.\Delta\boldsymbol{x}_0^{(\ell)})\big]\mathcal{L}^{\mathrm{P},u}(\hat{s};\mathcal{C}_x^{(\ell-1)}) \tag{17b}$$

(with $\Delta\boldsymbol{x}_0^{(\ell)} = \boldsymbol{x}_0^{(\ell-1)} - \boldsymbol{x}_0^{(\ell)}$). Similarly to Step 2, application of identity (17a) and (17b) requires an inverse extrapolation procedure (Chaillat et al., 2008 and Darve, 2000) furnishing the values of $\mathcal{L}_k^{\mathrm{S},u}$ and $\mathcal{L}^{\mathrm{P},u}$ at the level-ℓ quadrature points from those at the level-$(\ell-1)$ quadrature points.

5. When the leaf level $\ell = \bar{\ell}$ is reached, all local expansions have been computed. The far contribution $(\mathcal{K}u)^{\mathrm{FM}}(\boldsymbol{x})$ is evaluated using

$$\begin{aligned}(\mathcal{K}u)_k^{\mathrm{FM}}(\boldsymbol{x}) \approx \sum_{q=1}^{Q} w_q^{(\bar{\ell})}\Big[& e^{-ik_{\mathrm{S}}\hat{s}_q^{(\bar{\ell})}.(\boldsymbol{x}-\boldsymbol{x}_0^{(\bar{\ell})})}\mathcal{L}_k^{\mathrm{S},u}(\hat{s}_q;\mathcal{C}_x^{(\bar{\ell})}) \\ & + e^{-ik_{\mathrm{P}}\hat{s}_q.(\boldsymbol{x}-\boldsymbol{x}_0^{(\bar{\ell})})}(\hat{s}_q^{(\bar{\ell})})_k\mathcal{L}^{\mathrm{P},u}(\hat{s}_q;\mathcal{C}_x^{(\bar{\ell})})\Big],\end{aligned} \tag{18}$$

and the near contribution is evaluated for all level-$\bar{\ell}$ (leaf) cells $\mathcal{C}_x$ according to

$$(\mathcal{K}u)_k^{\mathrm{near}}(\boldsymbol{x}) = c_{ik}(\boldsymbol{x})u_i(\boldsymbol{x}) + \sum_{\mathcal{C}_y^{(\bar{\ell})}\in\mathcal{A}(\mathcal{C}_x^{(\bar{\ell})})} (\mathrm{P.V.})\int_{\partial\Omega\cap\mathcal{C}_y^{(\bar{\ell})}} u_i(\boldsymbol{y})T_i^k(\boldsymbol{x},\boldsymbol{y})dS_y. \tag{19}$$

The computation of the right-hand side (3) follows the same steps, with the multipole moments $\mathcal{R}_k^{\mathrm{S},u}$, $\mathcal{R}^{\mathrm{P},u}$ and local expansions $\mathcal{L}_k^{\mathrm{S},u}$, $\mathcal{L}^{\mathrm{P},u}$ replaced with their counterparts $\mathcal{R}_k^{\mathrm{S},t}$, $\mathcal{R}^{\mathrm{P},t}$ and $\mathcal{L}_k^{\mathrm{S},t}$, $\mathcal{L}^{\mathrm{P},t}$. The above steps are found (Chaillat et al. 2008) to have a complexity of at most $O(N\log N)$, with the exception of the direct and inverse extrapolations in steps 2 and 4, whose complexity is $O(N^{3/2})$.

The near-field contribution (19) involves (i) CPV-singular, (ii) weakly-singular and (iii) non-singular element integrals. CPV-singular integrals are split into (singular) integrals involving the static Kelvin traction kernel and (nonsingular) complementary integrals. The former are then evaluated analytically, taking advantage of the fact that three-noded triangular elements, which have constant unit normal and Jacobian, are used. Weakly-singular integrals (which feature the kernel $U_i^k(\boldsymbol{x},\boldsymbol{y})$) and non-singular integrals are computed using numerical Gaussian quadrature (the weak singularity being first cancelled by means of a suitable change of coordinates). Finally, when $\partial\Omega$ presents an edge or corner at $\boldsymbol{x}$, the free-term $c_{ij}(\boldsymbol{x})$ is evaluated using the method of Mantic (1993).

A detailed account of the implementation of this single-region elastodynamic FMM is described in (Chaillat et al. 2008), wherein analytical and numerical verifications of the algorithmic complexity of single-level and multi-level versions are also addressed.

3.2 Multi-Region FM-BEM

The above FM-BEM formulation can be naturally extended to multi-domain configurations. Such problems involve displacement and traction degrees of freedom associated with interfacial displacements $\boldsymbol{u}^{ij}$ or $\boldsymbol{t}^{ij}$ on the common interface Γ_{ij} separating subregions Ω_i and Ω_j (with $\boldsymbol{t}^{ij}$ conventionally defined in terms of the unit normal $\boldsymbol{n}^{ij}$ directed from Ω_i to Ω_j). Invoking the perfect bonding conditions

$$\boldsymbol{u}^{ij} = \boldsymbol{u}^{ji}, \qquad \boldsymbol{t}^{ij} = -\boldsymbol{t}^{ji} \tag{20}$$

the interfacial quantities $\boldsymbol{u}^{ij}, \boldsymbol{t}^{ij}$ with $i < j$ are retained as independent unknowns. For each subdomain Ω_i, the governing integral equation (1) is discretized by collocation at the displacement nodes and the element centres. This defines for each subregion a *rectangular*, overdetermined system of BEM equations. The FM-BEM then evaluates for each Ω_i the corresponding matrix-vector product. To define a square global system of equations, equations resulting from collocation at interface element nodes or centres arising from both adjacent subdomains are linearly combined, an operation that is performed externally on the matrix-vector products generated by the FM-BEM. A detailed presentation of this BEM-BEM coupling approach, including numerical experiments allowing to select suitable values of the weighting coefficients used in the foregoing equation combinations, is given in Chaillat et al. (2009).

4 Numerical Examples

Following are a few sample numerical results obtained on test problems using the present elastodynamic single-region or multi-region FM-BEMs. Other examples are presented in Chaillat et al. (2008, 2009).

4.1 Scattering of SV Waves by a Semi-Spherical Canyon

This example is concerned with the perturbation by a semi-spherical canyon of radius a of an oblique incident plane SV-wave of unit amplitude traveling in a elastic half space (Fig. 3) characterized by $\nu = 1/3$. This problem has been previously studied in Eshraghi and Dravinski (1989) by means of a wave function expansion and, for low frequencies, in Reinoso et al. (1997) using a standard BEM.

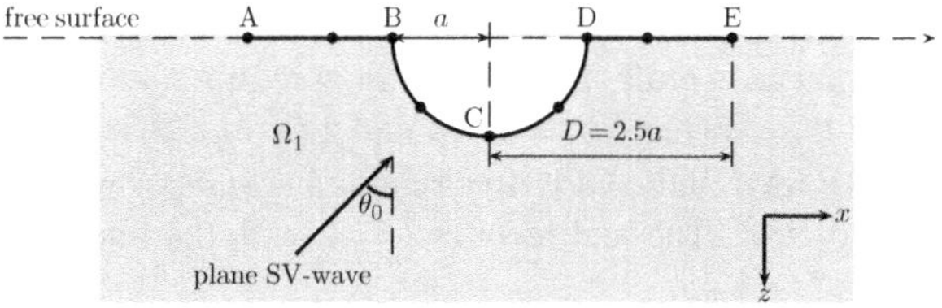

Fig. 3 Scattering of SV waves by a semi-spherical canyon: geometry and notation

Results obtained for a low frequency such that $k_S a = 0.75\pi$ are seen to agree well with published results from Eshraghi and Dravinski (1989) ($\theta = 0$, Fig. 4) and Eshraghi and Dravinski (1989) and Reinoso et al. (1997) ($\theta = \pi/6$, Fig. 5). The free surface is here meshed within a truncation radius $D = 2.5a$, a relatively small value chosen so as to reproduce the conditions used in Eshraghi and

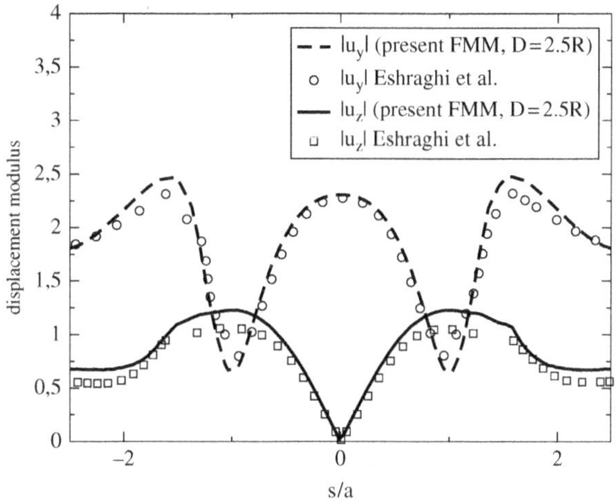

Fig. 4 Scattering of SV waves by a semi-spherical canyon: horizontal and vertical computed displacement on line ABCDE (with points A, B, C, D, E defined on Fig. 3) plotted against normalized arc-length coordinate s/a along ABCDE ($\theta = 0,\ k_S a = 0.75\pi$)

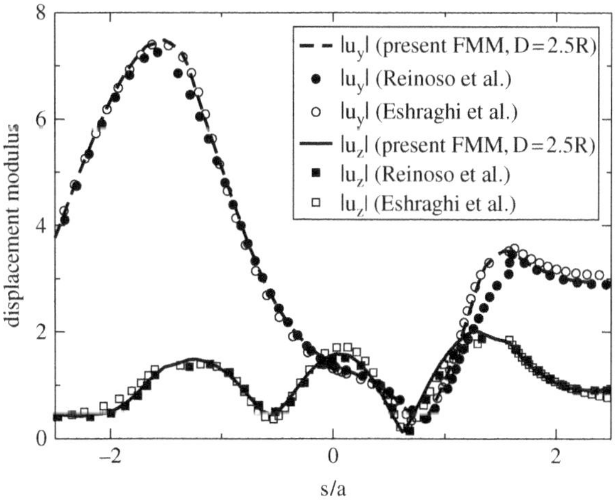

Fig. 5 Scattering of SV waves by a semi-spherical canyon: horizontal and vertical computed displacement on line ABCDE (with points A, B, C, D, E defined on Fig. 3) plotted against normalized arc-length coordinate s/a along ABCDE ($\theta = \pi/6,\ k_S a = 0.75\pi$)

Dravinski (1989) and Reinoso et al. (1997), and the mesh features $N = 7,602$ nodal unknowns overall. The computation required 8 (resp. 11) GMRES iterations for the case $\theta = 0$ (resp. $\theta = \pi/6$) and 1.5s of CPU per iteration on a single-processor PC (RAM: 3GB, CPU frequency: 3.40 GHz), with the GMRES relative tolerance set to 10^{-3}. The leaf level is $\bar{\ell} = 3$, with the linear size $d^{\min}$ of leaf cells such that $k_S d^{\min} \approx 1.45$.

4.2 Test Problem for the Multi-Region FMM

The test problem of a spherical cavity subjected to an internal time-harmonic uniform pressure P, surrounded by two concentric spherical layers Ω_1, Ω_2 embedded in an unbounded elastic medium Ω_3, is considered (Fig. 6). The radii of the cavity and two surrounding interfaces are a_1, a_2, a_3, respectively The mechanical properties of the respective media, in arbitrary units, are $(\rho_1, \mu_1, \nu_1) = (3, 4, 0.25)$, $(\rho_2, \mu_2, \nu_2) = (6, 5, 0.25)$, $(\rho_3, \mu_3, \nu_3) = (2, 1, 1/3)$. A closed-form exact solution is available for this test problem. Results in terms of RMS errors between numerical solutions computed using the present multi-domain FM-BEM and their analytical counterparts, given in Table 1, show that the formulation achieves satisfactory accuracy (the GMRES relative tolerance being again set to 10^{-3}).

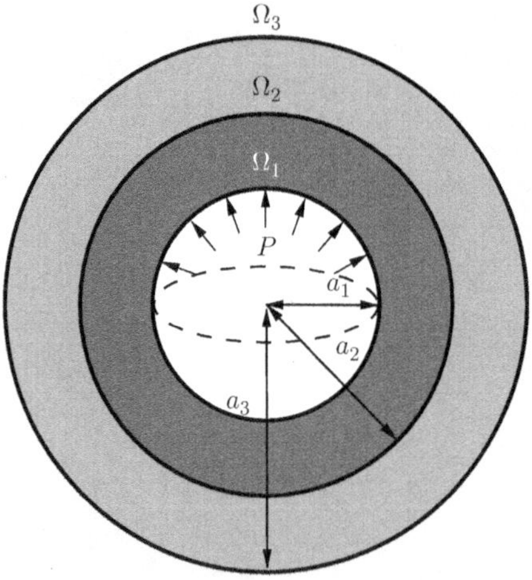

Fig. 6 Test problem: pressurized spherical cavity surrounded by concentric layers

Table 1 Pressurized spherical cavity surrounded by concentric layers: relative RMS errors

N	$k_S^1 d^{\min}$	$k_S^1 a_1$	$\bar{\ell}_i$	$E(u^1)$	$E(u^{12})$	$E(t^{12})$	$E(u^{23})$	$E(t^{23})$	Iters.
55,778	0.82	2.17	3;3;3	$3.0\,10^{-2}$	$1.4\,10^{-2}$	$2.2\,10^{-2}$	$1.3\,10^{-2}$	$2.8\,10^{-2}$	59
215,058	1.88	4.93	3;3;4	$1.0\,10^{-2}$	$1.3\,10^{-2}$	$1.0\,10^{-2}$	$1.4\,10^{-2}$	$1.4\,10^{-2}$	43

4.3 Wave Amplification in a Semi-Spherical Basin

This example is again concerned with the perturbation of an oblique incident plane SV-wave of unit amplitude traveling in a elastic half space, this time by a semi-spherical filled basin (Fig. 7). This configuration, has been studied in the frequency domain in Mossessian and Dravinski (1990) using a standard indirect BEM (using the half-space Green's functions). The mechanical parameters are defined through $\mu_1 = c_{S1} = 1$, $c_{P1} = 2$, $\mu_2 = 1/6$, $c_{S2} = 1/2$ and $c_{P2} = 1$. Results obtained for an incidence angle $\theta = \pi/6$ and a frequency such that $k_S a = 2\pi_1$ are presented, in terms of the x, y and z components of the surface displacement, in Figs. 8, 9 and 10. The free surface is here meshed within a truncation radius $D = 5a$, and the mesh features $N = 143,451$ nodal unknowns (discretized versions of the displacement $\boldsymbol{u}^1$ on the truncated free surface and the interfacial displacements $\boldsymbol{u}^{12}$ and tractions $\boldsymbol{t}^{12}$). The computation required 484 GMRES iterations and 36s of CPU per iteration.

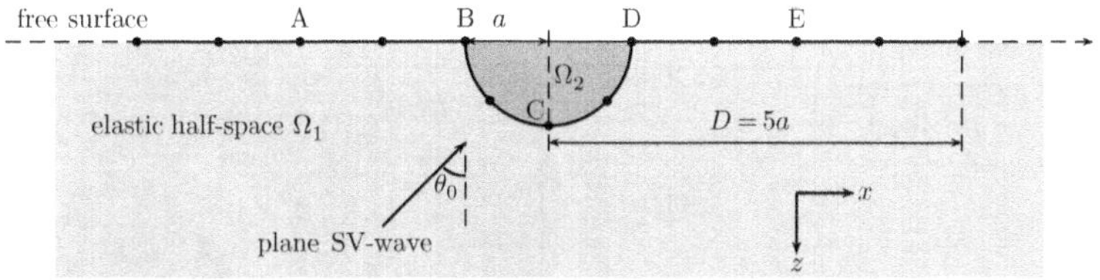

Fig. 7 Propagation of an oblique incident plane SV-wave in a semi-spherical basin: notation

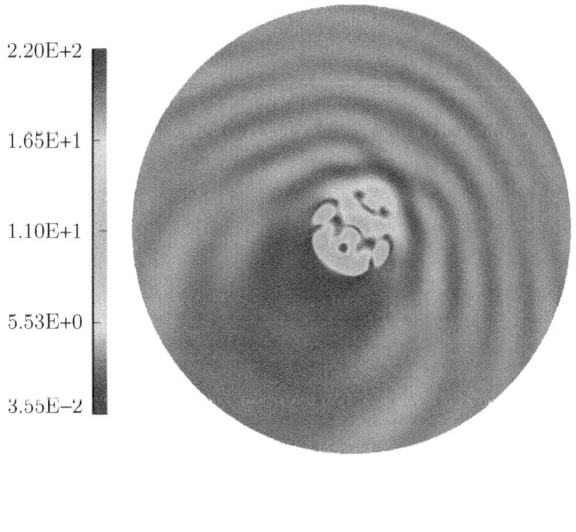

Fig. 8 Diffraction of an oblique incident SV plane wave by a semi-spherical basin: computed x component of displacement on basin surface and meshed part of free surface (normalized frequency $k_S a = 2\pi$)

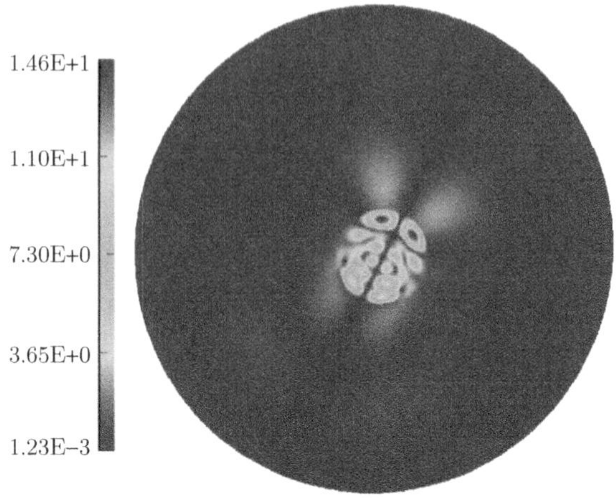

Fig. 9 Diffraction of an oblique incident SV plane wave by a semi-spherical basin: computed y component of displacement on basin surface and meshed part of free surface (normalized frequency $k_S a = 2\pi$)

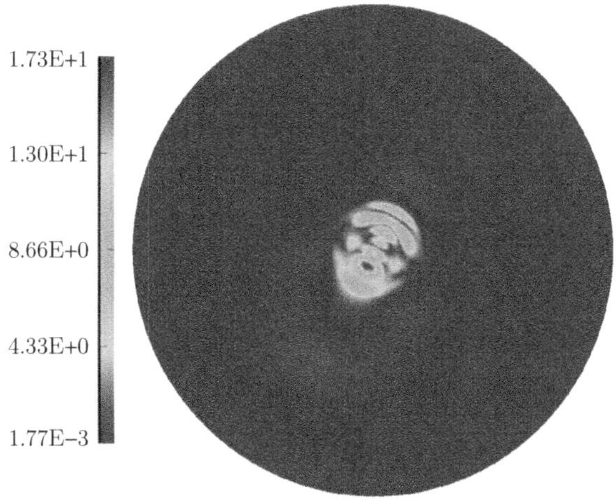

Fig. 10 Diffraction of an oblique incident SV plane wave by a semi-spherical basin: computed z component of displacement on basin surface and meshed part of free surface (normalized frequency $k_S a = 2\pi$)

5 Conclusions

In this contribution, the Fast Multipole Method has been succesfully extended to $3D$ single-region and multi-region elastodynamics in the frequency domain. Combined with the BEM formulation, it permits to reduce the computational burden, in both CPU time and memory requirements, for the analysis of elastic wave propagation (e.g. seismic), and allows to run BEM models of large size on an ordinary PC. Comparisons with analytical or previously published numerical results show the efficiency and accuracy of the present elastodynamic FM-BEM.

Applications of the present FM-BEM to realistic cases in seismology are under way. Moreover, a natural extension of this work consists in formulating multipole expansions of other fundamental solutions, with the half-space elastodynamic fundamental solution being currently investigated. Also, extending the formulation to complex wavenumbers will allow more realistic modelling where viscoelastic constitutive properties are assumed for the propagation medium. Finally, improving the efficiency of the elastodynamic FM-BEM also requires further research into refined (direct/inverse) extrapolation techniques (for lowering the $O(N^{3/2})$ of this step) and a well-chosen preconditioning strategy (for reducing the GMRES iteration count).

Acknowledgments This work is part of the project Quantitative Seismic Hazard Assessment (QSHA, http://qsha.unice.fr) funded by the French National Research Agency (ANR)

References

Beskos, D. Boundary element methods in dynamic analysis. *Appl. Mech. Rev.*, **40**:1–23 (1987).
Beskos, D. Boundary element methods in dynamic analysis, Part. II. *Appl. Mech. Rev.*, **50**:149–197 (1997).
Bonnet, M. *Boundary Integral Equation Method for Solids and Fluids*. Wiley, New York (1999).
Chaillat, S., Bonnet, M., Semblat, J. F. A multi-level fast multipole BEM for 3-D elastodynamics in the frequency domain. *Comp. Meth. Appl. Mech. Eng.*, **197:** 4233–4249 (2008).
Chaillat, S., Bonnet, M., Semblat, J. F. A new fast multi-domain BEM to model seismic wave propagation and amplification in 3D geological structures. *Geophys. J. Int.*, in press (2009).
Dangla, P., Semblat, J. F., Xiao, H., Delépine, N. A simple and efficient regularization method for 3D BEM: application to frequency-domain elastodynamics. *Bull. Seism. Soc. Am.*, **95**: 1916–1927 (2005).
Darve, E. The fast multipole method : numerical implementation. *J. Comp. Phys.*, **160**:195–240 (2000).
Epton, M. A., Dembart, B. Multipole translation theory for the three-dimensional Laplace and Helmholtz equations. *SIAM J. Sci. Comp.*, **16**:865–897 (1995).
Eringen, A. C., Suhubi, E. S. *Elastodynamics*, vol. II-linear theory. Academic Press, New York (1975).
Eshraghi, H., Dravinski, M. Scattering of plane harmonic SH, SV, P and Rayleigh waves by non-axisymmetric three-dimensional canyons: a wave function expansion approach. *Earthquake Eng. Struct. Dyn.*, **18**:983–998 (1989).
Fujiwara, H. The fast multipole method for solving integral equations of three-dimensional topography and basin problems. *Geophys. J. Int.*, **140**:198–210 (2000).
Gumerov, N. A., Duraiswami, R. *Fast Multipole Methods for the Helmholtz Equation in Three Dimensions*. Elsevier, Amsterdam (2005).
Guzina, B. B., Pak, R. Y. S. On the analysis of wave motions in a multi-layered solid. *Quart. J. Mech. Appl. Math.*, **54**:13–37 (2001).
Mantic, V. A new formula for the C-matrix in the somigliana identity. *J. Elast.*, **33**:191–201 (1993).
Mossessian, T. K., Dravinski, M. Amplification of elastic waves by a three dimensional valley. Part 1: steady state response. *Earthquake Eng. Struct. Dyn.*, **19**:667–680 (1990).
Nishimura, N. Fast multipole accelerated boundary integral equation methods. *Appl. Mech. Rev.*, **55**(4) (2002).
Reinoso, E., Wrobel, L. C., Power, H. Three-dimensional scattering of seismic waves from topographical structures. *Soil. Dyn. Earthquake Eng.*, **16**:41–61 (1997).
Rokhlin, V. Diagonal forms of translation operators for the Helmholtz equation in three dimensions. *Appl. Comp. Harmonic Anal.*, **1**:82–93 (1993).
Saad, Y., Schultz, M.H. GMRES: a generalized minimal residual algorithm for solving nonsymmetric linear systems. *SIAM J. Sci. Stat. Comput.*, **7**:856–869 (1986).
Sánchez-Sesma, F. J. Diffraction of elastic waves by 3D surface irregularities. *Bull. Seism. Soc. Am.*, **73**:1621–1636 (1983).
Sylvand, G. *La méthode multipôle rapide en éléctromagnétisme : performances, parallélisation, applications*. Ph.D. thesis, ENPC, Paris, France (2002).

A Semi-Analytical Approach for Boundary Value Problems with Circular Boundaries

Jeng-Tzong Chen, Ying-Te Lee and Wei-Ming Lee

Abstract In this paper, a semi-analytical approach is developed to deal with problems including multiple circular boundaries. The boundary integral approach is utilized in conjunction with degenerate kernel and Fourier series. To fully utilize the circular geometry, the fundamental solutions and the boundary densities are expanded by using degenerate kernels and Fourier series, respectively. Both direct and indirect formulations are proposed. This approach is a semi-analytical approach, since the error stems from the truncation of Fourier series in the implementation. The unknown Fourier coefficients are easily determined by solving a linear algebraic system after using the collocation method and matching the boundary conditions. Five goals: (1) free of calculating principal value, (2) exponential convergence, (3) well-posed algebraic system, (4) elimination of boundary-layer effect and (5) meshless, of the formulation are achieved. The proposed approach is extended to deal with the problems containing multiple circular inclusions. Finally, the general-purpose program in a unified manner is developed for BVPs with circular boundaries. Several examples including the torsion bar, water wave and plate vibration problems are given to demonstrate the validity of the present approach.

1 Introduction

Most of engineering phenomena are simulated by the mathematical models of boundary value problems. In order to solve the boundary value problems, researchers and engineers have paid more attention on the development of boundary integral equation method (BIEM), boundary element method (BEM) and meshless method than domain type methods, finite element method (FEM) and finite difference method (FDM). Although BEM has been involved as an alternative numerical method for solving engineering problems, some critical issues exist, e.g. singular and hypersingular integrals, boundary-layer effect, ill-posed matrix system and mesh generation.

J.-T Chen (✉)
Department of Harbor and River Engineering, National Taiwan Ocean University, Keelung, Taiwan
Department of Mechanical and Mechatronic Engineering,
e-mail: jtchen@mail.ntou.edu.tw

G.D. Manolis, D. Polyzos (eds.), *Recent Advances in Boundary Element Methods*,
DOI 10.1007/978-1-4020-9710-2_3,

It is well known that BEM is based on the use of fundamental solutions to solve partial differential equations. These functions are two-point functions which are singular as the source and field points coincide. Most of the efforts have been focused on the singular boundary integral equation for problems with ordinary boundaries. In some situations, the singular boundary integral equation is not sufficient to ensure a unique solution, e.g. degenerate boundary, fictitious frequency and spurious eigenvalue. Therefore, the hypersingular equation is required. The applications of hypersingular equations have been summarized in the review article of Chen and Hong (1999).

Unlike the conventional BEM and BIEM, Waterman (1965) introduced first the so-called T-matrix method for electromagnetic scattering problems. Various names, null-field approach or extended boundary condition method (EBCM), have been coined. The null-field approach or T-matrix method was widely used for obtaining numerical solutions of acoustics (Bates and Wall, 1977), elastodynamics (Waterman, 1976) and hydrodynamics (Martin, 1981). Boström (1982) introduced a new method of treating the scattering of transient fields by a bounded obstacle in the three-dimensional space. He defined new sets of time-dependent basis functions, and used these to expand the free-space Green's function and the incoming and scattered fields. The method is a generalization to the time domain of the null-field approach first given by Waterman (1965). A crucial advantage of the null-field approach or T-matrix method consists in the fact that the influence matrix can be computed easily since the singular and hypersingular integrals are avoided. However, they may result in an ill-posed matrix.

In the Fredholm integral equations, the degenerate kernel (or the so-called separate kernel) plays an important role. However, its applications in practical problems seem to have taken a back seat to other methods. The degenerate kernel can be seen as one kind of approximation for the fundamental solution, i.e., the kernel function is expressed as finite sums of products by two linearly independent functions. Kress (1989) proved that the integral equations of the second kind in conjunction with degenerate kernels have the convergence rate of exponential order instead of the linear algebraic order of conventional BEM. Recently, Chen et al. applied the degenerate kernels in conjunction with null-field integral equations to solve many boundary value problems including the Laplace (Chen et al., 2005), Helmholtz (Chen et al., 2007a), biharmonic (Chen et al., 2006) and biHelmholtz (Lee et al., 2007a) problems with holes and/or inclusions. The main gain of their approach is to avoid the improper integrals and free of mesh. They also linked the two numerical methods, Trefftz method and method of fundamental solutions, by using degenerate kernels (Chen et al., 2007b). Therefore, these two methods can be seen to be mathematically equivalent. The similar viewpoint was discussed by Schaback (2007). However, Schaback claimed that MFS is closely connected to the Trefftz method but they are not fully equivalent, since the source points on the far-away filed yield a trial space that is a space of harmonic polynomials (Schaback, 2007). In a word, the degenerate kernel has the property of transferring the integral equation to a linear algebraic system, since the kernel functions in the integral equations are approximated by using two linearly independent functions.

In this paper, we will apply the semi-analytical approach to deal with engineering problems containing multiple circular boundaries. The boundary integral approach is utilized in conjunction with degenerate kernels and Fourier series. Both direct and indirect formulations will be considered. To fully utilize the circular geometry, the fundamental solutions and the boundary densities are expanded by using degenerate kernels and Fourier series, respectively. This approach is a semi-analytical approach, since the error stems from the truncation of Fourier series in the implementation. Five advantages: (1) free of calculating principal value, (2) exponential convergence, (3) well-posed algebraic system (4) elimination of boundary-layer effect and (5) meshless, of the formulation are the main concern. It will also be extended to deal with the problems containing multiple circular inclusions. Finally, the general-purpose program in a unified manner will be developed for BVPs with circular boundaries. Several examples including the torsion bar, water wave and plate vibration problems are given to see the validity of the present approach.

2 Methods of Solution

2.1 Problem Statements

Suppose a boundary value problem has a domain D which is enclosed with the circular boundary B_k $(k = 0, 1, 2, \ldots, H)$

$$B = \bigcup_{k=0}^{H} B_k \tag{1}$$

as shown in Fig. 1. For the infinite plane problem, the radius a_0 in Fig. 1 is infinite. The governing equation can be expressed by

$$\mathcal{L}u(\mathbf{x}) = 0, \quad \mathbf{x} \in \Omega, \tag{2}$$

where $u(\mathbf{x})$ is the potential function, Ω is the domain of interest and $\mathcal{L}$ denotes the operator for the corresponding problems as shown below:

$$\mathcal{L}u(\mathbf{x}) = \begin{cases} \nabla^2 u(\mathbf{x}) & : \text{Laplacian operator}, \\ (\nabla^2 + k^2)u(\mathbf{x}) & : \text{Helmholtz operator}, \\ (\nabla^4 - \zeta^4)u(\mathbf{x}) & : \text{biHelmholtz operator}, \end{cases} \tag{3}$$

where ∇^2 is the Laplacian operator, k is the wave number which is the angular frequency over the speed of sound and ζ is the frequency parameter.

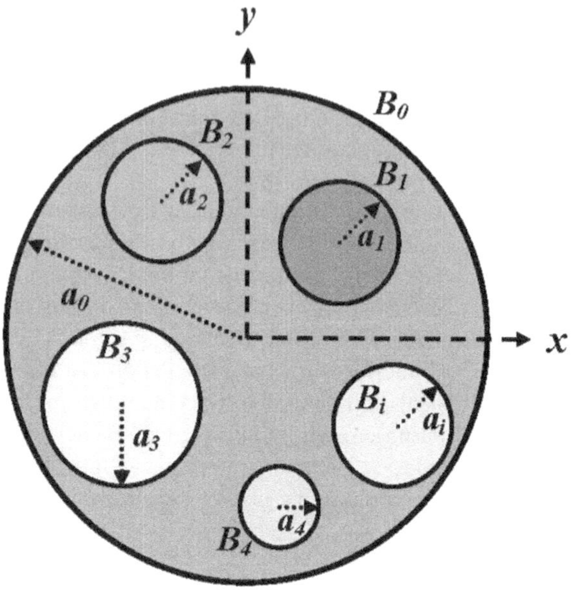

Fig. 1 Sketch of a circular boundary with circular holes and/or inclusions

2.2 Dual Null-Field Integral Formulation – the Conventional Version

Based on the dual boundary integral formulation for the domain point, we have

$$2\pi u(x) = \int_B T(s,x)u(s)dB(s) - \int_B U(s,x)t(s)\,dB(s), \quad x \in \Omega, \tag{4}$$

$$2\pi \frac{\partial u(x)}{\partial n_x} = \int_B M(s,x)u(s)\,dB(s) - \int_B L(s,x)t(s)\,dB(s), \quad x \in \Omega, \tag{5}$$

where s and x are the source and field points, respectively, B is the boundary, n_x denotes the outward normal vector at field point x, and the kernel function $U(s,x)$ is the fundamental solution which satisfies

$$\mathcal{L}\{U(x,s)\} = \delta(x-s), \tag{6}$$

in which $\delta(x-s)$ denotes the Dirac-delta function. The other kernel functions are

$$\begin{aligned} T(s,x) &\equiv \frac{\partial U(s,x)}{\partial n_s}, \; L(s,x) \equiv \frac{\partial U(s,x)}{\partial n_x}, \\ M(s,x) &\equiv \frac{\partial^2 U(s,x)}{\partial n_s \partial n_x}, \end{aligned} \tag{7}$$

where n_s denotes the outward normal vector of the source point s. By moving the field point x to the boundary, the dual boundary integral equations for the boundary point can be obtained as follows:

$$\pi u(x) = C.P.V. \int_B T(s,x)u(s)dB(s) - R.P.V. \int_B U(s,x)t(s)dB(s), \quad x \in B, \tag{8}$$

$$\pi \frac{\partial u(x)}{\partial n_x} = H.P.V. \int_B M(s,x)u(s)dB(s) - C.P.V. \int_B L(s,x)t(s)dB(s), \quad x \in B, \tag{9}$$

where the $R.P.V.$ is the Riemann principal value, $C.P.V.$ is the Cauchy principal value and $H.P.V.$ is the Hadamard (or called Mangler) principal value. The dual null-field integral equations are

$$0 = \int_B T(s,x)u(s)dB(s) - \int_B U(s,x)t(s)dB(s), \quad x \in \Omega^c, \tag{10}$$

$$0 = \int_B M(s,x)u(s)dB(s) - \int_B L(s,x)t(s)dB(s), \quad x \in \Omega^c, \tag{11}$$

when the field point x is moved to the complementary domain, and the superscript "c" denotes the complementary domain.

2.3 Dual Null-Field Integral Formulation – the Present Version (Direct BIEM)

By introducing the degenerate kernels, the collocation point can be exactly located on the real boundary free of facing singularity. Therefore, the representations of integral equations including the boundary point can be written as

$$2\pi u(x) = \int_B T^E(s,x)u(s)\,dB(s) - \int_B U^E(s,x)t(s)\,dB(s), \quad x \in \Omega \cup B, \tag{12}$$

$$2\pi \frac{\partial u(x)}{\partial n_x} = \int_B M^E(s,x)u(s)\,dB(s) - \int_B L^E(s,x)t(s)\,dB(s), \quad x \in \Omega \cup B, \tag{13}$$

and

$$0 = \int_B T^I(s,x)u(s)dB(s) - \int_B U^I(s,x)t(s)dB(s), \quad x \in \Omega^c \cup B, \tag{14}$$

$$0 = \int_B M^I(s,x)u(s)dB(s) - \int_B L^I(s,x)t(s)dB(s), \quad x \in \Omega^c \cup B, \tag{15}$$

once each kernel is expressed in terms of an appropriate degenerate form (I and E superscripts denote the difference). Here, we used the formulation of potential problem to illustrate the validity of our approach. The more detailed formulation for plate can be consulted with Lee et al. (2007a).

2.4 Indirect Boundary Integral Formulation

Based on the indirect boundary integral formulation, the potential and its derivative can be represented by

$$u(x) = \int_B P(s,x)\phi(s)dB(s), \tag{16}$$

$$\frac{\partial u(x)}{\partial n_x} = \int_B \frac{\partial P(s,x)}{\partial n_x}\phi(s)dB(s), \tag{17}$$

where $U(s,x)$ or $T(s,x)$ is chosen as $P(s,x)$, $\phi(s)$ is the unknown fictitious density distribution. By matching the boundary condition, the unknown fictitious density can be obtained. Therefore, the potential in the field is determined by using the Eq. (16). The extended application to plate problems can be found in Lee et al. (2007a). The comparison of direct and indirect approaches is shown in Table 1. It must be noted that the null-field equation is not available in the indirect formulation.

Table 1 Comparison of direct and indirect BIEMs

	Direct BIEM	Indirect BIEM
Null field	○	N. A.
Approach	UT (singular equation) LM (hypersingular equation)	$\begin{pmatrix} U \\ L \end{pmatrix}$ (single layer) $\begin{pmatrix} T \\ M \end{pmatrix}$ (double layer)
Singularity	Disappear	Disappear

2.5 Expansions of the Fundamental Solution and Boundary Density

The closed-form fundamental solutions as mentioned above are

$$U(\mathbf{s},\mathbf{x}) = \begin{cases} \frac{1}{2\pi}\ln r & \text{for the Laplace problem} \\ \frac{i}{4}H_0^{(1)}(kr) & \text{for the Helmholtz problem} \\ \frac{1}{8\zeta^2}[Y_0(\zeta r) + iJ_0(\zeta r) + \frac{2}{\pi}(K_0(\zeta r) + iI_0(\zeta r))] & \text{for the biHelmholtz problem} \end{cases} \tag{18}$$

where $r \equiv |s - x|$ is the distance between the source point and the field point, i is the imaginary number ($i^2 = -1$), $H_0^{(1)}$ is the first kind Hankel function of the zeroth order, $J_0(\zeta r)$ and $Y_0(\zeta r)$ are the first kind and second kind Bessel functions of the zeroth order, respectively, $I_0(\zeta r)$ and $K_0(\zeta r)$ are the first kind and second kind modified Bessel functions of the zeroth order, respectively. To fully utilize the property of circular geometry, the mathematical tools, separable kernel (or so-called degenerate kernel) and Fourier series, are utilized for analytical integrations.

2.5.1 Degenerate (Separable) Kernel for Fundamental Solutions

By employing the separating technique for source point and field point, the kernel function $U(s, x)$ can be expanded in terms of degenerate (separable) kernel in a series form as shown below:

$$U(s,x) = \begin{cases} U^I(s,x) = \sum_{j=0}^{\infty} A_j(s)B_j(x), & |x| \le |s|, \\ U^E(s,x) = \sum_{j=0}^{\infty} A_j(x)B_j(s), & |x| > |s|, \end{cases} \tag{19}$$

where the superscripts "I" and "E" denote the interior and exterior cases, respectively. The other kernels in the boundary integral equation can be obtained by utilizing the operators of Eq. (7) with respect to the kernel $U(s, x)$. When the degenerate kernel is used, we choose two linearly independent sets of $\{A_j\}$ and $\{B_j\}$. In the computation, the degenerate kernel can be expressed as finite sums of products of functions of s alone and functions of x alone. Equation (19) is valid for one, two and three dimensional cases as shown in Fig. 2. In this paper, we focus on two-dimensional problems. The degenerate kernels for the fundamental solutions of the three operators are shown in Chen et al. (2005, 2007), and Lee et al. (2007a).

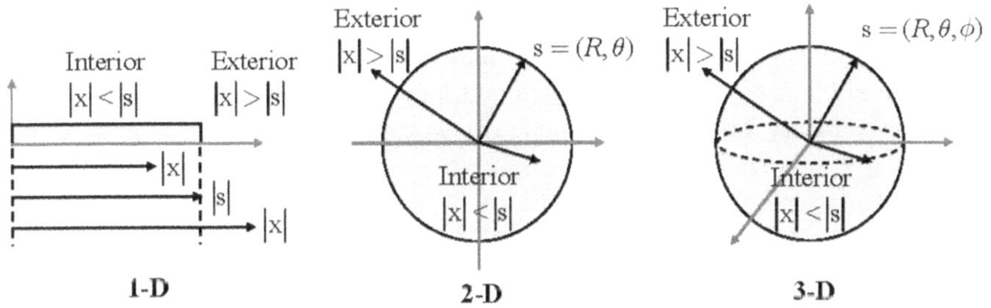

Fig. 2 Degenerate kernel for the one, two and three dimensional problems

2.5.2 Fourier Series Expansion for Boundary Densities

We apply the Fourier series expansion to approximate the boundary potential and its normal derivative as expressed by

$$u(s) = a_0 + \sum_{n=1}^{\infty} a_n \cos n\theta + b_n \sin n\theta, \quad s \in B, \tag{20}$$

$$t(s) = p_0 + \sum_{n=1}^{\infty} p_n \cos n\theta + q_n \sin n\theta, \quad s \in B, \tag{21}$$

where a_n, b_n, p_n and q_n $(n = 0, 1, 2, \ldots)$ are the Fourier coefficients and θ is the polar angle. In the real computation, the integrations can be analytically calculated by employing the orthogonal property of Fourier series and only the finite M number of terms is used in the summation. The present method belongs to one kind of semi-analytical methods since error only attributes to the truncation of Fourier series.

2.6 Adaptive Observer System

After collocating points in the null-field integral equation of Eq. (14), the boundary integrals through all the circular contours are required. Since the boundary integral equations are frame indifferent (i.e. rule of objectivity), the origin of the observer system can be adaptively located on the center of the corresponding boundary contour under integration. Adaptive observer system is chosen to fully employ the circular property by expanding the kernels into degenerate forms. Fig. 3 shows the boundary integration for the circular boundaries in the adaptive observer system. The dummy variable in the circular contour integration is the angle (θ) instead of radial coordinate (R). By using the adaptive system, all the boundary integrals can be determined analytically free of principal value senses.

2.7 Linear Algebraic Equation

In order to calculate the Fourier coefficients, N $(N = 2M + 1)$ boundary nodes for each circular boundary are needed and they are uniformly collocated on each circular boundary. From Eqs. (14) and (15), we have

$$0 = \sum_{j=0}^{H} \int_{B_j} T(s, x)u(s)\, dB(s) - \sum_{j=0}^{H} \int_{B_j} U(s, x)t(s)\, dB(s), \quad x \in D^c \cup B, \tag{22}$$

$$0 = \sum_{j=0}^{H} \int_{B_j} M(s, x)u(s)\, dB(s) - \sum_{j=0}^{H} \int_{B_j} L(s, x)t(s)\, dB(s), \quad x \in D^c \cup B. \tag{23}$$

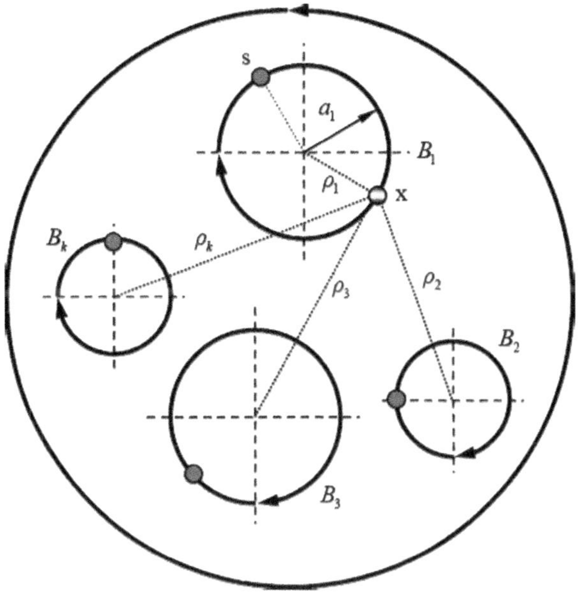

Fig. 3 The adaptive observer system

It is noted that the integration path is counterclockwise on the outer boundary, others are clockwise. For the B_j integral of the circular boundary, the kernels ($U(s, x)$, $T(s, x)$, $L(s, x)$ and $M(s, x)$) are expressed by degenerate kernels when the circle of observation located is the same as that of integral path. The boundary densities ($u(s)$ and $t(s)$) are substituted by using the Fourier series. The linear algebraic system is obtained

$$[U]\{t\} = [T]\{u\}, \tag{24}$$

where $[U]$ and $[T]$ are the influence matrices with a dimension of $(H+1)\times(2M+1)$ by $(H+1)\times(2M+1)$, $\{u\}$ and $\{t\}$ denote the column vectors of Fourier coefficients with a dimension of $(H + 1) \times (2M + 1)$ by 1. All the unknown coefficients can be calculated by using the linear algebraic equations. Then the unknown boundary data can be determined and the potential is obtained by substituting the boundary data into Eq. (12).

2.8 Transformation of Tensor Components

In order to determine the field of potential gradient, the normal and tangential derivatives should be calculated with care. For the non-concentric case, special treatment for the potential gradient should be taken care as the source and field points locate on different circular boundaries. The detailed tensor transformation is shown in Lee et al. (2007a).

3 Illustrative Examples

Example 1: A circular bar with an eccentric inclusion

A circular bar of radius R_0 with an eccentric circular inclusion of radius R_1 is shown in Fig. 4. The ratio of R_1/R_0 and e_x/R_0 are 0.3 and 0.6, respectively. The torsional rigidity G of cross section is expressed as

$$G = \sum_{k=0}^{H} \mu_k \left[\int_{\Omega} (x^2 + y^2)\, d\Omega - \int_{B_k} \varphi \frac{\partial \varphi}{\partial n} dB_k \right], \tag{25}$$

where μ_k is the shear modulus of kth inclusion and φ is the warping function. It is found that the solution converges fast by using only fourteen terms of Fourier series. The results of torsional rigidity for various values of μ_1/μ_0 are shown in Table 2. For verifying our results, the exact solution of Muskhelishvili (1953) and

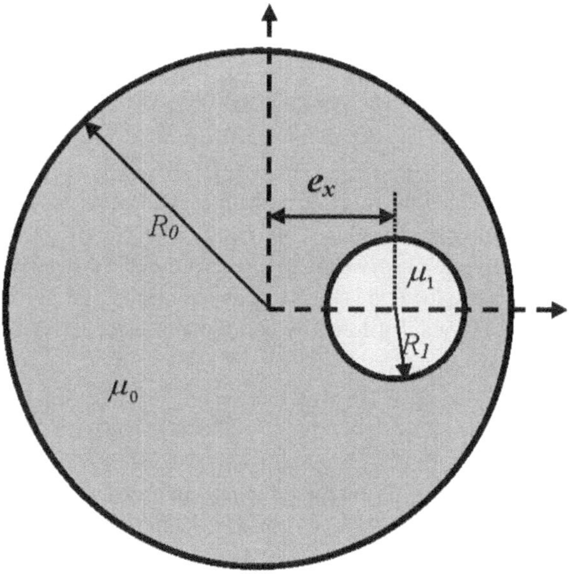

Fig. 4 Sketch of an eccentric circular inclusion problem

Table 2 Torsional rigidity of a circular bar with an eccentric inclusion

μ_1/μ_0	$2G/\pi\mu_0 R_0^4$ Muskhelishvili (1953)	Tang (1996)	Present (M=20)
0	0.82370	0.82377	0.82370
0.2	0.89180	0.89181	0.89180
0.6	0.96246	0.96246	0.96246
1.0	1.00000	1.00000	1.00000
5.0	1.10800	1.10794	1.10800
20.0	1.25224	1.25181	1.25224
1000	9.19866	N/A	9.19866

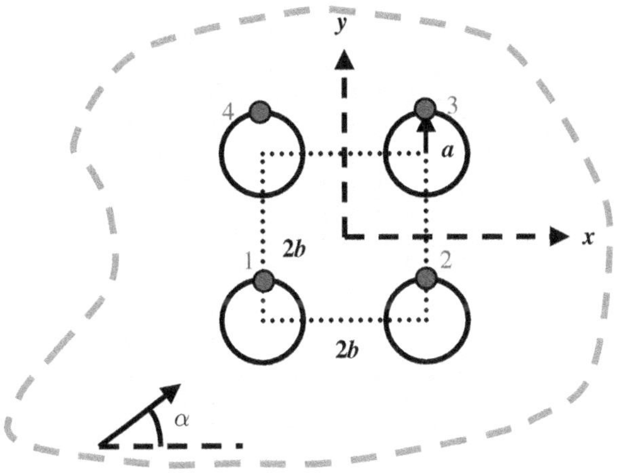

Fig. 5 Interaction of an incident water wave with four cylinders

the result of integral equation formulation by Tang (1996) are also shown in Table 2 for comparison. The present results match very well with the exact solution derived by Muskhelishvil and are better than those of Tang (1996).

Example 2: Water wave impinging four cylinders

In this example, we consider a water wave problem by an array of four bottom-mounted vertical rigid circular cylinders with the same radius a located at the vertices of a square $(-b, -b)$, $(-b, b)$, $(b, -b)$, (b, b), respectively, as shown in Fig. 5. Consider the incident wave in the direction of 45 degree ($\alpha = 45^\circ$). The first-order force for four cylinders in the direction of the incident wave is shown in Fig. 6. It is found that the force effect on the cylinder 2 and the cylinder 4 is identical as expected due to symmetry. The maximum free-surface elevation amplitude is plotted in Fig. 7. Also, the results of potentials at the north pole of each cylinder are also compared well with the approximate series solution given by Linton and

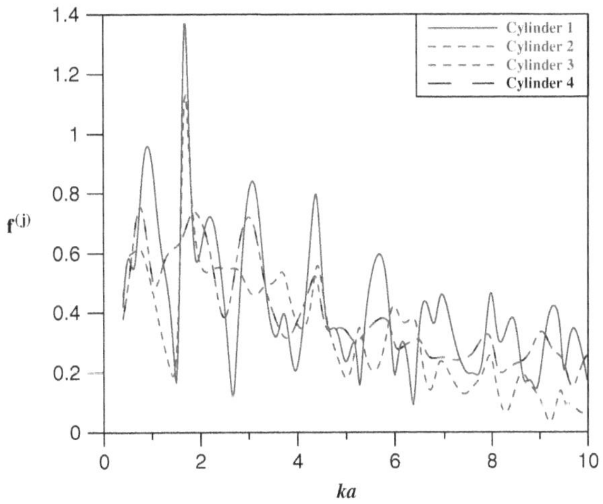

Fig. 6 The first-order force for four cylinders using the proposed method

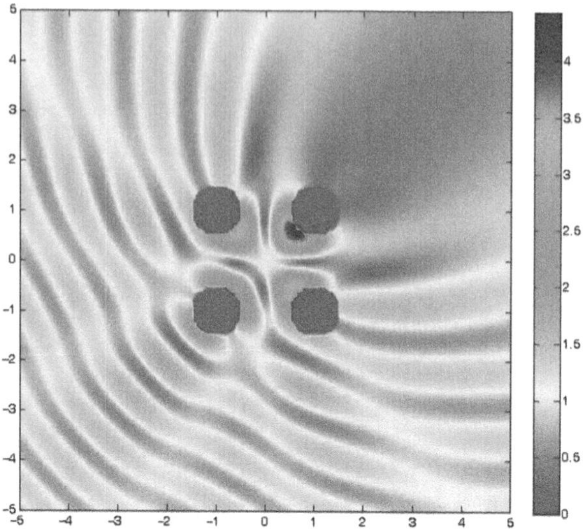

Fig. 7 Contour of the maximum free-surface elevation amplitude

Evans (1990) and the BEM data by Perrey-Debain et al. (2003) as shown in Table 3. The results agree well with those of Perrey-Debain et al.

Table 3 Potential (ϕ) at the north pole of each cylinder ($ka = 1.7$)

	Present method	Perrey-Debain et al.	Linton & Evan
Cylinder 1	−2.418395851	−2.418395682	−2.418395683
	+0.753719467 *i*	+0.753719398 *i*	+0.753719398 *i*
Cylinder 2	2.328927362	2.328927403	2.328927400
	−0.310367580 *i*	−0.310367705 *i*	−0.310367707 *i*
Cylinder 3	0.350612027	0.350611956	0.350611956
	−0.198852116 *i*	−0.198852086 *i*	−0.198852086 *i*
Cylinder 4	−0.383803194	−0.383803273	−0.383803272
	+1.292792513 *i*	+1.292792457 *i*	+1.292792455 *i*

Example 3: A circular plate with an eccentric hole

A circular plate weakened by an eccentric hole is considered. The offset distance e of the eccentric hole is 0.45 m ($e/a = 0.45$) as shown in Fig. 8. The FEM model of the ABAQUS used 8217 elements and 8404 nodes. The former six natural frequency parameters and modes by using FEM (Khurasia, 1978), ABAQUS and the present method are shown in Fig. 9. The results of the present method match well with those of FEM using ABAQUS. In the results of Khurasia (1978), the first mode was not reported while the second and fourth modes are lost. A little deviation is also shown in the results reported by Khurasia and Rawtani due to the coarse mesh. Owing to the lack of stiffness of the clamped boundary condition in reality, it is expected that

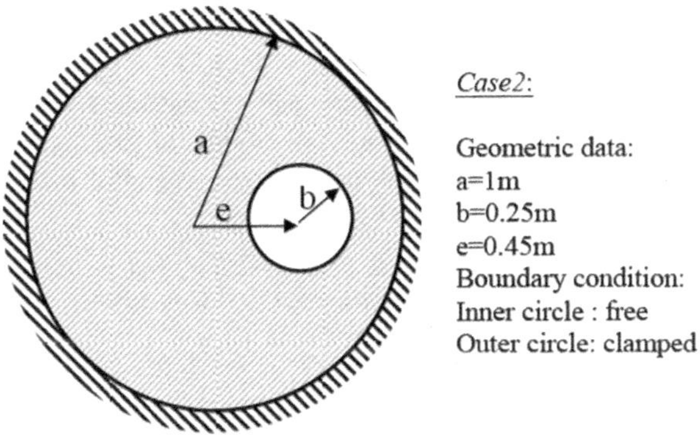

Fig. 8 A circular plate with an eccentric hole subject to clamped-free boundary conditions

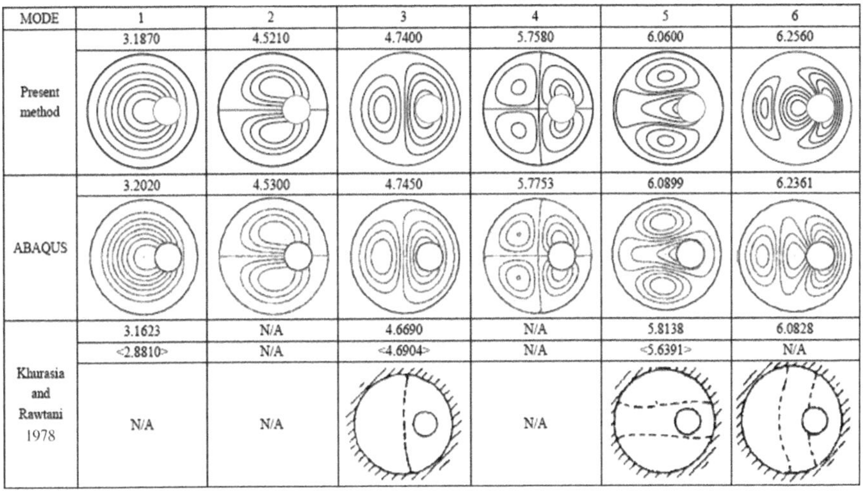

Fig. 9 The former six natural frequency parameters and modes of a circular plate with an eccentric hole where <> denotes the experimental data

the experimental data (Khurasia, 1978) are less than those obtained by using the other methods

4 Conclusions

For the boundary value problems with circular boundaries, we have proposed a null-field BIEM formulation by using degenerate kernels, the null-field integral equation and Fourier series in companion with adaptive observer system and vector decomposition. Three operators for Laplace, Helmholtz and biHelmholtz problems were all considered. This method is a semi-analytical approach for the problems with circular boundaries since only truncation error in the Fourier series is involved. The method shows great generality and versatility for the problems including circular boundaries. Not only the torsion problem but also the water wave as well as plate vibration problems were solved. Five goals: (1) free of calculating principal value, (2) exponential convergence, (3) well-posed algebraic system, (4) elimination of boundary-layer effect and (5) meshless, of the formulation were achieved. A general-purpose program for solving the problems with arbitrary number, various size and any location of circular cavities and/or inclusions was developed.

Acknowledgments Financial support from the National Science Council under Grants No. NSC-94-2211-E-019-009 and MOE CMBB-97-G-A-601 for National Taiwan Ocean University is gratefully acknowledged.

References

Bates RHT, Wall DJN (1977) Null field approach to scalar diffraction. I. General method. Philos. Trans. R. Soc. Lond. 287: 45–78

Boström A (1982) Time-dependent scattering by a bounded obstacle in three dimensions. J. Math. Phys. 23: 1444–1450

Chen JT, Hong H-K (1999) Review of dual boundary element methods with emphasis on hypersingular integrals and divergent series. ASME Appl. Mech. Rev. 52: 17–33

Chen JT, Shen WC, Wu AC (2005) Null-field integral equations for stress field around circular holes under anti-plane shear. Eng. Anal. Bound. Elem. 30: 205–217

Chen JT, Hsiao CC, Leu SY (2006) Null-field integral equation approach for plate problems with circular boundaries. ASME J. Appl. Mech. 73: 679–693

Chen JT, Chen CT, Chen PY, Chen IL (2007a) A semi-analytical approach for radiation and scattering problems with circular boundaries. Comput. Meth. Appl. Mech. Eng. 196: 2751–2764

Chen JT, Wu CS, Lee YT, Chen KH (2007b) On the equivalence of the Trefftz method and method of fundamental solution for Laplace and biharmonic equations. Comput. Math. Appl. 53: 851–879

Khurasia HB, Rawtani S (1978) Vibration analysis of circular plates with eccentric hole. ASME J. Appl. Mech. 45: 215–217

Kress R (1989) Linear integral equations. Springer, Berlin

Lee WM, Chen JT, Lee YT (2007a) Free vibration analysis of circular plates with multiple circular holes using indirect BIEMs. J. Sound Vib. 304: 811–830

Lee YT, Chen JT, Chou KS (2007b) Revisit of two classical elasticity problems using the null-field BIE. in APCOM'07 in conjunction with EPMESC XI, Kyoto, Japan

Linton CM, Evans DV (1990) The interaction of waves with arrays of vertical circular cylinders. J. Fluid Mech. 215: 549–569

Martin PA (1981) On the null-field equations for water-wave radiation problems. J. Fluid Mech. 113: 315–332

Muskhelishvili NI (1953) Some basic problems of the mathematical theory of elasticity. Noordhoff, Groningen

Perrey-Debain E, Trevelyan J, Bettess P (2003) Plane wave interpolation in direct collocation boundary element method for radiation and wave scattering: numerical aspects and applications. J. Sound Vib. 261: 839–858

Schaback R (2007) Adaptive Numerical Solution of MFS Systems, in ICCES Special Symposium on Meshless Methods, Patras, Greece

Tang RJ (1996) Torsion theory of the crack cylinder. Shanghai Jiao Tong University Publisher, Shanghai (in Chinese)

Waterman PC (1965) Matrix formulation of electromagnetic scattering. Proc. IEEE. 53: 805–812

Waterman PC (1976) Matrix theory of elastic wave scattering. J. Acoust. Soc. Am. 60: 567–580

The Singular Function Boundary Integral Method for Elliptic Problems with Boundary Singularities

Evgenia Christodoulou, Christos Xenophontos and Georgios Georgiou

Abstract We review the Singular Function Boundary Integral Method (SFBIM) for solving two-dimensional elliptic problems with boundary singularities. In this method the solution is approximated by the leading terms of the asymptotic expansion of the local solution. The unknowns to be calculated are the singular coefficients, i.e. the coefficients of the local asymptotic expansion, also called generalized stress intensity factors. The discretized Galerkin equations are reduced to boundary integrals by means of the divergence theorem. The Dirichlet boundary conditions are then weakly enforced by means of Lagrange multipliers, the values of which are additional unknowns. In the case of two-dimensional Laplacian problems, we have shown that this method converges exponentially with respect to the number of singular functions. This is demonstrated via several benchmark applications, including ones involving the biharmonic operator which can be viewed as an extension of the theory.

1 Introduction

Planar elliptic boundary value problems with boundary singularities have been extensively studied in the last few decades. Many different methods have been proposed for the solution of such problems, ranging from special mesh-refinement schemes to sophisticated techniques that incorporate, directly or indirectly, the form of the local asymptotic expansion, which is known in many occasions. These methods aim to improve the accuracy and resolve the convergence difficulties that are known to appear in the neighborhood of such singular points.

The local solution, centered at the singular point, in polar coordinates (r, θ) is of the general form:

$$u(r, \theta) = \sum_{j=1}^{\infty} a_j r^{\mu_j} f_j(\theta), \tag{1}$$

G. Georgiou (✉)
Department of Mathematics and Statistics, University of Cyprus, 1678 Nicosia, Cyprus
email: georgios@ucy.ac.cy

A chapter in honor of Dimitri Beskos' 65th birthday

G.D. Manolis, D. Polyzos (eds.), *Recent Advances in Boundary Element Methods*,
DOI 10.1007/978-1-4020-9710-2_4,

where μ_j are the eigenvalues and f_j are the eigenfunctions of the problem, which are uniquely determined by the geometry and the boundary conditions along the boundaries sharing the singular point. The singular coefficients α_j, also known as generalized stress intensity factors (Szabó and Babuška 1991) or flux intensity factors (Arad et al. 1998), are determined by the boundary conditions in the rest of the boundary. Knowledge of the singular coefficients is of importance in many engineering applications, especially in fracture mechanics.

In the past few years, Georgiou and co-workers (Georgiou et al. 1996, 1997; Elliotis et al. 2002, 2005a, 2005b, 2006, 2007; Li 2006; Xenophontos et al. 2006) developed the Singular Function Boundary Integral Method (SFBIM), in which the unknown singular coefficients are calculated directly. The solution is approximated by the leading terms of the local asymptotic solution expansion and the Dirichlet boundary conditions are weakly enforced by means of Lagrange multipliers. The method has been tested on standard Laplacian and biharmonic problems, yielding extremely accurate estimates of the leading singular coefficients, and exhibiting exponential convergence with respect to the number of singular functions.

In the present paper, the SFBIM is reviewed and its convergence is discussed. The method is presented in Section 2. In Section 3, some convergence results for Laplacian problems are provided. Numerical results for Laplacian and biharmonic problems are presented and discussed in Sections 4 and 5, respectively. Finally, Section 6 contains the conclusions and briefly discusses our current efforts for extending the method to three-dimensional Laplacian problems with edge singularities.

2 The Singular Function Boundary Integral Method

We consider the Laplacian problem with a boundary singularity, as depicted in Fig. 1. Find u such that:

$$\nabla^2 u = 0 \qquad \text{in} \quad \Omega, \tag{2}$$

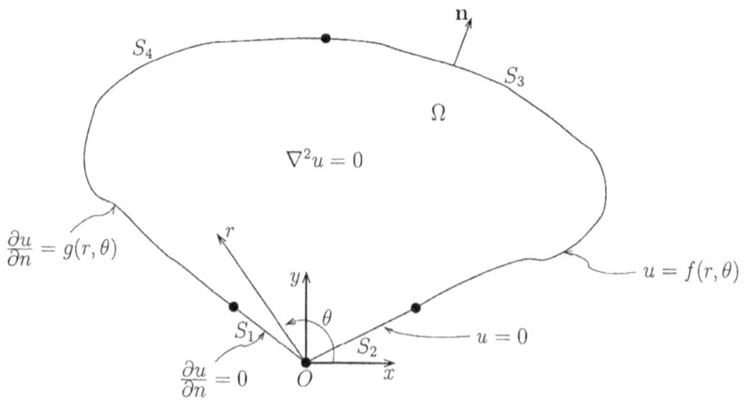

Fig. 1 A two-dimensional Laplace equation problem with a boundary singularity

with

$$\left.\begin{array}{ll}\dfrac{\partial u}{\partial n}=0 & on\ S_1\\ u=0 & on\ S_2\\ u=f(r,\theta) & on\ S_3\\ \dfrac{\partial u}{\partial n}=g(r,\theta) & on\ S_4\end{array}\right\} \tag{3}$$

where Ω has a smooth boundary, $\partial\Omega = S_1 \cup S_2 \cup S_3 \cup S_4$, with the exception of a boundary singularity at the corner O, formed by the straight boundary segments S_1 and S_2. In the remaining parts of the boundary, either Dirichlet or Neumann boundary conditions apply and the given functions f and g are such that no other boundary singularity is present.

In general, the asymptotic expansion of the solution in polar co-ordinates (r, θ) centered at the singular point, is given by (Grisvard 1995)

$$u(r,\theta)=\sum_{i=1}^{\infty}\alpha_i r^{\mu_i} f_i(\theta), \qquad (r,\theta)\in\Omega, \tag{4}$$

where α_i are the unknown singular coefficients, μ_i are the singularity powers arranged in ascending order, and the functions $f_i(\theta)$ represent the θ-dependence of the eigensolution.

The SFBIM is based on the approximation of the solution by the leading terms of the local solution expansion:

$$\bar{u}=\sum_{i=1}^{N_\alpha}\bar{\alpha}_i W^i \tag{5}$$

where N_α is the number of singular functions used, which are defined by

$$W^i \equiv r^{\mu_i} f_i(\theta). \tag{6}$$

Note that this approximation is valid only if Ω is a subset of the convergence domain of expansion (4).

By applying Galerkin's principle, the problem is discretized as follows:

$$\int_{\Omega} W^i \nabla^2 \bar{u}\, dV = 0, \qquad i=1,2,\ldots,N_\alpha, \tag{7}$$

By double application of Green's second identity, the above volume integral becomes:

$$\int_{\partial\Omega} W^i \frac{\partial \bar{u}}{\partial n} dS - \int_{\partial\Omega} \bar{u}\frac{\partial W^i}{\partial n} dS + \int_{\Omega} \bar{u}\nabla^2 W^i dV = 0, \qquad i=1,2,\ldots,N_\alpha, \tag{8}$$

and since the singular functions W^i, are harmonic, the above volume integral is reduced to a boundary one, as follows:

$$\int_{\partial\Omega} \left(\frac{\partial \bar{u}}{\partial n} W^i - \bar{u} \frac{\partial W^i}{\partial n} \right) dS = 0, \qquad i = 1, 2, \ldots, N_\alpha. \tag{9}$$

Thus, the dimension of the problem is reduced by one, leading to a considerable reduction of the computational cost. Since, now, W^i exactly satisfy the boundary conditions along S_1 and S_2, the above integral along these boundary segments is identically zero. Therefore, for $i = 1, 2, \ldots, N_a$ we have:

$$\int_{S_3} \left(\frac{\partial \bar{u}}{\partial n} W^i - \bar{u} \frac{\partial W^i}{\partial n} \right) dS + \int_{S_4} \left(\frac{\partial \bar{u}}{\partial n} W^i - \bar{u} \frac{\partial W^i}{\partial n} \right) dS = 0, \quad i = 1, 2, \ldots, N_\alpha. \tag{10}$$

It should be noted that the integrands in Eq. (10) are non-singular and all integrations are carried out far from the boundaries causing the singularity.

To impose the Neumann condition along S_4, we simply substitute the normal derivative by the known function g (Eq. (3)). The Dirichlet condition along S_3 is imposed by means of a Lagrange multiplier function λ, replacing the normal derivative. The function λ is expanded in terms of standard, polynomial basis functions M^j as

$$\lambda = \frac{\partial \bar{u}}{\partial n} = \sum_{j=1}^{N_\lambda} \lambda_j M^j, \tag{11}$$

where N_λ represents the total number of the unknown discrete Lagrange multipliers (or, equivalently, the total number of Lagrange-multiplier nodes) along S_4. The basis functions M^j are used to weight the Dirichlet condition along the corresponding boundary segment S_3. We thus obtain the following system of $N_\alpha + N_\lambda$ discretized equations:

$$\int_{S_3} \left(\lambda W^i - \bar{u} \frac{\partial W^i}{\partial n} \right) dS - \int_{S_4} \bar{u} \frac{\partial W^i}{\partial n} dS$$
$$= - \int_{S_4} W^i g(r, \theta) dS, \quad i = 1, 2, \ldots, N_\alpha, \tag{12}$$

$$\int_{S_3} \bar{u} M^j dS = \int_{S_3} f(r, \theta) M^j dS, \quad j = 1, 2, \ldots, N_\lambda. \tag{13}$$

The above system can be written in the following block form:

$$\begin{bmatrix} K_1 & K_2 \\ K_2^T & O \end{bmatrix} \begin{bmatrix} \mathrm{A} \\ \Lambda \end{bmatrix} = \begin{bmatrix} F_1 \\ F_2 \end{bmatrix} \tag{14}$$

where A is the vector of the unknown singular coefficients α_i, Λ is the vector of the unknown Lagrange multipliers λ_j, submatrices K_1 and K_2 contain the coefficients of the unknowns (obviously, K_1 is symmetric), and vectors F_1 and F_2 contain the right hand side contributions of Eqs. (12) and (13), respectively. It is easily shown that the system (14) is symmetric and nonsingular, provided $N_\alpha > N_\lambda$.. The "optimal" relationship between these two parameters will be discussed in the next section.

3 Convergence Analysis

In this section we briefly present results from Xenophontos et al. (2006) which show that the method converges at an exponential rate. To this end let

$$H^1(\Omega) = \left\{ u \in L^2(\Omega) : \nabla u \in L^2(\Omega) \right\}$$

denote the usual Sobolev space, with $\|\cdot\|_{1,\Omega}$ denoting its norm, and set

$$H_*^1(\Omega) = \left\{ u \in H^1(\Omega) : u|_{S_2} = 0 \right\}.$$

The space

$$H^{1/2}(\partial\Omega) = \left\{ u \in H^1(\Omega) : u|_{\partial\Omega} \in L^2(\partial\Omega) \right\}$$

is referred to as the trace space of functions in $H^1(\Omega)$ and its norm will be denoted by $\|\cdot\|_{1/2,\Omega}$. Finally, the dual space of $H^{1/2}(\partial\Omega)$, denoted by $H^{-1/2}(\partial\Omega)$, with norm $\|\cdot\|_{-1/2,\Omega}$, will also be used (see Xenophontos et al. 2006 for more details). The approximate solution u will be chosen from the finite dimensional space $V_\alpha \subset H_*^1(\Omega)$, defined by

$$V_\alpha = span\left\{ W^i \right\}_{i=1}^{N_\alpha}. \tag{15}$$

The Lagrange multiplier function λ will be chosen from the finite dimensional space and $V_\lambda \subset H^{-1/2}(S_3)$, which is defined as follows. Let S_3 be divided into quasi-uniform sections $\Gamma_i, i = 1, \ldots, n$ such that $S_3 = \cup_{i=1}^n \Gamma_i$. Let $h_i = |\Gamma_i|$ and set $h = \max_{1 \le i \le n} h_i$. We assume that for each segment Γ_i there exists an invertible mapping $F : I \to \Gamma_i$ which maps the interval $I = [-1, 1]$ to Γ_i, and define

$$V_\lambda = \left\{ \lambda_h : \lambda_h|_{\Gamma_i} \circ F_i^{-1} \in P_p(I), \quad i = 1, \ldots, n \right\}, \tag{16}$$

where $P_p(I)$ is the set of polynomials of degree $\leq p$ on $I = [-1, 1]$. In other words, the Lagrange multiplier function λ is a linear combination of piecewise polynomials of degree p, defined on a quasiuniform partition of S_3 characterized by the meshwidth h. Note that the number of Lagrange multipliers N_λ satisfies

$$N_\lambda = O(1/h).$$

In Xenophontos et al. (2006) it was shown that if u and $\partial u/\partial n$ are approximated by u and λ, given by Eqs. (5) and (11) respectively, then there exists a positive constant C, independent of N_α and N_λ, such that

$$\|u - \bar{u}\|_{1,\Omega} + \left\| \frac{\partial u}{\partial n} - \lambda \right\|_{-1/2,S_3} \leq C \left\{ \inf_{v \in V_\alpha} \|u - v\|_{1,\Omega} + \inf_{\eta \in V_\lambda} \|\lambda - \eta\|_{-1/2,S_3} \right\}.$$

Using the above best approximation result it was further shown that if $\partial u/\partial n \in H^k(S_3)$ for some $k \geq 1$, then there exist positive constants C and $\beta \in (0, 1)$, independent of N_α and N_λ, such that

$$\|u - \bar{u}\|_{1,\Omega} + \left\| \frac{\partial u}{\partial n} - \lambda \right\|_{-1/2,S_3} \leq C \left\{ \sqrt{N_\alpha} \beta^{N_\alpha} + h^m p^{-k} \right\}, \tag{17}$$

where $m = \min\{k, p+1\}$. Moreover, since the error between the exact coefficients α_i and approximate coefficients $\bar{\alpha}_i$ satisfies

$$|\alpha_i - \bar{\alpha}_i| \leq C \|u - \bar{u}\|_{L^2(\Omega)},$$

we have

$$|\alpha_i - \bar{\alpha}_i| \leq C\beta^{N_\alpha}, \tag{18}$$

which shows that the method approximates these coefficients at an exponential rate, as $N_\alpha \to \infty$.

Based on the error estimate (17) one may obtain the "optimal" matching between the parameters N_α and h, i.e. the relationship between the number of singular functions and the number of Lagrange multipliers used in the method, by choosing them in such a way so that the error in Eq. (17) is balanced. For example, in the case when p is kept fixed and $h \to 0$ (or equivalently $N_\lambda \to \infty$) we take $h^{p+1} \approx \sqrt{N_\alpha}\beta^{N_\alpha}$. This leads to the following approximate expression for N_α:

$$N_\alpha \approx (p+1) \left| \ln \frac{2p}{N_\lambda - 1} / \ln \beta \right|. \tag{19}$$

In practice, Eq. (19) is used as follows: We pick a value for N_λ and solve the linear system (14) for several values of $N_\alpha > N_\lambda$ concentrating only on the calculation of the first approximate coefficient $\bar{\alpha}_1$, which we record. Once we reach a value for N_α

which yields an approximate $\bar{\alpha}_1$ with, say, 16 converged significant digits, we then use Eq. (19) to calculate the constant β using the values for N_α and N_λ which gave us the converged coefficient $\bar{\alpha}_1$. With β now known, we can compute subsequent "optimal" pairs of N_α and $N_{\lambda.}$.

4 Numerical Results for Laplacian Problems

Results for the cracked-beam problem (Georgiou et al. 1997) and a Laplacian problem over an L-shaped domain (Elliotis et al. 2005a) are presented in this section. The former problem is defined in Fig. 2. A singularity arises at $x = y = 0$, where the boundary condition suddenly changes from $u = 0$ to $\partial u/\partial y = 0$. The local solution is given by

$$u = \sum_{j=1}^{\infty} \alpha_j r^{(2j-1)/2} \cos\left[\left(\frac{2j-1}{2}\right)\theta\right]. \tag{20}$$

The system of discretized equations, resulting from the application of the SFBIM, consists of two equation sets as follows:

$$-\int_{S_3} \bar{u}\frac{\partial W^i}{\partial x}dy + \int_{S_4}\left(\lambda W^i - \bar{u}\frac{\partial W^i}{\partial y}\right)dx + \int_{S_5} \bar{u}\frac{\partial W^i}{\partial x}dy = 0, \quad i = 1, \ldots, N_\alpha, \tag{21}$$

$$\int_{S_4} uM^j dx = 0.125 \int_{S_4} M^j dx = 0, \quad j = 1, \ldots, N_\lambda. \tag{22}$$

The interval $[-1/2, 1/2]$, corresponding to the boundary segment S_4, is subdivided uniformly into quadratic elements and, thus, the Lagrange multiplier function is approximated locally by quadratic polynomials. All numerical integrations were performed using Gaussian quadrature with 15 nodes on each subinterval.

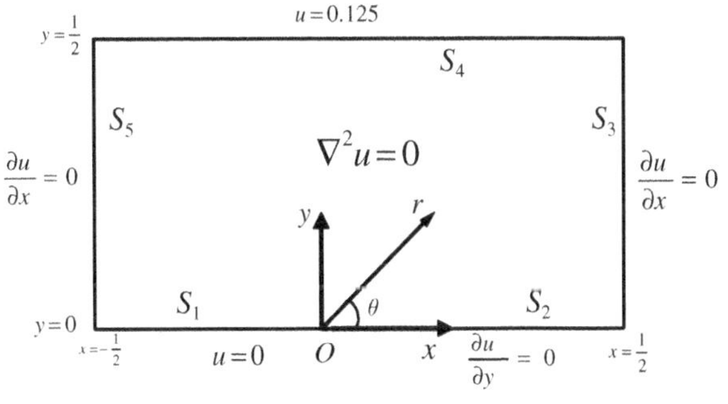

Fig. 2 The cracked-beam problem

Table 1 The coefficients $\bar{\alpha}_l$ calculated with $N_\alpha = 50$ and $N_\lambda = 25$ (cracked-beam problem)

l	$\bar{\alpha}_l$
1	0.191118631972
2	−0.118116071967
3	0.000000000000
4	0.000000000000
5	−0.01254698598
6	−0.01903340371

Using Eq. (19), we find that the "optimal" values for N_α and N_λ are 50 and 25, respectively. In Table 1, the first 6 approximate coefficients, calculated using these "optimal" values, are given.

Let us now consider the second Laplacian problem which is depicted in Fig. 3. This is equivalent to a Poisson equation problem, $\nabla^2 u = -1$, over an L-shaped domain, with homogeneous Dirichlet boundary conditions along the whole boundary. Note that along boundary parts S_2 and S_3 essential boundary conditions are applied. Due to symmetry, only half of the domain is considered. The local solution is given by

$$u = \sum_{j=1}^{\infty} \alpha_j r^{2(2j-1)/3} \sin\left[\frac{2}{3}(2j-1)\theta\right]. \tag{23}$$

This problem is very important in fracture mechanics and the so-called "generalized stress intensity factor", defined by $2\alpha_1/3$, is of great significance (Arad et al. 1998). It should be noted that two sets of Lagrange multiplier functions, denoted by λ_A and λ_B, are now required. Thus, the Dirichlet boundary conditions along S_2 and S_3 are replaced by:

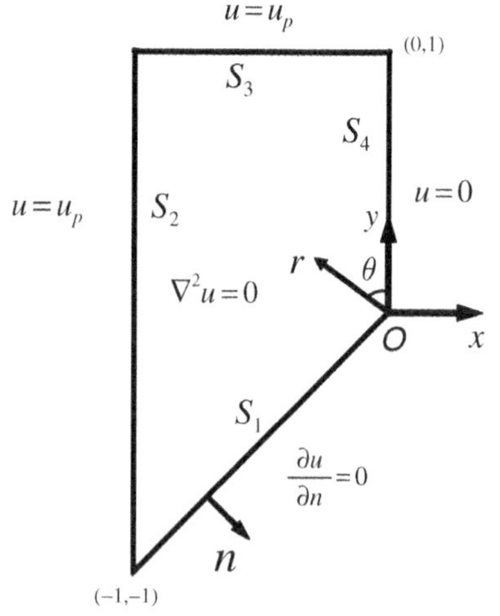

Fig. 3 Geometry of the second Laplacian problem: due to symmetry only half of the L-shaped domain is used

$$\lambda_A = \frac{\partial \bar{u}}{\partial x} = \sum_{j=1}^{N_{\lambda_A}} \lambda_A^j M^j \qquad \text{and} \qquad \lambda_B = \frac{\partial \bar{u}}{\partial y} = \sum_{j=1}^{N_{\lambda_B}} \lambda_B^j M^j, \tag{24}$$

where N_{λ_A} and N_{λ_B} are the numbers of nodes along S_2 and S_3, respectively. The following system of $N_\alpha + N_{\lambda_A} + N_{\lambda_B}$ equations is thus obtained:

$$\begin{aligned} &-\int_{S_2} \left(\lambda_A W^i - \bar{u}\frac{\partial W^i}{\partial x} \right) dy \\ &+ \int_{S_3} \left(\lambda_B W^i - \bar{u}\frac{\partial W^i}{\partial y} \right) dx = 0, \quad i = 1, \ldots, N_\alpha, \end{aligned} \tag{25}$$

$$-\int_{S_2} \bar{u} M^j dy = \int_{S_2} u_p M^j dy, \quad j = 1, \ldots, N_{\lambda_A}, \tag{26}$$

$$\int_{S_3} \bar{u} M^k dx = -\int_{S_3} u_p M^k dx, \quad k = 1, \ldots, N_{\lambda_B}. \tag{27}$$

As before, we use quadratic polynomials to approximate the Lagrange multiplier functions λ_A and λ_B over boundaries S_2 and S_3 respectively, and all numerical integrations are performed using Gaussian quadrature with 15 nodes on each subinterval. The "optimal" values for N_α and $N_\lambda = N_{\lambda_A} + N_{\lambda_B}$ are found via Eq. (19) to be $N_\alpha = 90$ and $N_\lambda = 38$. The computed leading singular coefficients are listed in Table 2.

Table 2 Converged values of the leading singular coefficients with N_α =90 and N_λ =38 (second Laplacian problem)

i	$\bar{\alpha}_i$	i	$\bar{\alpha}_i$
1	0.40193103	9	−0.000719
2	0.09364829	10	−0.000565
3	−0.0093830	11	−0.000395
4	−0.0298851	12	−0.000296
5	−0.0083588	13	−0.000219
6	−0.0047302	14	−0.000173
7	−0.0015451	15	−0.000138
8	−0.001098		

5 The SFBIM for Biharmonic Problems

In this section we describe the extension of the SFBIM to biharmonic problems arising in fracture mechanics. Even though no convergence analysis is available at present, we found that the method converges fast with respect to the number of singular functions, yielding very satisfactory results (Elliotis et al. 2005a, 2005b,

2006, 2007). We consider a two-dimensional solid elastic plate containing a single edge crack subjected to a uniform inplane load normal to the two edges parallel to the crack, while the remaining edges are stress free, as illustrated in Fig. 4. The resulting boundary value problem is to find u such that:

$$\nabla^4 u = 0 \quad \text{in} \quad \Omega = (-1, 1) \times (0, 1), \tag{28}$$

with

$$\left.\begin{array}{llll} u = 0, & \dfrac{\partial u}{\partial y} = 0, & on & S_A \\ \dfrac{\partial u}{\partial y} = 0, & \dfrac{\partial^3 u}{\partial y^3} = 0, & on & S_B \\ u = 2, & \dfrac{\partial u}{\partial x} = 2, & on & S_C \\ u = \dfrac{1}{2}(x+1)^2, & \dfrac{\partial u}{\partial y} = 0, & on & S_D \\ u = 0, & \dfrac{\partial u}{\partial x} = 0, & on & S_E \end{array}\right\} \tag{29}$$

where $\partial\Omega = S_A \cup S_B \cup S_C \cup S_D \cup S_E$.

The asymptotic expansion for u in the neighborhood of the singular point O is given by:

$$u(r, \theta) = \sum_{j=1}^{\infty} \left[c_j W_1^j(r, \theta) + d_j W_2^j(r, \theta) \right], \tag{30}$$

where (r, θ) are the polar coordinates centered at O, and c_j, d_j correspond to the even and odd SIFs, respectively (see also Schiff et al. 1979). The two sets of singular

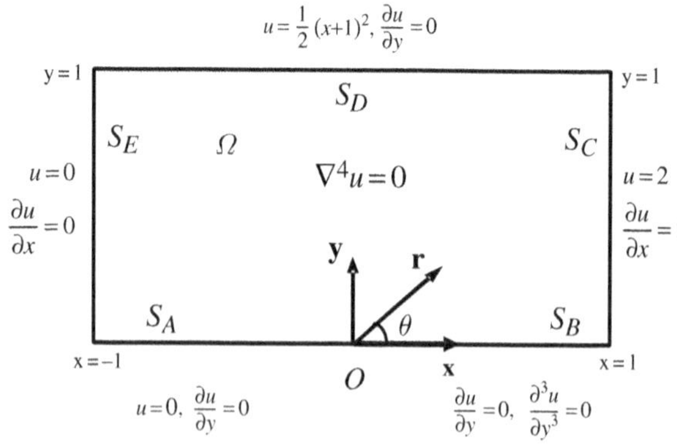

Fig. 4 The model fracture problem. Due to symmetry only half of the domain is used; the crack corresponds to S_A

functions W_k^j, $j = 1, 2, \ldots, N_\alpha$, $k = 1, 2$, are given by

$$W_k^j \equiv r^{\mu_j+1} f_k(\theta, \mu_j), \tag{31}$$

where

$$f_1(\theta, \mu_j) = \cos(\mu_j - 1)\theta - \cos(\mu_j + 1)\theta, \qquad \mu_j = j \tag{32}$$

and

$$f_2(\theta, \mu_j) = \cos(\mu_j - 1)\theta - \frac{\mu_j - 1}{\mu_j + 1}\cos(\mu_j + 1)\theta, \qquad \mu_j = j - \frac{1}{2}. \tag{33}$$

Note that the singular functions W_k^j satisfy the PDE (28) and the boundary conditions on S_A and S_B.

As usual, the solution will be approximated by the leading terms of the asymptotic expansion. The approximate solution u is given by:

$$\bar{u} = \sum_{i=1}^{N_\alpha} \bar{c}_i\, W_1^i + \sum_{i=1}^{N_\alpha} \bar{d}\, W_2^i, \tag{34}$$

where $\bar{c}_i$ and $\bar{d}_i$ are the approximations to the SIFs. Applying Galerkin's principle, the governing equation is weighted by the singular functions, which gives the following set of discretized equations:

$$\int_\Omega W_k^i \nabla^4 \bar{u} dV = 0, \quad i = 1, 2, \ldots, N_\alpha, k = 1, 2. \tag{35}$$

Next, applying Green's theorem twice and since the singular functions satisfy the governing biharmonic Eq. (28), the above integrals are reduced to:

$$\int_{\partial\Omega} \left(\frac{\partial \bar{u}}{\partial n} \nabla^2 W_k^i - \bar{u} \frac{\partial(\nabla^2 W_k^i)}{\partial n} \right) dS + \int_{\partial\Omega} \left(\frac{\partial\left(\nabla^2 \bar{u}\right)}{\partial n} W_k^i - \nabla^2 \bar{u} \frac{\partial W_k^i}{\partial n} \right) dS = 0, \tag{36}$$

for $i = 1, 2, \ldots, N_\alpha$, $k = 1, 2$. Now, since W_k^j satisfy exactly the boundary conditions along S_A and S_B, the above integral along these boundary segments is identically zero. Therefore, we have:

$$\int_{S} \left(\frac{\partial \bar{u}}{\partial n} \nabla^2 W_k^i - \bar{u} \frac{\partial(\nabla^2 W_k^i)}{\partial n} \right) dS + \int_{S} \left(\frac{\partial\left(\nabla^2 \bar{u}\right)}{\partial n} W_k^i - \nabla^2 \bar{u} \frac{\partial W_k^i}{\partial n} \right) dS = 0, \tag{37}$$

where $i = 1, 2, \ldots, N_\alpha, k = 1, 2$ and $S = S_C \cup S_D \cup S_E$. As usual, the Dirichlet boundary conditions are imposed by means of Lagrange multipliers. In the case of Laplacian problems, the Lagrange multipliers replace the normal derivative $\partial \bar{u}/\partial n$. In the case of biharmonic problems, another option for the Lagrange multipliers is to replace $\partial \nabla^2 \bar{u}/\partial n$, which is the choice made here. In the present problem, Dirichlet boundary conditions appear along the three boundary parts of interest, i.e. S_C, S_D and S_E, where the normal derivative of the solution is also specified. Therefore, Lagrange multipliers have been chosen to replace $\partial \nabla^2 \bar{u}/\partial n$ on boundary parts S_C, S_D and S_E. These are partitioned into three-node elements and the corresponding Lagrange multipliers, denoted respectively by λ_C, λ_D and λ_E, are expanded in terms of quadratic basis functions M^j as:

$$\lambda_C = \frac{\partial \left(\nabla^2 \bar{u}\right)}{\partial x} = \sum_{j=1}^{N_{\lambda_C}} \lambda_C^j M^j \text{ on } S_C, \tag{38}$$

$$\lambda_D = \frac{\partial \left(\nabla^2 \bar{u}\right)}{\partial y} = \sum_{j=1}^{N_{\lambda_D}} \lambda_D^j M^j \text{ on } S_D, \tag{39}$$

and

$$\lambda_E = \frac{\partial \left(\nabla^2 \bar{u}\right)}{\partial x} = \sum_{j=1}^{N_{\lambda E}} \lambda_E^j M^j \text{ on } S_E, \tag{40}$$

where N_{λ_C}, N_{λ_D} and N_{λ_E} are the numbers of the discrete Lagrange multipliers λ_C^j, λ_D^j and λ_E^j along the corresponding boundaries. The discrete Lagrange multipliers appear as additional unknowns in the problem. The required $N_{\lambda_C} + N_{\lambda_D} + N_{\lambda_E}$ additional equations are obtained by weighting the Dirichlet boundary conditions along S_C, S_D and S_E by the quadratic basis functions M^j in the Galerkin sense. The following linear system of $2N_\alpha + N_{\lambda_C} + N_{\lambda_D} + N_{\lambda_E}$ discretized equations is thus obtained:

$$\begin{aligned} &\int_{S_C} \left(\lambda_C W_k^i - \bar{u} \frac{\partial(\nabla^2 W_k^i)}{\partial x} - \nabla^2 \bar{u} \frac{\partial W_k^i}{\partial x} \right) dy \\ &+ \int_{S_D} \left(\lambda_D W_k^i - \bar{u} \frac{\partial(\nabla^2 W_k^i)}{\partial y} - \nabla^2 \bar{u} \frac{\partial W_k^i}{\partial y} \right) dx \\ &+ \int_{S_E} \left(-\lambda_E W_k^i + \bar{u} \frac{\partial(\nabla^2 W_k^i)}{\partial x} + \nabla^2 \bar{u} \frac{\partial W_k^i}{\partial x} \right) dy \\ &= - \int_{S_C} 2\nabla^2 W_k^i dy, i = 1, \ldots, N_\alpha, k = 1, 2, \end{aligned} \tag{41}$$

$$\int_{S_C} \bar{u} M^j dy = \int_{S_C} 2M^j dy, \qquad j = 1, 2, \ldots, N_{\lambda_C}, \tag{42}$$

$$\int_{S_D} \bar{u} M^j dx = \int_{S_D} \left[\frac{1}{2}(x+1)^2\right] M^j dx, \qquad j = 1, 2, \ldots N_{\lambda_D}, \tag{43}$$

$$-\int_{S_E} \bar{u} M^j dy = 0, \qquad j = 1, 2, \ldots N_{\lambda_E}. \tag{44}$$

As in the case of Laplacian problems, the integrands in Eqs. (41), (42), (43) and (44) are non-singular and all integrations are carried out far from the boundaries causing the singularity. Also, the stiffness matrix is symmetric and becomes singular if $N_\lambda > 2N_\alpha$ where $N_\lambda = N_{\lambda_C} + N_{\lambda_D} + N_{\lambda_E}$.

As already mentioned, no theory is available at this time for biharmonic problems. As a result, the "optimal" values of N_α and N_λ are found by systematic runs. We found that the choice $N_\alpha = 47$, $N_\lambda = 39 (= 7 + 25 + 7)$ produces very accurate results. These are converged in the sense that they are not affected by moderate changes of N_α and N_λ (see Elliotis et al. 2006 for more details). Table 3 shows the approximate coefficients $\bar{c}_i, \bar{d}_i, i = 1, 2, \ldots, 10$ obtained with this choice of parameters, along with the results from (Li et al. 2004) for comparison. It appears that the SFBIM can be effective for biharmonic problems as well.

Table 3 Comparison of converged values of the SIFs with those reported by Li et al. (2004) using the Collocation Trefftz method

Approximate SIFs	Collocation Trefftz	SFBIM
d_1	2.12751351	2.1275134
d_2	−1.0366925	−1.036692
d_3	0.0371711	0.037170
d_4	0.117749	0.11775
d_5	−0.122728	−0.12273
d_6	−0.109909	−0.10991
d_7	−0.002255	−0.00226
d_8	0.006863	0.00686
d_9	−0.005936	−0.00594
d_{10}	−0.011032	−0.01103
c_1	0.1667621	0.166762
c_2	0.0624433	0.062444
c_3	−0.1324738	−0.132474
c_4	−0.010221	−0.01022
c_5	0.105846	0.10585
c_6	0.031153	0.03115
c_7	−0.007149	−0.00714
c_8	−0.001684	−0.00169
c_9	0.009484	0.00950
c_{10}	0.004281	0.00426

6 Conclusions

The SFBIM for planar Laplacian and biharmonic problems with boundary singularities has been reviewed. The convergence of the method has been demonstrated theoretically for Laplacian problems and numerical applications has been presented for two Laplacian and a biharmonic elasticity test problems. The formulation of the SFBIM for a certain three-dimensional Laplace problem with a straight-edge singularity is currently investigated. The extraction of the leading singular functions for this problem is based on the work of Yosibash et al. (2002).

References

Arad N, Yosibash Z, Ben-Dor G, Yakhot A (1998) Comparing the flux intensity factors by a boundary element method for elliptic equations with singularities. Comm. Numer. Meth. Eng. 14:657–670

Elliotis M, Georgiou G, Xenophontos C (2002) The solution of a Laplacian problem over an L-shaped domain with a singular function boundary integral method. Comm. Numer. Methods Eng.18: 213–222

Elliotis M, Georgiou G, Xenophontos C (2005a) Solving Laplacian problems with boundary singularities: A comparison of a singular function boundary integral method with the p/hp version of the finite element method. Appl. Math. Comp. 169: 485–499

Elliotis M, Georgiou G, Xenophontos C (2005b) Solution of the planar Newtonian stick-slip problem with the singular function boundary integral method. Int. J. Numer. Meth. Fluids 48:1001–1021

Elliotis M, Georgiou G, Xenophontos C (2006) The singular function boundary integral method for a two-dimensional fracture problem. Eng. Anal. Bound. Elem. 80:100–106

Elliotis M, Georgiou G, Xenophontos C (2007) The singular function boundary integral method for biharmonic problems with crack singularities. Eng. Anal. Bound. Elem. 31: 209–215

Georgiou GC, Boudouvis A, Poullikkas A (1997) Comparison of two methods for the computation of singular solutions in elliptic problems. J. Comput. Appl. Math. 79: 277–290

Georgiou GC, Olson LG, Smyrlis Y (1996). A singular function boundary integral method for the Laplace equation. Commun. Numer. Meth. Eng. 12: 127–134

Grisvard P (1995) Elliptic Problems in Nonsmooth Domains. Pitman Publishers, London

Li ZC, Chen YL, Georgiou G, Xenophontos C (2006) Special boundary approximation methods for Laplace equation problems with boundary singularities. Comp. Math. Appl. 51:115–142

Li ZC, Lu TT, Hu HY (2004) The collocation Trefftz method for biharmonic equations with crack singularities. Eng. Anal. Bound. Elem. 28:79–96

Szabó B, Babuška I (1991) Finite Element Analysis. John Wiley & Sons, New York

Schiff BD, Fishelov D, Whiteman JR (1979) Determination of a stress intensity factor using local mesh refinement, in: J.R. Whiteman (Ed.), The Mathematics in Finite Elements and Applications III. Academic Press, London

Xenophontos C, Elliotis N, Georgiou G (2006) The singular function boundary integral method for elliptic problems with singularities. J. Sci. Comp. 28:517–532

Yosibash Z, Actis R, Szabó B (2002) Extracting edge flux intensity factors for the Laplacian. Int. J. Numer. Methods Eng. 53: 225–2420

Fast Multipole BEM and Genetic Algorithms for the Design of Foams with Functional-Graded Thermal Conductivity

Marco Dondero, Adrián P. Cisilino, Alexis Rodriguez Carranza and Georgios Stavroulakis

Abstract A numerical tool for the design of foam-like microstructures with customized heat conduction properties is presented in this paper. The spatial variation of the local material properties is achieved via the optimization of the spatial distribution of the gas cells. A Genetic Algorithm (GA) is used for this purpose. Although very effective and versatile GAs are computationally expensive. This problem is tackled in two ways: a Fast Multipole Boundary Element Method algorithm is used for the thermal analysis of the microstructures, and the GA is implemented in parallel using a cluster of PCs. Two examples illustrate the performance of the developed implementation.

1 Introduction

Functionally graded materials (FGMs) are two-component composites with a compositional gradient from one component to the other in order to obtain a given spatial variation of the local material properties. FGMs are very promising in applications where the operating conditions are severe. For example, wear-resistant linings, rocket heat shields, heat exchanger tubes, heat-engine components, plasma facings for fusion reactors, and electrical insulating metal/ceramic joints. The FGM concept originated in Japan in 1984 during the space plane project, in the form of a thermal barrier material capable of withstanding a surface temperature of 2000°K and a temperature gradient of 1000°K within a section equal to 10 mm. Since then, FGM thin films have been comprehensively researched and they are almost a commercial reality (Ruys et al. 2001).

The design of foams with customized heat conduction properties is proposed and tested in this work. Foams can be assimilated to a micro-heterogeneous material consisting in a thermoplastic polymer matrix containing small cells filled with gas that may constitute an important fraction of the total volume (Kazarian 2000).

M. Dondero (✉)
Welding and Fracture Division – INTEMA, Faculty of Engineering,
University of Mar del Plata, Mar del Plata, Argentina
e-mail: mdondero@fi.mdp.edu.ar

G.D. Manolis, D. Polyzos (eds.), *Recent Advances in Boundary Element Methods*,
DOI 10.1007/978-1-4020-9710-2_5, © Springer Science+Business Media B.V. 2009

The spatial distribution of the gas cells can be adjusted to obtain customized heat conduction properties.

One of the difficulties usually encountered in the computational design of random micro-heterogeneous materials is that the solution space is non-convex and the objective functions are not continuously differentiable. A numerical approach to simulate and accelerate the associated design process is due to Zohdi and Wriggers (2005) who proposed the utilization of Genetic Algorithms (GA).

Following Zohdi and Wriggers (2005), a high-performance numerical tool for the design of foam-like microstructures with customized heat conduction properties is presented in this work. Computations are carried on using representative volume elements (RVE) which size is determined after a homogenization analysis. A GA is used as the optimization method, with the spatial distribution of the gas cells as design variables and the temperature profile along the sample as objective function. The GA is implemented in parallel using a PC cluster and the RVEs are solved using a Fast Multipole Boundary Element Method.

2 The Fast Multipole Boundary Element Method and Modeling Considerations

Genetic Algorithms are robust global optimizers but they exhibit high computational cost resulting from the repetitive evaluation of the fitness function. As it will be seen in the following sections, the evaluation of the fitness function for this work requires the solution of thermal fields for computationally expensive RVE models containing a considerable number of gas cells. The Boundary Element Method (BEM) was selected for the solution of the thermal problem due to simplicity in the generation of the required data (the model discretization is restricted only to the boundaries) and the accuracy of the method (Aliabadi and Wrobel 2002).

The Fast Multipole Boundary Element Method (FMBEM) is used to reduce the computational cost in terms of both, operations and memory requirements of the direct BEM formulations. FMBEM reduces the computational cost of the direct BEM, from an order of $O(N^3)$ to a quasi-linear. This reduction is achieved by multilevel clustering of the boundary elements into cells, the use of the multipole series expansion for the evaluation of the fundamental solution in the far field and the use of an efficient iterative solver. The FMBEM used in this work is based on the work by Liu and Nishimura (2006). It uses a constant element discretization, analytical integration of the kernels and a preconditioned GMRES solver.

A typical model for the present application is illustrated in Fig. 1. It consists in a two-dimensional idealization of the foam microstructure with the gas cells assimilated to isolated circular holes. Boundary conditions are specified in order to induce a one-dimensional heat flux in the y-direction. The model discretization strategy was devised after a convergence analysis. With this purpose a reference solution was computed using direct BEM for a problem similar to that of Fig. 1, but containing 100 holes of radius $r = L/31$ (L being the specimen dimension)

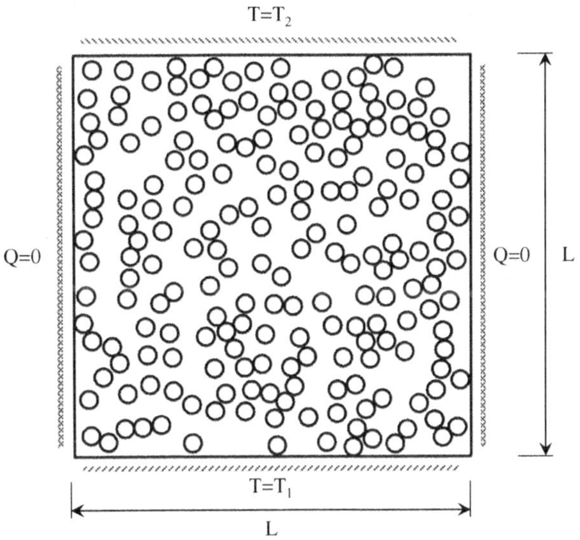

Fig. 1 Representative volume element with a void fraction $f = 0.3$

arranged on a regular square-array resulting in a void volume fraction (the ratio of the hole volume to the total sample volume) $f = 0.327$. The independency of the BEM solution with respect to the size of the element was explored by means of the total potential energy U. The number of elements for the model was progressively increased and the results compared. Obtained results are illustrated in Fig. 2 (dark symbols), where the potential energy values are normalized with respect to the total potential energy of a homogeneous hole-free specimen, U_0. From this analysis it was concluded that a model discretization with 4400 elements (40 elements per hole perimeter, 100 element along the sample side) provides mesh-independent results.

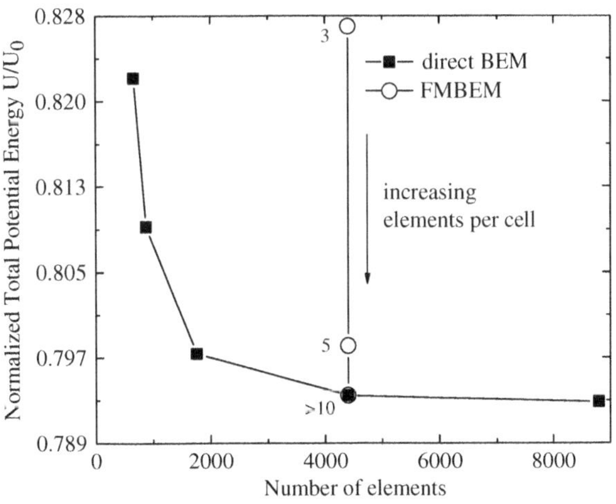

Fig. 2 Normalized total potential energy as a function of the number of elements

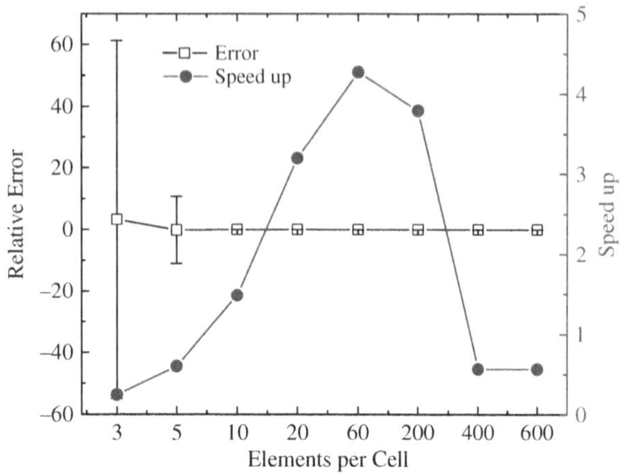

Fig. 3 Relative error and FMBEM speed up vs. elements per cell for a 4400 element mesh

Afterwards, the FMBEM algorithm was "tuned-up" for optimum performance and accuracy for the problem under analysis. In this process it was necessary to adjust the *elements per cell* parameter which determines the extent of the near and far fields for the collocation points (Liu and Nishimura 2006). This parameter affects the quality of the solution and the efficiency of the algorithm. Figure 3 illustrates the deviation of the FMBEM results and the algorithm speed up with respect to the direct BEM as a function of the *elements per cell* parameter. Error computations were done by comparing the solutions for the two models on an element by element basis and the results are presented in terms of the mean error value and its standard deviation (error bars in the figure). Results in Fig. 3 show that for values greater than 10 elements per cell, the relative error reduces noticeably and its dispersion vanishes. Similarly, the CPU time decreases for higher number of elements per cell, but it rapidly increases when this number is greater than 200. This change in the tendency occurs when the size of the near field is too big and thus the algorithm ends up working like a direct BEM (one may imagine, as the extreme case, a picture consisting in a single cell enclosing the complete model in the near field). The convergence rates of the FMBEM and BEM solutions are also plotted in Fig. 2. Figure 4 depicts the speed up of the algorithm as a function of the number of elements. Based on the above results the number of *elements per cell* was selected equal to 200 for further computations.

3 The Representative Volume Element

Foams have a hole-matrix like microstructure which is assumed to be two dimensional in this work. In addition, the holes are considered circular and randomly distributed. In order to study the macroscopic response of a heterogeneous material

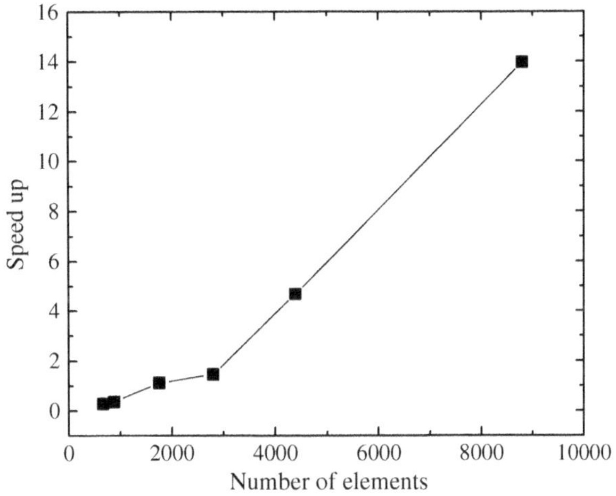

Fig. 4 FMBEM speed up as a function of the number of elements

the size of the sample must be independent of its size. This means that the sample must be big enough to hold a representative number of heterogeneities. The sample that satisfies these requirements is named Representative Volume Element (RVE).

In order to size the RVE, a series of FMBEM analysis were performed for samples containing an increasing number of randomly distributed holes and the total potential energy was computed in each case. In every case the model boundary conditions are that illustrated in Fig. 1. The following hole per sample sequence was used to study the dependence of the effective responses on the sample size: 10, 30, 60, 100, 150, 200 and 300. In order to get more reliable response data, tests were performed 20 times for each hole number set (each time with a different hole distribution) and the responses were averaged. Three constant void fractions were studied $f = 0.1, 0.3$ and 0.45. Results for the case $f = 0.3$ are illustrated in Fig. 5. Similar

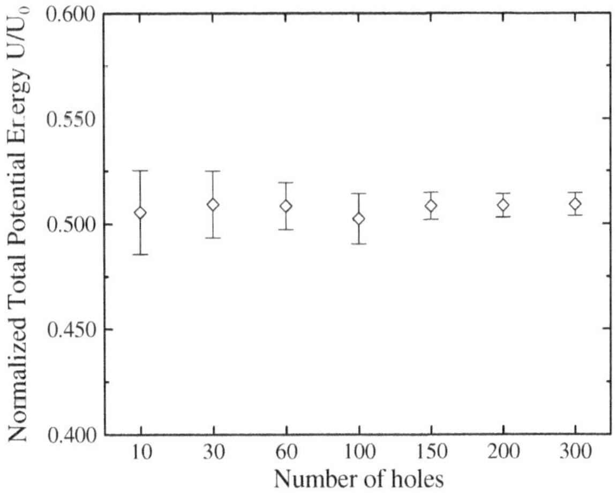

Fig. 5 Potential energy as a function of the number of holes for $f = 0.3$ void fraction. Error bars indicate result dispersion

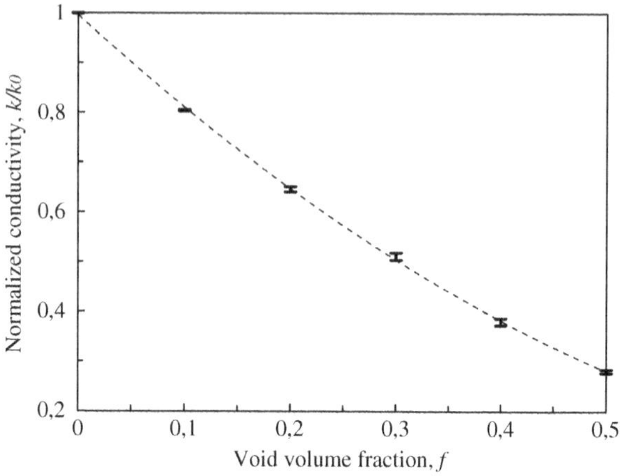

Fig. 6 Normalized thermal conductivity as a function of the void volume fraction

results were obtained for $f = 0.1$ and 0.45. Justified by the somewhat ad-hoc fact that for two successive enlargements of the number of holes the responses differed from one another, on average, by less than 0.5%, the 200 hole samples were selected as RVE for further tests.

RVEs were used to compute the sample conductivity in terms of the void volume fraction, *k*(*f*). This result will be used later in the examples section. Figure 6 illustrates the overall normalized conductivity k/k_0 for a series or RVE with void volume fractions in the range $0 \leq f \leq 0.5$ computed as $k(f)/k_0 = q \cdot L/(\Delta T \cdot k_0)$, k_0 being the conductivity of the matrix material and q the total one-dimensional flux in the y-direction. Each of the points in the plot is the mean value of 4 computations. The error bars indicate the dispersion of the results, which are below 1% in every case. The results were fitted using a polynomial approximation as follows:

$$k(f)/k_0 = 1.0514 \cdot f^2 - 1.9553 \cdot f + 1 \tag{1}$$

which has correlation coefficient $R^2 = 0.9997$.

4 Genetic Algorithms

Genetic Algorithms (hereafter "GA") simulate the natural evolution; hence their components are the chromosomes, the genetic material that dictates unique properties of the individuals (Goldberg 1999). A GA emulates the phenomena that take place during reproduction of species making use of the genetic operators. The latest are natural selection, pairing and mutation. Individuals live in an environment determined by the *objective (or fitness) function*, where they compete for survival and only the best succeed. The GA code used in this work is based on

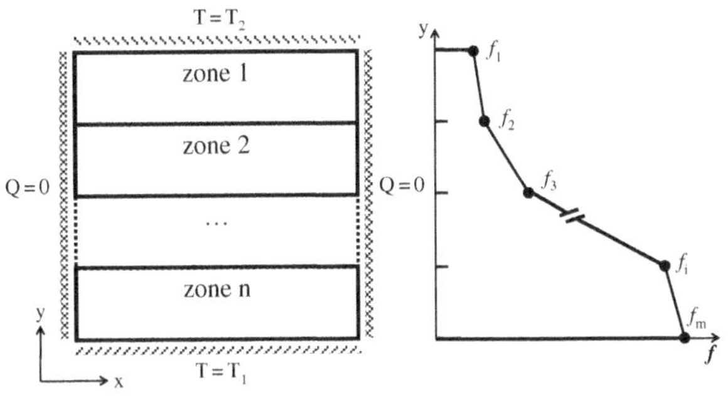

Fig. 7 Domain division, boundary conditions and piecewise linear void fraction distribution of the RVE sample

PIKAIA, a self-contained, genetic-algorithm-based optimization subroutine developed by Charbonneau and Knapp (1995), and available in public domain in the internet (Charbonneau and Knapp).

The GA is used to optimize the spatial distribution of the gas cells (holes) in the foam microstructure in order to obtain a given temperature distribution in the y-direction, $T(y)$ (the objective function). The optimization problem is solved by dividing the model domain into n zones (parallel bands in Fig. 7) of equal length with linear distribution of the void volume fraction. This approach results in a piecewise linear interpolation of the void volume fraction, $f(y)$, which is defined in terms of $m = n + 1$ discrete f_i values. The f_i are selected as design variables for the GA and they are codified into a chromosome

$$\text{chromosome} = [f_1, f_2, \ldots, f_\text{m}]. \tag{2}$$

The chromosome representation is done in binary format.

The fitness of the individuals (the fitness function) is the deviation of its temperature field from the objective temperature field, $T(y)$. This is assessed using a least-squares scheme for the differences between the FMBEM results and $T(y)$ for a set of p internal points evenly distributed over the complete model domain:

$$\mathit{fitness}\,(\text{individual } i) = \frac{\sqrt{\sum_{j=1}^{p}\left[T(y_j) - t_j\right]^2}}{p} \tag{3}$$

where t_j is the temperature solution at the jth internal point. In order to make the fitness value independent of the number of evaluation points the definition of the fitness function implies an average. The number of evaluations points can not be guaranteed constant for every model due to the constant change in the void volume fraction during the optimization process and the random nature of the foam microstructure.

In every case the fitness function evaluation is performed using a model with dimensions greater than that of the RVE (see Section 2). The random distribution

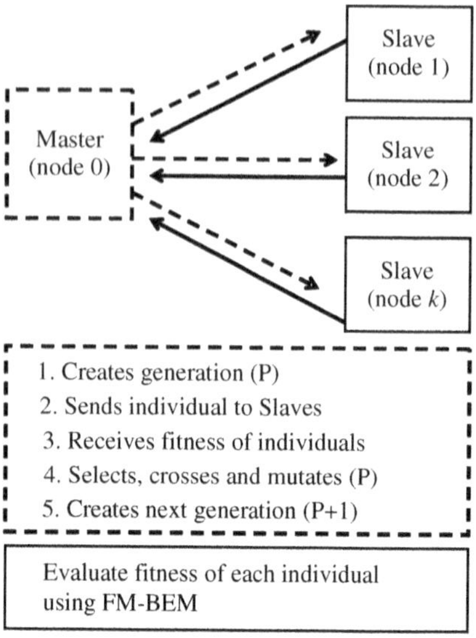

Fig. 8 PAGA master/slave parallel implementation: master (*broken lines*) and slave (*continuous lines*) tasks and communication scheme (messages through MPI)

for the hole positions are generated automatically by using the rejection method with the piece-wise definition for $f(y)$ as distribution function (Press et al. 1992).

The critical issue in the implementation of a GA is the computational cost of the evaluation of the fitness function, which must be performed hundreds of even thousands of times for the solution of a single problem. In order to accelerate the computations a parallel version of the GA was developed. The GA are relatively easy to implement in parallel due to natural independence in the evaluation of the fitness function for each individual. The developed algorithm uses a master-slave scheme where the master node is in charge of the management of the GA (creating and populating each generation) and the slave nodes are dedicated to the evaluation of the fitness of the individuals by solving the FMBEM models (see Fig. 8). This was implemented by incorporating MPI routines to PIKAIA. The parallel version of the GA runs on a Beowulf cluster of 8 PCs with GNU/ Linux.

5 Examples

There are presented in this section two benchmark examples used for the validation of the proposed implementation. In both cases the objective function are temperature distributions, $T(y)$, corresponding to given thermal conductivities, $k(y)$. All computations are performed using a RVE with dimensions $L \times L = 60\,\text{mm} \times 60\,\text{mm}$ with hole radius $r = 1$ mm and matrix thermal conductivity $k_0 = 1$ W/mm·°C.

5.1 Piece-Wise Temperature Distribution

The first example consists in a square 60 mm × 60 mm sample with a piece wise linear objective temperature field (see Fig. 9)

$$T(y) = 4.2 \cdot y - 62, \ \ 0\,\text{mm} \le y \le 30\,\text{mm}$$
$$T(y) = 1.2 \cdot y + 28, \ 30\,\text{mm} < y \le 60\,\text{mm} \tag{4}$$

Boundary conditions are $T_2 = 100^{\circ}$C along the side $y = 60$ mm and prescribed flux $q = -1.25\,\text{W/mm}^2$ along the side $y = 0$ mm. The lateral sides are insulated (see Fig. 1).

It is easy to see that the above objective function is the temperature-field solution for the problem consisting in a two-zone sample with normalized thermal conductivities $[k(f)/k_0]_1 = 1$ and $[k(f)/k_0]_2 = 0.2852$. Note that the thermal conductivities for the zones are the limit values of equation (1), which correspond to void volume fractions $f = 0$ and $f = 0.5$ respectively.

The optimization domain is divided into 8 zones, resulting in nine design variables f_i. The chromosome was codified using 6 significant digits (i.e. the number of genes) per design variable. Parameters for the GA are: population size 24 individuals, 50 generation, crossover probability 0.85, one point mutation mode with adjustable rate, initial mutation rate 0.005, minimum mutation rate 0.001, maximum mutation rate 0.0185, and full generational replacement reproduction plan with elitism.

Figure 10 illustrates the evolution of the fitness function in terms of the generation number for a typical computer run. Results are plotted for the best individual

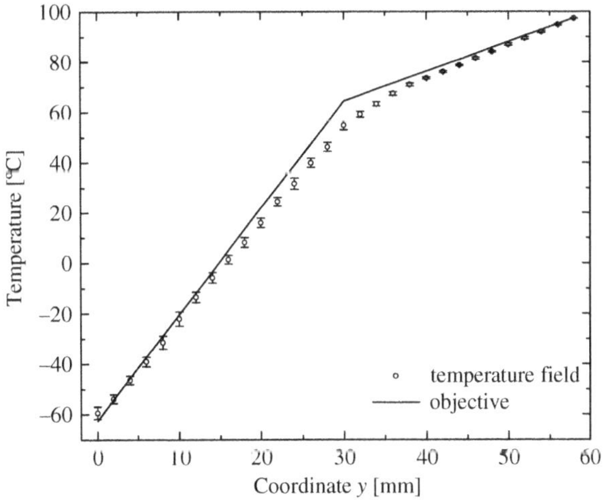

Fig. 9 Objective and resultant temperature fields for the first example

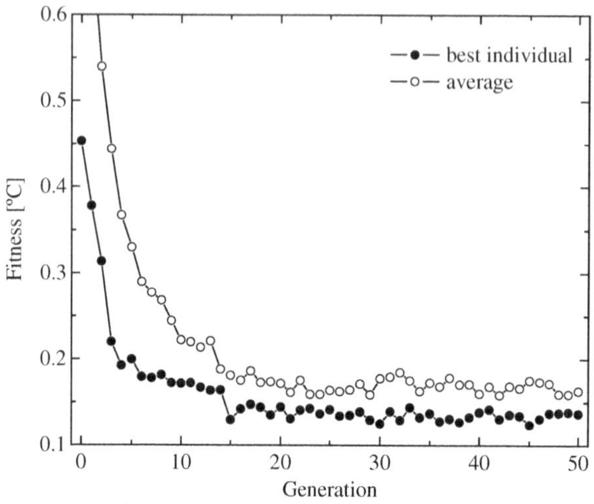

Fig. 10 Fitness evolution for the first example: best-of-generation and average results

(the most fitted) in each generation and the average value for all the individuals. A minimum is achieved after approximately 30 generations. It is worth noting that the convergence for the best individual is not monotonous, but there are small occasional increments in the fitness function as the optimization procedure progresses. These increments are a consequence of the random nature of the foam microstructure. Sample geometries are generated for each generation following the design variables values f_i. But due to the random nature of the microstructure the same set of design values does not produce the same model geometries and consequently the fitness function varies. This means that even for the case when no improvement is made by the GA in a generation and so the best fitted individual is kept, the fitness function value could be reduced or augmented. This phenomenon is reduced by enlarging the RVE size, indicating that the fluctuations in the fitness function evolution are negligible with respect to the characteristic temperature of the example. Note that in the present example the fluctuations in the fitness function can be estimated in approximately 0.01°C, and they are negligible in comparison with the overall temperature difference across the sample (162°C).

The optimized microstructure is shown in Fig. 11 together with the contour plot of the temperature field and the optimal void volume fraction ($f_i = 0.445, 0.465, 0.458, 0.486, 0.405, 0.022, 0.032, 0.023, 0.009$). Besides the final temperature distribution is plotted in Fig. 9 with the error bars indicating the dispersion of the results. It can be seen that the maximum difference between the objective function and the optimized result occurs in the neighborhood of $y = L/2 = 30\,\text{mm}$, the position where the objective function presents an abrupt change in the slope due to the discontinuity in the conductivity. As it was expected the optimization procedure fails to reproduce this transition because it has been designed to produce smooth variations in the void volume fraction.

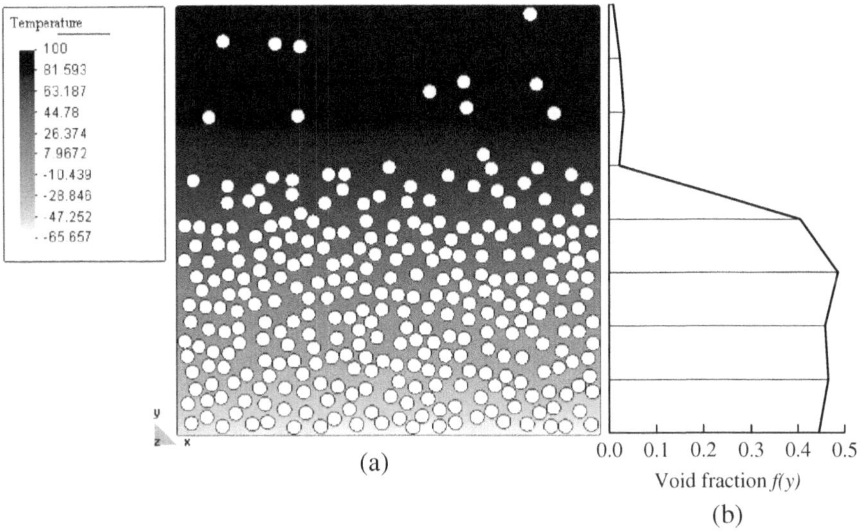

Fig. 11 First example: (**a**) optimized microstructure with temperature-map, and (**b**) void fraction solution

5.2 Smooth Temperature Distribution

The second example deals with a more challenging problem. In this case the sample conductivity varies continuously in the y-direction. The variation is chosen to have Gaussian-like shape (see Fig. 12) and it is written in terms of Padé polynomials (Jones and Thron, 1980) as follows:

$$\frac{k(y)}{k_0} = k_{\min} + \frac{A}{(y-\mu)^2/\sigma + 2} \tag{5}$$

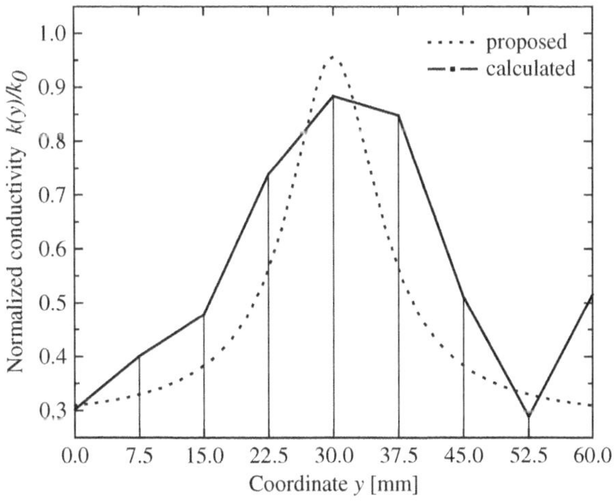

Fig. 12 Proposed and obtained conductivity variations along the sample

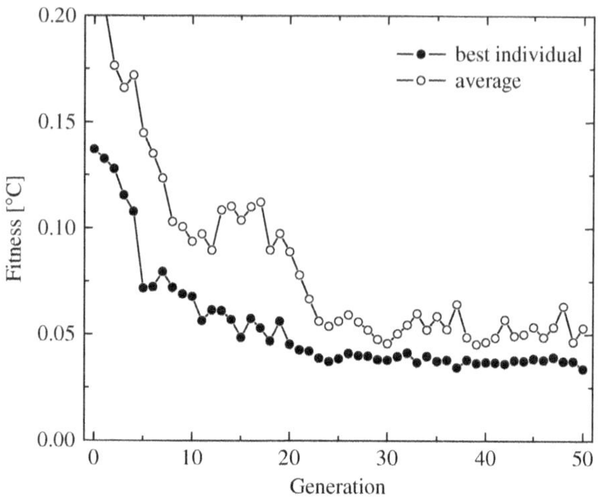

Fig. 13 Fitness evolution for the second example: best-of-generation and average results

where the constants were set $k_{min} = 0.2805$, $A = 3\sqrt{2}/\pi$, $\mu = L/2 = 30$, $\sigma = 20$. It is worth noting that like in the previous example the maximum and minimum conductivity values correspond to the void volume fractions $f = 0$ and $f = 0.5$ respectively.

The problem boundary conditions are similar to those of the first example but with prescribed temperatures $T_1 = 20°C$ and $T_2 = 100°C$ along the sides $y = 0$ mm $y = 60$ mm respectively. The resulting objective temperature field is

$$T(y) = 0.3133 + 1.989 \cdot y - 16.41 \cdot \tan^{-1}(0.0857 \cdot y - 2.5690)\,. \qquad (6)$$

The GA parameters were set same to that of the first example. The evolution of the fitness function with the generations is plotted in Fig. 13 for 2 computer runs. The behavior of the fitness function is in general similar to that of the first example. The minimum is achieved after approximately 20 generations and there exist small fluctuations due to the random nature of the microstructure.

Figure 14 depicts the microstructure for one of the best fitted individuals together with the contour plot for the temperature field and the optimal void volume fraction ($f_i = 0.477, 0.383, 0.319, 0.143, 0.059, 0.079, 0.295, 0.492, 0.292$). In accordance with the maximum conductivity values, the minimum void volume fraction occurs in the central part of the sample.

The comparison between the objective function and the resulting temperature field is plotted in Fig. 15. The computed solutions are in excellent agreement with the objective function.

Finally, the void volume fraction results in Fig. 14b were correlated to the conductivity in equation 5 by means of equation 1. The result is plotted in Fig. 12. It can be seen that the resulting conductivity posses the same general trend of that used for the formulation of the problem, with the maximum value in the sample central zone.

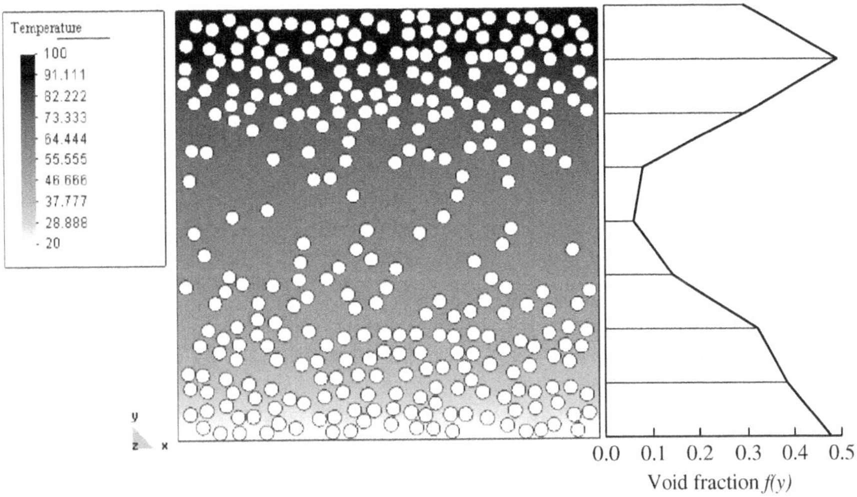

(a) (b)

Fig. 14 Second example: (**a**) optimized microstructure with temperature-map, and (**b**) void fraction solution

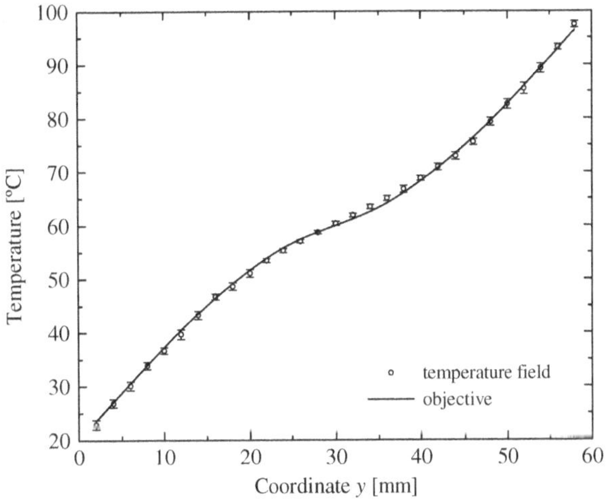

Fig. 15 Objective and resultant temperature fields for the second example

6 Conclusions

It has been presented in this work an efficient numerical tool for the design of foam-like microstructures with functional-graded thermal conductivity. The devised methodology is based on a parallel Genetic Algorithm (GA) as optimization method with a Fast Multipole Boundary Element Method (FMBEM) code for the evaluation of the fitness function using representative volume elements (RVE).

The FMBEM is specially suited for the optimization method: the boundary-only discretization strategy makes the model data generation a simple task, while the fast multipole formulation results in important savings in computing time when compared to direct BEM.

Two validation examples demonstrate the performance of the implementation. The proposed methodology is general and robust and can be easily extended to optimize more complex microstructures with the distribution of inclusions and/or holes, their shapes, orientations and material properties as design variables.

Acknowledgments This work has been partially supported by the projects ALFA II ELBENET "Europe Latin America Boundary Element Network" of the European Union, PROSUL 040/2006 "Optimization of microstructures of polymer-matrix composites" (CNPq-Brazil) and PICT 12-14114 of the Agencia Nacional de Investigaciones Científicas y Técnologicas of Argentina.

References

Aliabadi M.H. and Wrobel L.C. "The Boundary Element Method" John Wiley & Sons, Chichister, UK (2002)

Charbonneau P. and Knapp B. "A User's Guide to PIKAIA 1.0" NCAR Technical Note TN-418+IA (1995)

Charbonneau P. and Knapp B. "PIKAIA: Optimization (Maximization) of User-Supplied 'Fitness' Function ff Over n-Dimensional Parameter Space x Using a Basic Genetic Algorithm Method." High Altitude Observatory, National Center for Atmospheric Research, Boulder CO 80307-3000, US. (http://download.hao.ucar.edu/archive/pikaia)

Goldberg D. E. "Genetic Algorithms in search, optimization & machine learning", Addison-Wesley, New York, USA (1999)

Jones W.B. and Thron W.J. "Continued Fractions: Theory and Applications". Addison-Wesley Publishing Company, Reading, MA, 185–197 (1980)

Kazarian S.G. "Polymer Processing with Supercritical Fluids" Poly. Sci. Ser. C, 42(1), 78–101 (2000)

Liu Y.J. and Nishimura N. "The Fast Multipole Boundary Element Method for Potential Problems: A Tutorial". Eng. Anal. Bound. Elem., 30(5), 371–381 (2006)

Press W.H., Teukolsky S.A., Vetterling W.T. and Flannery B.P. "Numerical Recipes in Fortran 77", Second Edition. Cambridge University Press, Cambridge, UK (1992)

Ruys A.J., Popov E.B., Sun D., Russell J.J. and Murray C.C.J. "Functionally Graded Electrical/Thermal Ceramic Systems". J. Eur. Ceram. Soc., 21(10-11), 2025–2029 (2001)

Zohdi T.I. and Wriggers P. "Introduction to Computational Micromechanics" Lecture Notes in Applied and Computational Mechanics, Volume 20, Springer-Verlag, Berlin (2005)

An Integral Equation Formulation of Three-Dimensional Inhomogeneity Problems

C.Y. Dong, F. Xie and E. Pan

Abstract A displacement integral equation formulation of three-dimensional infinite isotropic matrix with inhomogeneities of arbitrary shapes is derived based on the assumption that both the inhomogeneity and matrix have the same Poisson's ratio. Compared to the conventional boundary integral equation formulation which requires both the tractions and displacements on the interface between the inhomogeneity and matrix, the present displacement integral formulation only contains the unknown interface displacements. Therefore, its numerical implementation can easily be carried out since the handling of corners in any irregular shaped inhomogeneity is avoided. Thus, through the interface discretization using quadrilateral boundary elements, the resulting system of equations can be formulated so that the interface displacements can be obtained. Stresses at any point of interest can also be obtained by using the corresponding stress integral equation formulation which contains only the inhomogeneity-matrix interface displacements. Numerical results from the present approach are in excellent agreement with existing ones.

1 Introduction

Since Eshelby's classic work on the problem of an elastic ellipsoidal inhomogeneity embedded in an infinite elastic medium (1957, 1959), various inclusion/inhomogeneity problems have been extensively investigated (Mura, 1987; Ting, 1996). The interaction between the inhomogeneity and matrix can be analyzed using various numerical methods, e.g. the finite element method (FEM) and boundary element method (BEM).

The FEM has been used to solve various inclusion/inhomogeneity problems, e.g. Thomson and Hancock (1984), Ghosh and Mukhopadhyay (1993) and Nakamura and Suresh (1993). For the irregular shaped inhomogeneity with random space

C.Y. Dong (✉)
Department of Mechanics, School of Science, Beijing Institute of Technology, Beijing 100081, China
e-mail: cydong218@yahoo.com.cn

A chapter in honor of Dimitri Beskos' 65th birthday

G.D. Manolis, D. Polyzos (eds.), *Recent Advances in Boundary Element Methods*, DOI 10.1007/978-1-4020-9710-2_6, © Springer Science+Business Media B.V. 2009

distribution, the finite element discretization could be difficult. Interaction among various three-dimensional (3D) inclusions/inhomogeneities was investigated using a domain integral equation method by Dong et al. (2003a) in which only the isotropic fundamental solution is required even for anisotropic materials so that the complicated fundamental solution for anisotropic medium can be avoided. However, the drawback is that the inhomogeneity needs to be discretized into finite elements. The BEM can also be used to solve various inhomogeneity problems based on the inhomogeneity-matrix interface conditions, i.e. displacement continuity and traction equilibrium (Dong et al., 2003b). However, for inhomogeneities of irregular shapes, special technique, e.g. discontinuous elements or double nodes in coupling the equations from the inhomogeneity to the matrix, is required to satisfy the displacement continuity condition and traction equilibrium at the corner. This demand is difficult for 3D inhomogeneity problems, especially for those inhomogeneities with irregular shapes.

In this paper, following the method proposed by Leite et al. (2003) for two-dimensional (2D) reinforced solid, a displacement integral equation formulation is developed to solve 3D inhomogeneity problems by assuming that the matrix and inhomogeneity have the same Poisson's ratio. The corresponding stress integral equation formulation is also derived in this paper. Compared to the conventional BEM for the inhomogeneity problem, the present integral equation formulation only contains the interface displacements (no interface tractions). Similar to the domain integral equation approach (Dong et al., 2003a), the present integral equation approach can also overcome the corner singularity problems (due to different normal directions at the corner nodes). As such, various irregular inhomogeneity problems, especially in 3D, can be easily solved using the derived displacement integral formulation. Numerical examples are presented to show the validity and efficiency of the proposed method.

2 Basic Formulation

For an infinite isotropic matrix subjected to remote stresses, the displacement and stress integral equations at point P in the matrix can be given as follows (Dong et al., 2003a)

$$u_k(P) = u_k^0(P) + \int_\Gamma U_{ki}(P,q)t_i(q)d\Gamma(q) - \int_\Gamma T_{ki}(P,q)u_i(q)d\Gamma(q) \quad (1)$$

and

$$\sigma_{kl}(P) - \sigma_{kl}^0(P) = \int_\Gamma U_{kli}(P,q)t_i(q)d\Gamma(q) - \int_\Gamma T_{kli}(P,q)u_i(q)d\Gamma(q) \quad (2)$$

where q is the field point acting at the inhomogeneity-matrix interface Γ. u_k^0 and σ_{kl}^0 are, respectively, the displacements and stresses at point P caused by the remote stresses in an infinite homogeneous isotropic elastic matrix. U_{ki}, T_{ki}, U_{kli} and T_{kli} are the fundamental solutions to the infinite isotropic elastic medium (Brebbia and Dominguez, 1992).

When the source point P approaches the boundary point p being on the inhomogeneity-matrix interface, Eq. (1) becomes

$$c_{ki}u_i(p) = u_k^0(p) + \int_{\Gamma} U_{ki}(p,q)\,t_i(q)\,d\Gamma(q) - \int_{\Gamma}^{C} T_{ki}(p,q)\,u_i(q)\,d\Gamma(q) \quad (3)$$

where c_{ki} depends on the boundary geometry at the source point p. The symbol $\int^{C}$ denotes the Cauchy principal value integral.

For the I-th isotropic inhomogeneity, the corresponding displacement boundary integral equation can be given as (Brebbia and Dominguez, 1992)

$$c_{ki}^{I}(p)u_i^{I}(p) = \int_{\Gamma_I} U_{ki}^{I}(p,q)\,t_i^{I}(q)\,d\Gamma(q) - \int_{\Gamma_I}^{C} T_{ki}^{I}(p,q)\,u_i^{I}(q)\,d\Gamma(q) \quad (4)$$

where Γ_I represents the I-th inhomogeneity-matrix interface.

If we assume that the Poisson's ratio is the same for both the matrix and inhomogeneity, one can then find the following relationships (Leite et al., 2003)

$$U_{ki}^{I} = \frac{G}{G^{I}}U_{ki} \quad (5a)$$

$$T_{ki}^{I} = T_{ki} \quad (5b)$$

$$U_{kli}^{I} = U_{kli} \quad (5c)$$

$$T_{kli}^{I} = \frac{G^{I}}{G}T_{kli} \quad (5d)$$

Substituting (5a) and (5b) into Eq. (4), then adding Eqs. (3) and (4), and considering the interface conditions, one can obtain the following displacement boundary integral equation

$$c_{ki}\left(1+\frac{G^{I}}{G}\right)u_i(p) = u_i^0(p) - \int_{\Gamma_I}^{C}\left(1-\frac{G^{I}}{G}\right)T_{ki}(p,q)\,u_i(q)\,d\Gamma \quad (6)$$

For the matrix with multiple inhomogeneities, Eq. (6) can be extended to the following form

$$c_{ki}\left(1+\frac{G^{I}}{G}\right)u_i(p) = u_i^0(p) - \int_{\Gamma_I}^{C}\left(1-\frac{G^{I}}{G}\right)T_{ki}(p,q)\,u_i(q)\,d\Gamma - \sum_{J=1,\neq I}^{NI}\int_{\Gamma_J}\left(1-\frac{G^{J}}{G}\right)T_{ij}(p,q)\,u_j(q)\,d\Gamma \quad (7)$$

where NI denotes the number of the inhomogeneities. The source point p is acting on the I-th inhomogeneity-matrix interface.

Similarly, the stress integral equation for point P being in the inhomogeneity or matrix can be written as

$$\sigma_{kl}(P) = \sigma_{kl}^{0}(P) - \sum_{I=1}^{NI} \int_{\Gamma_I} \left(1 - \frac{G^I}{G}\right) T_{kli}(P, q) u_i(q) d\Gamma \qquad (8)$$

From Eqs. (7) and (8), one can find that the displacement and stress integral equations contain only the inhomogeneity-matrix interface displacements. Note that the tractions on the inhomogeneity-matrix interfaces disappear in these two integral equations. Therefore, unlike the conventional BEM in which the discontinuous element has to be used near the corner of the irregular inhomogeneities (Dong et al., 2003b), arbitrary inhomogeneity shapes can be easily treated using Eqs. (7) and (8). Therefore, Eqs. (7) and (8) can be considered as an extension of the 2D formulation of Leite et al. (2003) to the corresponding 3D elasticity.

In numerical implementation, quadratic quadrilateral boundary elements are used to discretize the inhomogeneity-matrix interfaces. Thus, the resulting system of equations can be written as

$$\mathbf{AU} = \widetilde{\mathbf{U}} \qquad (9)$$

where $\boldsymbol{A}$ is the related coefficient matrix from Eq. (7). $\mathbf{U}$ is the vectors of the inhomogeneity-matrix interface nodal displacements. $\widetilde{\mathbf{U}}$ is the vector of the displacements on the inhomogeneity-matrix interface due to the remote stresses.

Once the interface displacements are available, the stresses at point P in the inhomogeneity or matrix can be calculated using Eq. (8).

3 Numerical Examples

In this section, a couple of numerical examples are presented to show that the proposed formulation is accurate and efficient in analyzing 3D inhomogeneity problems. These include some benchmark problems where analytical solutions are available and problems that can only be solved numerically.

3.1 A Spherical Inhomogeneity Embedded in an Infinite Isotropic Medium (Matrix)

A spherical inhomogeneity is embedded in an infinite isotropic elastic medium subjected to the remote unit stresses, i.e. $\sigma_{xx}^{0} = \sigma_{yy}^{0} = \sigma_{zz}^{0} = \sigma^{0} = 1$, along x-, y-, and z-axes respectively. The Poisson's ratio is taken to be 0.3 for both the matrix and inhomogeneity. The radius of the spherical inhomogeneity is assumed to be $R = 1$.

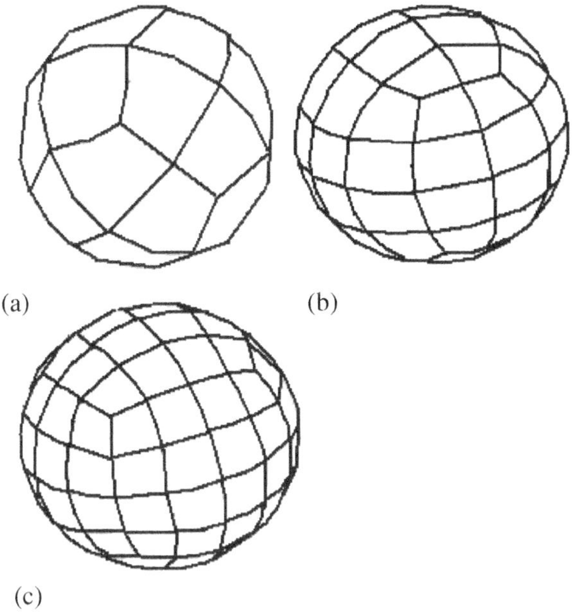

Fig. 1 Discretization of the spherical surface using quadratic quadrilateral boundary elements of 24 in (**a**), 64 in (**b**), and 96 in (**c**)

The interface between the matrix and inhomogeneity is discretized into 24, 64 and 96 quadratic quadrilateral boundary elements, respectively, as shown in Fig. 1. Assuming that the ratio of the Young's moduli between the matrix and inhomogeneity is $E_M/E_I = 1/2$ in which E_M and E_I denote the Young's moduli of the matrix and inhomogeneity, respectively. Due to the symmetry of the geometry and loadings, only the stress σ_{xx} along z-axis within the spherical inhomogeneity is displayed in Fig. 2. It can be found that with increasing elements the results from the present method are in excellent agreement with existing analytical solutions (Dong et al., 2003a).

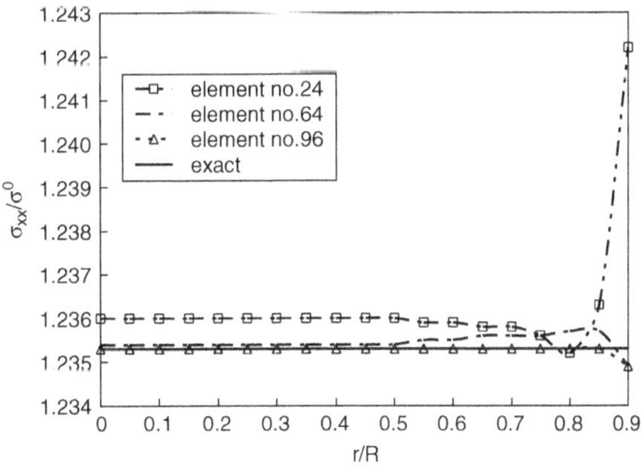

Fig. 2 Variation of the normalized stress σ_{xx}/σ^0 along z-axis within the spherical inhomogeneity (r/R is the normalized distance of the observation point to the spherical center)

3.2 A Cylindrical Inhomogeneity Embedded in an Infinite Isotropic Elastic Medium

A cylindrical inhomogeneity with radius $r = 1$ and height $h = 0.5$ is embedded in an infinite isotropic elastic medium subjected to the remote loading $\sigma_{xx}^0 = \sigma_{yy}^0 = \sigma_{zz}^0 = \sigma^0 = 1$. The ratio of the Young's moduli between the matrix and inhomogeneity is taken to be $E_M/E_I = 1/2$, and the corresponding Poisson's ratios are chosen as $\nu_M = \nu_I = 0.3$. It is noted again that the symbols with subscripts M and I denote the values related to the matrix and inhomogeneity, respectively.

The surface of the cylindrical inhomogeneity is discretized with 52 quadratic quadrilateral boundary elements as shown in Fig. 3. For comparison, the FEM is also employed to solve the same problem. The infinite medium is approximated by a cylindrical body with radius $R = 20$ and height $h = 20$. The corresponding FEM solutions are obtained by using 46337 tetrahedron 4-node elements as shown in Fig. 4 (symmetric property has been considered). The results at selected points A (0, 0, 0), B (0, 0, −0.25), C (0, 0, −0.5) and D (0, 0.081, 0.0077) are given in Table 1. It is observed clearly from Table 1 that the results from the present method are in good agreement with those from FEM.

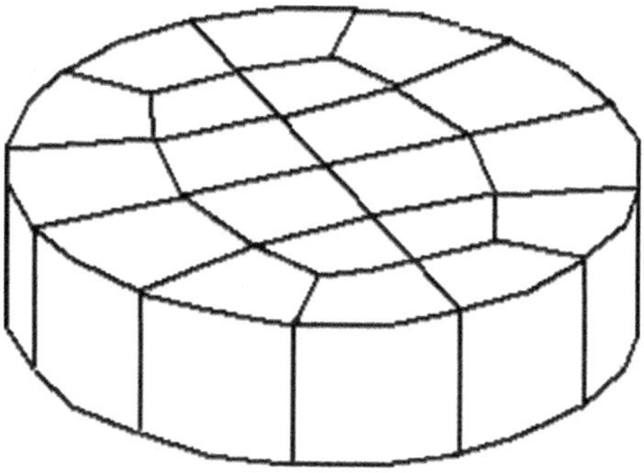

Fig. 3 Boundary element meshes of the cylindrical inhomogeneity

3.3 Two Spherical Inhomogeneities Embedded in an Infinite Isotropic Medium

Two spherical inhomogeneities with radii R_1 and R_2 and material parameters E_1, ν_1 and E_2, ν_2 are embedded in an infinite isotropic elastic matrix with material parameters E_M, ν_M (Fig. 5). The matrix is subjected to the remote loading $\sigma_{xx}^0 = \sigma_{yy}^0 = \sigma_{zz}^0 = \sigma^0 = 1$. The Poisson's ratio of the inhomogeneity and matrix are again taken to be 0.3. The distance between the centers of the two spheres is assumed to be d.

In the numerical calculation, 96 quadratic quadrilateral boundary elements are used to discretize the surface of each inhomogeneity as shown in Fig. 1(c). For

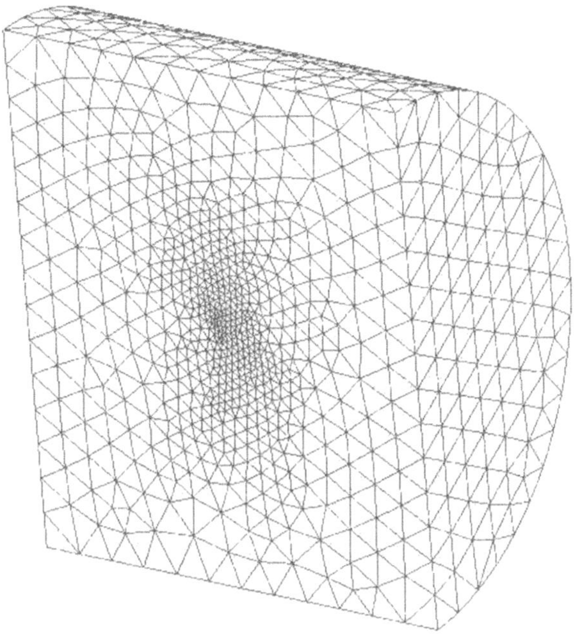

Fig. 4 Finite element meshes of one half of the cylindrical inhomogeneity

Table 1 Comparison of the results from BEM and FEM

	Point A		Point B		Point C		Point D	
	BEM	FEM	BEM	FEM	BEM	FEM	BEM	FEM
σ_{xx}/σ^0	1.376	1.409	1.373	1.398	1.367	1.387	0.9703	0.9623
σ_{yy}/σ^0	1.082	1.130	1.085	1.100	1.098	1.070	1.025	1.024
σ_{zz}/σ^0	1.376	1.407	1.371	1.399	1.354	1.390	0.9703	0.9652

$R_1 = R_2 = 1, d = 1.5, E_1 = E_2$ and $E_1/E_M = 0.5$, the stresses from the proposed formulation along y- and z-axes are displayed, respectively, in Figs. 6 and 7. It can be found that $\sigma_{xx} = \sigma_{zz}$ along y-axis due to the symmetry of problem. For $R_1 = R_2 = 1$, $d = 1.5$, $E_2/E_M = 2$ and $E_1/E_M = 0.5$, the stresses along y- and z-axes are

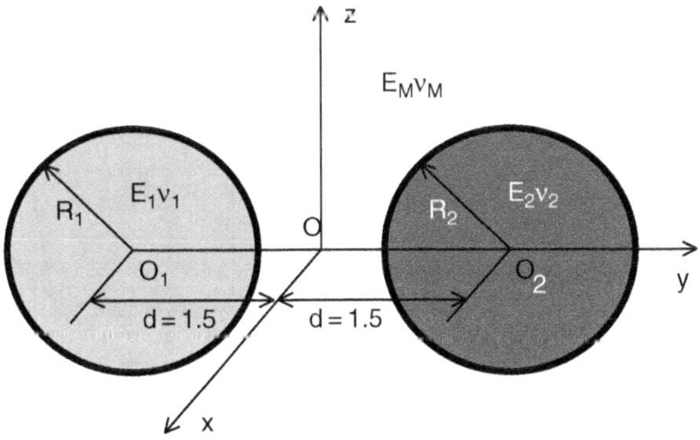

Fig. 5 Two spherical inhomogeneities in an infinite matrix

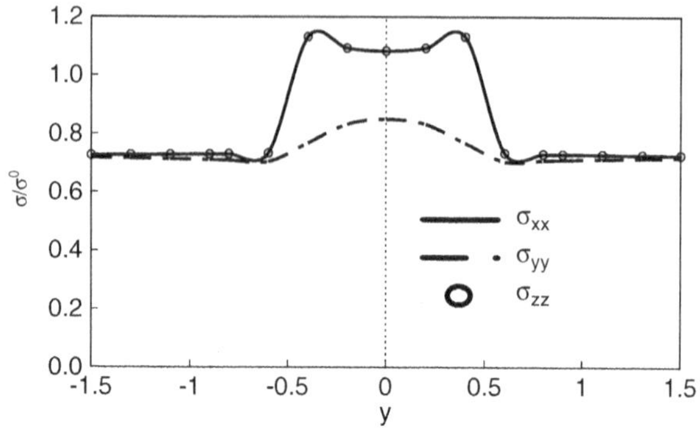

Fig. 6 Variation of the normalized stresses σ/σ^0 along y-axis ($E_1/E_M = 0.5$, $E_1 = E_2$)

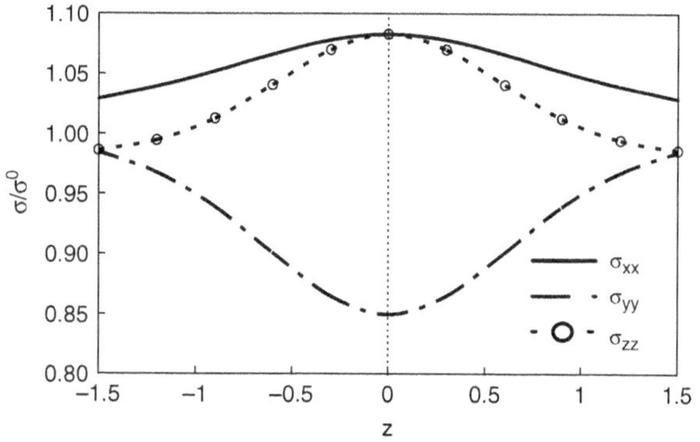

Fig. 7 Variation of the normalized stresses σ/σ^0 along z-axis ($E_1/E_M = 0.5$, $E_1 = E_2$)

displayed, respectively, in Figs. 8 and 9. Since the two spherical inhomogeneities now have different material properties, the stress distribution along y-axis is no longer symmetric about the vertical z-axis. However, the stress along z-axis is still symmetric about the vertical z-axis.

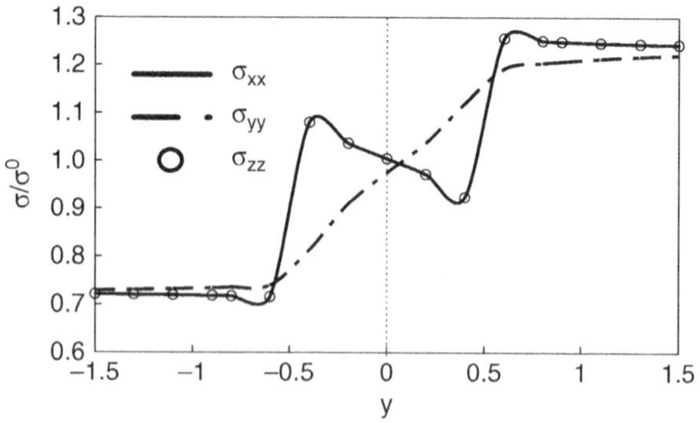

Fig. 8 Variation of the normalized stresses σ/σ^0 along y-axis ($E_1/E_M = 0.5$, $E_2/E_M = 2$)

Fig. 9 Variation of the normalized stresses σ/σ^0 along z-axis ($E_1/E_M = 0.5$, $E_2/E_M = 2$)

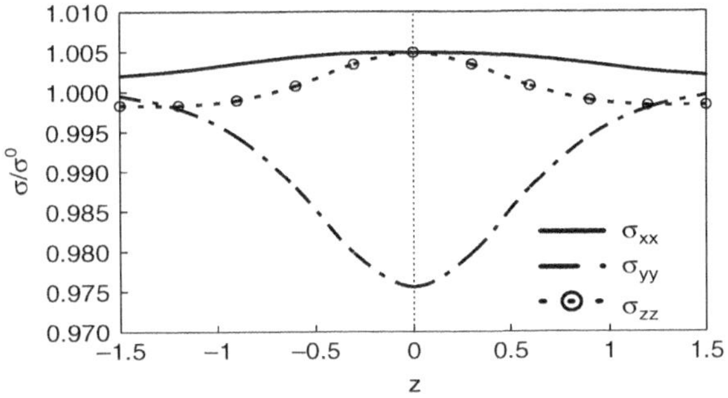

4 Conclusions

An integral equation formulation for 3D inhomogeneity problems is derived based on the assumption that the matrix and inhomogeneity have the same Poisson's ratio. The proposed integral equation formulation only contains the displacements on the inhomogeneity-matrix interface. Therefore, the present formulation can easily be applied to solve various irregular inhomogeneity problems because it does not contain the interface-matrix traction (so that the corner singularity issue can be avoided). Three inhomogeneity problems are studied using the present formulation. The obtained results are in good agreement with those available in literature, and they can be considered as benchmark solutions for future investigation.

Acknowledgments This work was supported by the National Natural Science Foundation of China under Grant No. 10772030.

References

Brebbia, CA, Dominguez, J (1992) Boundary Elements – An Introduction Course, Computational Mechanics Publications, Southampton

Dong CY, Lo SH, Cheung YK (2003a) Numerical solution of 3D elastostatic inclusion problems using the volume integral equation method. Comput. Methods Appl. Mech. Eng. 192: 95–106

Dong CY, Lo SH, Cheung YK (2003b) Stress analysis of inclusion problems of various shapes in an infinite anisotropic elastic medium. Comput. Methods Appl. Mech. Eng. 192: 683–696

Eshelby JD (1959) The elastic field outside an ellipsoidal inclusion. Proc. R. Soc. Lond. A. 252: 561–569

Eshelby JD (1957) The determination of the elastic field of an ellipsoidal inclusion and related problems. Proc. R. Soc. Lond. A. 241: 376–396

Ghosh S, Mukhopadhyay SN (1993) A material based finite element analysis of heterogeneous media involving Dirichlet tessellations. Comput. Methods Appl. Mech. Eng. 104: 211–247

Leite LGS, Coda HB, Venturini WS (2003) Two-dimensional solids reinforced by thin bars using the boundary element method. Eng. Anal. Bound. Elem. 27: 193–201

Mura T (1987) Micromechanics of Defects in Solids. Martinus Nijhoff Publishers, Dordrecht, Boston, Lancaster

Nakamura T, Suresh S (1993) Effects of thermal residual stresses and fiber packing on deformation of metal-matrix composites. Acta Metall. Mater. 41: 1665–1681
Thomson RD, Hancock JW (1984) Local stress and strain fields near a spherical elastic inclusion in a plastically deforming matrix. Int. J. Frac. 24: 209–228
Ting TCT (1996) Anisotropic Elasticity: Theory and Applications. Oxford University Press, New York

Energy Flux Across a Corrugated Interface of a Basin Subjected to a Plane Harmonic SH Wave

Marijan Dravinski

Abstract Energy flux across a corrugated interface of a basin subjected to a plane harmonic SH wave is investigated using an indirect boundary integral equation approach. The corrugation is assumed to be in the form of an ellipse perturbed in a sinusoidal manner. The results clearly demonstrate that wide range of frequencies presence of the corrugation produced overall reduction in the interface energy density flux for both deep and shallow basins when compared with the corresponding smooth basin case. The reduction of the interface energy density flux is found to nonuniformly increase with the increase of the corrugation amplitude and it is influenced by the corrugation period as well. Finally, the general reduction in the interface energy density flux is found to lead to the smaller interface motion.

1 Introduction

Sedimentary basins are generally exposed to larger surface motion amplification and longer duration of shaking than the sites on crystalline rocks (Aki, 1988, Bard and Bouchon, 1985). The response of the basins with smooth interfaces is well understood at this point in time (e.g., Wong, 1982, Dravinski, 1983, Sanchez-Sesma, 1983, Bard and Bouchon, 1985, and Bouchon et al., 1996). The assumption in such modelling is that the basin's interface is known exactly. In practice, the basin's interface is probably known only at a limited number of sites leading to a considerable uncertainty about the exact interface shape. So the lack of basin's geophysical data requires often that the available interface data are fitted by a smooth surface. This means that the actual basin's interface may vary about the smooth shape used for the modelling of the surface response. This, of course, will introduce certain error in the results of the simulated basin's response. Recently, the author (Dravinski, 2007) investigated the role of the basin interface irregularities on the local surface ground motion amplification. It was shown that the presence of the

M. Dravinski (✉)
USC Viterbi School of Engineering, University of Southern California, Los Angeles, CA 90089-1453, USA
e-mail: mdravins@usc.edu

G.D. Manolis, D. Polyzos (eds.), *Recent Advances in Boundary Element Methods*,
DOI 10.1007/978-1-4020-9710-2_7,

interface corrugation may result in a significant reduction of peak ground motion atop the basin when compared to the corresponding smooth basin response. In addition, it was demonstrated that the presence of the corrugation will in general affect the fundamental frequency of the basin as well when compared to the corresponding smooth basin value.

The present paper investigates how the energy flux across the interface is affected by the presence of the corrugation. The aim is to gain further understanding of the wave propagation mechanism of motion amplification within basins with a corrugated interface.

Scattering of elastic waves by the irregularities of the crystal waveguides have been investigated by several researchers (e.g., Fu et al. 2002 and Fu, 2005). For a detailed review of the literature on this topic the reader is referred to the article by Fu (2005).

2 Statement of Problem

The problem model depicted by Fig. 1 consists of a sedimentary basin, $\mathbf{x} \in D_2$, perfectly embedded within an elastic half-space, $\mathbf{x} \in D_1 : \{|x_1| \leq \infty; x_3 \geq 0\}$. The material of the media is assumed to be linearly elastic, homogeneous and isotropic. The Cartesian coordinate system is denoted by $\{x_i\}; i = 1, 2, 3$. All the field variables are assumed to be independent of the coordinate x_2 thus $\partial()/\partial x_2 \equiv 0$ and $\mathbf{x} = (x_1, 0, x_3)$.

The basin is subjected to a plane harmonic incident SH wave

$$u_2^{inc} = e^{ik_1(x_1 \sin\theta_0 - x_3 \cos\theta_0) - i\omega t} \tag{1}$$

where k_1, θ_0 and ω denote the half-space wave number, the off-vertical angle of incidence and the circular frequency, respectively.

For the antiplane strain model at hand the motion takes place only along the x_2 – axes and the displacement field is defined by $\mathbf{u} = (0, u_2, 0)$. The basin's

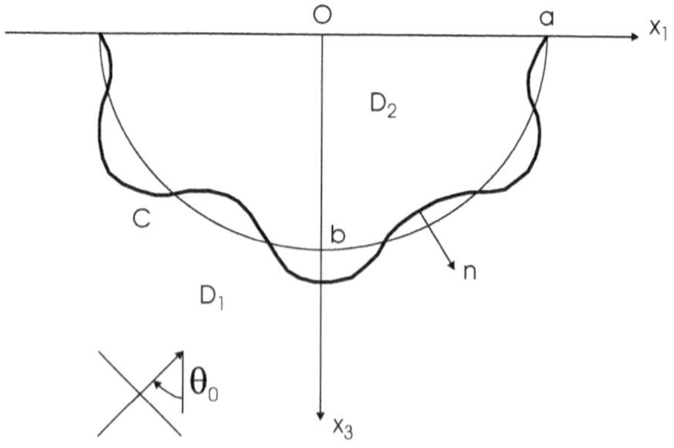

Fig. 1 Problem model of a basin with a corrugated interface C subjected to a plane harmonic incident SH wave. Here a and b denote the principal axes, **n** represents the unit normal on C and θ_0 is the angle of incidence

interface C is assumed to be irregular but sufficiently smooth without any sharp corners and of the form (Dravinski, 2007)

$$C(x_1, x_3) = C_0(x_1, x_3) + C_\varepsilon(x_1, x_3) \tag{2}$$

where C_0 denotes the smooth fundamental elliptic interface defined by

$$C_0 : \left\{ \frac{x_1^2}{a^2} + \frac{x_3^2}{b^2} = 1 \right\} \tag{3}$$

with a, b being the principal axes and C_ε denoting the interface perturbation. The surface perturbation C_ε is assumed to be of the form

$$\begin{aligned} &C_\varepsilon : \{x_1 = \varepsilon \sin \frac{2\pi\theta}{\Lambda} \cos\theta; x_3 = \varepsilon \sin \frac{2\pi\theta}{\Lambda} \sin\theta\} \\ &0 \leq \varepsilon < 1; 0 \leq \theta \leq \pi \end{aligned} \tag{4}$$

where ε and Λ represent the amplitude and the period of the perturbation, respectively and θ denotes the angle $\tan^{-1} x_3/x_1$.

It should be noted that for $\Lambda = \pi/(2m)$, m being an integer, the average of the perturbation surface C_ε over the interface C_0 is zero, i.e.,

$$\int_{C_0} C_\varepsilon dC_0 = 0 \tag{5}$$

The condition (5) simply follows from the procedures often used in the numerical modeling of the basin response in which the irregular interface of the basin is replaced by a smooth one in an "average sense". It should be noted that while the fundamental surface C_0 is symmetric with respect to the plane $x_1 = 0$, the perturbed interface C is not.

As the incident wave strikes the basin it will generate the unknown scattered waves. Therefore, the motion within the half-space and the basin is of the form

$$\begin{aligned} \boldsymbol{u}_2^{(1)} &= \boldsymbol{u}_2^{ff} + u_2^{s(1)}; \quad \mathbf{x} \in D_1 \\ \boldsymbol{u}_2^{(2)} &= \boldsymbol{u}_2^{s(2)}; \quad \mathbf{x} \in D_2 \end{aligned}$$

where the superscripts ff and s denote the free-field and the scattered waves, respectively. Furthermore, the superscripts (1) and (2) indicate the fields corresponding to the domains D_1 and D_2, respectively. The free-field represents the sum of the incident and reflected waves in the absence of the basin.

By imposing stress free boundary conditions along the surface of the half-space and the continuity of displacement and traction along the basin's interface C the total displacement field can be determined using an indirect boundary integral equation approach (e.g., Dravinski, 1983). In this paper the focus is on the energy flux along the basin interface.

The rate of energy flow across the basin interface C is defined by (Pao and Mow, 1973)

$$\dot{E} = -\int_C \mathbf{T}^n \cdot \dot{\mathbf{u}}\, dA \tag{6}$$

where $\mathbf{T}^n$ and $\dot{\mathbf{u}}$ denote the traction vector on the interface C with unit normal $\mathbf{n}$ and the particle velocity vector, respectively and $\cdot$ represents the dot product. Then the time average of the energy flow over one period $T = 2\pi/\omega$ defined as

$$< \dot{E} >= \frac{1}{T}\int_0^T \dot{E}\, dt \tag{7}$$

becomes

$$< \dot{E} >= \frac{-i\omega}{4}\int_C (\overline{T_2^n}\,\overline{u_2}^* - \overline{T_2^n}^*\,\overline{u_2})dA \tag{8}$$

where

$$\mathbf{T}^n(\mathbf{x},\omega) = \overline{\mathbf{T}^n}(\mathbf{x})e^{-i\omega t}$$
$$\mathbf{u}(\mathbf{x},\omega) = \overline{\mathbf{u}}(\mathbf{x})\mathbf{e}^{-i\omega t}$$

and the superscript $*$ denotes the complex conjugate. Using the fact that $\mathbf{T}^n = \sigma \cdot \mathbf{n}$ it follows from Eq. (8) that

$$< \dot{E} >= \frac{-i\omega}{4}\int_C \left\{n_1(\overline{\sigma_{12}u_2}^* - \overline{\sigma_{12}}^*\overline{u_2}) + n_3(\overline{\sigma_{32}u_2}^* - \overline{\sigma_{32}}^*\overline{u_2})\right\} dA \tag{9}$$

where σ_{ij} are the components of the stress tensor and $\mathbf{n} = (n_1, 0, n_3)$ is the unit normal on the interface C. The energy flux density or energy flux per unit area at any point of the interface C is given then by the integrand of the last equation, i.e.,

$$e(\mathbf{x},\omega) = -\frac{i\omega}{4}\left\{n_1(\overline{\sigma_{12}u_2}^* - \overline{\sigma_{12}}^*\overline{u_2}) + n_3(\overline{\sigma_{32}u_2}^* - \overline{\sigma_{32}}^*\overline{u_2})\right\} \mathbf{x} \in C \tag{10}$$

It should be noted that the energy flux density defined by the last equation is a real number.

3 Numerical Results

Throughout the numerical calculations the following parameters are used:

$$0 \le \varepsilon \le 0.2;\ \Lambda = \frac{\pi}{2m};\ m = 1-4$$
$$a = 1;\ b = 1, 0.7$$
$$\mu_1 = \rho_1 = 1;\ \mu_2 = 1/6;\ \rho_2 = 2/3$$

where μ_1 and ρ_1 denote the half-space shear modulus and material density, respectively while μ_2 and ρ_2 are the same parameters for the basin. The BIE parameters of the problem can be found in the article by Dravinski (2007).

For convenience, a dimensionless frequency is introduced as

$$\eta = \frac{2a}{\lambda^{inc}} \tag{11}$$

where λ^{inc} denotes the wavelength of the incident wave.

3.1 Interface Energy Flux Density

Figure 2 depicts the energy flux density along the interface of semicircular type of basins with different amplitudes of corrugation subjected to a vertical incident wave. Clearly, the presence of the corrugation resulted in reduction of the energy flux density for most points on the interface C. Especially significant reduction of the energy flux density can be observed for larger values of the corrugation amplitude ε.

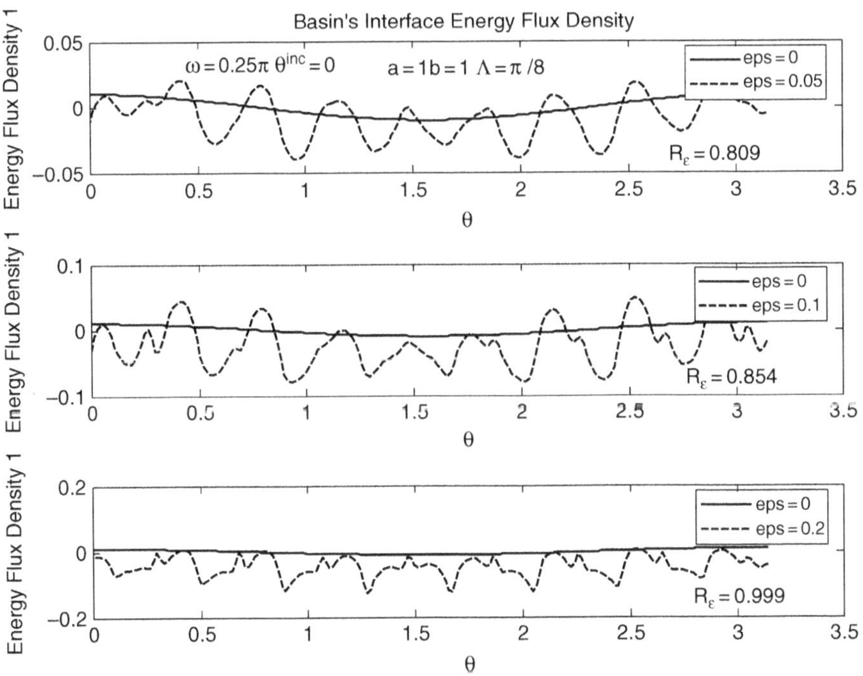

Fig. 2 Interface energy flux density for semicircular type of corrugated basins with different corrugation amplitude ε and a vertical incidence. $a = b = 1$, $\eta = 1/4$, $\Lambda = \pi/8$, $\theta_0 = 0$

In order to quantify more precisely the reduction of the interface energy density flux due to presence of the corrugation the following flux ratio is introduced

$$R_\varepsilon = \frac{A^+}{A^+ - A^-} \tag{12}$$

where

$$A^+ = \sum_i \int_{\theta_i}^{\theta_{i+1}} \{e|_{\varepsilon=0} - e|_{\varepsilon\neq 0}\} d\theta; e|_{\varepsilon=0} - e|_{\varepsilon\neq 0} > 0; \mathbf{x} \in C \tag{13}$$

$$A^- = \sum_i \int_{\theta_i}^{\theta_{i+1}} \{e|_{\varepsilon=0} - e|_{\varepsilon\neq 0}\} d\theta; e|_{\varepsilon=0} - e|_{\varepsilon\neq 0} < 0; \mathbf{x} \in C \tag{14}$$

$$i = 0 : N - 1; \theta_0 = 0; \theta_N = \pi$$

Here A^+ denotes the sum of the areas under the curve $e|_{\varepsilon=0} - e|_{\varepsilon\neq 0}$ for which $e|_{\varepsilon=0} > e|_{\varepsilon\neq 0}$, A^- represents the same sum for $e|_{\varepsilon=0} < e|_{\varepsilon\neq 0}$ while $\theta_i, i = 1 : N - 2$ are the zeros of the function $e|_{\varepsilon=0} - e|_{\varepsilon\neq 0}; \mathbf{x} \in C$ over the range $0 \leq \theta \leq \pi$.

It is easy to show that the value of the flux ratio falls within the range $0 \leq R_\varepsilon \leq 1$. Furthermore, if $A^+ > |A^-|$, then $R_\varepsilon > \frac{1}{2}$ which implies overall reduction of the energy density flux along the interface due to the corrugation. On the other hand, if $A^+ < |A^-|$, then $R_\varepsilon < \frac{1}{2}$ which indicates overall increase of the energy density flux along C caused by the corrugation.

It is evident from the flux ratios depicted by Fig. 2 that the presence of the corrugation caused overall reduction of the interface energy density flux for all three corrugation amplitudes. For this particular case the energy density flux reduced uniformly with the increase of the corrugation amplitude ε.

By increasing the frequency from $\eta = 1/4$ to $\eta = 1$ the interface energy flux densities are depicted by Fig. 3. Again, increase in the corrugation amplitude resulted in greater reduction of the interface energy flux density.

By changing the peak depth of the basin from $b = 1$ to $b = 0.7$ the energy flux densities at the interface C are depicted at two frequencies by Figs. 4 and 5. As before, the presence of the corrugation resulted in the reduction of the interface energy flux density for all cases considered here, i.e., the flux ratio $R_\varepsilon > \frac{1}{2}$. Although these results are similar to those of Figs. 2 and 3 however, for the deeper basin results in Fig. 3 the reduction of the interface energy density, measured by the flux ratio R_ε, uniformly increases with greater corrugation amplitude ε. This is not case for the shallower basin results of Fig. 5. Here the largest reduction of the energy flux density (or maximumR_ε) can be observed for the corrugation amplitude $\varepsilon = 0.1$ and not for $\varepsilon = 0.2$. Therefore, the reduction of the energy flux with increase of the corrugation amplitude is not uniform.

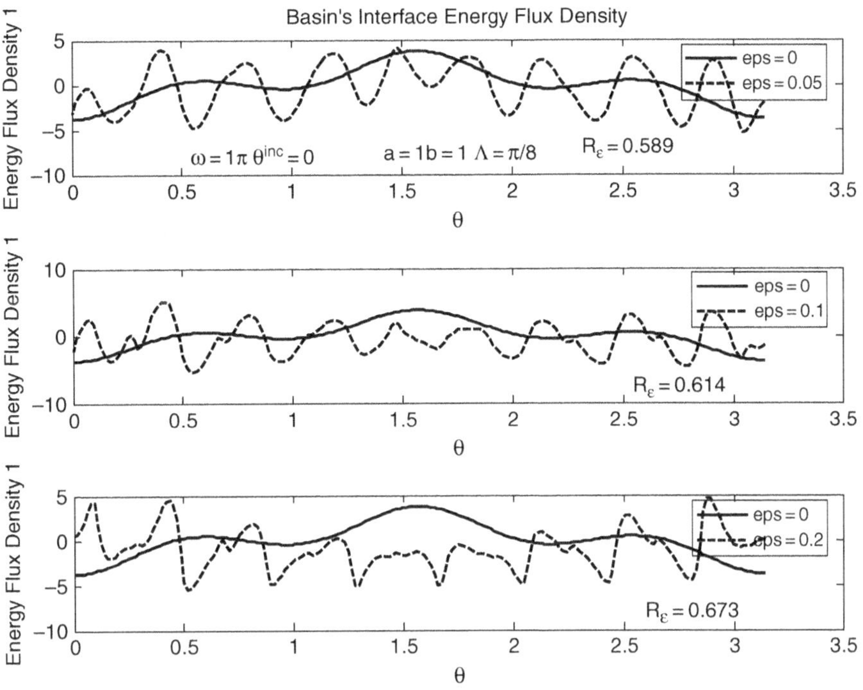

Fig. 3 Interface energy flux density for semicircular type of corrugated basins with different corrugation amplitude ε and a vertical incidence. $a = b = 1$, $\eta = 1$, $\Lambda = \pi/8$, $\theta_0 = 0$

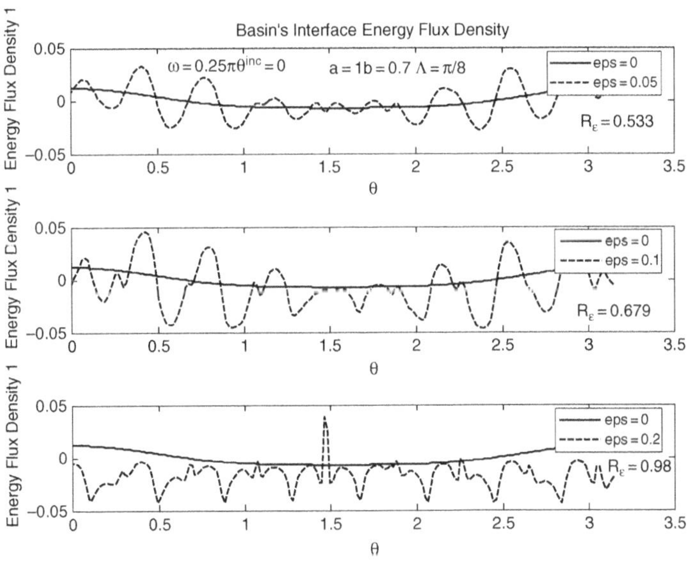

Fig. 4 Interface energy flux density for semi elliptic type of corrugated basins with different corrugation amplitude ε and a vertical incidence. $a = 1$, $b = 0.7$, $\eta = 1/4$, $\Lambda = \pi/8$, $\theta_0 = 0$

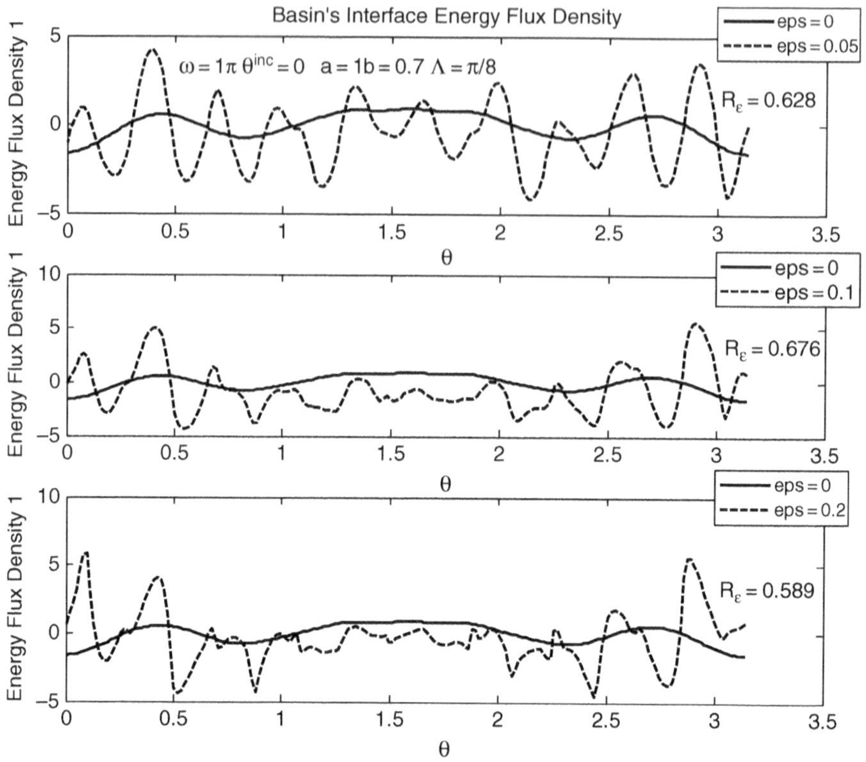

Fig. 5 Interface energy flux density for semi elliptic type of corrugated basins with different corrugation amplitude ε and a vertical incidence. $a = 1, b = 0.7, \eta = 1, \Lambda = \pi/8, \theta_0 = 0$

Change in the corrugation period from $\Lambda = \pi/8$ to $\Lambda = \pi/4$ resulted in the interface energy density flux depicted for deep and shallower basins by Figs. 6 and 7, respectively. In both cases presence of the corrugation produced an overall decrease in the interface energy density flux (i.e., $R_\varepsilon > \frac{1}{2}$). Similar results have been obtained for different frequencies but are omitted in order to reduce the number of figures.

Based on the results of Figs. 2 through 7 the following conclusions can be made:

1. Presence of the corrugation produced an overall reduction of the interface energy density flux for both deep and shallow basins at various frequencies of the input motion.
2. For most of the cases considered in this study, larger corrugation amplitude ε in general produced a more pronounced reduction of the interface energy density flux. However, this is not valid uniformly for all models considered in this study.

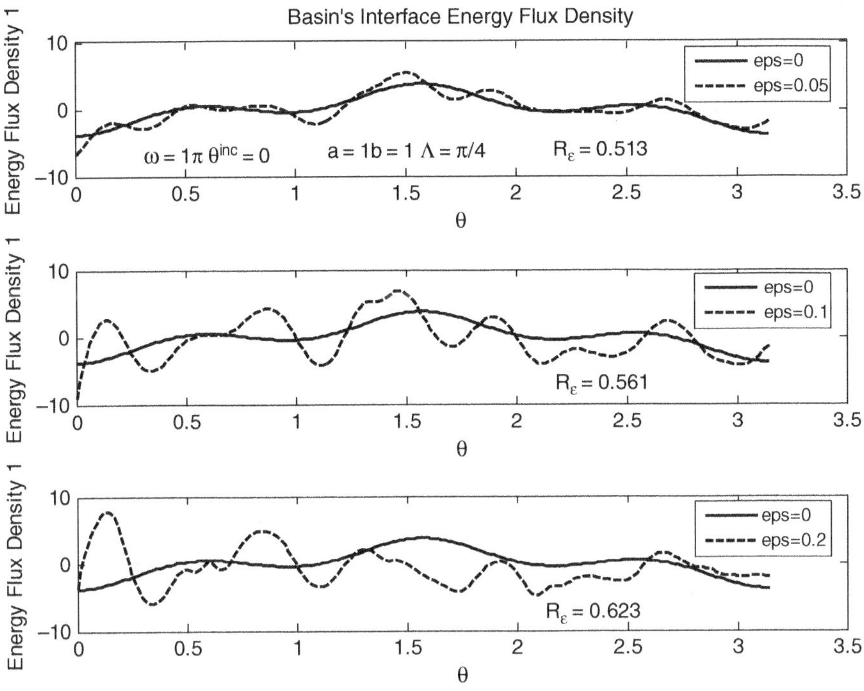

Fig. 6 Interface energy flux density for semi elliptic type of corrugated basins with different corrugation amplitude ε and a vertical incidence. $a = 1, b = 1, \eta = 1, \Lambda = \pi/4, \theta_0 = 0$

3. The reduction of the interface energy density flux is affected by the nature of the corrugation described in terms of the corrugation amplitude ε, and the period Λ.

3.2 Interface Motion

At this point it is of interest to investigate the role of corrugation upon the basin interface motion. For that purpose the interface displacement has been evaluated for a wide range of parameters present in the problem (e.g., frequency, amplitude and period of corrugation). Two typical results, shown by Figs. 8 and 9, correspond to the energy flux calculations depicted earlier by Figs. 3 and 5.

It can be clearly seen that the presence of the corrugation produced overall significant reduction in the interface displacement amplitude for both deep and shallow basins when compared with the interface motion of the corresponding smooth basin. Thus it appears that the reduction in the interface energy density flux observed in Figs. 3 and 5 is associated with overall smaller interface motion as well.

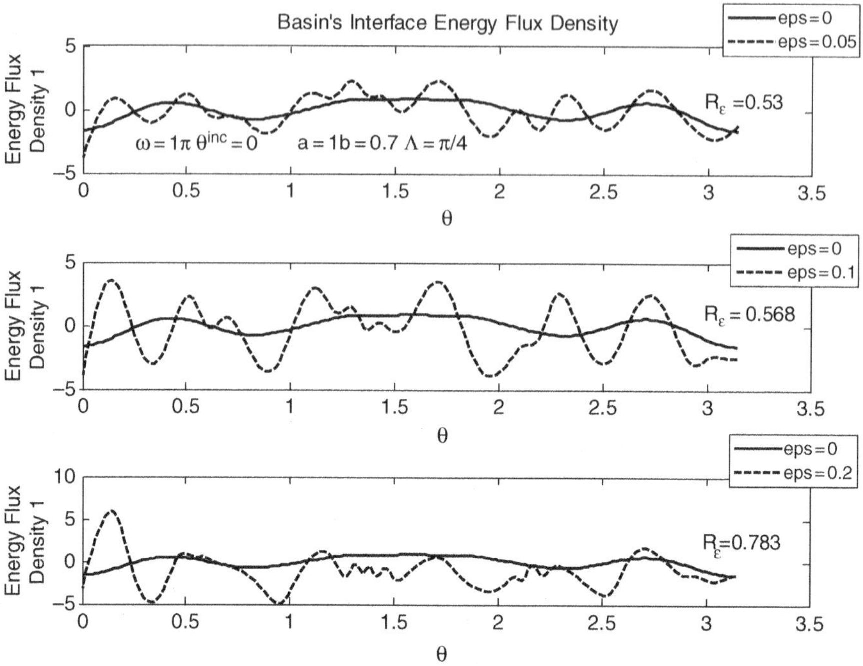

Fig. 7 Interface energy flux density for semicircular type of corrugated basins with different corrugation amplitude ε and a vertical incidence. $a = 1$, $b = 0.7$, $\eta = 1$, $\Lambda = \pi/4$, $\theta_0 = 0$

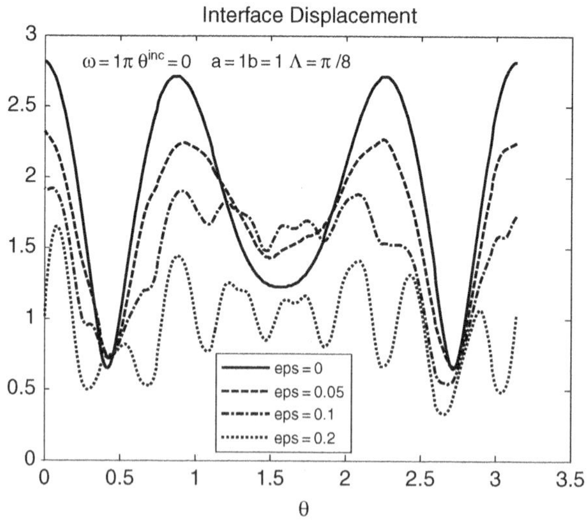

Fig. 8 Displacement amplitude spectra along the interface C of semicircular type of basin subjected to a vertical incident wave. Parameters: $\eta = 1$, $a = 1$, $b = 1$, $\Lambda = \pi/8$, $\theta_0 = 0$

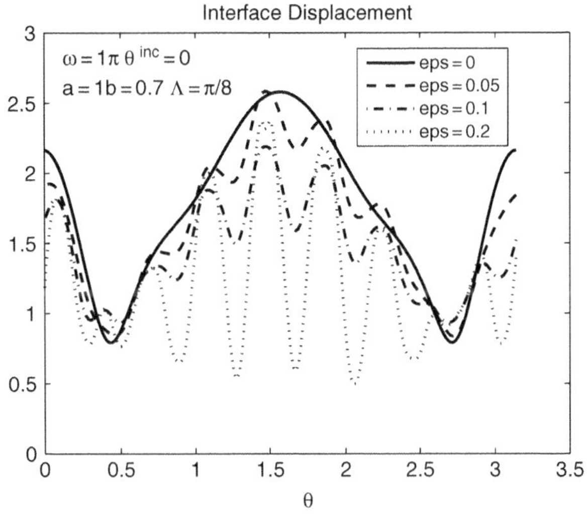

Fig. 9 Displacement amplitude spectra along the interface *C* of semi elliptical type of basins subjected to a vertical incident wave. Parameters: $\eta = 1$, $a = 1, b = 0.7, \Lambda = \pi/8$, $\theta_0 = 0$

4 Summary and Conclusions

Energy flux across a corrugated interface of a basin subjected to a plane harmonic SH wave is investigated for linearly elastic and homogeneous media. The corrugation is assumed to be in the form of an ellipse perturbed in a sinusoidal manner. The problem is solved by using an indirect boundary integral equation approach. The interface energy flux density was evaluated for a wide range of parameters present in the problem (e.g., frequency, basin depth, amplitude and period of the corrugation). The reduction of the interface energy density is defined in terms of the flux ratios.

The presented results clearly demonstrate that presence of the corrugation produced overall reduction in the interface energy density flux for both deep and shallow basins at different frequencies. For majority of the cases studied in this work, larger corrugation amplitude produced greater reduction in the interface energy density flux. However, this reduction appears not be uniform for all the models. The reduction of the interface energy density flux is influenced by the corrugation amplitude and period. Furthermore, it was shown that decrease in the interface energy density flux leads to smaller interface motion.

References

Aki, K. (1988). Local site effects and strong ground motion. In *Proceedings of the Special Conference on Earthquake Engineering and Soil Dynamics 2, Am. Soc. Civil Eng., Park City, Utah.*

Bard, P. Y., and M. Bouchon (1985). The two-dimensional resonance of sediment-filled valleys, *Bull. Seism. Soc. Am.*, **75,** 519–541.

Bouchon, M., C. A. Shultz, and M. N. Toksöz (1996). Effect of three dimensional topography on seismic motion, *J. Geophys. Res.*, **101,** 5835–5846.

Dravinski, M. (1983). Scattering of plane harmonic SH waves by dipping layers of arbitrary shape, *Bull. Seism. Soc. Am.,* **73,** 1303–1319.

Dravinski, M. (2007). Scattering of waves by a sedimentary basin with a corrugated interface, *Bull. Seism. Soc. Am.,* **97,** 256–264.

Fu, L. Y. (2005). Rough surface scattering: Comparison of various approximation theories for 2D SH waves, *Bull. Seism. Soc. Am.,* **95,** 646–663.

Fu, L. Y., R. S. Wu, and M. Campillo (2002). Energy partition and attenuation of regional phases by random free surface, *Bull. Seism. Soc. Am.,* **92,** 1992–2007.

Pao, Y.-H., and C.-C. Mow (1973). *Diffraction of Elastic Waves and Dynamic Stress Concentrations,* Crane Russak, New York.

Sanchez-Sesma, F. J. (1983). Diffraction of elastic waves by three dimensional irregularities, *Bull. Seism. Soc. Am.,* **73,** 1621–1636.

Wong, H. L. (1982). Diffraction of P, SV, and Rayleigh waves by surface topographies, *Bull. Seism. Soc. Am.,* **72,** 1167–1184.

Boundary Integral Equations and Fluid-Structure Interaction at the Micro Scale

Attilio Frangi

Abstract Providing an estimate of gas damping in Micro-Electro-Mechanical-Systems (MEMS) is a complex task since MEMS are fully three dimensional micro-structures which cannot in general be reduced to simple 1D or 2D models and since the gas cannot be treated as a continuum medium at the microscale. Moreover, the range of working pressures is extremely large, starting from standard conditions down to almost vacuum. We show that integral equations are the ideal tool for addressing this issue both at standard conditions and at near-vacuum. In the former case robust formulations and industrial codes have already been developed for large scale problems employing Fast Solvers and implementing a linear, quasi-static, incompressible Stokes formulation with slip boundary conditions. In the latter situation, on the contrary, the development is technically less mature but very promising. The tools investigated herein are eventually applied to the analysis of an industrial MEMS sensor produced and tested by STMicroelectronics.

1 Introduction

The specific class of inertial polysilicon Micro-Electro-Mechanical-Systems (MEMS) sensors (e.g. accelerometers and gyroscopes) is now experiencing an increasing commercial success but still poses several scientific challenges. These MEMS typically consist of collections of fixed parts and vibrating shuttles separated by gaps which are few microns wide.

Estimating mechanical dissipation in MEMS seems, for several reasons, to be an ideal application for Boundary Integral Methods (BIE). First, the micromechanical structures, a typical example of which is the sensor of Fig. 1, are innately three-dimensional and too geometrically complicated to analyze analytically. Therefore, a numerical approach is needed. Second, the mechanical dissipation is primarily due to pressure and drag forces generated by the air surrounding the mechanical

A. Frangi (✉)
Department of Structural Engineering, Politecnico of Milano, Milano, Italy
e-mail: attilio.frangi@polimi.it

A chapter in honor of Dimitri Beskos' 65th birthday

G.D. Manolis, D. Polyzos (eds.), *Recent Advances in Boundary Element Methods*,
DOI 10.1007/978-1-4020-9710-2_8,

structure. Hence, even though the exterior domain is effectively infinite in extent, the only quantities of interest are velocities and forces on the structure surface. Surface-only integral equations, if they can be formulated, have a dimensional advantage over volume methods in such a setting. Finally, since resonating frequencies are generally low (see e.g. Frangi et al. (2006) and the example of Section 4) and the shuttle velocity is always small with respect to the mean molecular velocity, linear formulations can be developed, as explained in detail in the following sections.

Unfortunately, at the length scales of interest rarefaction effects in the flow are always a key factor. These are conventionally quantified by means of the Knudsen number, defined as $\mathrm{Kn} = \lambda/\ell$, where ℓ is a characteristic length of the flow and λ the mean free path of gas molecules. Based on the Knudsen number, gas flows can be qualitatively classified as continuum ($\mathrm{Kn} < 0.01$), slip ($0.01 < \mathrm{Kn} < 0.1$), transition ($0.1 < \mathrm{Kn} < 10$), and free-molecule ($\mathrm{Kn} > 10$) (Chapman and Cowling, 1960; Cercignani, 1988; Bird, 1994; Gad-el-Hak, 1999; Karniadakis and Beskok, 2002). Since λ increases when package pressure decreases, and since the range of working pressures for MEMS is extremely large, starting from standard conditions down to almost vacuum, all the flow types defined above need to be properly analysed by using, in general, different numerical tools. However, all of them could be conveniently addressed by BIE, at least in principle.

Indeed, there have been a number of experimentally-verified successes in evaluating gas damping for air-packaged MEMS (continuum and slip flows) using fast integral equation solvers (Wang et al., 2002, 2006; Ye et al., 2003; Frangi and Tausch, 2005; Frangi et al., 2006; Yuhong et al., 2006; Liu and Shen, 2007).

The situation is however less developed and understood in the transition regime. Even if a Boundary-Volume Integral Equation approach has been formulated theoretically in Cercignani (1988), no numerical implementation has been reported so far. Different techniques of deterministic and statistical nature are generally preferred (Fukui and Kaneko, 1987; Cho et al., 1993; Cai et al., 2000; Aoki et al., 2002; Emerson et al., 2007), even if their application to realistic 3D low-speed MEMS is still under investigation (Mieussens, 2000; Frangi et al., 2007).

On the contrary, the collisionless or free-molecule flow lends itself again to the development of robust and competitive BIE approaches. Based on the formal classification given above, the free-molecule flow represents the limiting case where the Knudsen number tends to infinity and collisions between molecules can be neglected. In inertial MEMS this applies typically at pressures in the range of few mbars and below. The numerical technique generally employed for estimating the quality factor of MEMS in this range is the Test Particle Monte Carlo (TPMC) method in the various forms analysed e.g. in Bird (1994); Kadar et al. (1996); Bao et al. (2002); Hutcherson and Ye (2004). However, this approach is stochastic in nature which is a major drawback for the extremely low-speed applications at hand, often implying long runs in order to obtain reliable averages, while in the design and optimization phases of MEMS fast and agile tools are preferred. In this chapter it is shown that a different technique based on integral equations is indeed very competitive with respect to the TPMC for the typical working conditions of inertial MEMS since it is deterministic, fast and robust.

In the next sections of this chapter, we review many of the issues associated with integral formulation, discretization and fast solution of the Stokes and free-molecule solvers. Finally Section 4 collects numerical results for an industrial MEMS produced by STMicroelectronics and presents a comparison with available experimental data in the free-molecule regime.

Although not strictly necessary, in the sequel it will be assumed that the displacement $\mathbf{s}(\mathbf{x}, t)$ and velocity $\mathbf{w}(\mathbf{x}, t)$ of movable parts in a MEMS can be expressed as:

$$\mathbf{s}(\mathbf{x}, t) = \mathbf{g}(\mathbf{x})q(t), \qquad \mathbf{w}(\mathbf{x}, t) = \mathbf{g}(\mathbf{x})\dot{q}(t) \tag{1}$$

where vector $\mathbf{g}(\mathbf{x})$ is a function of position $\mathbf{x}$ to be specified and corresponds to a modal shape of shuttle and springs. This implies a sort of decoupling between the structural response and the fluid action, the latter never affecting the shape of the displacement field but only the time dependent weight $q(t)$. The case in which the global MEMS movement is the linear combination of several terms like those in eqn. (1) requires only a straightforward extension of the techniques analysed herein and will not be addressed. Rather frequently dissipation originates from portions of the shuttle where $\mathbf{g}(\mathbf{x})$ corresponds to a rigid body motion. In this case further simplifications are allowed and advisable. Moreover, in the typical working conditions of inertial MEMS, as discussed later, the surface force density exerted by the shuttle on the gas flow is $\mathbf{t}(\mathbf{x}, t) = \boldsymbol{\sigma}(\mathbf{x}, t) \cdot \mathbf{n}(\mathbf{x})$ ($\mathbf{n}(\mathbf{x})$ is the surface normal pointing outside the fluid domain) and admits the form $\mathbf{t}(\mathbf{x}, t) = \mathbf{p}(\mathbf{x})\dot{q}(t)$, where $\mathbf{p}(\mathbf{x})$ is a real vector function. The global analysis can thus be reduced to the 1D model $M\ddot{q}(t) + B\dot{q}(t) + Kq(t) = F(t)$ where, if S_{SH} denotes the surface of the shuttle, the damping coefficient B is:

$$B = \int_{S_{\mathrm{SH}}} \mathbf{g}(\mathbf{x}) \cdot \mathbf{t}(\mathbf{x})\mathrm{d}S \tag{2}$$

If, as customary, the electrostatic forcing is sinusoidal with angular frequency ω_0, $q(t)$ admits the simple form $q(t) = A\sin(\omega_0 t + \phi)$. This model predicts the quality factor $Q = (2/\pi)E/D = \omega_0 M/B$ where E is the maximum elastic energy stored in the system $1/2KA^2$ and D is the dissipation per unit-cycle $D = \pi BA^2\omega_0$.

2 Quasi-Static Stokes Flow with Slip Boundary Conditions

The most mature and best validated of the fast solvers for estimating gas damping in MEMS are based on solving integral formulations of the 3D incompressible quasi-static Stokes equations (Wang et al., 2006, 2002; Ye et al., 2003; Frangi et al., 2006). For quasi-static Stokes to be a good model of the gas surrounding a micromachined structure, the gas should be incompressible, sufficiently viscous, moving slowly, and not too rarefied. This list of assumptions hold for several kinds of inertial sensors like the biaxial accelerometer analysed in Frangi et al. (2006), the rotational accelerometer of Section 4 and gyroscopes. In this background section the quasi-static Stokes model is given, and the standard integral formulation and the common numerical discretizations described. In later sections techniques for extending the model to include the effects of rarefaction will be described.

2.1 Governing Equations

For an isotropic Newtonian fluid, if the fluid velocity, $\mathbf{v}$, is divergence-free, then the relation between $\mathbf{v}$ and the stress tensor, $\boldsymbol{\sigma}$, is given by

$$\boldsymbol{\sigma}(\mathbf{x},t) = -p(\mathbf{x},t)\mathbb{1} + \eta\left(\nabla\mathbf{v}(\mathbf{x},t) + \nabla^T\mathbf{v}(\mathbf{x},t)\right) \tag{3}$$

where η is the fluid viscosity and p denotes pressure.

The quasi-static Stokes equations are derived from combining incompressibility, conservation of mass, and conservation of momentum yielding the well-known quasi-static Stokes equations

$$\nabla p(\mathbf{x},t) - \eta\Delta\mathbf{v}(\mathbf{x},t) = \mathbf{0} \qquad \nabla\cdot\mathbf{v}(\mathbf{x},t) = 0 \quad \text{in} \quad \Omega \tag{4}$$

where Ω denotes the volume occupied by the fluid. Given the linearity of the governing equations and taking into account eqn. (1), the velocity field has the form:

$$\mathbf{v}(\mathbf{x},t) = \mathbf{u}(\mathbf{x})\dot{q}(t) \tag{5}$$

where $\mathbf{u}(\mathbf{x})$ is the time independent solution of the Stokes model when the velocity $\mathbf{g}(\mathbf{x})$ is imposed to solid surfaces.

The standard boundary conditions for the classical quasi-static Stokes equation are no-slip conditions. For the case of several interacting micromachined structures surrounded by fluid, the no-slip condition implies that for each structure surface point, the fluid velocity must match the structure's velocity. More precisely, for the MEMS problem, the domain of the fluid, Ω, is defined as the domain exterior to the micromachined structures. For each point $\mathbf{x}$ on surface S, where S is defined to be the union the structure surfaces,

$$\mathbf{u}(\mathbf{x}) = \mathbf{g}(\mathbf{x}) \tag{6}$$

Note also that typically the velocity distant from the micromachined structures is assumed to approach zero. Nonzero background velocities can be treated by perturbation.

Given $\mathbf{u}$ which satisfies Eqs. (4) and (6), the vector force density the structure exerts on the fluid can be computed from the product of the stress tensor with the surface normal pointing out of Ω.

2.2 Integral Formulation

Either Green's identities or Lorentz reciprocity can be used to derive an integral formulation that relates the Stokes flow generated traction forces to the surface velocities. Specifically, if **x** is a point on a smooth region of in S,

$$\frac{1}{2}\mathbf{u}(\mathbf{x}) = \int_S \{\mathcal{V}(\mathbf{r}) \cdot \mathbf{t}(\mathbf{y}) - [\mathcal{K}(\mathbf{r}) \cdot \mathbf{n}(\mathbf{y})] \cdot \mathbf{u}(\mathbf{y})\}\, \mathrm{d}S_y \tag{7}$$

where $\mathbf{r} \equiv \mathbf{x} - \mathbf{y}$, the kernels (also referred to as Greens functions) $\mathcal{V}$ and $\mathcal{K}$ are given by

$$V_{ik}(\mathbf{r}) = \frac{1}{8\pi\eta}\left(\frac{\delta_{ik}}{r} + \frac{r_i r_k}{r^3}\right)$$

$$K_{iqk}(\mathbf{r}) = -\frac{3}{4\pi}\frac{1}{r^5} r_i r_q r_k\,,$$

and the integrals of the strongly singular kernel, $\mathcal{K}$, should be interpreted in the Cauchy principal value sense. The kernels $\mathcal{V}$ and $\mathcal{K}$ are often interpreted as generating the velocities associated with stokeslet and stresslet point sources, but these kernels also coincide with the Kelvin kernels used in the displacement equation for incompressible elasticity (Bonnet, 1999).

Combining the no-slip boundary conditions with eqn. (7) yields a first kind integral equation which relates the known structure velocities to the unknown fluid traction forces

$$\frac{1}{2}\mathbf{g}(\mathbf{x}) = \int_S \{\mathcal{V}(\mathbf{r}) \cdot \mathbf{t}(\mathbf{y}) - [\mathcal{K}(\mathbf{r}) \cdot \mathbf{n}(\mathbf{y})] \cdot \mathbf{g}(\mathbf{y})\}\, \mathrm{d}S_y. \tag{8}$$

In the common case where the structure velocities correspond to rigid body motion, the second integral in eqn. (8) vanishes, resulting in

$$\frac{1}{2}\mathbf{g}(\mathbf{x}) = \int_S \mathcal{V}(\mathbf{r}) \cdot \mathbf{t}(\mathbf{y})\mathrm{d}S_y. \tag{9}$$

The formulation in eqn. (9) is common in the literature on integral equations for Stokes flow, but there are alternatives that can have superior properties. Of immediate concern is the fact that eqn. (9) does not have a unique solution, an issue that will be addressed in the next section.

2.3 Null Space Problem

The differential form of Stokes equation (4) only involves the gradient of the pressure and is therefore insensitive to spatially constant shifts in pressure. Since the pressure is a component of **t**, and **t** is the unknown in eqn. (9), this can not have a unique solution. The matter is even more problematic when the surface S corresponds to N unconnected structures. In that case,

$$[\mathcal{V}\mathbf{t}] = \int_S \boldsymbol{\mathcal{V}}(\mathbf{r}) \cdot \mathbf{t}(\mathbf{y}) \mathrm{d}S_y$$

has null space $\mathcal{N}(\mathcal{V})$ of dimension N whose basis is given by

$$\mathbf{t}^\alpha(\mathbf{x}) = \begin{cases} \mathbf{n}(\mathbf{x}), & \mathbf{x} \in S^\alpha \\ 0 & \text{elsewhere} \end{cases}, \qquad 1 \leq \alpha \leq N. \tag{10}$$

Many techniques exist for dealing with the null space of eqn. (9) as discussed in Frangi and Di Gioia (2005). Unfortunately, standard BEM approaches are ill-conditioned when applied to complex structures even if the null space is filtered out exactly.

In order to cure the issue of ill-conditioning, a new boundary element method, the Mixed Velocity Traction approach (MVT), has been recently proposed in Frangi and Tausch (2005) and extended to large scale problems in Frangi et al. (2006) using fast solvers.

The second tool required for setting up the MVT is the traction integral equation, an integral identity which can be obtained through careful differentiation of eqn. (7), for a smooth $\mathbf{x} \in S$:

$$\frac{1}{2}\mathbf{t}(\mathbf{x}) = \int_S \{-[\mathbf{n}(\mathbf{x}) \cdot \boldsymbol{\mathcal{K}}(\mathbf{r})] \cdot \mathbf{t}(\mathbf{y}) - [\mathbf{n}(\mathbf{x}) \cdot \boldsymbol{\mathcal{W}}(\mathbf{r}) \cdot \mathbf{n}(\mathbf{y})] \cdot \mathbf{u}(\mathbf{y})\} \, \mathrm{d}S_y \tag{11}$$

where $\boldsymbol{\mathcal{W}}$ is a fourth order two-point kernel with components:

$$\begin{aligned} W_{qiks}(\mathbf{r}) =& \frac{\eta}{4\pi}\frac{1}{r^3}\Big[2\delta_{sk}\delta_{iq} + \frac{3}{r^2}\left(\delta_{ik}r_q r_s + \delta_{kq}r_i r_s + \delta_{is}r_k r_q + \delta_{sq}r_i r_k\right) \\ & - 30\frac{r_i r_k r_q r_s}{r^4}\Big] \end{aligned}$$

Equation (11) also contains an hypersingular integral interpreted here in the finite-part sense. Again, when the structure velocities correspond to rigid body motion the integral of $[\mathbf{n}(\mathbf{x}) \cdot \boldsymbol{\mathcal{W}}(\mathbf{r}) \cdot \mathbf{n}(\mathbf{y})] \cdot \mathbf{g}(\mathbf{y})$ over a closed surface vanishes. The MVT simply consists in enforcing a linear combination of eqns. (9) and (11):

$$\mathbf{g}(\mathbf{x}) - \frac{\gamma}{\eta}\frac{1}{2}\mathbf{t}(\mathbf{x}) = \int_S \Big\{\boldsymbol{\mathcal{V}}(\mathbf{r}) \cdot \mathbf{t}(\mathbf{y}) + \frac{\gamma}{\eta}[\mathbf{n}(\mathbf{x}) \cdot \boldsymbol{\mathcal{K}}(\mathbf{r})] \cdot \mathbf{t}(\mathbf{y})\Big\} \mathrm{d}S_y \tag{12}$$

where γ is a length scale to be calibrated. The benefits of this formulation have been pointed out in Frangi and Tausch (2005) where it has been shown that the mixed formulation is well posed when $\gamma > 0$ and that not only does it filter out automatically all the exact null space of the velocity equation, but also considerably improves the condition number which is crucial for the iterative solvers employed.

2.4 Numerical Implementation

The numerical solutions of eqn. (9) require, as usual, the discretization of S (in this case we choose a triangulation $\mathcal{T}$ with M flat triangles) and the choice of the space X_h to which the interpolation of $\mathbf{t}$ belongs. Since $\mathbf{t}$ represents tractions which are typically discontinuous along edges and corners, we choose X_h as the space of piecewise constant functions. At this stage, different alternative procedures can be employed. In the Galerkin approach eqn. (12) is contracted with a traction test field $\tilde{\mathbf{t}}(\mathbf{x}) \in X_h$, integrated over S and enforced for any choice of $\tilde{\mathbf{t}}(\mathbf{x}) \in X_h$.

Let τ_β and τ_γ be generic triangles of the mesh with $\mathbf{x} \in \tau_\beta$ and $\mathbf{y} \in \tau_\gamma$, and let A_β be the area of τ_β. If $\mathbf{T}_\gamma$ is the traction on τ_γ, the discretized Galerkin approach yields the linear system:

$$\begin{aligned}
&\int_{\tau_\beta} \mathbf{g}(\mathbf{x})\mathrm{d}S_x - A_\beta \frac{\gamma}{2\eta}\mathbf{T}_\beta \qquad\qquad \forall\beta = \{1 : M\} \\
&= \sum_{\gamma=1}^{M} \left[\int_{\tau_\beta} \int_{\tau_\gamma} \left(\mathcal{V}(\mathbf{r}) + \frac{\gamma}{\eta}\mathbf{n}(\mathbf{x}) \cdot \mathcal{K}(\mathbf{r}) \right) \mathrm{d}S_y \mathrm{d}S_x \right] \cdot \mathbf{T}_\gamma
\end{aligned} \tag{13}$$

However, though accurate, this approach requires the evaluation of double surface integrals. This issue has been studied at length, but still represents a considerable obstacle when computing time becomes an issue and hence, typically, in large scale problems associated to MEMS.

The classical and faster collocation approach consists in enforcing eqn. (12) at the center of each triangle. It is worth stressing that the collocation approach can be recovered from eqn. (13) by employing a single point quadrature rule for the integration over $\mathbf{x}$, where $\mathbf{x}_\beta$ represents the center of mass of the τ_β triangle.

Recently, however, it has been shown in Frangi and Tausch (2005) that, in the context of this mixed velocity approach, a different approximate numerical scheme requiring only single surface integrals can be advantageously applied: qualocation (Tausch and White, 1998). In this approach, eqn. (13) is applied with a one point Gauss–Hammer rule for the integration over $\mathbf{y}$ (the inner integration) while a standard numerical rule or an analytical approach is applied for the outer integral. Hence qualocation is, in a sense, the dual of collocation with respect to the Galerkin approach. Numerical examples presented in Frangi and Tausch (2005) show that qualocation is less sensitive to the value of the coupling parameter than collocation. For instance, it can be proved that, as $\gamma \to \infty$, the field $\mathbf{t}$ obtained from collocation vanishes, while this is not the case with qualocation.

As is clear from eqn. (12), the discretized integral equations generate systems of equations that are *dense*. If direct factorization is used to solve memory required to store the matrix will grow like n^2 and the matrix solve time will increase like n^3. If instead, a preconditioned Krylov-subspace method like GMRES is used to solve the system, then it is possible to reduce the solve time to order n^2 but the memory requirement will not decrease.

In order to develop algorithms that use memory and time that grows more slowly with problem size, it is essential *not* to form the matrix explicitly. Instead, one can exploit the fact that Krylov-subspace methods for solving systems of equations only require matrix-vector products and not an explicit representation of the matrix and this can be accomplished in nearly order n operations (Barnes and Hut, 1986; Greengard and Rokhlin, 1997). Several researcher simultaneously observed the powerful combination of discretized integral equations, Krylov-subspace methods, and fast matrix-vector products (Greengard and Rokhlin, 1997; Hackbusch and Nowak, 1989; Borup and Gandhi, 1985). Such methods are now referred to, somewhat pejoratively, as fast solvers. Among these techniques, two in particular have been applied to the analysis of gas damping in MEMS: the Fast Multipole Method and the Precorrected FFT (Ye et al., 2003). Acceleration is achieved by computing the far-field interactions in an approximate way, while near-field interaction still resorts to classical technique for numerical integration of nearly-singular or singular integrals. In particular FMM represent nowadays a well established technique applied successfully in different fields of mechanics. A complete explanation of FMM is however beyond the scope of this paper and reference is made to Greengard and Rokhlin (1997); Nishimura (2002). For the details of FMM applied in the context of the MVT approach of interest herein, the reader is referred to Frangi (2005); Frangi et al. (2006).

2.5 Extension to the Slip Flow Regime

Working pressures of MEMS are spread over a large range (1 bar – 10^{-6} bar). This issue, associated to the micro-scale at hand, promotes rarefaction effects which, at low pressures, have to be dealt with using techniques of rarefied gas-dynamics (Cercignani, 1988; Gad-el-Hak, 1999).

As briefly recalled in the Introduction, in the range $0.01 < \mathrm{Kn} < 0.1$ an accurate prediction of the flow properties can be obtained by applying a continuum approach with slip boundary conditions (slip BC). Since the dimensions typical of MEMS are of a few microns, the flow mainly develops in the slip regime even at ambient pressure.

Let $\mathbf{t}^S$ denote the surface components of tractions:

$$\mathbf{t}^S(\mathbf{x}) = [\mathbb{1} - \mathbf{n}(\mathbf{x}) \otimes \mathbf{n}(\mathbf{x})] \cdot \mathbf{t}(\mathbf{x}),$$

where $\mathbb{1} - \mathbf{n}(\mathbf{x}) \otimes \mathbf{n}(\mathbf{x})$ is the surface projector tensor. As pointed out recently (Yuhong et al., 2006), first order boundary slip conditions should be expressed in terms of $\mathbf{t}^S$:

$$\mathbf{u}(\mathbf{x}) = \mathbf{g}(\mathbf{x}) - c_t \mathbf{t}^S(\mathbf{x}) \qquad c_t := \frac{2-\sigma}{\sigma}\frac{\lambda}{\eta} \tag{14}$$

where σ is the tangential momentum accommodation coefficient (Karniadakis and Beskok, 2002).

Since eqn. (14) operates linearly on $\mathbf{t}$, the BEM formulation eqn. (12) can be easily adapted, yielding (Frangi et al., 2006):

$$\mathbf{g}(\mathbf{x}) - \frac{c_t}{2}\mathbf{t}^S(\mathbf{x}) - \frac{\gamma}{\eta}\frac{1}{2}\mathbf{t}(\mathbf{x}) = \int_S \Big\{ \mathcal{V}(\mathbf{r}) \cdot \mathbf{t}(\mathbf{y}) + c_t\,[\mathcal{K}(\mathbf{r}) \cdot \mathbf{n}(\mathbf{y})] \cdot \mathbf{t}^S(\mathbf{y})$$
$$+ \frac{\gamma}{\eta}\Big([\mathbf{n}(\mathbf{x}) \cdot \mathcal{K}(\mathbf{r})] \cdot \mathbf{t}(\mathbf{y}) - c_t\,[\mathbf{n}(\mathbf{x}) \cdot \mathcal{W}(\mathbf{r}) \cdot \mathbf{n}(\mathbf{y})] \cdot \mathbf{t}^S(\mathbf{y})\Big) \Big\} \mathrm{d}S_y. \quad (15)$$

3 Gas Structure Interaction in the Free Molecule Regime

As discussed in Section 1, in the free molecule regime the flow is so rarefied that collisions between molecules can be reasonably neglected and the attention is focused on the interaction of the flow with solid surfaces. Also in this case Boundary Integral Equations (BIE) approaches prove to be very competitive.

For the MEMS addressed in this work the velocity of shuttle surfaces $\mathbf{w}$ is generally small with respect to the average thermal molecular speed, hence $|\tilde{\mathbf{w}}| = |\mathbf{w}|/\sqrt{2RT} \ll 1$. For a sinusoidal motion $q(t) = A\sin(\omega_0 t + \phi)$, this condition can be written as $c_L \ll 1$, where $c_L := \omega_0 A/(\sqrt{2RT})$.

Thanks to the absence of molecular collisions, the flow impinging on a shuttle surface can be decomposed in the contribution of molecules coming from the far field region with a given distribution function and a contribution from molecules reflected by other solid surfaces. The former can indeed be evaluated analytically. As far as the latter is concerned, if L denotes a typical distance between visible surfaces, and if the average time required by a molecule to bounce back to the shuttle is small with respect to the duration of one oscillation, i.e. $c_Q := \omega_0 L/\sqrt{\rho RT} \ll 1$, the properties of the flow can be computed assuming quasi-static conditions.

Clearly the two conditions $c_L \ll 1$, $c_Q \ll 1$ need to be accurately checked in the applications. However, for the class of inertial resonators of interest herein, they are generally largely met. As an example, in the case of the rotational accelerometer of Section 4, $c_Q \simeq 6\,10^{-4}$, $c_L \simeq 10^{-5}$.

According to the kinetic theory of gases (Chapman and Cowling, 1960; Cercignani, 1988; Bird, 1994), all the macroscopic quantities of interest, like mean velocity $\mathbf{v}$, density ρ, temperature T and stress tensor $\boldsymbol{\sigma}$, can be expressed as moments of the distribution function $f(\mathbf{x}, \boldsymbol{\xi})$ in the velocity space:

$$\rho(\mathbf{x}, t) = \int_{\mathbb{R}^3} f(\mathbf{x}, \boldsymbol{\xi}, t)\mathrm{d}\boldsymbol{\xi} \quad (16)$$

$$\rho(\mathbf{x}, t)\mathbf{v}(\mathbf{x}, t) = \int_{\mathbb{R}^3} f(\mathbf{x}, \boldsymbol{\xi}, t)\boldsymbol{\xi}\mathrm{d}\boldsymbol{\xi} \quad (17)$$

$$T(\mathbf{x}, t) = \frac{1}{3R\rho(\mathbf{x}, t)} \int_{\mathbb{R}^3} f(\mathbf{x}, \boldsymbol{\xi}, t)|\boldsymbol{\xi} - \mathbf{v}(\mathbf{x}, t)|^2\mathrm{d}\boldsymbol{\xi} \quad (18)$$

$$\boldsymbol{\sigma}(\mathbf{x}, t) = \int_{\mathbb{R}^3} f(\mathbf{x}, \boldsymbol{\xi}, t)\,(\boldsymbol{\xi} - \mathbf{v}(\mathbf{x}, t)) \otimes (\boldsymbol{\xi} - \mathbf{v}(\mathbf{x}, t))\,d\boldsymbol{\xi} \quad (19)$$

where R is the universal gas constant divided by the molar mass.

The gas interacts with both fixed and movable surfaces of the MEMS. As a matter of fact, silicon surfaces originating from etching procedures are very rough so that, as a first approximation, diffuse reflection from the solid surface will be assumed in the sequel. If $\mathbf{n}(\mathbf{x})$ denotes the unit normal vector to the solid surface pointing inside the fluid domain, diffuse reflection (Cercignani, 1988) means that molecules are re-emitted from a given solid surface S_R according to the distribution function

$$f(\mathbf{x}, \boldsymbol{\xi}, t) = \frac{\rho_w(\mathbf{x}, t)}{(2\pi RT_w(\mathbf{x}, t))^{3/2}} \exp\left(-\frac{|\boldsymbol{\xi} - \mathbf{w}(\mathbf{x}, t)|^2}{2RT_w(\mathbf{x}, t)}\right) \qquad \text{for} \quad (\boldsymbol{\xi} - \mathbf{w}) \cdot \mathbf{n} > 0 \tag{20}$$

where T_w is the wall temperature and $\mathbf{w}$ is the wall velocity. The function ρ_w is proportional to the flux of incoming molecules:

$$\rho_w(\mathbf{x}, t) = \left(\frac{2\pi}{RT_w(\mathbf{x}, t)}\right)^{1/2} \int_{\mathbb{R}^3, (\boldsymbol{\xi}-\mathbf{w})\cdot\mathbf{n}<0} (|\boldsymbol{\xi} - \mathbf{w}(\mathbf{x}, t)| \cdot \mathbf{n}(\mathbf{x}))\, f(\mathbf{x}, \boldsymbol{\xi}, t) \mathrm{d}\boldsymbol{\xi} \tag{21}$$

and is defined in such a way that the global net flux of incoming and outgoing molecules vanishes according to eqn. (20).

Moreover it will be assumed that the flow is nearly isothermal, with far field temperature T_0 and wall temperature $T_w = T_0$.

If the analysis domain extends to infinity in some direction, the distribution function for incoming molecules is $f = f_0$, where f_0 is the far-field equilibrium Maxwellian defined by temperature T_0, density ρ_0 and zero average velocity:

$$f(\mathbf{x}, \boldsymbol{\xi}, t) = f_0(\boldsymbol{\xi}) = \frac{\rho_0}{(2\pi RT_0)^{3/2}} \exp\left(-\frac{|\boldsymbol{\xi}|^2}{2RT_0}\right) \tag{22}$$

3.1 Boundary Integral Equation Approach

Since intermolecular collisions are neglected, the total (material) derivative of f along the molecular trajectories vanishes:

$$\frac{\partial f}{\partial t} + \boldsymbol{\xi} \cdot \nabla f = 0 \tag{23}$$

The condition $c_L \ll 1$ discussed in the previous section implies that the deviation from the equilibrium state of the gas is small and f can be expressed as

$$f(\mathbf{x}, \boldsymbol{\xi}, t) = f_0(\boldsymbol{\xi})(1 + f_1(\mathbf{x}, \boldsymbol{\xi}, t)),$$

where $f_0 f_1$ is a perturbation of the equilibrium distribution f_0 eqn. (22) and all the governing equations can be linearized with respect to f_1. Boundary conditions simply to:

$$f_1(\mathbf{x}, \boldsymbol{\xi}, t) = \rho_{w1}(\mathbf{x}, t) + 2\tilde{\boldsymbol{\xi}} \cdot \tilde{\mathbf{w}}(\mathbf{x}, t) \quad \text{for} \quad \xi_n \geq 0 \tag{24}$$

with $\tilde{\boldsymbol{\xi}} = \boldsymbol{\xi}_w/\sqrt{2RT}$ and

$$\rho_{w1}(\mathbf{x}, t) = \sqrt{\pi}\tilde{\mathbf{w}}(\mathbf{x}, t) \cdot \mathbf{n}(\mathbf{x}) - \frac{2}{\pi} \int_{\mathbb{R}^3, \boldsymbol{\xi}\cdot\mathbf{n}<0} (\tilde{\boldsymbol{\xi}} \cdot \mathbf{n}) \exp(-|\tilde{\boldsymbol{\xi}}|^2) f_1(\mathbf{x}, \boldsymbol{\xi}, t) \mathrm{d}\tilde{\boldsymbol{\xi}} \tag{25}$$

on solid surfaces. In eqn. (25) ρ_{w1} is the linear term in the expansion (Frangi et al., 2007) of ρ_w eqn. (21).

Finally, for the flow of molecules impinging on the structures and coming from infinity, $f_1(\mathbf{x}, \boldsymbol{\xi}, t) = 0$.

The stresses in the fluid next to the solid surfaces can be linearized as well yielding:

$$\begin{aligned}\boldsymbol{\sigma}(\mathbf{x}, t) &= \int_{R^3} f(\boldsymbol{\xi} - \mathbf{v}) \otimes (\boldsymbol{\xi} - \mathbf{v}) \mathrm{d}\boldsymbol{\xi} \simeq \int_{R^3} f_0 \boldsymbol{\xi} \otimes \boldsymbol{\xi} \mathrm{d}\boldsymbol{\xi} + \int_{R^3} f_0 f_1 \boldsymbol{\xi} \otimes \boldsymbol{\xi} \mathrm{d}\boldsymbol{\xi} \\ &= \rho_0 R T_0 \left[\mathbb{1} + \frac{2}{\pi^{3/2}} \int_{R^3} \exp\left(-|\tilde{\boldsymbol{\xi}}|^2\right) \tilde{\boldsymbol{\xi}} \otimes \tilde{\boldsymbol{\xi}} f_1 \mathrm{d}\tilde{\boldsymbol{\xi}} \right]\end{aligned} \tag{26}$$

Typically, the output of interest from the analysis is the global viscous force exerted on the shuttle surfaces, which can be obtained through integration over the surfaces of the tractions $\mathbf{p}$.

The quasi-static condition $c_Q \ll 1$ implies that the time derivative in eqn. (23) can dropped and the velocity of the shuttle can be simply enforced as a stationary boundary condition. As a consequence, for any $\mathbf{y}$ in the analysis domain, the probability $f_1(\mathbf{y}, \boldsymbol{\xi}, t)$ remains always constant along the line $\mathbf{y} + \lambda\boldsymbol{\xi}$, with $\lambda > 0$ (i.e. along the trajectories of molecules emitted from $\mathbf{y}$). Thus, for any bounded and simply connected region of fluid Ω, for given $\mathbf{x} \in S$ and $\boldsymbol{\xi}$, there exist one and only one $\mathbf{y} \in S$ such that:

$$f_1(\mathbf{x}, \boldsymbol{\xi}, t) = f_1(\mathbf{y}, \boldsymbol{\xi}, t), \qquad \mathbf{x} = \mathbf{y} + \lambda\boldsymbol{\xi}, \quad \lambda > 0 \tag{27}$$

and $\mathbf{x}$ is visible from $\mathbf{y}$. Let us denote $\mathbf{r}(\mathbf{x}, \boldsymbol{\xi})$ the vector such that $\mathbf{y} = \mathbf{x} + \mathbf{r}$ and $\mathbf{x}$ and $\mathbf{y}$ are associated in the sense defined above. Hence we have:

$$\boldsymbol{\xi} = -\xi \frac{\mathbf{r}}{r} \tag{28}$$

where r is the length of $\mathbf{r}$ and ξ is the velocity module.

An idea frequently exploited in similar applications (Cercignani, 1988) is to transform the integration over the molecular velocity space into an integration over the surface S^+ visible from $\mathbf{x}$. Indeed, $\mathrm{d}\boldsymbol{\xi} = \xi^2 \mathrm{d}\xi \mathrm{d}S_{\mathbb{1}}$, where ξ is the velocity module and $S_{\mathbb{1}}$ is the surface of the unit sphere, and

$$\mathrm{d}S_{\mathbb{1}} = -\frac{1}{r^3} \mathbf{r}(\mathbf{x}, \boldsymbol{\xi}) \cdot \mathbf{n}(\mathbf{y}) \mathrm{d}S(\mathbf{y})$$

Account taken of eqn. (1) and of the linearity of the governing equations, the unknown function $\rho_{w1}(\mathbf{x}, t)$ admits the general form $\rho_{w1}(\mathbf{x}, t) = J(\mathbf{x})\dot{q}(t)$ and the final governing equation for $J(\mathbf{x})$ writes:

$$J(\mathbf{x}) = \sqrt{\pi}\tilde{\mathbf{g}}(\mathbf{x}) \cdot \mathbf{n}(\mathbf{x}) - \frac{1}{\pi}\int_{S^+} J(\mathbf{y})\big(\mathbf{r} \cdot \mathbf{n}(\mathbf{x})\big)\big(\mathbf{r} \cdot \mathbf{n}(\mathbf{y})\big)\frac{1}{r^4}\mathrm{d}S$$

$$+ \frac{3}{2}\frac{1}{\sqrt{\pi}}\int_{S^+} \big(\mathbf{r} \cdot \tilde{\mathbf{g}}(\mathbf{y})\big)\big(\mathbf{r} \cdot \mathbf{n}(\mathbf{x})\big)\big(\mathbf{r} \cdot \mathbf{n}(\mathbf{y})\big)\frac{1}{r^5}\mathrm{d}S \qquad (29)$$

where we have used the identities:

$$\int_{R^+} x^3 \exp(-x^2)\mathrm{d}x = \frac{1}{2} \qquad \int_{R^+} x^4 \exp(-x^2)\mathrm{d}x = \frac{3\sqrt{\pi}}{8}$$

It is worth stressing that eqn. (29) is the extension to moving boundaries of a classical integral identity Cercignani (1988) in the kinetic theory of rarefied gases.

The numerical implementation of the scalar equation eqn. (29) is straightforward and advantage is taken of the techniques developed in the last decade for the evaluation of boundary integral equations with high order singularities. In the implementation employed in this work the surface is discretized with three-node triangles and J is assumed to be constant over each triangle. These choices permit to adopt simple and fast integration strategies which suit well into the iterative solver employed.

Testing the visibility condition, as required by the occurrence of S^+ in eqn. (29), could be potentially a major issue but this is eventually mitigated for MEMS thanks to the very regular layout of the structures to analyse.

A similar procedure applied to eqn. (26) leads to an integral identity for $\mathbf{t}(\mathbf{x})$ on the MEMS surfaces:

$$\begin{aligned}\frac{\pi^{3/2}}{\rho_0 2RT}\mathbf{t}(\mathbf{x}) = {} & \left(1 + \frac{1}{2}J(\mathbf{x})\right)\frac{\pi^{3/2}}{2}\mathbf{n} + \pi\tilde{g}_n\mathbf{n} + \frac{\pi}{2}\tilde{g}_t \\ & - \frac{3\sqrt{\pi}}{8}\int_{S^+} \mathbf{r}\Big(\mathbf{r} \cdot \mathbf{n}(\mathbf{x})\Big)\Big(\mathbf{r} \cdot \mathbf{n}(\mathbf{y})\Big)\frac{1}{r^5}J(\mathbf{y})\mathrm{d}S \\ & + 2\int_{S^+} \mathbf{r}\Big(\mathbf{r} \cdot \tilde{\mathbf{g}}(\mathbf{y})\Big)\Big(\mathbf{r} \cdot \mathbf{n}(\mathbf{x})\Big)\Big(\mathbf{r} \cdot \mathbf{n}(\mathbf{y})\Big)\frac{1}{r^6}\mathrm{d}S \qquad (30)\end{aligned}$$

where $\tilde{g}_t$ and $\tilde{g}_n$ denote the nondimensional components of $\mathbf{g}$ in the tangential and normal directions with respect to the solid surfaces. Moreover the following identity has been exploited:

$$\int_{R^+} x^5 \exp(-x^2)\mathrm{d}x = 1$$

4 Numerical Application

Let us consider the out-of-plane rotational resonator of Fig. 1 produced by STMicroelectronics. It is basically a plane structure (plane $\mathbf{e}_1 - \mathbf{e}_2$) with uniform out-of-plane (along $\mathbf{e}_3$) thickness $t = 15\,\mu\text{m}$. The gap with the substrate is $g = 1.8\,\mu\text{m}$. The highly perforated central mass is attached to the substrate via a set of four flexible beams connect to the solid circle in the middle and it (almost) rigidly vibrates around the in plane axis $\mathbf{e}_2$ so that the whole structure can be approximated with a one degree-of-freedom model in terms of the rotation angle ϑ, with $\mathbf{s}(\mathbf{x}, t) = x_1\theta(t)\mathbf{e}_3$. Hence in eqn. (1) $q(t) = \theta(t)$ and $\mathbf{g}(\mathbf{x}) = x_1\mathbf{e}_3$. The first resonating frequency is $f \simeq 4000\,\text{Hz}$.

As can be appreciated from the enlargement in Fig. 2, all the holes are squares of constant size and are aligned along circles so as to have almost constant spacing all over the MEMS. Dissipation occurring in the comb-finger capacitors can be reasonably neglected. According to the assumptions of Section 1, gas dissipation is proportional to $\dot{\vartheta}$ through $\mathbf{t}(\mathbf{x})$ and the governing equation is:

$$\rho t \ddot{\vartheta}(t) \int_S x_1^2 \mathrm{d}S + \dot{\vartheta}(t) \int_{\partial V} x t_3(\mathbf{x}) \mathrm{d}V + K\vartheta(t) = V(t) \tag{31}$$

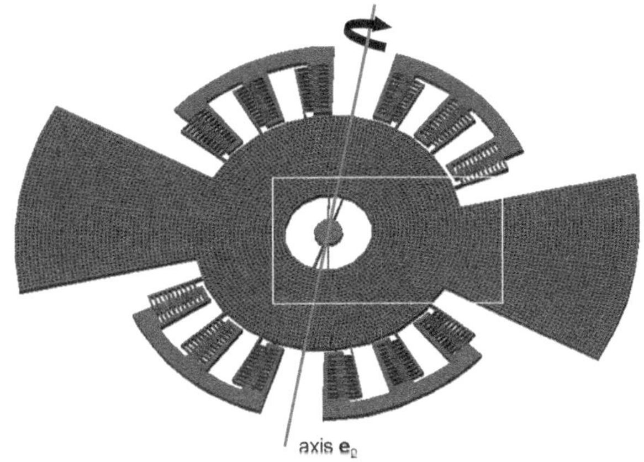

Fig. 1 Out-of-plane rotational resonator: global layout

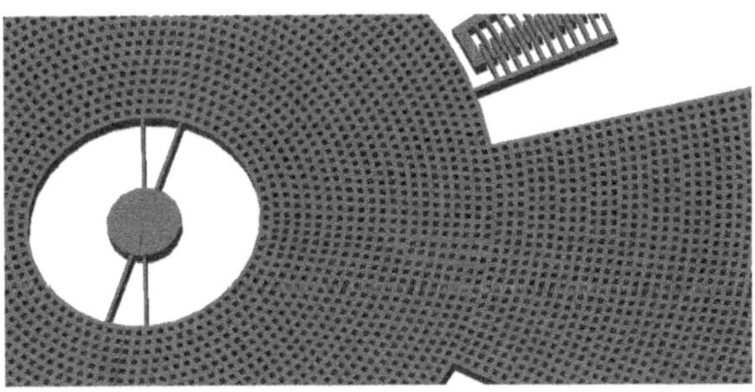

Fig. 2 Out-of-plane rotational resonator: detail of the perforated mass

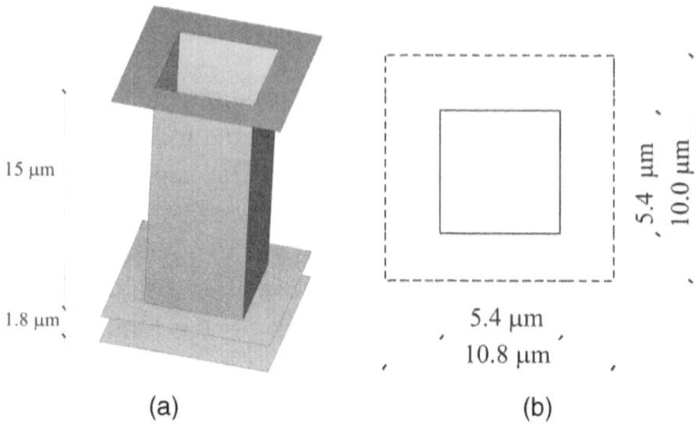

Fig. 3 Physical bounding surfaces of a fluid elementary cell: (**a**) 3D view; (**b**) in plane dimensions

where S is the in-plane distribution of masses in the MEMS, $V(t)$ accounts for the sinusoidal electrostatic forcing and K for the the elastic stiffness of the four central springs.

At this stage we introduce several realistic hypotheses in order to further simplify the problem at hand. The MEMS is assumed to be made by the assemblage of N identical units containing one hole each and having infinitesimal dimensions. Moreover the cells near the axis $\mathbf{e}_2$ contribute negligibly to the damping forces. Finally, if we denote by $\mathbf{X}_I$ the centre of the I-th cell, the distribution of tractions is expressed as the product of X_{I1} and of the same function $T_3(\mathbf{x} - \mathbf{X}_I)$ over any cell $t_3(\mathbf{x}) = X_{I1} T_3(\mathbf{x} - \mathbf{X}_I)$. It should be remarked that T_3 is the out-of-plane damping force exerted on the surface of the unit due to the enforced velocity $\mathbf{e}_3$. The integrals in eqn. (31) can now be simplified with good accuracy since:

$$\rho t \int_S x_1^2 \mathrm{d}S \simeq M\left(\sum_{I=1} X_{I1}^2\right) := Mb$$

$$\int_{\partial V} t_3(\mathbf{x})\, x_1 \mathrm{d}S \simeq \int_{\partial U} T_3(\mathbf{y}) \mathrm{d}S \left(\sum_{I=1} X_{I1}^2\right) := Cb$$

where M is the mass of the typical unit, ∂U the surface of the unit, and $b = \sum_{I=1} X_{I1}^2$.

Finally:

$$Mb\, \ddot{\vartheta}(t) + Cb\, \dot{\vartheta}(t) + K\, \vartheta(t) = T(t) \quad \text{or} \quad \ddot{\vartheta}(t) + 2\xi\, \omega_0\, \dot{\vartheta}(t) + \omega_0^2\, \vartheta(t) = \frac{T(t)}{Mb}$$

with:

$$\omega_0 = \sqrt{\frac{K}{Mb}} \qquad \frac{C}{M} = 2\xi\omega_0$$

The integral equation approach described in Section 3 has been employed to compute ξ in the free-molecule regime after implementing small modifications in order to account for symmetry conditions. The results obtained with a mesh of $\simeq$ 6000 elements are presented in Fig. 4 and compared with experimental tests conducted by ST Microelectronics on several different resonators showing surprisingly good agreement.

Even though no experimental data are available yet at higher pressures, a set of analyses have been performed also with the MVT approach of Section 2. MVT has been anyway validated in the past with different sets of experiments (Frangi et al., 2006; Frangi, 2005) always displaying exceptional accuracy. The results obtained in this case are represented by the crosses in Fig. 5.

One interesting feature of the combined analyses at high and low pressure is that the damping coefficient $\xi(p)$ can eventually be extrapolated to the whole

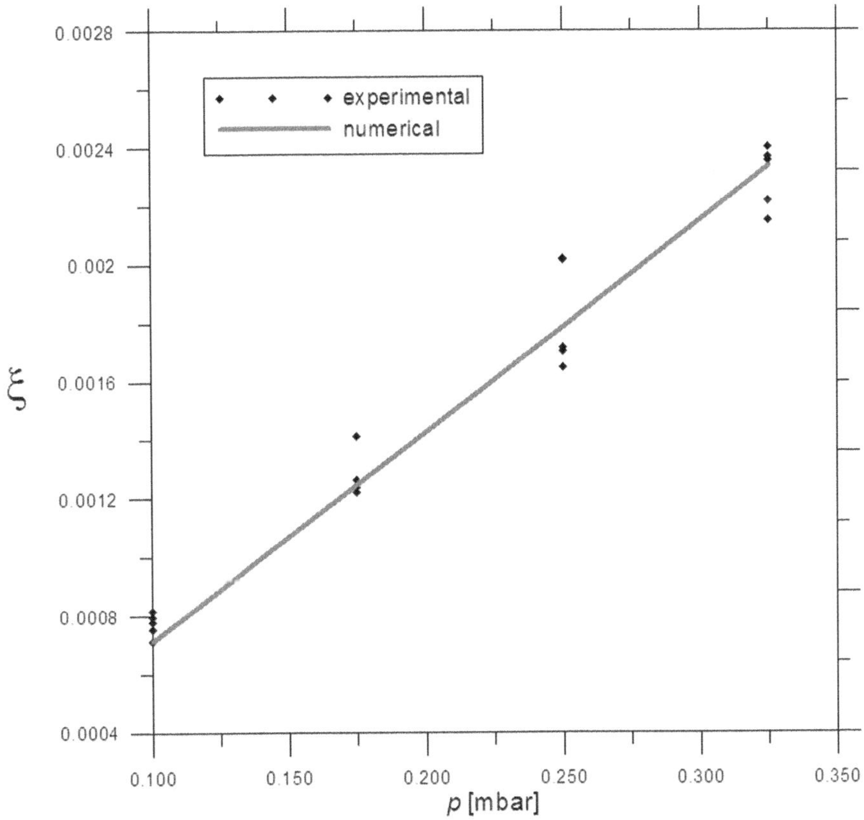

Fig. 4 Damping coeffecients: comparison of experiments and simulations in the free-molecule regime

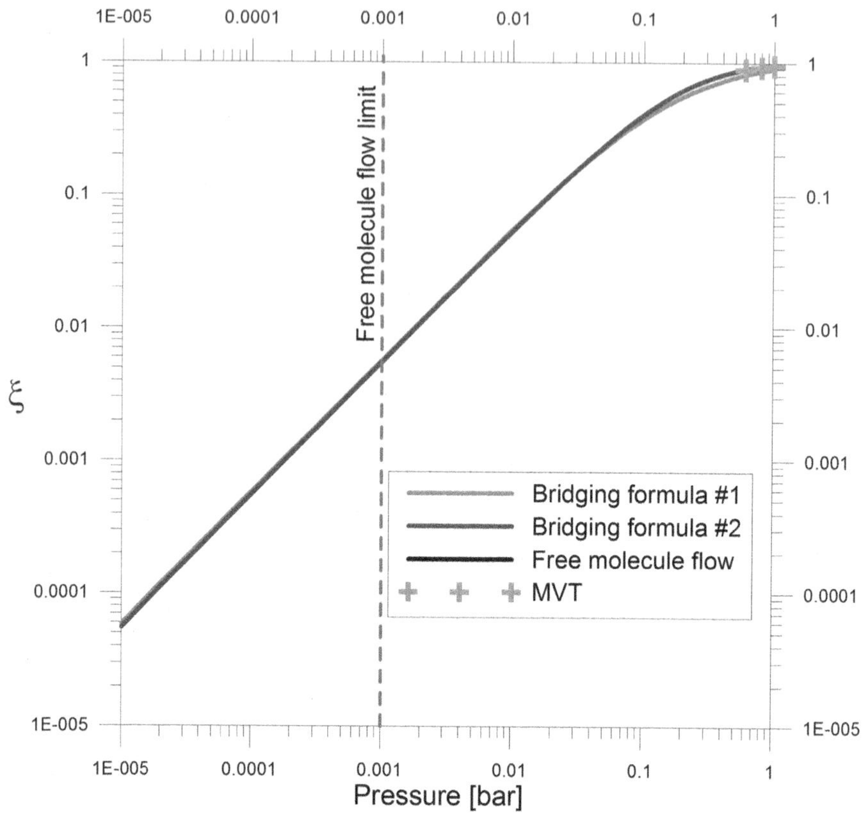

Fig. 5 Bridging curves connecting estimates of ξ coefficient at low and high pressures

pressure range by adopting bridging formulas which are very promising for applications. Indeed, let α denote the inverse of the slope of the $\xi(p)$ curve in the free molecule flow regime and let ξ_0 be the limit value with stick boundary condition (corresponding ideally to $p \to \infty$ and $Kn \to 0$).

Hence any bridging formula must satisfy at least the following two relations:

$$\lim_{p \to 0} \xi(p) = \alpha \qquad \lim_{p \to \infty} \xi(p) = \xi_0 \tag{32}$$

For the present problem we have $\alpha = 1.82578 \cdot 10^{-1}$ bar, $\xi_0 = 1.0644$.

Clearly several options are available and two of them are considered herein:

$$\xi^{\#1} = \xi_0 p \, \frac{1}{\alpha + p} \tag{33}$$

$$\xi^{\#2} = \xi_0 p \, \frac{1 + ap}{\alpha + bp + ap^2} \tag{34}$$

Formula (1) is simpler but necessarily induces larger errors even in the slip regime. Formula (2) on the contrary, requires an identification procedure to estimate the best possible parameters a and b. This can be easily achieved by performing a limited number of analyses with MVT (typically two or three) near $p \sim 1$ bar where slip boundary conditions apply, as reported in Figure 5 where $a = 2.17860\ \text{bar}^{-1}$, $b = 0.96621$. A validation of these formulas with kinetic approaches suitable for the transition regime is currently underway.

5 Conclusions

We have focused here on a specific but rather diffused class of MEMS inertial sensors like accelerometers, gyroscopes and other resonators which consist of a fixed stator and a suspended shuttle which vibrates at a certain frequency. We have shown that, for two different operating conditions, namely standard air pressure and near-vacuum, fluid damping can be conveniently addressed by means of a linear Boundary Integral Equations (BIEs). BIEs an ideal and robust tool since they reduce the dimensionality of the domain to be analysed allowing for the analysis of full-scale 3D structures and since they are expressed in terms of surface unknowns yielding directly and accurately the viscous force of interest. In the latter case a validation with experimental results on a real MEMS has been performed showing excellent agreement.

Acknowledgments Financial support from MIUR and Fondazione Cariplo is gratefully acknowledged.

References

Aoki K., Takata S., Nakanishi T. : Poiseuille-type flow of a rarefied gas between two parallel plates driven by a uniform external force, *Phys. Rev.*, 65, 026315-1–22 (2002)

Bao M., Yang H., Yin H., Sun Y. : Energy transfer model for squeeze-film air damping in low vacuum, *J. Micromech. Microeng.*, 12, 341–346 (2002)

Barnes J. and Hut P., : A hierarchical $O(N \log N)$ force-calculation algorithm, *Nature*, 324, 446–449 (1986).

Bird G.A. : *Molecular Gas Dynamics and the Direct Simulation of Gas Flows*, Clarendon Press, (1994)

Bonnet M. : *Boundary Integral Equation Methods for Solids and Fluids*, Wiley, 1999

Borup D.T. and Gandhi O.P. : Calculation of high-resolution SAR distributions in biological bodies using the FFT algorithm and conjugate gradient method, *IEEE Trans. Microw. Theor. Tech.*, 33, 417–419 (1985)

Cai C., Boyd I., Fan J., Candler G. : Direct simulation methods for low-speed microchannel flows, *J. Thermophys. Heat Transfer*, 14, 368–380 (2000)

Cho Y., Kwak B., Pisano A.P., Howe R.T. : Slide film damping in laterally driven microstructures, *Sensors and Actuators A*, 40, 31–39 (1993)

Chapman S., Cowling T.G. The Mathematical Theory of Non-Uniform Gases, Cambridge University Press, Cambridge (1960)
Cercignani C. The Boltzmann Equation and Its Applications, Springer, New York 1988
Ding J., Ye W. : A fast integral approach for drag force calculations to oscillatory slip Stokes flows., *Int. J. Numer. Methods Eng.*, 60, 1535–1567 (2004)
Emerson D.R., Gu X.J., Stefanov S.K, Sun Y.H., Barber R.W. : Nonplanar oscillatory shear flow: from the continuum to the free-molecular regime, *Phys. Fluids*, 19, 107–105 (2007)
Epton M.A., Dembart B. : Multipole translation theory for the three-dimensional Laplace and Helmholtz equations, *SIAM J. Sci. Comput.*, 16, 865–897 (1995)
Frangi A., Di Gioia A. : Multipole BEM for the evaluation of damping forces on MEMS, *Comput. Mech.*, 37, 24–31 (2005)
Frangi A, Tausch J. : A qualocation enhanced approach for the Dirichlet problem of exterior Stokes flow, *Eng. Anal. Bound. Elem.*, 29, 886–893 (2005)
Frangi A. : A fast multipole implementation of the qualocation mixed-velocity-traction approach for exterior Stokes flows, *Eng. Anal. Bound. Elem.*, 29, 1039–1046 (2005)
Frangi A, Spinola G., Vigna B. : On the evaluation of damping in MEMS in the slip-flow regime, *Int. J. Numer. Methods Eng.*, 68, 1031–1051 (2006)
Frangi A., Frezzotti A., and Lorenzani S. : On the application of the BGK kinetic model to the analysis of gas-structure interactions in MEMS, *Comput. Struct.*, 85, 810–817 (2007)
Fukui S., Kaneko R. : Analysis of ultra-thin gas film lubrication based on the linearized Boltzmann equation, *JSME Int. J.*, 30, 1660–1666 (1987)
Gad-el-Hak M. : The fluid mechanics of microdevices - the Freeman scholar lecture, *J. Fluids Eng.*, 121, 5–33 (1999)
Greengard L., Rokhlin V. : A new version of the fast multipole algorithm for the Laplace equation in three dimensions, *Acta Numerica*, 6, 229–269 (1997)
Hackbusch W. and Nowak Z.P. : On the fast matrix multiplication in the boundary element method by panel clustering, N*umer. Math.*, 54, 463–491 (1989)
Hutcherson S., Ye W. : On the squeeze-film damping of micro-resonators in the free-molecule regime, *J. Micromech. Microeng.*, 14, 1726–1733 (2004)
Kadar Z., Kindt W., Bossche A., Mollinger J. : Quality factor of torsional resonators in the low-pressure region, *Sens. Actuators A*, 53, 299–303 (1996)
Karniadakis G.E. and Beskok A. Micro Flows, Fundamentals and Simulation, Springer, New York (2002)
Liu Y.J., Shen L. : A dual BIE approach for large-scale modelling of 3-D electrostatic problems with the fast multipole boundary element method, *Int. J. Numer. Methods Eng.*, 71, 837–855 (2007)
Lu C.C. and Chew W.C. : Fast algorithms for solving hybrid integral equations, *IEEE Proc.-H*, 140, 455–460 (1993)
Mieussens L. : Discrete velocity models and numerical schemes for the Boltzmann-BGK equation in plane and axisymmetric geometries, *J. Comput. Phys.*, 162, 429–466 (2000)
Mukherjee S., Telekunta S., Mukherjee Y.X. : BEM modeling of damping forces on MEMS with thin plates, *Eng. Anal. Bound. Elem.*, 29, 1000–1007 (2005)
Nishimura N., Yoshida K., Kobayashi S. : Application of fast multipole Galerkin boundary integral equation method to elastostatic crack problems in 3D, *Int. J. Numer. Methods Eng.*, 50, 525–547 (2001)
Nishimura N. : Fast multipole accelerated boundary integral equation methods, *Appl. Mech. Rev.*, 55, 299–324 (2002)
Park J.H., Bahukudumbi P., Beskok A. : Rarefaction effects on shear driven oscillatory gas flows: A direct simulation Monte Carlo study in the entire Knudsen regime, *Phys. Fluids*, 16, 317–330 (2004)
Tausch J. and White J. : Second kind integral formulations of the capacitance problem, *Adv. Comput. Math.*, 9, 217–232 (1998)
Wang X., Judy M. and White J. : Validating fast simulation of air damping in micromachined devices, *Proc.* MEMS '02, (2002)

Wang X., Kanapka J., Ye W., Aluru N., White J. : Algorithms in FastStokes and its application to micromachined device simulation, *IEEE Trans. Comput. Aided Des. Integr. Circ. Syst.*, 25, 248–257 (2006)

Ye W., Wang X., Hemmert W., Freeman D., White J. : Air damping in lateral oscillating micro-resonators: a numerical and experimental study, *J. MEMS*, 12, 557–566 (2003)

Yuhong S., Barber R.W, Emerson D.R. : Inverted velocity profiles in rarefied cylindrical Couette gas flow and the impact of the accommodation coefficient, *Phys. Fluids*, 17, 047102–7 (2006)

A 2D Time-Domain BEM for Dynamic Crack Problems in Anisotropic Solids

F. García-Sánchez, Ch. Zhang, J. Sládek and V. Sládek

Abstract This chapter presents a time-domain boundary element method (BEM) for transient dynamic crack analysis in two-dimensional, homogeneous, anisotropic and linear elastic solids. Strongly singular displacement boundary integral equations (DBIEs) are applied on the external boundary of the cracked body, while hypersingular traction boundary integral equations (TBIEs) are used on the crack-faces. The quadrature formula of Lubich is used for approximating the convolution integrals and a collocation method is adopted for the spatial discretization of the time-domain boundary integral equations (BIEs). By means of a suitable change of variable an efficient regularization technique is applied to compute the strongly singular and hypersingular integrals arising in the time-domain BEM. Discontinuous quadratic quarter-point elements are implemented at the crack-tips to capture the local square-root behavior of the crack-opening-displacements (CODs) properly. Numerical examples for computing the dynamic stress intensity factors (SIFs) are shown and discussed to demonstrate the robustness, the accuracy and the efficiency of the present time-domain BEM.

1 Introduction

Boundary element method (BEM) or boundary integral equation method (BIEM) has been successfully applied to static and dynamic crack analysis since many years, at least for homogeneous, isotropic and linear elastic solids (e.g., Aliabadi, 1997, 2002; Balas et al., 1989; Beskos, 1997; Bonnet, 1999 and Domínguez, 1993). However, its extension to homogeneous, anisotropic and linear elastic solids is not straight-forward. The main reason lies in the fact that the corresponding elastodynamic fundamental solutions for generally anisotropic solids are much more complicated than that for isotropic solids. In contrast to isotropic case, they cannot

F. García-Sánchez (✉)
Department of Civil Engineering, of Materials and Manufacturing,
University of Málaga, 29013 Málaga, Spain
e-mail: fgsanchez@uma.es

G.D. Manolis, D. Polyzos (eds.), *Recent Advances in Boundary Element Methods*,
DOI 10.1007/978-1-4020-9710-2_9,

be given explicitly in closed and simple forms. In three-dimensional (3D) case, dynamic fundamental solutions can be expressed as surface-integrals over a unit-sphere, while in two-dimensional (2D) case they can be represented by line-integrals over a unit-circle. This fact may affect an easy and efficient numerical implementation of the BEM. Recently, frequency-domain BEM for 2D dynamic crack problems in anisotropic solids under harmonic loading has been presented by Dineva et al. (2005) and García-Sánchez et al. (2004). Time-domain BEM for dynamic crack problems in anisotropic solids has been reported in Tan et al. (2005a, b), Wünsche et al. (2005), Zhang (2000, 2002a, b, 2005), Zhang and Savaidis (2003). In these works either Laplace-domain or time-domain fundamental solutions have been implemented.

To avoid the use of the complicated dynamic fundamental solutions, Albuquerque et al. have developed a so-called dual reciprocity BEM, either in the time-domain (Albuquerque et al. 2002, 2004) or in the Laplace-domain (Albuquerque et al. 2003). A special feature of the dual reciprocity BEM is that it uses static instead of dynamic fundamental solutions. Fortunately, static fundamental solutions for anisotropic solids have closed-form expressions.

From the mathematical point of view one of the main characteristics in the numerical solution of crack problems is the degeneration of the boundary integral equations (BIEs) when they are applied to both crack-surfaces (3D) or crack-faces (2D) at the same position. Several possibilities have been suggested in the past years to overcome this difficulty. An overview on the topic can be found, e.g., in Zhang Ch (2002b). One possibility among others is the dual BEM approach, which uses the displacement BIEs (DBIEs) over one crack-face and the traction BIEs (TBIEs) over the other one. Another way is the combined application of the TBIEs over one crack-face with the crack-opening-displacements (CODs) as unknown quantities, and the DBIEs over other boundaries. This approach is presented in this chapter.

A collocation method is adopted for the spatial discretization, and the convolution quadrature of Lubich (1988a, b) is used for the temporal discretization. This allows us to use Laplace-domain fundamental solutions instead of time-domain fundamental solutions. In the present time-domain BEM, time-dependent solutions are obtained directly without any inverse Laplace-transform. Difficulties in computing strongly singular and hypersingular integrals arising in the present time-domain BEM are overcome by using a special regularization technique based on a suitable variable change (García et al., 2004; García-Sánchez et al., 2006; Sladek and Sladek, 1998). Moreover, to ensure an accurate computation of the dynamic stress intensity factors (SIFs), discontinuous quarter-point elements are applied to capture the local square-root behavior of the CODs at the crack-tips. To show the accuracy and the efficiency of the present time-domain BEM, numerical examples for computing transient dynamic SIFs are presented and discussed.

A comparative study of three different dynamic BEM approaches for 2D dynamic crack analysis in infinite, homogeneous, anisotropic and linear elastic solids has been recently presented by García-Sánchez and Zhang (2007).

2 Problem Statement and BIEs

Let us consider a crack in a 2D, homogeneous, anisotropic and linear elastic solid with a bounded domain of arbitrary shape. In the absence of body forces, the cracked solid satisfies the followings equations of motion (Achenbach, 1980)

$$\sigma_{ij,j} = \rho \ddot{u}_i, \tag{1}$$

Hooke's law

$$\sigma_{ij} = C_{ijkl} u_{k,l}, \tag{2}$$

the initial conditions

$$u_i(\mathbf{x}, t) = \dot{u}_i(\mathbf{x}, t) = 0, \qquad \text{for} \qquad t = 0, \tag{3}$$

and the boundary conditions

$$u_i(\mathbf{x}, t) = u_i(\mathbf{x}, t), \qquad \mathbf{x} \in \Gamma_u. \tag{4}$$

$$p_i(\mathbf{x}, t) = \sigma_{ij}(\mathbf{x}, t) n_j(\mathbf{x}) = \bar{p}_i(\mathbf{x}, t), \qquad \mathbf{x} \in \Gamma_t, \tag{5}$$

$$p_i(\mathbf{x}, t) = \sigma_{ij}(\mathbf{x}, t) n_j(\mathbf{x}) = 0, \qquad \mathbf{x} \in \Gamma_c. \tag{6}$$

In Eqs. (1), (2), (3), (4), (5) and (6), $\mathbf{x} = (x_1, x_2)$ represents a generic field point, σ_{ij} and u_i denote the stress and the displacement components, C_{ijkl} is the fourth order elasticity tensor, $p_i(\mathbf{x}, t)$ is the traction vector, n_j is the outward unit normal vector, ρ is the mass density, $\bar{u}_i(\mathbf{x}, t)$ and $\bar{p}_i(\mathbf{x}, t)$ are the prescribed displacements and tractions on the external boundary parts $\Gamma_u$and Γ_p ,and $\Gamma_c = \Gamma_c^+ + \Gamma_c^-$ represents the upper and the lower crack-faces, respectively.

Throughout the analysis, a comma after a quantity represents partial derivatives with respect to spatial variables, and superscript dots stand for the time differentiations of the quantity. Unless otherwise stated, the conventional summation rule over double indices is implied, and the indices i and j take the values 1 and 2.

On the external boundary $\Gamma_E = \Gamma_u + \Gamma_t$ of the cracked solid, the following displacement boundary integral equations (DBIEs) are applied

$$c_{ij}(\xi) u_j(\xi, t) + \fint_{\Gamma_E} p_{ij}^*(\xi, \mathbf{x}, t) * u_j(\mathbf{x}, t) d\Gamma + \int_{\Gamma_c^+} p_{ij}^*(\xi, \mathbf{x}, t) * \Delta u_j(\mathbf{x}, t) d\Gamma$$
$$= \int_{\Gamma_E} u_{ij}^*(\xi, \mathbf{x}, t) * p_j(\mathbf{x}, t) d\Gamma, \quad \xi \in \Gamma_E, \tag{7}$$

where $\fint$ stands for the Cauchy-principal value integral, $\mathbf{x} = (x_1, x_2)$ and $\xi = (\xi_1, \xi_2)$ represent the field and the source points, u_{ij}^* and p_{ij}^* are, respectively, displacement and traction fundamental solutions, which are related by

$$p_{ij}^* = C_{jkln} \frac{\partial u_{il}^*}{\partial x_n} n_k(\mathbf{x}), \tag{8}$$

$\Delta u_j(\mathbf{x}, t)$ are the crack-opening-displacements (CODs) defined by

$$\Delta u_i(\mathbf{x}, t) = u_i(\mathbf{x} \in \Gamma_c^+, t) - u_i(\mathbf{x} \in \Gamma_c^-, t), \tag{9}$$

and an asterisk $*$ denotes Riemann convolution which is defined by

$$g(\mathbf{x}, t) * h(\mathbf{x}, t) = \int_0^t g(\mathbf{x}, t - \tau) h(\mathbf{x}, \tau) d\tau. \tag{10}$$

The term $c_{ij}(\xi)$ in Eq. (7) depends on the smoothness of the boundary. For smooth boundaries one has $c_{ij}(\xi) = 0.5\delta_{ij}$.

Substituting Eq. (7) into Hooke's law (2) we obtain the traction boundary integral equations (TBIEs) as

$$\begin{aligned} p_j(\xi, t) &+ \int_{\Gamma_E} s_{ij}^*(\xi, \mathbf{x}, t) * u_j(\mathbf{x}, t) d\Gamma + \fint_{\Gamma_c^+} s_{ij}^*(\xi, \mathbf{x}, t) * \Delta u_j(\mathbf{x}, t) d\Gamma \\ &= \int_{\Gamma_E} d_{ij}^*(\xi, \mathbf{x}, t) * p_j(\mathbf{x}, t) d\Gamma, \qquad \xi \in \Gamma_c^+, \end{aligned} \tag{11}$$

which are applied to one of the crack-faces. In Eq. (11), $\fint$ denotes the Hadamard-finite-part integral,

$$s_{ij}^* = C_{ikln} \frac{\partial p_{lj}^*}{\partial \xi_n} n_k(\xi), \qquad d_{ij}^* = C_{ikln} \frac{\partial u_{lj}^*}{\partial \xi_n} n_k(\xi), \tag{12}$$

and $n_r(\xi)$ represents the outward unit normal vector to the boundary at the collocation point.

3 Numerical Solution Procedure

3.1 Spatial Discretization

For the spatial discretization of the external boundaries and the crack-faces, different quadratic elements are adopted, which are summarized in Fig. 1.

In the standard continuous quadratic element both the geometry and the boundary quantities u_k and p_k are described by the same shape functions, i.e, $\phi_{1G} = \phi_1$, $\phi_{2G} = \phi_2$ and $\phi_{3G} = \phi_3$ (see Fig. 2)

$$\phi_{1G} = \phi_1 = \frac{1}{2}\zeta(\zeta - 1), \phi_{2G} = \phi_2 = 1 - \zeta^2, \phi_{3G} = \phi_3 = \frac{1}{2}\zeta(\zeta + 1), \tag{13}$$

where $\zeta \in [-1, +1]$ is the local coordinate of the element.

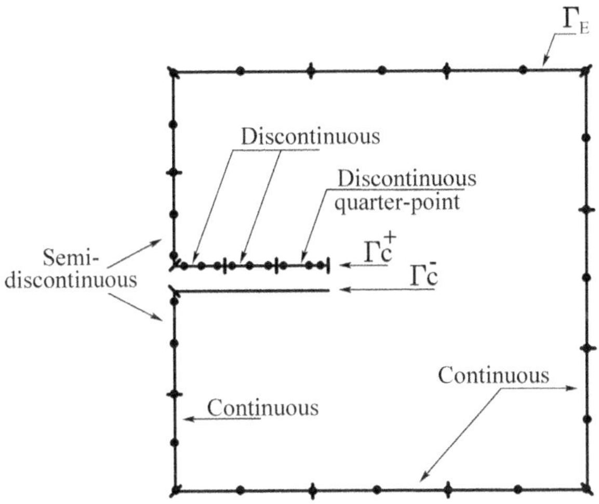

Fig. 1 Type of elements used in the BEM

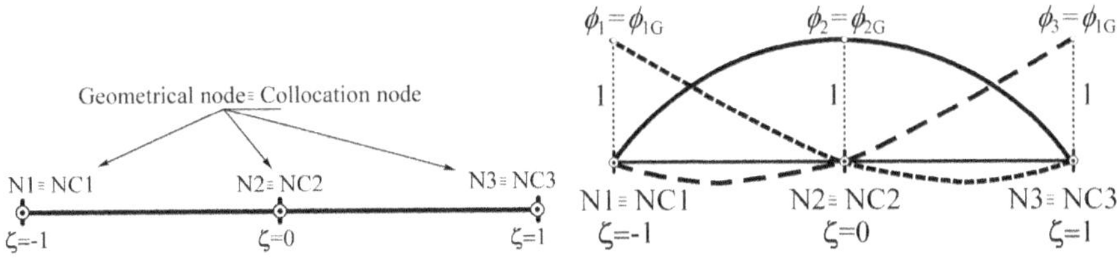

Fig. 2 Standard continuous quadratic element and its shape functions

In the discontinuous quadratic element, the collocation nodes are not at the ends of the element but inside the element, thus they do not coincide with the geometrical nodes (see Fig. 3).

In this case, the geometry of the elements is represented by same shape functions ϕ_{1G}, ϕ_{2G} and ϕ_{3G} of the standard continuous quadratic element, while the boundary quantities Δu_k are represented in terms of their values at the collocation nodes NC1, NC2 and NC3. The shape functions ϕ_1, ϕ_2 and ϕ_3 are given by

$$\phi_1 = \frac{(\zeta - \zeta_2)(\zeta - \zeta_3)}{(\zeta_1 - \zeta_2)(\zeta_1 - \zeta_3)}, \; \phi_2 = \frac{(\zeta - \zeta_1)(\zeta - \zeta_3)}{(\zeta_2 - \zeta_1)(\zeta_2 - \zeta_3)}, \; \phi_3 = \frac{(\zeta - \zeta_1)(\zeta - \zeta_2)}{(\zeta_3 - \zeta_1)(\zeta_3 - \zeta_2)}. \quad (14)$$

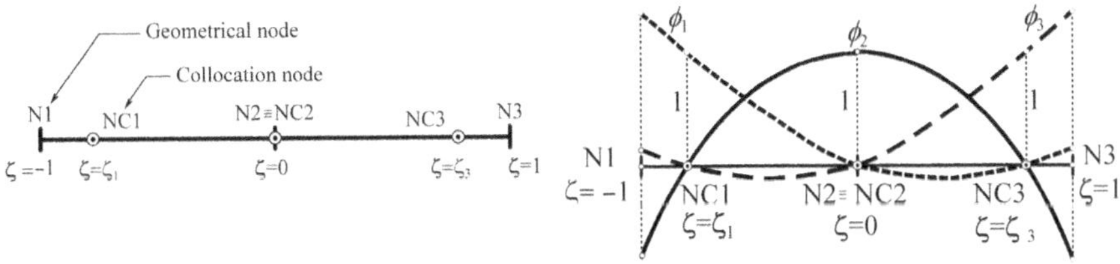

Fig. 3 Standard discontinuous quadratic element and its shape functions

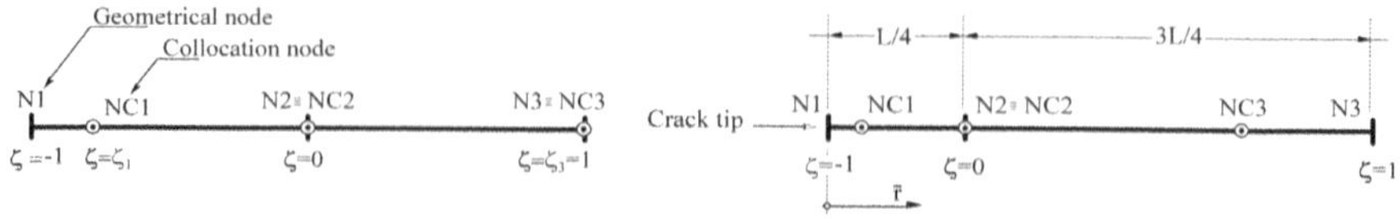

Fig. 4 Semi-discontinuous (*left*) and discontinuous quarter-point (*right*) elements

In the semi-discontinuous quadratic elements, the collocation node at only one end of the element is moved towards the mid-node (see Fig. 4). As in the discontinuous quadratic elements, the geometry of the elements is represented by the shape functions ϕ_{1G}, ϕ_{2G} and ϕ_{3G}, while the boundary quantities u_k and p_k are represented by the shape functions ϕ_1, ϕ_2 and ϕ_3.

In the discontinuous quarter-point element, the same shape functions are used as in the discontinuous quadratic element, i.e., Eqs. (13) and (14). In this element the mid-node is not at the central position but shifted to a quarter-length position of the element towards the crack-tip (see Fig. 4). Due to this location one has for the natural coordinate

$$\zeta = 2\sqrt{\frac{\bar{r}}{L}} - 1, \tag{15}$$

with $\bar{r}$ being the distance from the crack-tip and L the element-length.

In the present time-domain BEM, the collocation nodes NC1, NC2 and NC3 are located at

$$\begin{aligned}
\zeta_1 &= -3/4 \rightarrow \bar{r}_1 = L/64 \quad \text{for NC1,}\\
\zeta_2 &= 0 \quad\;\;\; \rightarrow \bar{r}_2 = L/4 \quad\;\; \text{for NC2,}\\
\zeta_3 &= +3/4 \rightarrow \bar{r}_3 = 49L/64 \text{ for NC3.}
\end{aligned} \tag{16}$$

As mentioned before, the use of the discontinuous quadratic quarter-point element ensures a proper description of the local $\sqrt{r}$-behavior of CODs at the crack-tip.

3.2 Time Discretization and Time-Stepping Scheme

In the present time-domain BEM, the Riemann convolution integral

$$f(t) = g(t) * h(t) = \int_0^t g(t-\tau)h(\tau)d\tau \tag{17}$$

is approximated by the convolution quadrature formula of Lubich (1988a, b), which is given by

$$f(n\,\Delta t) = \sum_{j=0}^{n} \omega_{n-j}(\Delta t)h(j\,\Delta t), \quad n = 0, 1, 2, \ldots, N, \tag{18}$$

where the time-interval t is divided into N equal time-steps Δt, and the weights $\omega_n(\Delta t)$ are defined by

$$\omega_n(\Delta t) = \frac{r^{-n}}{N} \sum_{m=0}^{N-1} \bar{g}\left(\frac{\delta(\zeta_m)}{\Delta t}\right) e^{-2\pi inm/N}. \tag{19}$$

In Eq. (19), $\bar{g}(\cdot)$ is the Laplace-transform of $g(t)$, $\delta(\zeta_m) = \sum_{j=1}^{2}(1-\zeta_m)^j/2$, $\zeta_m = re^{2\pi im/N}$, and $r = \varepsilon^{1/2N}$, with ε being the numerical error in computing the Laplace-transform $\bar{g}(\cdot)$. Equations (18) and (19) imply that the present formulation needs Laplace-domain instead time-domain fundamental solutions unlike the conventional time-domain BEM, e.g., Tan et al. (2005a, b), Wünsche et al. (2005) and Zhang (2002b).

After spatial and temporal discretizations of the DBIEs and the TBIEs, a system of linear algebraic equations for the discrete boundary quantities can be obtained as

$$\sum_{j=0}^{n} \boldsymbol{H}^{n-j} \cdot \boldsymbol{u}^j = \sum_{j=0}^{n} \boldsymbol{G}^{n-j} \cdot \boldsymbol{p}^j, \tag{20}$$

where $\boldsymbol{G}^{n-j}$ and $\boldsymbol{H}^{n-j}$ are the time-domain system matrices, $\boldsymbol{u}^j$ is the vector containing the discrete boundary displacements and the CODs, and $\boldsymbol{p}^j$ is the vector containing the discrete boundary tractions. According to Eq. (19), the system matrix at the $(n-1)$-th time step can be obtained by

$$\boldsymbol{G}^{n-j} = \frac{r^{-(n-j)}}{N} \sum_{m=0}^{N-1} \bar{\boldsymbol{G}}(s_m)\, e^{-2\pi i(n-j)m/N}, \tag{21}$$

$$\boldsymbol{H}^{n-j} = \frac{r^{-(n-j)}}{N} \sum_{m=0}^{N-1} \bar{\boldsymbol{H}}(s_m)\, e^{-2\pi i(n-j)m/N}, \tag{22}$$

where$s_m = \delta(\zeta_m)/\Delta t$, and the Laplace-domain system matrices are given by

$$\bar{\boldsymbol{G}}(s_m) = \begin{cases} \sum_{e=1}^{E} \int_{\Gamma_e} \bar{u}_{ij}^*(\xi, \mathbf{x}, s_m)\, \phi_q(\mathbf{x})\, d\Gamma, & \text{for DBIEs,} \\ \sum_{e=1}^{E} \int_{\Gamma_e} \bar{d}_{ij}^*(\xi, \mathbf{x}, s_m)\, \phi_q(\mathbf{x})\, d\Gamma, & \text{for TBIEs,} \end{cases} \tag{23}$$

$$\bar{\boldsymbol{H}}(s_m) = \begin{cases} \sum_{e=1}^{E} \int_{\Gamma_e} \bar{p}_{ij}^*(\xi, \mathbf{x}, s_m)\, \phi_q(\mathbf{x})\, d\Gamma, & \text{for DBIEs,} \\ \sum_{e=1}^{E} \int_{\Gamma_e} \bar{s}_{ij}^*(\xi, \mathbf{x}, s_m)\, \phi_q(\mathbf{x})\, d\Gamma, & \text{for DBIEs.} \end{cases} \tag{24}$$

When the integration is along the boundary element which contains the collocation point, Eq. (23) involves weakly singular and strongly singular integrals of the type $1/r$, while Eq. (24) contains strongly singular and hypersingular integrals of the type $1/r^2$. A special regularization technique is applied for evaluating the strongly singular and the hyper-singular integrals, for more details see García F et al., (2004), García-Sánchez et al., (2006) and García-Sánchez and Zhang (2007). Weakly singular and regular integrals can be computed numerically by using standard Gaussian quadrature.

By considering the boundary conditions (4), (5) and (6), Eq. (20) can be rearranged as

$$\sum_{j=0}^{n} A^{n-j} \cdot x^j = y^j \,, \tag{25}$$

where $\boldsymbol{A}^{n-j}$ is the rearranged system matrix, $\boldsymbol{x}^j$ is the vector containing the unknown boundary quantities, and $\boldsymbol{y}^j$ is the vector containing the prescribed or known boundary quantities. By invoking the zero initial conditions (3), Eq. (25) leads to the following explicit time-stepping scheme (Zhang, 2000, 2002a, 2005)

$$\boldsymbol{x}^n = \left(\boldsymbol{A}^0\right)^{-1} \cdot \left(\boldsymbol{y}^n - \sum_{j=1}^{n-1} \boldsymbol{A}^{n-j} \cdot \boldsymbol{x}^j\right) \tag{26}$$

for computing the unknown CODs at the n-th time-step. In Eq. (26), $\left(\boldsymbol{A}^0\right)^{-1}$ is the inverse of the system matrix $\boldsymbol{A}^0$ at the time-step $n = 0$.

3.3 Computation of Dynamic SIFs

As mentioned above, discontinuous quadratic quarter-point elements are applied at the crack-tips to capture the local $\sqrt{r}$-behavior of the CODs around these zones. This allows us to compute the dynamic SIFs very efficiently and accurately. For this purpose, the following relationship between the dynamic SIFs and the CODs is applied

$$\begin{bmatrix} K_I \\ K_{II} \end{bmatrix} = \sqrt{\frac{\pi}{8d}} \cdot \frac{1}{\Delta} \begin{bmatrix} B_{11} & B_{12} \\ B_{21} & B_{22} \end{bmatrix} \begin{bmatrix} \Delta u_1 \\ \Delta u_2 \end{bmatrix}, \tag{27}$$

where

$$B_{ij} = \begin{bmatrix} \mathrm{Im}\left(\frac{q_1-q_2}{\mu_1-\mu_2}\right) & \mathrm{Im}\left(\frac{p_2-p_1}{\mu_1-\mu_2}\right) \\ \mathrm{Im}\left(\frac{\mu_1 q_2-\mu_2 q_1}{\mu_1-\mu_2}\right) & \mathrm{Im}\left(\frac{\mu_2 p_1-\mu_1 p_2}{\mu_1-\mu_2}\right) \end{bmatrix}, \tag{28}$$

$$\Delta = B_{11}B_{22} - B_{12}B_{21} \,. \tag{29}$$

In Eqs. (27), (28) and (29), Δu_1 and Δu_2 are the CODs at the collocation point closest to the crack-tip, d is the distance between this collocation point and the crack-tip ($d = L/64$ in this work, see Fig. 4 and Eq. (16) for details), K_I and K_{II} are the mode-I and the mode-II SIFs, $\mu_1$and$\mu_2$are the complex roots of the characteristic equation of the material with positive imaginary part and

$$p_m = b_{11}\mu_m^2 + b_{12} - b_{16}\mu_m, \; q_m = (b_{12}\mu_m^2 + b_{22} - b_{26}\mu_m)/\mu_m \,. \tag{30}$$

In Eq. (30), b_{ij} $(i, j = 1, 2, 6)$ are the materials compliance coefficients.

4 Numerical Examples

The accuracy, the robustness and the stability of the present time-domain BEM are analyzed by using several examples in this section, where different crack configurations, material anisotropy, spatial and temporal discretizations have been considered. For convenience, normalized SIFs are introduced as

$$\overline{K}_I(t) = K_I(t)/\sigma_0\sqrt{\pi a}, \qquad \overline{K}_{II}(t) = K_{II}(t)/\sigma_0\sqrt{\pi a}, \tag{31}$$

where σ_0 is the stress amplitude of the loading, and σ_0 is the half-length of the crack in the case of a central crack and the crack-length in the case of an edge-crack.

4.1 Central Crack Problem

In the first example, we consider a rectangular plate with a central crack in plane strain condition as depicted in Fig. 5. The size of the cracked plate is defined by $h = 20$ mm and $2a = 4.8$ mm. The plate is discretized into 34 elements: 24 equal

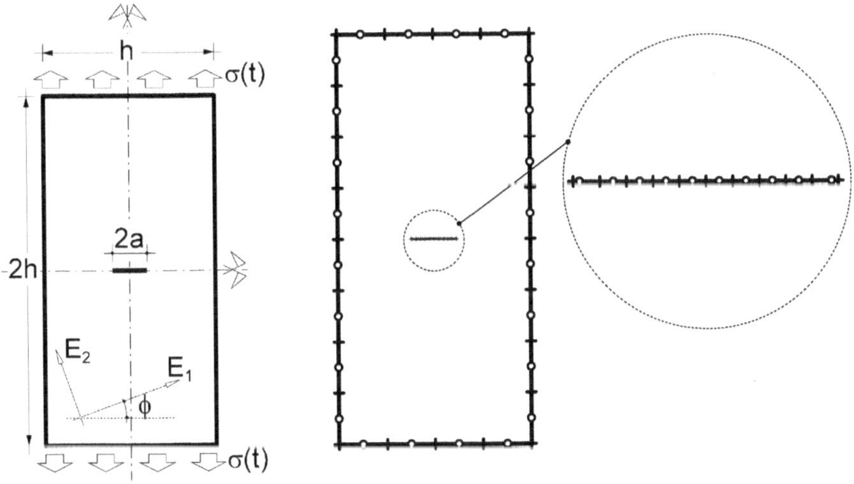

Fig. 5 Configuration and mesh for a central crack in an anisotropic plate

elements for the external boundary and 10 equal elements for the crack. A tensile impact loading of the form $\sigma(t) = \sigma_0 \cdot H(t)$ is applied to the top and the bottom of the plate, where σ_0 is the loading amplitude and $H(t)$ is the Heaviside-function. The following orthotropic material constants are used in the numerical calculations

$$E_1 = 118.30GPa, \quad E_2 = 54.80GPa, \quad G_{12} = 8.79GPa,$$
$$\nu_{12} = 0.083, \quad \rho = 1900kg/m^3.$$

A time-step $c_L\Delta t = h/50$ is selected and the normalized mode-I dynamic SIF is plotted in Fig. 6, where c_L is the wave velocity along the material-axis with E_2. In Fig. 6, numerical results of Albuquerque et al. (2004) and Wünsche et al. (2005) obtained by different BEM approaches and the FEM results obtained via ANSYS are also included for comparison purposes.

It can be easily observed that our numerical results agree very well with that of Albuquerque et al. (2004) and Wünsche et al. (2005) except the slight discrepancies around the peaks. These discrepancies do not exist between our numerical results and the FEM results.

Next, fully anisotropic material properties in plane stress are investigated by using four different inclination angles ϕ between the principal material axis, 1, and the crack-line, see Fig. 5 for details. The same material constants as in the previous orthotropic case are used. Through the transformation of the elasticity tensor, the inclination angle ϕ determines the degree of the material anisotropy. Also here, a time-step $c_Lt = h/50$ is chosen. The normalized mode-I and mode-II dynamic SIFs

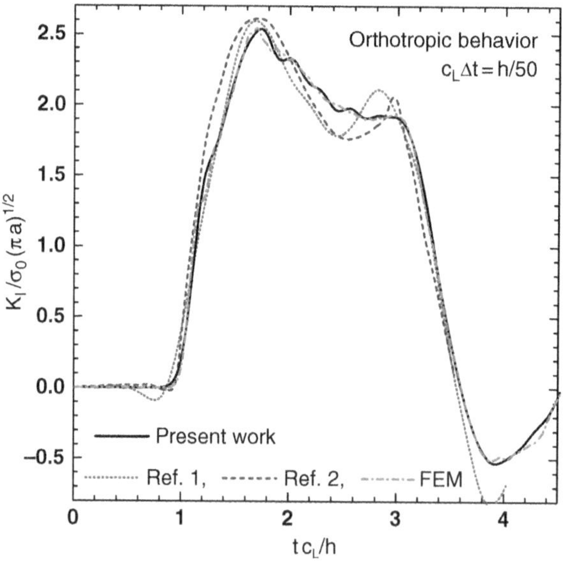

Fig. 6 Results for a central crack in an orthotropic plate (Albuquerque et al., 2004 and Wünsche et al. 2005)

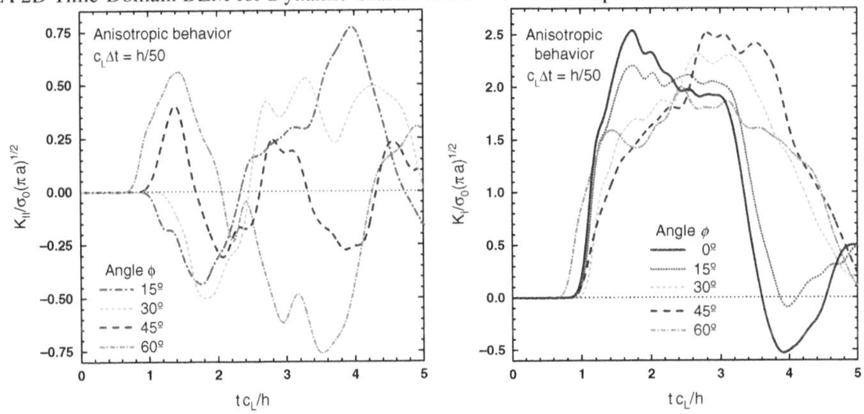

Fig. 7 Influence of the material anisotropy on the dynamic SIFs for a central crack in an anisotropic plate

versus the dimensionless time are presented in Fig. 7, which shows the influence of the material anisotropy on the normalized dynamic SIFs.

In order to make a sensitivity analysis of the present time-domain BEM to the used time-steps and spatial meshes, several values for the time-steps have been applied for a fixed BEM mesh, while BEM meshes have been used for a given time-step. The used meshes have 4β elements for the upper and the lower boundaries, 8β elements for the lateral ones, and 5α for the crack. In this way, we can refine the mesh by using the parameters α and β. In all investigated cases uniform meshes have been used. The values for α and β have been taken as 1, 2 and 3. The time-step is defined by $c_L\Delta t = h/\gamma$ and the values of 50, 60, 80 and 100 have been used for γ. Figures 8 and 9 show the influences of the time-steps and the BEM meshes on the accuracy and the stability of the present time-domain BEM for the case $\phi = 30°$. It can be concluded from Figs. 8 and 9 that the present time-domain BEM is quite insensitive to the used time-steps and spatial meshes.

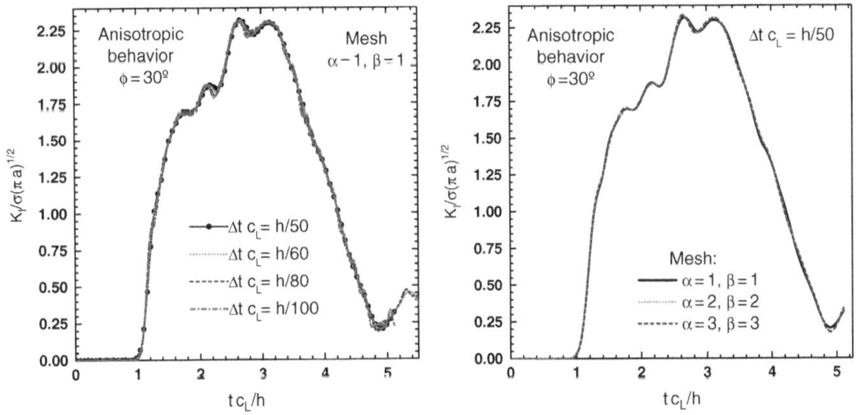

Fig. 8 Mode-I SIF for several time-steps and meshes for a central crack in an anisotropic plate

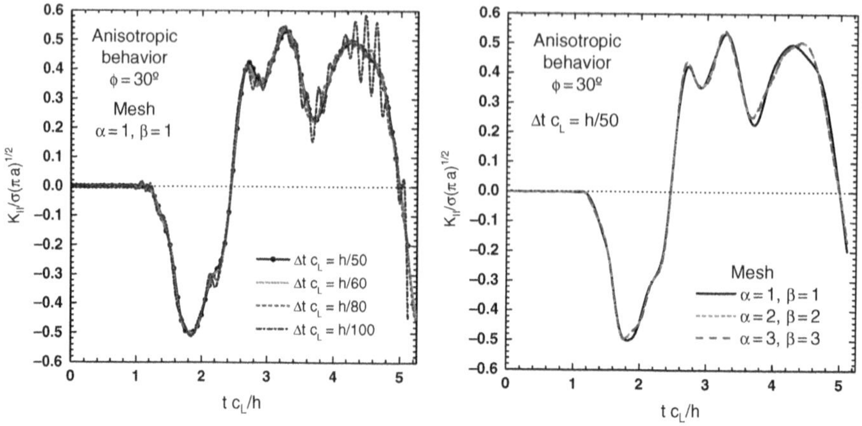

Fig. 9 Mode-II SIF for several time-steps and meshes for a central crack in an anisotropic plate

4.2 Edge-Crack Problem

An edge-crack of length a in a homogeneous, orthotropic and linear elastic plate as depicted in Fig. 10 is considered now. The crack is parallel to the upper and the bottom boundaries of the plate, and the geometry of the cracked plate is determined by $h = 20\,\text{mm}$, $b = 52\,\text{mm}$ and $a = 12\,\text{mm}$. The BEM mesh is also shown in Fig. 10, where 21 elements for the external boundary and 8 elements for the crack are used. The same orthotropic material properties as that used for the central crack are considered, and plane stress condition is assumed. Two different time-steps $c_L \Delta t = a/3$ and $c_L \Delta t = a/10$ have been selected.

Numerical results for the normalized mode-I dynamic SIF versus the dimensionless time are presented in Fig. 11, and they are compared with the FE results by using ANSYS.

The good agreement between our results and the FEM results verifies the accuracy of the present time-domain BEM. In addition, it can be observed that the

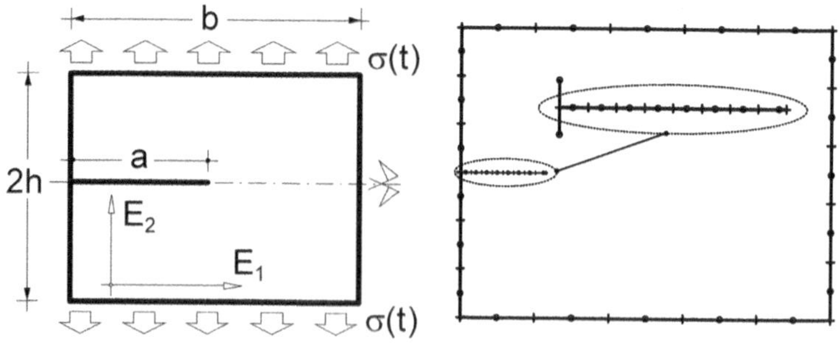

Fig. 10 Configuration and mesh for an edge-crack in an orthotropic plate

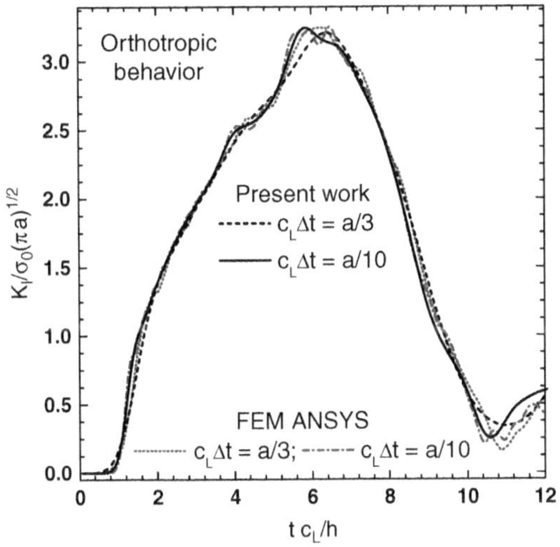

Fig. 11 Mode-I SIF for an edge-crack in an orthotropic plate

behavior of the BEM results is slightly better than that of the FEM results in view of the stability of the numerical results.

To check the stability of the present time-domain BEM for the edge-crack problem, different meshes and time-steps have been considered for an inclination angle of 30° between the principal axis of the material, 1, and the crack-line.

The number of the equal-length elements for the BE mesh have been defined as follows: 8β equal-length elements for the left boundary, 6β for the right boundary, 5β for each of the top and the bottom boundary, and 5α for the crack. The size of the elements is governed by the parameters α and β. The corresponding numerical results are presented in Figs. 12 and 13.

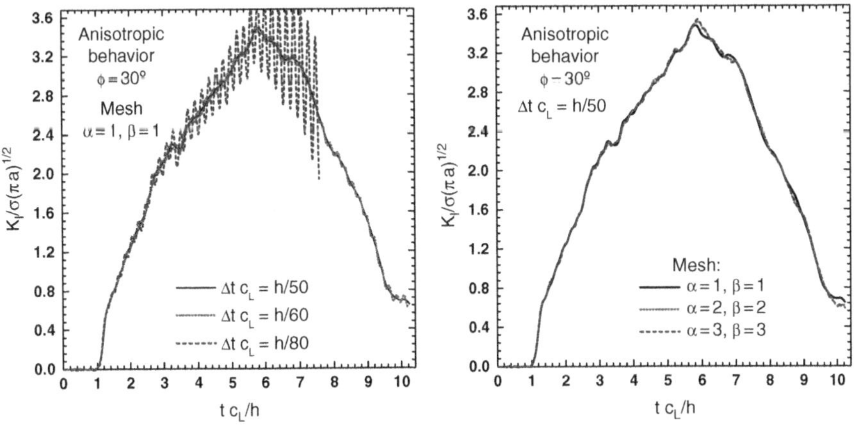

Fig. 12 Mode-I SIF of an edge-crack for different time-steps and meshes

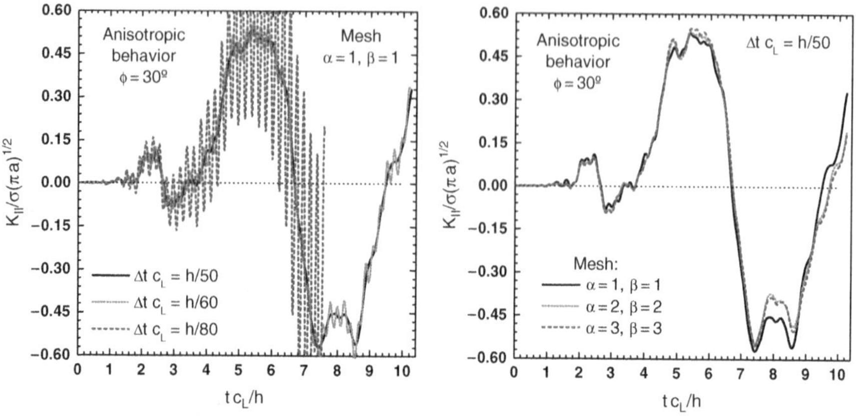

Fig. 13 Mode-II SIF of an edge-crack for different time-steps and meshes

By comparing Figs. 12 and 13 with Figs. 8 and 9 it can be seen that the numerical results for this problem are more sensitive and less stable with respect to the used time-steps than the central crack problem. However, similar to the central crack problem, the numerical results for the edge-crack problem as shown in Figs. 12 and 13 are still quite insensitive to the used mesh-sizes.

4.3 Inclined Edge-Crack Problem

In the last example, we consider a slanted edge-crack of length a in a homogeneous, orthotropic and linear elastic plate as shown in Fig. 14. The crack has an inclination angle of 45° with respect to the vertical plate boundary. The geometry of the cracked plate is defined by $h = 22$ mm, $w = 32$ mm, $c = 6$ mm and $a = 22.63$ mm. The BE mesh consists of 24 elements for the external boundary and 4 elements for the crack. A tensile impact loading of the form $\sigma = \sigma_0 \cdot H(t)$ is applied on the upper boundary of the plate while on all other boundaries the normal displacements are impeded. Plane stress condition is assumed.

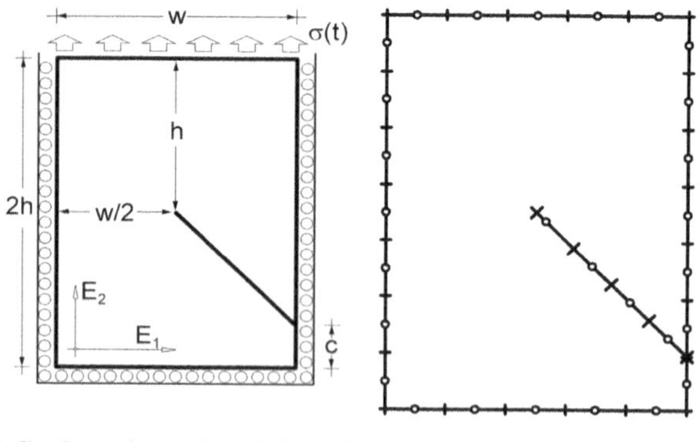

Fig. 14 Configuration and mesh for a slanted edge-crack in an orthotropic plate

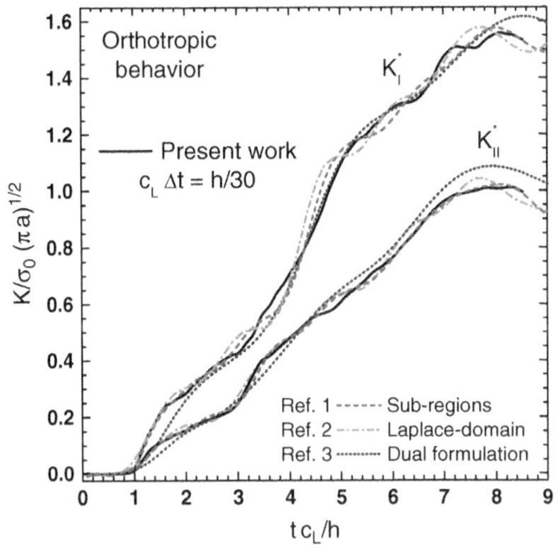

Fig. 15 Results for a slanted edge-crack in an orthotropic plate (Albuquerque et al., 2002, 2003, 2004)

Numerical results for this problem have been presented by Albuquerque et al. (2002, 2003, 2004) using different BEM formulations and the following orthotropic material constants

$$E_1 = 82.4GPa, \quad E_2 = 164.80GPa, \quad G_{12} = 29.4GPa,$$
$$\nu_{12} = 0.4006, \quad \rho = 2450kg/m^3.$$

The corresponding numerical results for the normalized mode-I and mode-II dynamic SIFs are presented in Fig. 15. Here, a time-step $c_L \Delta t = h/30$ is used. Here again, our numerical results agree very well with that of references (Albuquerque et al., 2002, 2003, 2004).

5 Conclusions

In this chapter, a time-domain BEM for transient dynamic crack analysis in 2D, homogeneous, generally anisotropic and linear elastic solids is presented. The quadrature formula of Lubich is applied for the temporal discretization, which requires Laplace-domain instead of time-domain fundamental solutions. A collocation method is adopted for the spatial discretization. At crack-tips, quarter-point elements are implemented for a proper description of the local behavior of the CODs at the crack-tips. Dynamic stress intensity factors are obtained directly from the numerically computed CODs.

The arising strongly singular and hypersingular integrals in the time-domain BEM are computed via a special regularization technique based on a suitable variable change. The time-domain BEM presented in this chapter is general and can be applied to both straight and curved cracks.

Numerical examples show that the present time-domain BEM is very accurate and robust, and it is valid even for isotropic material properties as a special case. In addition, the present time-domain BEM is less sensitive and more stable with respect to the used time-steps in comparison to the classical time-stepping scheme using time-domain fundamental solutions.

Acknowledgments This work is supported by the German Research Foundation (DFG), which is gratefully acknowledged.

References

Achenbach JD (1980) Wave Propagation in Elastic Solids. North Holland, Amsterdam

Albuquerque EL, Sollero P, Aliabadi MH (2002) The boundary element method applied to time dependent problems in anisotropic materials. Int J Solids Struct 39:1405–1422

Albuquerque EL, Sollero P, Fedelinski P (2003) Dual reciprocity boundary element method in Laplace domain applied to anisotropic dynamic problems. Computers & Structures 81:1703–1713

Albuquerque EL, Sollero P, Fedelinski P (2004) Dual boundary element method for anisotropic dynamic fracture mechanics. Int J Numer Meth Eng 59:1187–1205

Aliabadi MH (1997) Boundary element formulations in fracture mechanics. Appl Mech Rev 50:83–96

Aliabadi MH (2002) The Boundary Element Method, Vol. 2. John Wiley & Sons Ltd, New York

Balas J, Sladek J, Sladek V (1989) Stress Analysis by Boundary Element Methods. Elsevier, Amsterdam

Beskos DE (1997) Boundary element methods in dynamic analysis: Part II (1986–1996). Appl Mech Rev 50:149–197

Bonnet M (1999) Boundary Integral Equation Methods for Solids and Fluids. John Wiley & Sons Ltd, New York

Dineva P, Rangelov T, Gross D (2005) BIEM for 2D steady-state problems in cracked anisotropic materials. Eng Anal Bound Elem 29:689–698

Domínguez J (1993) Boundary Elements in Dynamics. Computational Mechanics Publications, Southampton

García F, Sáez A, Domínguez J (2004) Traction boundary elements for cracks in anisotropic solids. Eng Anal Bound Elem 28:667–676

García-Sánchez F, Sáez A, Domínguez J (2006) Two-dimensional time-harmonic BEM for cracked anisotropic solids. Eng Anal Bound Elem 30:88–99

García-Sánchez F, Zhang Ch (2007) A comparative study of three BEM for transient dynamic crack analysis of 2-D anisotropic solids. Comput Mech 40:753–769

Lubich C (1988a) Convolution quadrature and discretized operational calculus, Part I. Numer Math 52:129–145

Lubich C (1988b) Convolution quadrature and discretized operational calculus, Part II. Numer Math 52:413–425

Sladek V, Sladek J (1998) Singular Integrals in Boundary Element Methods. CMP, Southampton

Tan A, Hirose S, Zhang Ch (2005a) A time–domain collocation-Galerkin BEM for transient dynamic crack analysis in anisotropic solids. Eng Anal Bound Elem 29:1025–1038

Tan A, Hirose S, Zhang Ch, Wang C-Y (2005b) A time-domain BEM for transient wave scattering analysis by a crack in anisotropic solids. Eng Anal Bound Elem 29:610–623

Wünsche M, Zhang Ch, Sládek J, Sládek V, Hirose S (2005) 2-D dynamic crack analysis in anisotropic solids by a hypersingular time-domain BEM. In: Proceedings of the 10th International Conference on Numerical Methods in Continuum Mechanics & 4th Workshop on Trefftz Methods (CD-ROM), August 22–25, 2005. Zilina, Slovak Republic

Zhang Ch (2000) Transient elastodynamic antiplane crack analysis of anisotropic solids. Int J Solids Struct 37:6107–6130
Zhang Ch (2002a) A 2-D time–domain BIEM for dynamic analysis of cracked orthotropic solids. Comp Model Eng & Sci 3:381–398
Zhang Ch (2002b) A 2D hypersingular time–domain traction BEM for transient elastodynamic crack analysis. Wave Motion 35:17–40
Zhang Ch (2005) Transient dynamic crack analysis in anisotropic solids. In: Ivankovic A, Aliabadi MH (Ed.) Crack Dynamics, WIT Press, Southampton, Boston
Zhang Ch, Savaidis A (2003) 3-D transient dynamic crack analysis by a novel time-domain BEM. Comp Model Eng & Sci 4:603–618

Simulation of Elastic Scattering with a Coupled FMBE-FE Approach

Lothar Gaul, Dominik Brunner and Michael Junge

Abstract In this paper scattering problems with elastic obstacles that are hit by an incident acoustic wave are discussed. Underwater acoustics mainly differs from air acoustics in the fact that a strong coupling scheme between the structural part and the acoustic domain is necessary. Such a scheme is discussed, using the fast multilevel multipole boundary element method (FMBEM) to model the exterior acoustic fluid. The structural part is modeled with the finite element method (FEM). To obtain a high flexibility, an interface to a commercial FE package is established. For a high efficiency, an iterative solver with preconditioning is applied. The numerical results are compared with an analytical solution for a model problem.

1 Introduction

Fluid-structure interaction deals with the mutual influence of an acoustic and a structural domain. If air is assumed as acoustic fluid, the influence of the surrounding fluid on the vibration behavior of the structure can usually be neglected. In contrast, this is typically not the case if the fluid is water. Due to the high density, the feedback of the acoustic pressure onto the structure has to be taken into account (Fahy, 2007). As a consequence fully coupled simulation schemes have to be applied which are computationally more expensive, since a structural problem and an acoustic problem have to be solved simultaneously. In this paper, a scattering problem is discussed. Here, the structure is excited by an incident wave.

For geometrically simple structures, analytical solutions exist (Junger and Feit, 1986). Typically, the structure of engineering problems are more complex so that numerical schemes have to be used. The finite element method is usually applied for the structural part (Zienkiewicz and Taylor, 2000). For exterior acoustic fluid domains, the application of the boundary element method is advantageous, since

L. Gaul (✉)
Institute of Applied and Experimental Mechanics, Pfaffenwaldring 9,
70550 Stuttgart, Germany
e-mail: gaul@iam.uni-stuttgart.de

G.D. Manolis, D. Polyzos (eds.), *Recent Advances in Boundary Element Methods*,
DOI 10.1007/978-1-4020-9710-2_10,

the Sommerfeld radiation condition is automatically fulfilled (Wu, 2000; Estorff, 2000 and Gaul et al., 2003). It can also handle scattering problems (Seybert et al., 1985; Wu, 2000 and Ochmann et al., 2003). There are fast methods to overcome the drawback of the fully populated matrices (Rokhlin, 1993; Greengard and Rokhlin, 1987; Gyure and Stalzer, 1998 and Rjasanow and Steinbach, 2007). For instance, the fast multipole method (FMM) helps to solve large scale problems in acoustics with some thousand degrees of freedom (Fischer, 2004; Schneider, 2003; Junge et al., 2005) and Shen and Liu, 2007. There are various formulations for the coupling of the FEM and BEM (Fischer, 2004; Moosrainer, 2000; Gaul and Fischer, 2006; Hughes and Chen, 2004; Everstine and Henderson, 1990 and Chen et al., 1998). Recently, the FMBEM was coupled with the FEM (Fischer, 2004; Brunner et al., 2007, 2009), allowing the simulation of large-scale coupled problems.

This paper starts with an introduction to the FEM and BEM. After this, the fast multipole method is applied. Then, a strong coupling scheme is presented and at the end a model problem with an incident plane wave is investigated.

2 FE Formulation for the Structural Domain

Due to it's simplicity, the finite element method is predominantly chosen for simulating linear elastodynamic systems. This section gives a brief derivation on the basis of the principle of virtual work. Throughout this paper, all field variables are assumed to be harmonic in time with the behavior $e^{-j\omega t}$. The imaginary unit is denoted by j and $\omega = 2\pi f$ is the angular frequency with the excitation frequency f. For the structural domain Ω_s (see Fig. 1), the elastodynamic equilibrium condition is written in the form

$$\text{div}\ \sigma + \omega^2 \rho_s \mathbf{u} = 0 \qquad \text{in} \quad \Omega_s, \tag{1}$$

where σ is the stress tensor, ρ_s is the density of the structure and $\mathbf{u}$ denotes the displacement. The surface stress vector $\mathbf{t}_p$ on the coupling interface Γ_i is given by

$$\mathcal{T}\mathbf{u} = \mathbf{t}_p \qquad \text{on} \quad \Gamma_i, \tag{2}$$

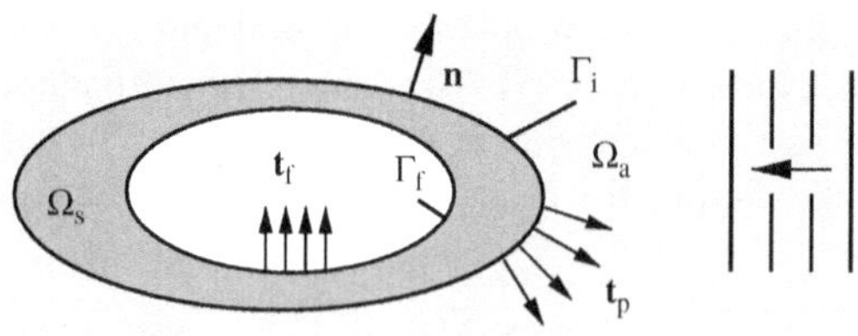

Fig. 1 Domains of the coupled problem. The exterior acoustic domain Ω_a is in contact with the structure Ω_s on the fluid-structure interface Γ_i. The surface stress vector is denoted by $\mathbf{t}_p$. Please note the incident plane wave on the right hand side

with the differential operator $\mathcal{T}$. In the following, we want to restrict to the case that the surface stress vector $\mathbf{t}_f$ on the Neumann boundary Γ_f vanishes. In case of a homogeneous and linear elastic structure, a weak formulation of (1) and (2) is (Zienkiewicz and Taylor, 2000)

$$\int_{\Omega_s} \delta\varepsilon : \sigma \, d\Omega = \omega^2 \int_{\Omega_s} \rho_s \delta\mathbf{u} \cdot \mathbf{u} \, d\Omega + \int_{\Gamma_i} \delta\mathbf{u} \cdot \mathbf{t}_p \, d\Gamma \,, \tag{3}$$

where $\delta\varepsilon$ is the virtual strain tensor. The surface stress vector $\mathbf{t}_p$ on the fluid-structure interface Γ_i is substituted by the acoustic pressure p and the normal $\mathbf{n}$

$$\mathbf{t}_p = -p\,\mathbf{n} \,. \tag{4}$$

Here, $\mathbf{n}$ is assumed to point out of the structural domain. The acoustic pressure p is thus coupled with the surface stress vector and influences the behavior of the structural domain.

In order to obtain an approximate solution the continuous system is discretized. The structural displacements $\mathbf{u}$, their corresponding virtual quantities as well as the acoustic pressure p are approximated with linear shape functions, $\mathbf{N}_u$ and $\mathbf{N}_p$, and nodal values

$$\mathbf{u} = \mathbf{N}_u \mathbf{u}^{(e)} \,, \quad \delta\mathbf{u} = \mathbf{N}_u \delta\mathbf{u}^{(e)}, \quad p = \mathbf{N}_p \mathbf{p}^{(e)} \,. \tag{5}$$

The index $(\,)^{(e)}$ stands for the concentrated nodal values of one element. Equation (3) must hold for an arbitrary non–zero virtual displacement $\delta\mathbf{u}^{(e)}$. In matrix notation, one obtains

$$\underbrace{\sum_e \int_{\Omega_s^{(e)}} (\mathcal{D}\mathbf{N}_u)^T \mathcal{E} \mathcal{D}\mathbf{N}_u \, d\Omega}_{\mathbf{K}_s} \; \mathbf{u} = \omega^2 \underbrace{\sum_e \int_{\Omega_s^{(e)}} \rho_s \mathbf{N}_u^T \mathbf{N}_u \, d\Omega}_{\mathbf{M}_s} \; \mathbf{u} - \mathbf{C}_{FE}\, \mathbf{p} \,, \tag{6}$$

with the differential operator matrix $\mathcal{D}$ and the elastic stiffness matrix $\mathcal{E}$. For isotropic materials its entries are characterized by Young's modulus E and Poisson's ratio ν. The sum over the structural elements e form the global stiffness matrix $\mathbf{K}_s$ and the global mass matrix $\mathbf{M}_s$ of the solid partition. The coupling matrix $\mathbf{C}_{FE}$ takes into account the fluid pressure and is given by

$$\mathbf{C}_{FE} = \sum_e \int_{\Gamma_i^{(e)}} \mathbf{N}_u^T \mathbf{n} \mathbf{N}_p \, d\Gamma \,. \tag{7}$$

Structural damping may be incorporated in the model for a more realistic system behavior. A simple approach is Rayleigh damping, where the damping matrix is proportional to the mass and stiffness matrix

$$\mathbf{D}_s = \alpha_R \, \mathbf{M}_s \, + \, \beta_R \, \mathbf{K}_s \,. \tag{8}$$

Finally, this leads to the FE formulation

$$\underbrace{\left(-\omega^2\mathbf{M}_\mathrm{s} - \mathrm{j}\omega\mathbf{D}_\mathrm{s} + \mathbf{K}_\mathrm{s}\right)}_{\mathbf{K}_\mathrm{FE}}\mathbf{u} + \mathbf{C}_\mathrm{FE}\mathbf{p} = 0\,. \tag{9}$$

In contrast to the boundary element method, which is discussed in the following section, the mass matrix and stiffness matrix of the FE formulation need only be computed once, since they are frequency independent. The damped dynamic stiffness matrix $\mathbf{K}_\mathrm{FE}$ is then assembled at low numerical cost for all considered frequencies.

To obtain a high flexibility, the commercial finite element package ANSYS is used to set up $\mathbf{K}_\mathrm{s}$ and $\mathbf{M}_\mathrm{s}$. This has the advantage, that complex structures as they are typical for engineering problems can easily be handled. Thin structures are often modeled by shell elements. These complex elements do not have to be implemented by one's own but the element library of ANSYS is used for this purpose. For the communication between ANSYS and the research code, a simple binary interface is established.

3 BE Formulation for the Fluid Domain

Below, the governing equations for the fluid are derived in the frequency domain for the exterior acoustic problem as depicted in Fig. 1. The normal on the boundary is always assumed to point outwards. On the interface Γ_i, the acoustic domain Ω_a is in contact with the vibrating structural domain Ω_s. The structure is excited by the acoustic field of an incident plane wave. First, the exterior acoustic problem as depicted in Fig. 2 (left) is investigated. After this, a generalization for scattering problems is presented.

Starting point of the fluid formulation is the linear time-harmonic Helmholtz equation (Kinsler et al., 1999)

$$\Delta p(\mathbf{x}) + \kappa^2\, p(\mathbf{x}) = 0 \qquad \mathbf{x} \in \Omega_\mathrm{a}\,, \tag{10}$$

where $\kappa = \frac{\omega}{c}$ is the circular wavenumber with the speed of sound of the acoustic fluid c. The acoustic pressure is denoted by p and $\mathbf{x}$ is an arbitrary point in the fluid. The Laplace operator is given by Δ.

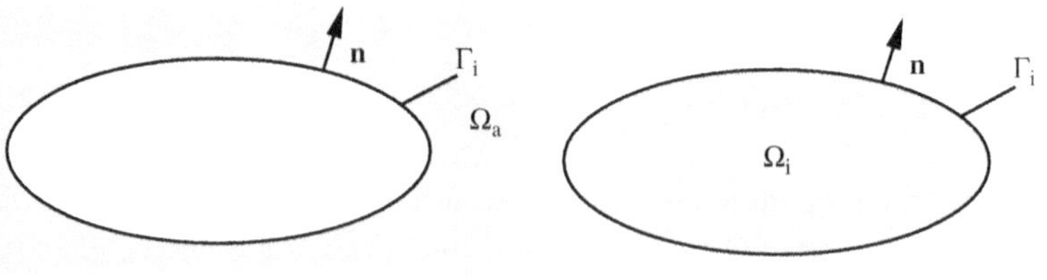

Fig. 2 Exterior acoustic domain Ω_a (*left*) and associated interior domain Ω_i (*right*). In both cases, the normal $\mathbf{n}$ is assumed to point outwards

A weak form of the Helmholtz equation is obtained by weighting with the fundamental solution

$$P^*(\mathbf{x},\mathbf{y}) = \frac{e^{j\kappa r}}{4\pi r}, \tag{11}$$

where $r = |\mathbf{x} - \mathbf{y}|$ denotes the distance between the field and the load point. Applying Green's second theorem yields the representation formula

$$p(\mathbf{x}) = -\int_{\Gamma_i} P^*(\mathbf{x},\mathbf{y}) \frac{\partial p(\mathbf{y})}{\partial \mathbf{n}_y} \mathrm{d}s_y + \int_{\Gamma_i} \frac{\partial P^*(\mathbf{x},\mathbf{y})}{\partial \mathbf{n}_y} p(\mathbf{y}) \mathrm{d}s_y, \qquad \mathbf{x} \in \Omega_a \tag{12}$$

which is valid for $\mathbf{x}$ within the acoustic domain. The boundary integral equation is obtained by shifting $\mathbf{x}$ onto the smooth boundary. With the definition of the acoustic flux $q(\mathbf{x}) := \frac{\partial p(\mathbf{x})}{\partial \mathbf{n}_x}$ one obtains

$$\frac{1}{2} p(\mathbf{x}) = -\underbrace{\int_{\Gamma_i} P^*(\mathbf{x},\mathbf{y})\, q(\mathbf{y}) \mathrm{d}s_y}_{(Vq)(\mathbf{x})} + \underbrace{\int_{\Gamma_i} \frac{\partial P^*(\mathbf{x},\mathbf{y})}{\partial \mathbf{n}_y} p(\mathbf{y}) \mathrm{d}s_y}_{(Kp)(\mathbf{x})}, \qquad \mathbf{x} \in \Gamma_i . \tag{13}$$

The single layer potential is denoted by V and the double layer potential by K, respectively. Analogously, the hypersingular boundary integral equation is derived by an additional derivative with respect to the normal $\mathbf{n}_x$

$$\frac{1}{2} q(\mathbf{x}) = -\underbrace{\int_{\Gamma_i} \frac{\partial P^*(\mathbf{x},\mathbf{y})}{\partial \mathbf{n}_x} q(\mathbf{y}) \mathrm{d}s_y}_{(K'q)(\mathbf{x})} + \underbrace{\int_{\Gamma_i} \frac{\partial^2 P^*(\mathbf{x},\mathbf{y})}{\partial \mathbf{n}_x \partial \mathbf{n}_y} p(\mathbf{y}) \mathrm{d}s_y}_{-(Dp)(\mathbf{x})}, \qquad \mathbf{x} \in \Gamma_i \tag{14}$$

where K' denotes the adjoint double layer potential and D is the hypersingular operator.

For exterior acoustic problems, neither the singular boundary integral equation (13) nor the hypersingular boundary integral equation (14) has an unique solution for all frequencies. One possibility to overcome the problem is the approach of Burton and Miller (1971), which uses a linear combination of both integral equations (13) and (14)

$$\left(-\frac{1}{2} I + K - \alpha D\right) p(\mathbf{x}) = \left(V + \frac{1}{2}\alpha I + \alpha K'\right) q(\mathbf{x}) \tag{15}$$

where typically $\alpha = j/\kappa$ is chosen for a good condition number. In case of a scattering problem, the resulting acoustic field is the superposition of the incident and the scattered field (Wu, 2000)

$$p = p^{\text{inc}} + p^{\text{s}} \qquad \text{and} \qquad q = q^{\text{inc}} + q^{\text{s}} . \tag{16}$$

The incident field $(\ )^{\mathrm{inc}}$ is caused by a plane wave and corresponds to the sound field in the absence of the structure. For simple sources it is given analytically (see Section 5). The scattered field is denoted by $(\ )^{\mathrm{s}}$. To derive a boundary integral representation for the scattering problem, first (15) is applied to p^{s} in Ω_{a}

$$\left(-\frac{1}{2}I + K - \alpha D\right)\, p^{\mathrm{s}}(\mathbf{x}) = \left(V + \frac{1}{2}\alpha I + \alpha K'\right) q^{\mathrm{s}}(\mathbf{x})\,. \tag{17}$$

Then the same is done for p^{inc} in the interior domain Ω_{i} (see Fig. 2 right) in absence of the structure

$$\left(\frac{1}{2}I + K - \alpha D\right)\, p^{\mathrm{inc}}(\mathbf{x}) = \left(V - \frac{1}{2}\alpha I + \alpha K'\right) q^{\mathrm{inc}}(\mathbf{x})\,. \tag{18}$$

Please note the modified leading signs of the mass terms. Adding (17) to (18) and using (16) yields

$$\left(-\frac{1}{2}I + K - \alpha D\right)\, p(\mathbf{x}) - \left(V + \frac{1}{2}\alpha I + \alpha K'\right) q(\mathbf{x}) = -I\; p^{\mathrm{inc}}(\mathbf{x}) - \alpha I\; q^{\mathrm{inc}}(\mathbf{x})\,. \tag{19}$$

There are various methods to derive an algebraic system of equations. Here, a Galerkin formulation is used. Equation (19) is tested on the interface Γ_{i} with linear trial functions ν, yielding

$$\begin{aligned}\int_{\Gamma_{\mathrm{i}}} \nu(\mathbf{x}) \left(-\frac{1}{2}I + K - \alpha D\right)\, p(\mathbf{x})\mathrm{d}s_x - \int_{\Gamma_{\mathrm{i}}} \nu(\mathbf{x}) \left(V + \frac{1}{2}\alpha I + \alpha K'\right) q(\mathbf{x})\mathrm{d}s_x = \\ - \int_{\Gamma_{\mathrm{i}}} \nu(\mathbf{x}) I\; p^{\mathrm{inc}}(\mathbf{x})\mathrm{d}s_x - \alpha \int_{\Gamma_{\mathrm{i}}} \nu(\mathbf{x}) I\; q^{\mathrm{inc}}(\mathbf{x})\mathrm{d}s_x\,.\end{aligned} \tag{20}$$

The pressures p and p^{inc} are interpolated with piecewise linear shape functions whereas constant shape functions are used for the fluxes q and q^{inc}. The resulting algebraic system of equations reads

$$\underbrace{\left(-\frac{1}{2}\mathbf{I} + \mathbf{K} - \alpha\mathbf{D}\right)}_{\mathbf{K}_{\mathrm{BE}}}\mathbf{p} - \underbrace{\left(\mathbf{V} + \frac{1}{2}\alpha\mathbf{I}' + \alpha\mathbf{K}'\right)}_{\mathbf{C}^{q}_{\mathrm{BE}}}\mathbf{q} = \underbrace{-\mathbf{I}\mathbf{p}^{\mathrm{inc}} - \alpha\mathbf{I}'\mathbf{q}^{\mathrm{inc}}}_{\mathbf{rhs}_{\mathrm{BE}}}\,. \tag{21}$$

A serious drawback of classical BE methods is, that setting up and storing the resulting fully populated matrices has a complexity of order $\mathcal{O}(N^2)$, where N denotes the number of degrees of freedom (DOF). If additionally a direct solver is used in combination with these fully populated matrices, the numerical expense is of order $\mathcal{O}(N^3)$. Obviously, standard BE methods are not practical for large scale problems having more than a thousand nodes. To overcome this drawback the fast multilevel multipole algorithm with a numerical complexity of order $\mathcal{O}(N \log^2 N)$ (Fischer, 2004) is applied together with a preconditioned GMRES solver.

3.1 Fast Multilevel Multipole Algorithm

For the introduced operators, one typically has to evaluate potentials of the type

$$\Phi(\mathbf{x}_b) = \sum_{a=1}^{A} \frac{e^{j\kappa|\mathbf{x}_b - \mathbf{y}_a|}}{|\mathbf{x}_b - \mathbf{y}_a|} q_a, \tag{22}$$

where q_a denotes the source strengths of A sources and $|\mathbf{x}_b - \mathbf{y}_a|$ is the distance between the field and load point. Standard BE methods typically consider the interaction of every combination of a load point with a field point. In contrast to this, the multipole algorithm sets up a clustering and sums up the contribution of all sources q_a in the center $\mathbf{z}_a$ of a cluster (see Fig. 3). At the next step, this so-called far-field signature is translated to the center $\mathbf{z}_b$ of the other clusters and from there finally distributed to $\mathbf{x}_b$.

From a mathematical point of view, the separation of the distance $|\mathbf{x}_b - \mathbf{y}_a|$ succeeds in the fundamental solution by using the diagonal form of the multipole expansion (Rokhlin, 1993)

$$\frac{e^{j\kappa|\mathbf{x}_b - \mathbf{y}_a|}}{|\mathbf{x}_b - \mathbf{y}_a|} = \frac{j\kappa}{4\pi} \sum_{l=0}^{\infty} (2l+1) j^l h_l^{(1)}(\kappa|\mathbf{D}|) \int_{\mathbb{S}^2} e^{j\kappa(\mathbf{d}_a + \mathbf{d}_b)\cdot\mathbf{s}} P_l\left(\mathbf{s}\cdot\hat{\mathbf{D}}\right) \mathrm{d}s\,, \tag{23}$$

with the Hankel functions h_l and the Legendre polynomials P_l. The vectors which are local to the clusters are denoted by $\mathbf{d}_a$ and $\mathbf{d}_b$ (see Fig. 3), whereas $\mathbf{D}$ is defined by the centers of two interacting clusters. The unit distance vector is defined by $\hat{\mathbf{D}} = \mathbf{D}/|\mathbf{D}|$, where $|\cdot|$ denotes the Euclidian norm. The integral over the unit sphere $\mathbb{S}^2$ is approximated by Gauss point quadrature using discrete values of the the far-field directions $\mathbf{s}$ (Rokhlin, 1993). Since one can not compute an infinite sum, the

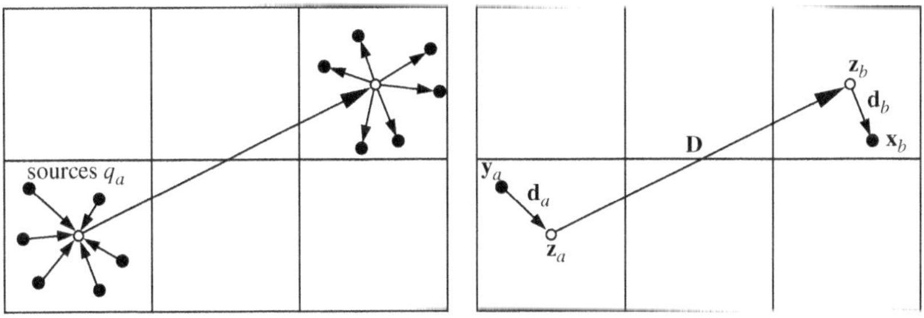

Fig. 3 Clustering of the sources q_a (*left*) and splitting up of the vector between load point and field point into three parts (*right*)

series has to be truncated. In this case the integration over the unit sphere $\mathbb{S}^2$ and the summation can be interchanged. Introducing the translation operator

$$M_L(\mathbf{s}, \mathbf{D}) = \sum_{\ell=0}^{L} (2\ell + 1) \mathrm{j}^\ell h_\ell^{(1)}(\kappa |\mathbf{D}|) P_\ell(\mathbf{s} \cdot \hat{\mathbf{D}}), \tag{24}$$

the original potential (22) can now be expressed in the form

$$\Phi(\mathbf{x}_b) = \frac{\mathrm{j}\kappa}{4\pi} \int_{\mathbb{S}^2} \mathrm{e}^{\mathrm{j}\kappa \mathbf{d}_b \cdot \mathbf{s}} M_L(\mathbf{s}, \mathbf{D}) \underbrace{\sum_{a=1}^{A} \mathrm{e}^{\mathrm{j}\kappa \mathbf{d}_a \cdot \mathbf{s}} q_a}_{F(\mathbf{s})} \, \mathrm{d}\mathbf{s}. \tag{25}$$

The choice of L in (24), which is called the *expansion length*, has a significant influence on the accuracy and the performance of the multipole algorithm. Proper choice helps to circumvent divergence of the series and will be discussed later in this section. The sum on the right hand of (25) is called the *far-field signature* $F(\mathbf{s})$. It is local to the cluster with the sources q_a, since only the vector $\mathbf{d}_a$ appears. In contrast to this, the translation operator M_L only depends on the vector $\mathbf{D}$ between two clusters' centers. Thus, if a regular cluster grid is used, the translation operators can be reused. Translating the far-field signature to another cluster using a translation operator forms the so called *near-field signature*. The solution is finally recovered by an exponential function of $\mathbf{d}_b$ and an integration over the unit sphere.

Since the multipole expansion is only valid for well separated load and field points, one has to split up the clusters into a near- and far-field. All clusters which fulfill the condition

$$|\mathbf{D}| < c_d \frac{d}{2} \tag{26}$$

form the near-field. Here, d denotes the cluster diameter and c_d is a constant. The arising near-field is represented by a sparse matrix. It has to be evaluated by classical BEM. All other clusters are in the far-field and form the so called *interaction list*.

3.1.1 Multilevel Algorithm

To obtain an optimal efficiency, a hierarchical multilevel cluster tree is used (see Fig. 4). It is set up by consecutive bisectioning such that a mother cluster is divided into two son clusters on the next level. The procedure starts with the *root cluster*, which is the smallest parallelepiped containing all elements of the model. The division is stopped if a specified number of elements per cluster is reached. These final clusters, which do not have any sons, are called *leaf clusters*. The interaction list of every cluster is formed by those clusters, which are in the near-field of the *mother cluster* but not in its own near-field (see Fig. 4).

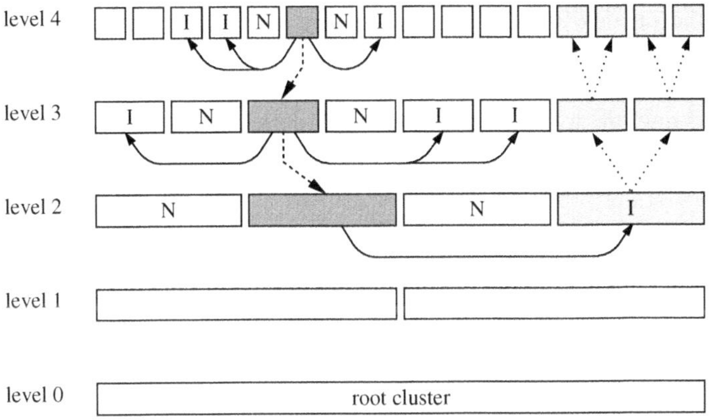

Fig. 4 Simplified model of a multilevel clustertree, which is created by bisectioning of the root cluster. Interaction clusters are denoted by I whereas near-field clusters are represented by N

Obviously, the far-field signature has to be translated to the interaction lists on different levels. Since the cluster diameters are different on every level, the expansion length L has to be adapted to every level, too. Typically the well-established semi-empirical rule

$$L(\kappa\, d_\ell) = \kappa d_\ell + c_e \log(\kappa\, d_\ell + \pi) \tag{27}$$

is used to estimate the number of series terms on level ℓ of the cluster tree (Coifman et al., 1993). The parameter c_e has to be chosen by the user and determines the desired accuracy. In order to maintain the accuracy of the multipole expansion when the cluster diameter increases on the next level, an interpolation and filtering strategy has to be applied. It is advantageous to use a fast Fourier transform for this purpose. This is because new far-field directions have to be added, which is only possible for the original form of the multipole expansion. A more detailed description can be found in (Gyure and Stalzer, 1998; Fischer, 2004 and Schneider, 2003).

The resulting *fast multipole method* (FMM) has a quasi linear complexity of order $\mathcal{O}(N \log^2 N)$. For a detailed investigation of the numerical complexity, the reader is referred to (Gaul and Fischer, 2006).

The evaluation of the matrix-vector product with the FMM algorithm is similar for all operators which will be needed for the coupling formulations. Only slight modifications are necessary in order to take into account the different test and shape functions. The general procedure can be summarized with the following steps:

1. Compute the near-field part by a sparse matrix-vector multiplication.
2. Evaluate the far-field signature $F(\mathbf{s})$ for every leaf cluster.
3. Translate the far-field signature to all interaction clusters by means of the translation operators (Eq. 24) and sum it up as the near-field signature $N(\mathbf{s})$ there (solid lines in Fig. 4).

4. Shift the far-field signature to the mother cluster and repeat step 3 until the interaction list is empty (dashed lines).
5. Go the opposite direction and shift the near-field signature $N(\mathbf{s})$ to the son clusters until the leaf clusters are reached (dotted lines).
6. Recover the solution by integration over the unit sphere.

4 Coupled FMBE/FE Formulation

For the coupling of (9) and (21) an additional relation between the flux $\mathbf{q}$ and the structural displacement $\mathbf{u}$ is needed. It is obtained from Euler's equation. In the following only reverberant surfaces are investigated, where the particle displacement is equal to the surface displacement $\mathbf{u}$ of the structure. The constant flux $\mathbf{q}$ on each element e is computed by the structural displacements of the adjacent nodes k of e in an extra step. For triangular elements this is done by

$$q^{(\mathrm{e})} = \frac{1}{3}\rho_{\mathrm{f}}\omega^2 \sum_{\mathrm{k}\in\mathrm{e}} \mathbf{u}^{\mathrm{k}} \cdot \mathbf{n}^{(\mathrm{e})}, \tag{28}$$

where ρ_{f} is the density of the fluid, $\mathbf{u}^{\mathrm{k}}$ is the displacement at node k and $\mathbf{n}^{(\mathrm{e})}$ is the normal of element e. The transformation is written in matrix notation as $\mathbf{q} = \mathbf{T}_{uq}\mathbf{u}$, where $\mathbf{q}$ denotes the vector with the flux on the elements und $\mathbf{u}$ are the nodal structural displacements. Matrix $\mathbf{T}_{uq}$ is sparse and cheap to compute. Together with the FE system (9) and the BE system (21), the coupled system of equations is

$$\underbrace{\begin{pmatrix} \mathbf{K}_{\mathrm{FE}} & \mathbf{C}_{\mathrm{FE}} \\ \mathbf{C}^q_{\mathrm{BE}}\mathbf{T}_{uq} & \mathbf{K}_{\mathrm{BE}} \end{pmatrix}}_{\mathbf{K}} \begin{pmatrix} \mathbf{u} \\ \mathbf{p} \end{pmatrix} = \begin{pmatrix} 0 \\ \mathbf{rhs}_{\mathrm{BE}} \end{pmatrix} . \tag{29}$$

It is worth mentioning, that two sparse near-field matrices have to be assembled for the BE part. One near-field for $\mathbf{K}_{\mathrm{BE}}$ and the other one for $\mathbf{C}^q_{\mathrm{BE}}$.

4.1 Iterative Solver

As discussed above, the fast multipole method helps to efficiently compute the BE matrix-vector product. It is especially well suited in combination with an iterative solver (van der Vorst, 2003). One possibility to solve the coupled system is to introduce the Schur complement $\mathbf{S}$ to replace (29) by the reduced system (Brunner et al., 2009)

$$\underbrace{\left(\mathbf{K}_{\mathrm{BE}} - \mathbf{C}^q_{\mathrm{BE}}\mathbf{T}_{uq}\mathbf{K}^{-1}_{\mathrm{FE}}\mathbf{C}_{\mathrm{FE}}\right)}_{\mathbf{S}} \mathbf{p} = \mathbf{rhs}_{\mathrm{BE}} , \tag{30}$$

with $\mathbf{p}$ as the vector of unknowns. A single GMRES is directly applied for $\mathbf{S}$. This formulation is especially well suited if the inverse of $\mathbf{K}_{\mathrm{FE}}$ can be computed in a cheap way for instance by factorization based solver for $\mathbf{K}_{\mathrm{FE}}^{-1}$ like a $\mathrm{LDL}^{\mathrm{T}}$ decomposition. The factorization has to be computed only once and can be reused at every iteration step with low effort. The structural displacements $\mathbf{u}$ are computed in a postprocessing step with the second equation of (29).

For preconditioning, only the near-field of $\mathbf{K}_{\mathrm{BE}}$ of $\mathbf{S}$ is taken into account. An imcomplete LU factorization (ILU) turned out to be efficient in the investigated frequency range (Brunner et al., 2009).

5 The Spherical Scatterer

To investigate the correct implementation of the proposed coupling scheme, a spherical shell structure as depicted in Fig. 5 is investigated, which is totally submerged in an acoustic fluid. An analytical solution for both a rigid sphere and an elastic sphere is available for an incident plane wave (Junger and Feit, 1986). Both cases are compared in the following. For the simulation of the rigid sphere, $\mathbf{C}_{\mathrm{BE}}^{q}$ in (29) is simply set to zero.

The elastic structure is modeled with the commercial FE package ANSYS with SHELL63 elements (12 △ and 3986 □) with six degrees of freedom at each node. The mean element size is 0.28 m and the model consists of 3994 nodes. The sphere has a radius r=5.00 m and a shell thickness t=0.05 m. Steel is assumed as linear elastic material (Young's modulus E=207 GPa, Poisson's ratio ν=0.3, density ρ_{s}=7669 kg/m³, Rayleigh damping coefficients: α_{R}=0 1/s, β_{R}=0 s). The material data of water (density ρ_{f}=1000 kg/m³, speed of sound c=1387 m/s) are applied for the simulation of the exterior acoustic fluid.

For the right hand side of (30) the pressure and flux of the incident field is needed. In case of a plane wave (see Fig. 5), p^{inc} is given at every node X by

$$p^{\mathrm{inc}}(\mathbf{x}) = \mathrm{e}^{\mathrm{j}\kappa(\mathbf{w}\cdot\mathbf{n}_{\mathrm{pw}})}\,, \tag{31}$$

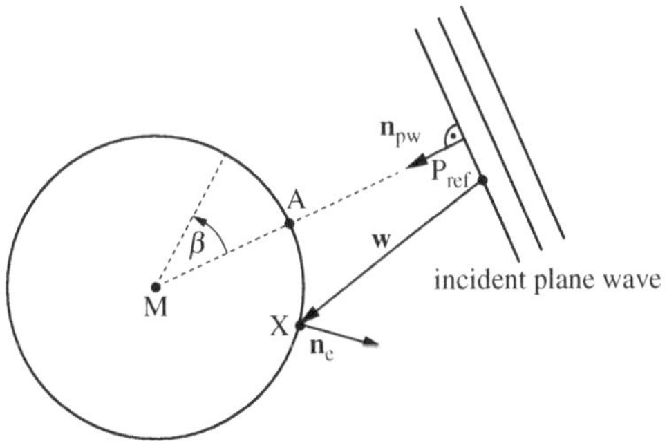

Fig. 5 Test example: Spherical scatterer. The incident wavefront is defined by a reference point $\mathrm{P}_{\mathrm{ref}}$ and the normal direction $\mathbf{n}_{\mathrm{pw}}$

where **w** denotes the vector from an arbitrary reference point P_{ref} to the node X and $\mathbf{n}_{pw}$ is the normal of the wave front. The corresponding flux at an arbitrary point **x** on an element is found by

$$q^{inc}(\mathbf{x}) = j\kappa \mathbf{n}_e \cdot \mathbf{n}_{pw} e^{j\kappa(\mathbf{w}\cdot\mathbf{n}_{pw})} , \tag{32}$$

with the normal of the element $\mathbf{n}_e$. The constant flux on each element is approximated by the value at the center of the element.

5.1 Numerical Results

First, the pressure and displacement amplitudes at node A (see Fig. 5) are investigated within the frequency range $f \in [5, 100\text{Hz}]$. Figure 6 compares the numerical results of the FMBE-FE approach with the analytical ones for both an elastic sphere and a rigid sphere. Obviously, the agreement between the numerical and the analytical results is quite good. Additionally, one can clearly see that the elasticity of the sphere has a strong influence on the pressure results, justifying the use of a fully coupled solution scheme. Only at very low frequencies $f < 10\,\text{Hz}$, both solutions tend to the same value. In this range, there are hardly any deformations and only rigid body motions occur.

In a second scenario, the directivity pattern along a semicircle on the surface of the sphere is computed which starts at point A (see Fig. 5) and is further defined by the angle β. Again, the numerical results are compared to the analytical ones for both the elastic and the rigid case. The simulations are performed at 75 Hz. The results are visualized in Fig. 7. An angle β=0° corresponds to the direction of the incident plane wave. Here, the ratio p^s/p^{inc} is the highest for both the elastic and rigid case. The elastic sphere additionally has two sidelobes which can not be observed in the rigid case. Generally, the values are higher in the elastic case than in the rigid

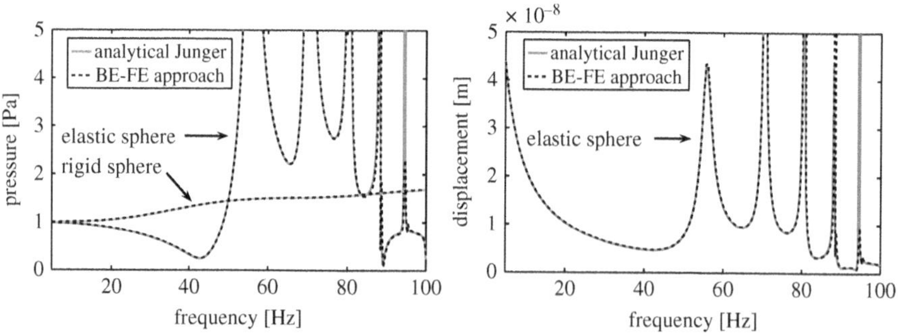

Fig. 6 Sphere: Pressure and displacement amplitudes at point A (see Fig. 5) for a frequency sweep between 5 and 100 Hz. Please note that the displacement is zero in case of a rigid sphere

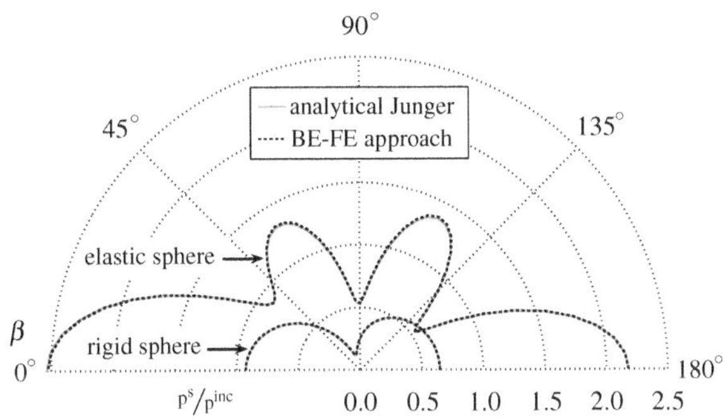

Fig. 7 Directivity pattern of the spherical shell structure for f=75 Hz. The dimensionless pressure p^s/p^{inc} is plotted for all points on a semicircle starting at point A (see Fig. 5)

case. Once again, the results of the FE-BE approach are almost identical with the analytical solution.

5.2 *Efficiency*

For the simulation of large scale problems, both the simulation time and the memory consumption should be moderate. Concerning the simulation time, most of the time is spent with setting up the BE near-field matrices and with solving the obtained linear system. Since an iterative GMRES solver is applied, the solution time depends directly on the number of iteration steps required to achieve convergence to a prescribed accuracy. A proper preconditioning has a strong influence on the convergence rate. In this paper, a ILU preconditioner on $\mathbf{K}_{\mathrm{BE}}$ is applied to accelerate the GMRES. The required number of iterations steps for a tolerance of 10^{-6} is visualized in Fig. 8 (left). It is observable, that the number of iteration steps is lower in case of the rigid sphere, where it is almost independent of the frequency. This is not astonishing, since the structural part of the Schur complement is neglected for preconditioning. With increasing frequency, the influence of the elasticity of the structure increases. For a lower number of iterations, the second part of the Schur complement would need to be considered as well.

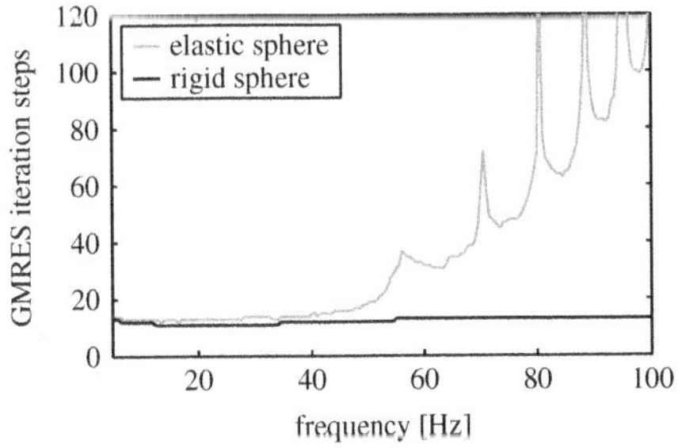

	memory [MB]
$\mathbf{K}_{\mathrm{BE}}$ (near–field)	31.4
$\mathbf{C}_{\mathrm{BE}}$ (near–field)	49.2
$\mathbf{K}_{\mathrm{FE}}$	7.6
$\mathbf{K}_{\mathrm{FE}}$ LDL^T factorization	57.8
$\mathbf{C}_{\mathrm{FE}}$	1.0
$\mathbf{T}_{uq}$	0.6

Fig. 8 Sphere: Number of GMRES iteration steps with ILU preconditioning (*left*) and memory consumption for the submatrices (*right*)

The memory consumption of all matrices is summarized in the table on the right hand side of Fig. 8. The sparse BE near-field matrices are more expensive than the sparsily populated FE system matrix. The memory consumption of $\mathbf{C}_{\text{FE}}$ and $\mathbf{T}_{uq}$ is negligible. The significant part of the memory consumption is the LDL^{T} factorization of the dynamic stiffness matrix $\mathbf{K}_{\text{FE}}$, which is symmetric and real valued in absence of structural damping. Theoretically, one could apply an iterative solver for $\mathbf{K}_{\text{FE}}^{-1}$ to avoid the factorization. But as a consequence the efficiency would decrease because of the resulting nested solution scheme. Additionally, proper preconditioning of $\mathbf{K}_{\text{FE}}$ which typically results from shell elements is a non-trivial task.

6 Conclusion

In this paper scattering problems with fluid-structure interaction are investigated. For this purpose, a conforming coupling formulation between the boundary element method and finite element method is presented. The finite element matrices are directly set up by a commercial finite element package. The research code has an interface in order to directly read the matrices. This leads to a high flexibility and even complex engineering problems can be handled. The boundary element part is accelerated by the fast multipole method. Thus, even large scale problems can efficiently be computed. Numerical results for a spherical shell structure with an incident plane wave are compared with an analytical solution. The comparison shows a good accuracy of the applied FE-BE approach. Additionally, the influence of the elasticity of the structure turned out to have a strong effect on the overall results. This necessitates the use of a strong coupling scheme. The overall computation time is influenced by proper preconditioning. Although an ILU preconditioner significantly improves the convergence, more advanced preconditioners need to be developed for an optimal performance.

Acknowledgments This research was financially supported by the German Research Foundation (DFG) under the transfer project SFB404/T3. The authors acknowledge valuable contributions of Germanischer Lloyd AG, Hamburg within this transfer project.

References

Brunner D, Junge M, Gaul L (2007) Strong coupling of the fast multilevel multipole boundary element method with the finite element method for vibro-acoustic problems. In: Proceedings of the 14th International Congress on Sound and Vibration (ICSV14)

Brunner D, Junge M, Gaul L (2009) A comparison of FE–BE coupling schemes for large scale problems with fluid-structure interaction. International Journal for Numerical Methods in Engineering 77 (5): 664–688

Burton AJ, Miller GF (1971) The application of integral equation methods to the numerical solution of some exterior boundary-value problems. Proceedings of the Royal Society of London 323:201–220

Chen ZS, Hofstetter G, Mang HA (1998) A Galerkin-type BE-FE formulation for elasto-acoustic coupling. Computer Methods in Applied Mechanics and Engineering 152:147–155

Coifman R, Rokhlin V, Wandzura S (1993) The fast multipole method for the wave equation: A pedestrian description. IEEE Antennas and Propagation Magazine 35:7–12
Estorff O von (ed) (2000) Boundary Elements in Acoustics: Advances and Applications. WIT Press, Southampton, UK
Everstine GC, Henderson FM (1990) Coupled finite element/boundary element approach for fluid-structure interaction. Journal of Sound and Vibration 87:1938–1947
Fahy F (2007) Sound and Structural Vibration. Academic Press, New York, 2nd edition
Fischer M (2004) The Fast Multipole Boundary Element Method and its Application to Structure-Acoustic Field Interaction. PhD Thesis, University of Stuttgart
Gaul L, Fischer M (2006) Large-scale simulations of acoustic-structure interaction using the fast multipole BEM. Zeitschrift für Angewandte Mathematik und Mechanik 86:4–17
Gaul L, Kögl M, Wagner M (2003) Boundary Element Methods for Engineers and Scientists. Springer Verlag, Berlin Heidelberg
Greengard L, Rokhlin V (1987) A fast algorithm for particle simulations. Journal of Computational Physics 73:325–348
Gyure MF, Stalzer MA (1998) A prescription for the multilevel Helmholtz FMM. IEEE Computational Science & Engineering 5:39–47
Hughes MD, Chen K (2004) An efficient preconditioned iterative solver for solving a coupled fluid structure interaction problem. International Journal of Computer Mathematics 81:583–594
Junge M, Fischer M, Maess M, Gaul L (July 11–14 2005) Acoustic simulation of an idealized exhaust system by coupled FEM and Fast Multipole BEM. In: Proceedings of the 12th International Congress on Sound and Vibration (ICSV12), Lisboa
Junger MC, Feit D (1986) Sound, Structures, and Their Interaction. The MIT Press, Cambridge, 2nd edition
Kinsler LE, Frey AR, Coppens A, Sanders J (1999) Fundamentals of Acoustics. John Wiley and Sons, Inc., New York
Moosrainer M (2000) Fluid-Struktur-Kopplung. PhD Thesis, Universität der Bundeswehr München
Ochmann M, Homm A, Makarov S, Semenov S (2003) An iterative GMRES-based boundary element solver for acoustic scattering. Engineering Analysis with Boundary Elements 27: 714–725
Rjasanow S, Steinbach O (2007) The Fast Solution of Boundary Integral Equations. Springer–Verlag, New York
Rokhlin V (1993) Diagonal forms of translation operators for the Helmholtz equation in three dimensions. Applied and Computational Harmonic Analysis, 1:82–93
Schneider S (2003) Efficient Usage of the Boundary Element Method for Solving the Time Harmonic Helmholtz Equation in Three Dimensions. PhD Thesis, Technische Universität Dresden
Seybert AF, Soenarko B, Rizzo FJ, Shippy DJ (1985) An advanced computational method for radiation and scattering of acoustic waves in three dimensions. Journal of the Acoustical Society of America, 77:362–368
Shen L, Liu YJ (2007) An adaptive fast multipole boundary element method for three–dimensional acoustic wave problems based on the Burton-Miller formulation. Computer Mechanical 40:461–472
van der Vorst HA (2003) Iterative Krylov Methods for Large Linear Systems. Cambridge University Press, Cambridge
Wu TW (2000) Boundary Element Acoustics: Fundamentals and Computer Codes. WIT Press, Southampton, UK
Zienkiewicz OC, Taylor RL (2000) The Finite Element Method, Vol. 1. Butterworth-Heinemann, Oxford, 5th edition

An Application of the BEM Numerical Green's Function Procedure to Study Cracks in Reissner's Plates

S. Guimarães and J. C. F. Telles

Abstract The Numerical Green's Function (NGF) technique, previously proposed by the present authors, is here extended to fracture mechanics problems involving Reissner's plate theory. The technique numerically produces a plate bending fundamental Green's function that automatically includes embedded cracks to be used in the classical boundary element method (BEM) to solve this class of problems. The applications discussed include torsion, bending moment and shear force loadings. In addition, also presented is a series of numerical results computed in terms of normalized stress intensity factors to illustrate the good accuracy of the procedure.

1 Introduction

The Linear Elastic Fracture Mechanics theory employs stresses as well as rotations and dislocations, in the vicinity of the crack tip, to obtain reliable stress intensity factor coefficients, essential in the prediction of tip behaviour and crack stability. There exist many techniques to model cracks using the boundary element method (BEM), these mainly differ in the numerical procedure adopted to include the cracks into the formulation (Telles and Guimarães, 2000). Sub-regions discretize the crack as the continuation of a fictitious interface using boundary elements for this. The so called dual and displacement discontinuity procedure avoid the interface discretization but keep the use of boundary elements to represent the existing crack. The Green's function approach, also including the adopted numerical Green's function (NGF) treated here, avoids boundary elements over the crack surfaces, since the fundamental solution removes boundary integration there. Concerning the NGF technique, some solutions have been presented in the literature for plates, as in Telles et al. (1995) and, recently, in Silveira et al. (2005) for crack propagation, but all restricted to in-plane loadings. In the present work, the ability of the hyper-singular formulation to represent a displacement discontinuity is used to develop a NGF solution for Reissner's plate (Reissner, 1947, Vander Weeën, 1982), i.e., for plates with

J. C. F. Telles (✉)
Programa de Engenharia Civil, COPPE/UFRJ, Caixa Postal 68506, CEP 21945-970, Rio de Janeiro, RJ, Brazil
e-mail: telles@coc.ufrj.br

G.D. Manolis, D. Polyzos (eds.), *Recent Advances in Boundary Element Methods*,
DOI 10.1007/978-1-4020-9710-2_11, © Springer Science+Business Media B.V. 2009

out of plane loadings also taking into account the transversal shear deformation. The numerical results presented here include bending moment, torsion and shear force loadings to confirm the good accuracy of the procedure.

2 The BEM Applied to Reissner's Plate Theory

The BEM integral equation for generalized displacements (i.e. rotations and linear out of plane displacement) in a Reissner's plate of boundary Γ is (Karam and Telles, 1988)

$$C_{ij}(\xi)u_j(\xi) = \fint_{\Gamma} \left(u^*_{ij}(\xi,x)p_j(x) - p^*_{ij}(\xi,x)u_j(x)\right)d\Gamma(x) + \int_{\Omega} \left(u^*_{i3}(\xi,x) - \frac{\nu}{(1-\nu)\lambda^2}u^*_{i\alpha,\alpha}(\xi,x)\right) q(x)d\Omega(x) \quad (1)$$

where ξ and x are the source and field points, respectively; $\fint$ stands for Cauchy's principal value, $C_{ij}(\xi)$ is a geometric coefficient at ξ, equal to δ_{ij} or $\delta_{ij}/2$ for either an internal point ξ or a point on a smooth part of the boundary; $u_j(x)$ and $p_j(x)$ stand for generalized displacements (two rotations and one deflection), and generalized tractions (moment, torsion and shear tractions), respectively; the fundamental solution, $u^*_{ij}(\xi,x)$ and $p^*_{ij}(\xi,x)$, is the standard (Vander Weeën, 1982) infinite domain for Reissner's plates considering a unit point load applied in i direction at ξ; ν is Poisson's ratio and λ is the Reissner constant, $\lambda = \sqrt{10}/h$, where h is the thickness of the plate. Throughout the paper, the Greek and the Latin indexes vary from 1, 2 and 1, 3, respectively.

The domain integral of Eq. (1) represents the transversal uniform distributed loading, $q(x)$, over the domain Ω of the plate. This domain term is not included in what follows (i.e. $q(x) = 0$ is assumed from now on) since it does not interfere with the numerical Green's function evaluation and can be included as a future development in the final formulation.

Suppose a crack as a boundary cavity in Eq. (1), denoted by $\Gamma^F = \Gamma^+ \cup \Gamma^-$ (+ and − standing for "upper" and "lower" surfaces of the crack). Consider a null generalized tractions over the crack boundary or, at the most, $p_j(x^+) = -p_j(x^-)$, what makes null any integrand multiplied by the component $(p_j(x^+) + p_j(x^-))$. Writing the integrals only over boundary Γ^-, the classical and hypersingular formulations, derived from Eq. (1) for $\xi \in \Omega$, are expressed as (Guimarães et al., 2000):

$$u_i(\xi) = \int_{\Gamma^-} p^*_{ij}(\xi,x)c_j(x)d\Gamma(x) \quad (2a)$$

$$p_i(\xi) = \int_{\Gamma^-} P^*_{ij}(\xi,x)c_j(x)d\Gamma(x) \quad (2b)$$

where $c_j(x) = u_j(x^+) - u_j(x^-)$ are the generalized crack openings, i.e., the discontinuities in rotations and in deflection at x, and $P^*_{ij}(\xi, x)$ are the properly manipulated derivatives of the standard fundamental solution (Karam and Telles, 1988), obtained following the traction boundary integral procedure, expressed as

$$\begin{aligned}P^*_{\alpha\gamma} &= \frac{D(1-\nu)}{4\pi r^2}\left\{(4A(z)+2zK_1(z)+1-\nu)\left(\delta_{\alpha\gamma}n_\beta m_\beta + n_\alpha m_\gamma\right)\right.\\&+(4A(z)+1+3\nu)\,n_\gamma m_\alpha - \left(16A(z)+6zK_1(z)+z^2K_0(z)+2-2\nu\right)\\&\left[\left(-n_\alpha r_{,m}+n_\beta m_\beta r_{,\alpha}\right)r_{,\gamma}+\left(-\delta_{\alpha\gamma}r_{,m}+m_\gamma r_{,\alpha}\right)r_{,n}\right]\\&-2\left(8A(z)+2zK_1(z)+1+\nu\right)\left(m_\alpha r_{,\gamma}r_{,n}-n_\gamma r_{,\alpha}r_{,m}\right)+\\&\left.-4\left(24A(z)+8zK_1(z)+z^2K_0(z)+2-2\nu\right)r_{,\alpha}r_{,\gamma}r_{,n}r_{,m}\right\}\end{aligned} \tag{3a}$$

$$\begin{aligned}P_{\alpha 3} &= \frac{D(1-\nu)\lambda^2}{4\pi r}\left\{(2A(z)+zK_1(z))\left(-n_\alpha r_{,m}+n_\beta m_\beta r_{,\alpha}\right)\right.\\&\left.+2\left(4A(z)+zK_1(z)\right)r_{,\alpha}r_{,n}r_{,m}+2A(z)m_\alpha r_{,n}\right\}\end{aligned} \tag{3b}$$

$$\begin{aligned}P^*_{3\gamma} &= -\frac{D(1-\nu)\lambda^2}{4\pi r}\left\{(2A(z)+zK_1(z))\left(m_\gamma r_{,n}+n_\beta m_\beta r_{,\gamma}\right)\right.\\&\left.+2\left(4A(z)+zK_1(z)\right)r_{,\gamma}r_{,m}r_{,n}-2A(z)\,n_\gamma r_{,m}\right\}\end{aligned} \tag{3c}$$

$$P^*_{33} = \frac{D(1-\nu)\lambda^2}{4\pi r^2}\left\{\left(z^2B(z)+1\right)n_\beta m_\beta+\left(z^2A(z)+2\right)r_{,m}r_{,n}\right\}, \tag{3d}$$

in which, $D = Eh^3/\left(12(1-\nu^2)\right)$ is the flexural rigidity of the plate, E the Young modulus; only in Eq. (3), n represents the direction of the outward boundary normal vector at x and n_α its components, m is the normal vector direction at the source point ξ and m_α its components; $A(z) = K_0(z) + \frac{2}{z}\left(K_1(z) - \frac{1}{z}\right)$ and $B(z) = K_0(z) + \frac{1}{z}\left(K_1(z) - \frac{1}{z}\right)$ are dependent on the modified Bessel's functions of the second kind $K_0(z)$ and $K_1(z)$, where $z = \lambda r$ and $r = \sqrt{r_\alpha \cdot r_\alpha}$ is the known distance from point x and ξ with components $r_\alpha = x_\alpha(x) - x_\alpha(\xi)$; the derivatives of r are $r_{,\alpha} = \frac{\partial r}{\partial x_\alpha} = \frac{r_\alpha}{r}$, $r_{,n} = \frac{\partial r}{\partial \mathbf{n}(x)} = r_{,\alpha}n_\alpha$ and $r_{,m} = \frac{\partial r}{\partial \mathbf{m}(x)} = -r_{,\alpha}m_\alpha$.

3 Numerical Green's Function Approach

Let us define ζ as the points over a single straight crack surface, $\Gamma^F(\zeta) = \Gamma^+(\zeta) \cup \Gamma^-(\zeta)$, embedded in an infinite medium under the action of a unit point load applied at ξ, $\left(\xi \notin \Gamma^F\right)$. The fundamental solution for this problem at x, $\left(x \notin \Gamma^F\right)$, also called the Green's function, can be written in terms of a superposition of the classical fundamental solution plus a complementary part, so that the sum of both provides satisfaction of the traction free condition on Γ^F (Telles et al., 1995):

$$\begin{aligned} u_{ij}^G(\xi, x) &= u_{ij}^*(\xi, x) - u_{ij}^C(\xi, x) \\ p_{ij}^G(\xi, x) &= p_{ij}^*(\xi, x) - p_{ij}^C(\xi, x) \end{aligned} \tag{4}$$

Superscripts G, $*$ and c, stand for Green's fundamental solution, classical Reissner's plate solution and complementary components of the respective fundamental displacements and tractions. Subscript i stands for the direction of the applied load at ξ and j for the respective component at x, $\left(x \notin \Gamma^F\right)$. Using Eq. (2), the complementary solutions are expressed by the following integrals,

$$u_{ij}^C(\xi, x) = \int_{\Gamma^-} p_{jk}^*(x, \zeta) c_{ik}(\xi, \zeta) d\Gamma(\zeta) \tag{5}$$

$$p_{ij}^C(\xi, x) = \int_{\Gamma^-} P_{jk}^*(x, \zeta) c_{ik}(\xi, \zeta) d\Gamma(\zeta), \tag{6}$$

defining the complementary generalized displacements and tractions at an internal point x, $\left(x \notin \Gamma^F\right)$, as a function of the generalized fundamental displacement discontinuities $c_{ik}(\xi, \zeta) = u_{ik}(\xi, \zeta^+) - u_{ik}(\xi, \zeta^-)$ associated to a unit point load at ξ. Note that the fundamental $c_{ik}(\xi, \zeta)$ is the solution of the complementary problem in which the load at the crack line is minus the standard full space tractions for a unit point load at ξ. Eqs. (5) and (6) can be numerically evaluated using Gaussian quadrature to obtain the fundamental Green's functions as,

$$\begin{aligned} u_{ij}^G(\xi, x) &= u_{ij}^*(\xi, x) + \sum_{n=1}^{N} p_{jk}^*(x, \zeta_n) c_{ik}(\xi, \zeta_n) J_n W_n \\ p_{ij}^G(\xi, x) &= p_{ij}^*(\xi, x) + \sum_{n=1}^{N} P_{jk}^*(x, \zeta_n) c_{ik}(\xi, \zeta_n) J_n W_n \end{aligned} \tag{7}$$

where J_n is the Jacobian of the transformation to the standard quadrature interval, W_n is the corresponding weight factor at the Gauss station n, N is the total number of integration points and ζ_n is the nth Gauss point over the crack surface. The final standard equation, assuming the fundamental Green's function is known, reads

$$\begin{aligned} C_{ij}(\xi) u_j(\xi) = &\fint_{\Gamma} \left(u_{ij}^G(\xi, x) p_j(x) - p_{ij}^G(\xi, x) u_j(x)\right) d\Gamma(x) \\ &- \int_{\Gamma^-} c_{ij}(\xi, \zeta) p_j(\zeta) d\Gamma(\zeta) \end{aligned} \tag{8}$$

The last integral over the crack surface exists only if prescribed crack tractions $p_j(x)$ are to be taken into consideration.

3.1 The $c_{ik}(\xi,\zeta)$ Computation: Fundamental Rotations and Deflection Discontinuities

To determine the displacement discontinuities $c_{ij}(\xi,\zeta)$ of the complementary problem, the limit of Eq. (6) is taken as $x \to \Gamma^-$, yielding

$$\begin{aligned}
-p^*_{i\alpha}(\xi,\bar{\zeta}) &= \mathop{\not{\int}}_{\Gamma^-} P^*_{\alpha\gamma}(\bar{\zeta},\zeta)c_{i\gamma}(\xi,\zeta)d\Gamma(\zeta) + \mathop{\text{⨍}}_{\Gamma^-} P^*_{\alpha 3}(\bar{\zeta},\zeta)c_{i3}(\xi,\zeta)d\Gamma(\zeta)\\
-p^*_{i3}(\xi,\bar{\zeta}) &= \mathop{\text{⨍}}_{\Gamma^-} P^*_{3\gamma}(\bar{\zeta},\zeta)c_{i\gamma}(\xi,\zeta)d\Gamma(\zeta) + \mathop{\not{\int}}_{\Gamma^-} P^*_{33}(\bar{\zeta},\zeta)c_{i3}(\xi,\zeta)d\Gamma(\zeta)
\end{aligned} \tag{9}$$

Note that the source point in Eq. (9) is $\bar{\zeta}$, ($\bar{\zeta} \in \Gamma^-$), and $-p^*_{ij}(\xi,\bar{\zeta})$ is the boundary condition of the complementary problem. Also, $\not{\int}$ stands for Hadamard's finite part integral.

Equation (9) can be solved using a standard weighted residual method such as the point collocation technique as discussed in detail elsewhere (Telles et al., 1995, Silveira et al., 2005). Adopting a standard Gauss quadrature procedure, with a total of N Gauss Points and M collocating source points, designated respectively by ζ_n and $\bar{\zeta}_m$, the following expressions can be written, provided a correcting term, $E_{ij}(\xi,\bar{\zeta}_m)$, is introduced to deal with the singularities:

$$\begin{aligned}
-p^*_{i\alpha}(\xi,\bar{\zeta}_m) &= \sum_{n=1}^{N} P^*_{\alpha k}(\bar{\zeta}_m,\zeta_n)c_{ik}(\xi,\zeta_n)J_n W_n - E_{i\alpha}(\xi,\bar{\zeta}_m)\\
&\qquad\qquad\qquad\qquad\qquad\qquad m = 1,..M\\
-p^*_{i3}(\xi,\bar{\zeta}_m) &= \sum_{n=1}^{N} P^*_{3k}(\bar{\zeta}_m,\zeta_n)c_{ik}(\xi,\zeta_n)J_n W_n - E_{i3}(\xi,\bar{\zeta}_m)
\end{aligned} \tag{10}$$

The correcting terms are defined as:

$$\begin{aligned}
E_{i\alpha}(\xi,\bar{\zeta}_m) &= c_{ik}(\xi,\bar{\zeta}_m)e^{(1)}_{\alpha k} + \frac{\partial c_{ik}(\xi,\bar{\zeta}_m)}{d\Gamma(\zeta)}e^{(2)}_{\alpha k}\\
E_{i3}(\xi,\bar{\zeta}_m) &= c_{ik}(\xi,\bar{\zeta}_m)e^{(1)}_{3k} + \frac{\partial c_{ik}(\xi,\bar{\zeta}_m)}{d\Gamma(\zeta)}e^{(2)}_{3k}
\end{aligned} \tag{11}$$

where, for the specific case of the bending plate problem,

$$e^{(1)}_{\alpha\beta} = -\mathop{\not{\int}}_{\Gamma^-} P^*_{\alpha\beta}(\bar{\zeta}_m,\zeta)d\Gamma(\zeta) + \sum_{n=1}^{N} P^*_{\alpha\beta}(\bar{\zeta}_m,\zeta_n)J_n W_n \tag{12}$$

$$e^{(1)}_{\alpha 3} = -\fint_{\Gamma^-} P^*_{\alpha 3}(\bar{\zeta}_m, \zeta) d\Gamma(\zeta) + \sum_{n=1}^{N} P^*_{\alpha 3}(\bar{\zeta}_m, \zeta_n) J_n W_n \tag{13}$$

$$e^{(2)}_{\alpha\beta} = -\fint_{\Gamma^-} P^*_{\alpha\beta}(\bar{\zeta}_m, \zeta)\left[\Gamma(\zeta) - \Gamma(\bar{\zeta}_m)\right] d\Gamma(\zeta) + \sum_{n=1}^{N} P^*_{\alpha\beta}(\bar{\zeta}_m, \zeta_n)\left[\Gamma(\zeta_n) - \Gamma(\bar{\zeta}_m)\right] J_n W_n \tag{14}$$

$$e^{(2)}_{\alpha 3} = -\int_{\Gamma^-} P^*_{\alpha 3}(\bar{\zeta}_m, \zeta)\left[\Gamma(\zeta) - \Gamma(\bar{\zeta}_m)\right] d\Gamma(\zeta) + \sum_{n=1}^{N} P^*_{\alpha 3}(\bar{\zeta}_m, \zeta_n)\left[\Gamma(\zeta_n) - \Gamma(\bar{\zeta}_m)\right] J_n W_n \tag{15}$$

$$e^{(1)}_{3\beta} = -\fint_{\Gamma^-} P^*_{3\beta}(\bar{\zeta}_m, \zeta) d\Gamma(\zeta) + \sum_{n=1}^{N} P^*_{3\beta}(\bar{\zeta}_m, \zeta_n) J_n W_n \tag{16}$$

$$e^{(1)}_{33} = -\fint_{\Gamma^-} P^*_{33}(\bar{\zeta}_m, \zeta) d\Gamma(\zeta) + \sum_{n=1}^{N} P^*_{33}(\bar{\zeta}_m, \zeta_n) J_n W_n \tag{17}$$

$$e^{(2)}_{3\beta} = -\int_{\Gamma^-} P^*_{3\beta}(\bar{\zeta}_m, \zeta)\left[\Gamma(\zeta) - \Gamma(\bar{\zeta}_m)\right] d\Gamma(\zeta) + \sum_{n=1}^{N} P^*_{3\beta}(\bar{\zeta}_m, \zeta_n)\left[\Gamma(\zeta_n) - \Gamma(\bar{\zeta}_m)\right] J_n W_n \tag{18}$$

$$e^{(2)}_{33} = -\fint_{\Gamma^-} P^*_{33}(\bar{\zeta}_m, \zeta)\left[\Gamma(\zeta) - \Gamma(\bar{\zeta}_m)\right] d\Gamma(\zeta) + \sum_{n=1}^{N} P^*_{33}(\bar{\zeta}_m, \zeta_n)\left[\Gamma(\zeta_n) - \Gamma(\bar{\zeta}_m)\right] J_n W_n \tag{19}$$

and $[\Gamma(\zeta_n) - \Gamma(\bar{\zeta}_m)]$ is the difference between the Γ coordinate of points ζ_n and $\bar{\zeta}_m$.

Equations (15) and (18), though having a built-in logarithmic behaviour, are theoretically integrable since the integrand limit when $\zeta \to \bar{\zeta}_m$ exists. These integrals, however, are still part of the correcting terms since Gauss quadrature integration would not provide adequate numerical accuracy. When the M collocation points are taken to be the same as the N Gauss points in Eq. (10), the integrand containing $P^*_{jk}(\bar{\zeta}_m, \zeta_n)$ at the singular point of the summation, when $n = m$, cannot be numerically evaluated. Therefore, the computation of the regular terms at this point is performed through a limit, as $n \to m$, and added to the Eq. (10) to produce

$$
\begin{aligned}
-p^*_{i\alpha}(\xi, \bar{\zeta}_m) = {} & \sum_{\substack{n=1 \\ n\neq 1}}^{N} P^*_{\alpha k}(\bar{\zeta}_m, \zeta_n) c_{ik}(\xi, \zeta_n) J_n W_n - E_{i\alpha}(\xi, \bar{\zeta}_m) \\
& + \frac{1}{2} \frac{\partial^2 c_{ik}(\xi, \bar{\zeta}_m)}{\partial \Gamma(\zeta)^2} \lim_{n \to m} \left\{ P^*_{\alpha k}(\bar{\zeta}_m, \zeta_n) \left[\Gamma(\zeta_n) - \Gamma(\bar{\zeta}_m) \right]^2 \right\} \\
& \qquad m = 1, \ldots N \qquad (20) \\
-p^*_{i3}(\xi, \bar{\zeta}_m) = {} & \sum_{\substack{n=1 \\ n\neq 1}}^{N} P^*_{3k}(\bar{\zeta}_m, \zeta_n) c_{ik}(\xi, \zeta_n) J_n W_n - E_{i3}(\xi, \bar{\zeta}_m) \\
& + \frac{1}{2} \frac{\partial^2 c_{ik}(\xi, \bar{\zeta}_m)}{\partial \Gamma(\zeta)^2} \lim_{n \to m} \left\{ P^*_{3k}(\bar{\zeta}_m, \zeta_n) \left[\Gamma(\zeta_n) - \Gamma(\bar{\zeta}_m) \right]^2 \right\}
\end{aligned}
$$

Equation (20) is used to generate a system of equations for $c_{ik}(\xi, \zeta_n)$ at the collocation points, represented as

$$\mathbf{S} \cdot \mathbf{c_i}(\xi) = -\mathbf{p}^*_\mathbf{i}(\xi) \qquad (21)$$

where the vectors $\mathbf{c_i}(\xi)$ and $-\mathbf{p}^*_\mathbf{i}(\xi)$ include, respectively, the unknown displacement discontinuities and the values of the standard Reissner fundamental solution at the crack lines, for one or more embedded cracks, in three generalized directions, $\boldsymbol{i}$, as a result of the unit applied load at ξ. $\mathbf{S}$ is a square matrix of dimension $3N$, that is computed only once for the complete analysis since is only a function of the crack geometries and is the same for any position of ξ. Therefore, as in the other applications of this technique (Telles et al., 1995, Silveira et al., 2005), the $\mathbf{S}$ matrix in Eq. (21) once computed is just subjected to repeated back substitutions for the other independent vectors $-\mathbf{p}^*_\mathbf{i}(\xi)$ functions of the general point ξ positions. The process is then independent of the external boundary points x, what makes the technique attractive.

4 Implementation Features

The present procedure adopts the collocation points at the straight crack lines to be the same as the Gauss station ones. The correcting terms of type $e^{(1)}_{\alpha k}$ and $e^{(1)}_{3k}$ are to be assembled at the diagonal sub-matrices of $\mathbf{S}$. The $e^{(2)}_{\alpha k}$ and $e^{(2)}_{3k}$ correcting terms are correlated to the first derivative of $c_{ik}(\xi, \zeta_n)$ at $\bar{\zeta}_m$, $\partial c_{ik}(\xi, \bar{\zeta}_m)/d\Gamma(\zeta)$, and its values are no longer just connected to the diagonal sub-matrices of **S**, as explained below. The first and second derivatives at the collocation points $\bar{\zeta}_m$, $\partial c_{ik}(\xi, \bar{\zeta}_m)/d\Gamma(\zeta)$ and $\partial^2 c_{ik}(\xi, \bar{\zeta}_m)/d\Gamma(\zeta)^2$, arising from the unabridged Eq. (20), are modelled as functions of the fundamental crack displacements $c_{ik}(\xi, \zeta_n)$, at the Gauss collocation points, using a Lagrange polynomial interpolation. For this reason, the corrections related to these derivatives affect all terms of **S** related to $c_{ik}(\xi, \zeta_n)$ of the singular crack line, which can represent the entire crack line if the integration is not subdivided into segments. All corrections, though, are performed over the entire crack, especially if subdivided to prevent against near-singular integration difficulties.

Another peculiarity of the implementation is the usage of a special procedure to modify the Lagrange interpolation, $L_n(\zeta)$, whenever the derivatives of displacement discontinuities is to be used over the segment adjoining the tip. In this case, instead of simply deriving the Lagrange interpolation function, the following expressions are used to force the $\sqrt{r'}$ type behaviour of $c_{ik}(\xi, \zeta_n)$ near the tip (r' is the distance from the tip to ζ):

$$\frac{\partial c_{ij}(\xi, \zeta)}{\partial \zeta} = \frac{\partial}{\partial \zeta}\left(\sqrt{a^2 - \zeta^2} \sum_{n=1}^{N} \frac{L_n(\zeta)}{\sqrt{a^2 - \zeta_n^2}}\right) c_{ij}(\xi, \zeta_n) \tag{22}$$

$$\frac{\partial^2 c_{ij}(\xi, \zeta)}{\partial \zeta^2} = \frac{\partial^2}{\partial \zeta^2}\left(\sqrt{a^2 - \zeta^2} \sum_{n=1}^{N} \frac{L_n(\zeta)}{\sqrt{a^2 - \zeta_n^2}}\right) c_{ij}(\xi, \zeta_n) \tag{23}$$

Where ζ_n and ζ represent, respectively, the coordinates $\Gamma(\zeta_n$ and $\Gamma\ (\zeta)$, and a is half the length of the crack line. Though the procedure does not affect much the overall displacement discontinuity results, it guarantees quicker convergence to the analytical values very near the tip. In this way accurate stress intensity factors are obtained without need to perform regressions, just using the displacement discontinuity values at any collocation point sufficiently near the tip.

The singular integrals, present in Eq. (12) to (19), can be computed by any efficient numerical procedure. A semi-analytical procedure is adopted here where the regular components of the fundamental solution, expressed in terms of Bessel Functions, are computed numerically while the singular terms are computed analytically. Therefore, for $z > 2$, the Gauss numerical integration can be adopted since all terms are regular. A semi-analytical integration over the domain $z \leq 2$ is adopted to cope with singularities, performing the numerical integration over regular terms and using analytical integration for singular ones. Still over the domain $z \leq 2$, some terms, though not theoretically singular, produce poor results when computed by the Gauss integration technique. Therefore, logarithm terms of the type $r^p\ ln(r)$, up to the power 5, are treated analytically. This has shown to be essential to the final convergence of results, especially when considering torsion and shear examples. In addition, whenever necessary, a cubic transformation of the type proposed by Telles (1987) is adopted to compute the Gauss numerical integration of the near singular terms.

The limits, present in Eq. (20), are used whenever $n = m$, or very near (tolerance adopted $r < 0.1h$), such limits have been computed analytically using symbolic mathematical programs and are appropriately added to the final **S** matrix.

The actual problem crack displacement discontinuities, $c_j(\zeta)$, can be computed as a pos-processing procedure, using the standard traction equation and the classical fundamental solution, as

$$-\int\!\!\!\!\!\!=_{\Gamma^-} P^*_{ij}(\bar{\zeta},\zeta)c_j(\zeta)d\Gamma(\zeta) = -p_i(\bar{\zeta}) + \int_{\Gamma} \Big(U^*_{ij}(\bar{\zeta},x)p_j(x) - P^*_{ij}(\bar{\zeta},x)u_j(x)\Big)\, d\Gamma(x) \quad \forall\bar{\zeta} \in \Gamma^-, \tag{24}$$

or, in matrix form, the same previously calculated **S** matrix appears for the post-processing of the problem being solved:

$$\mathbf{S}\cdot\mathbf{c_i} = -\mathbf{p_i} \tag{25}$$

For edge cracks, the external boundary is selected to bisect the embedded crack, what causes the actual crack opening displacements to be computed considering only the real part of the crack to compose the right- and left-hand sides of Eq. (24). In this case, matrix **S** of Eq. (25) has to be computed again.

5 Stress Intensity Factors

The force intensity factors and stress intensity factors are calculated at the Gauss points using the expressions below (Wearing and Ahmadi-Brooghami, 1998),

$$K_1 = \frac{Eh^3c_2}{48\sqrt{2r'}} \quad K_2 = \frac{Eh^3c_1}{48\sqrt{2r'}} \quad K_3 = \frac{5Ehc_3}{24(1+\nu)\sqrt{2r'}} \tag{26}$$

$$K_I = \frac{Ehc_2}{8\sqrt{2r'}} \quad K_{II} = \frac{Ehc_1}{8\sqrt{2r'}} \quad K_{III} = \frac{5Ec_3}{16(1+\nu)\sqrt{2r'}} \tag{27}$$

where $c_i(i = 1, 2, 3)$ represents the existing rotations and deflection discontinuities of the crack line, r' is the distance from the crack tip to the point where c_i is calculated, E is the Young modulus, ν is the Poisson's ratio and h is the thickness of the plate; K_1, K_2 and K_3 are, respectively, the torsion, bending and shear force intensity factors and, K_I, K_{II} and K_{III} are the corresponding stress intensity factors. Usually, the stress intensity factors and force intensity factors are normalized accordingly.

6 Examples

The following results confirm the accuracy of the procedure. The common features associated to the geometric modelling of the examples are stated below. The subdivision of the crack line into segments, for numerical integration purposes, is used in this work since it provides better results especially for shear or torsion loads. The crack line subdivision may be slightly different in each example but with 6 fixed Gauss points in each segment. The segments adjacent to the tips are fixed to a length of 0,01a where 6 Gauss points is enough to provide the numerical stress/force intensity factor limit as $r' \to 0$. In this way, the stress or force intensity factors for the NGF results are calculated directly at the second Gauss point nearest the tip without any regression or manipulation of the solution. The bending moment, torsion and shear distributed applied loads are designated as M, T and Q in the examples that follow.

6.1 Crack in Infinite Medium

Here, the NGF examples consider a crack embedded in an infinite medium, subjected to uniformly distributed moment, torsion and shear. First, two alternative loading positions are compared for they should lead to the same stress intensity factors: first, directly the NGF solution for **c** from the Eq. (25), **p** being the applied force on the crack line and second the solution for remotely loaded cracked plates simulating an infinite medium, considering a 1000a wide plate with a 2a crack length as depicted in Fig. 1. Both results, for the three loading cases of bending, torsion and shear, coincide graphically, as expected, the reason why there are only one curve representing the NGF solution for each ν in Figs. 2 to 6.

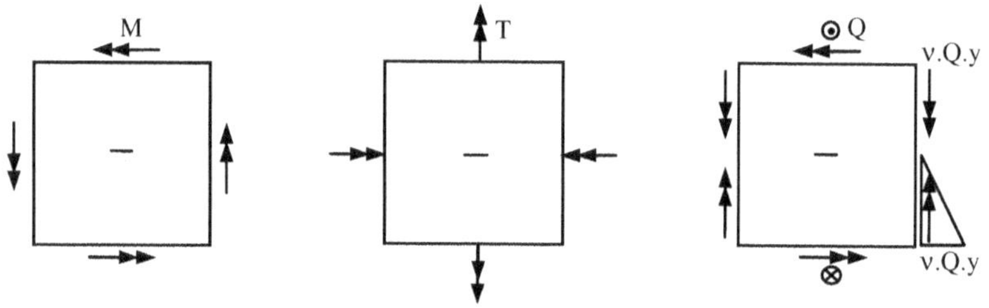

Fig. 1 Remotely loaded cracked plate, under bending moment, torsion and shear loadings, from *left* to *right*: a 1000a wide *square* plate for a 2a crack embedded and a coordinate system centred in crack

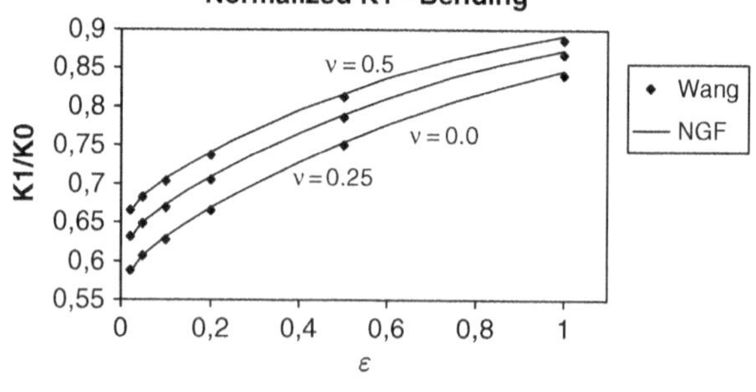

Fig. 2 Normalized moment intensity factor KI/K0= $K_{\text{I}} / M\sqrt{a}$ for bending problem, varying plate thickness ε and $\nu = 0.0$, 0.25 and 0.50

Fig. 3 Normalized torsion intensity factor K2/K0= $K_2/T\sqrt{a}$ for torsion problem, varying plate thickness ε and ν = 0.0, 0.25 and 0.50

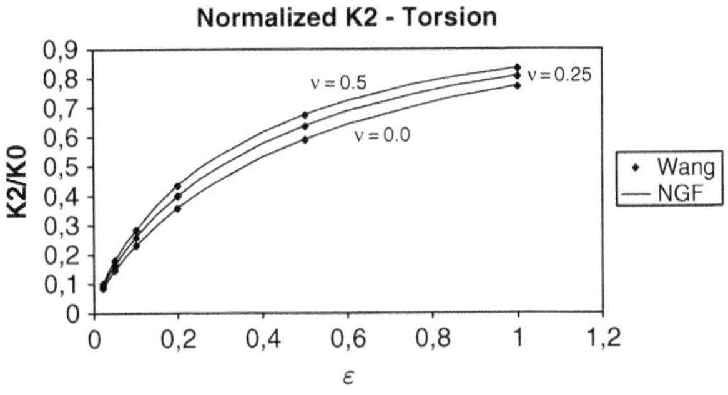

Fig. 4 Normalized shear intensity factor K3/K0= $K_3/\frac{T\sqrt{10a}}{h(1+\nu)}$ for torsion problem, varying plate thickness ε and ν = 0.0, 0.25 and 0.50

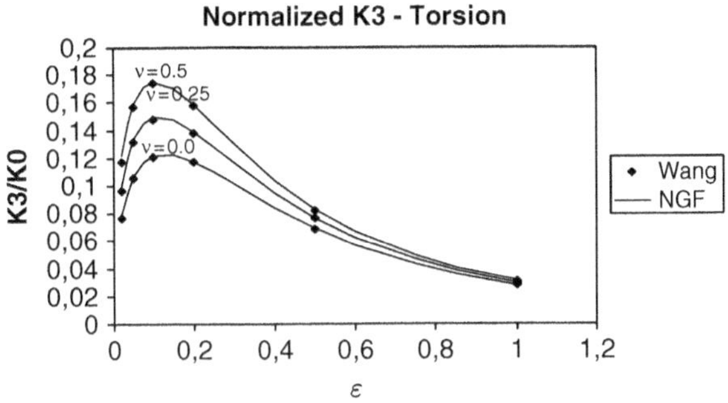

Fig. 5 Normalized torsion force intensity factor K2/K0= $K_2/\frac{Qa\sqrt{10a}}{h(1+\nu)}$ for shear problem, varying plate thickness ε and ν = 0.0, 0.25 and 0.50

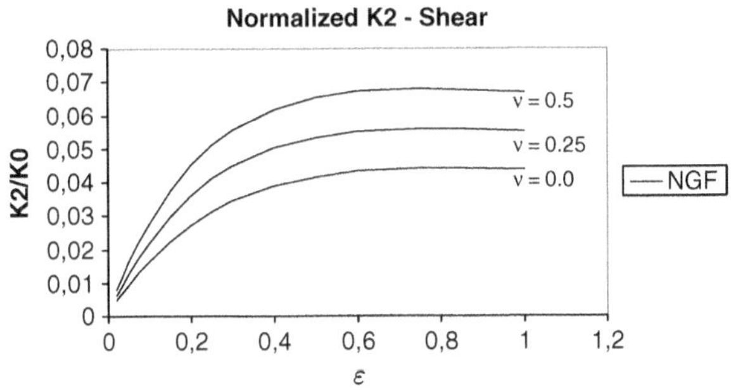

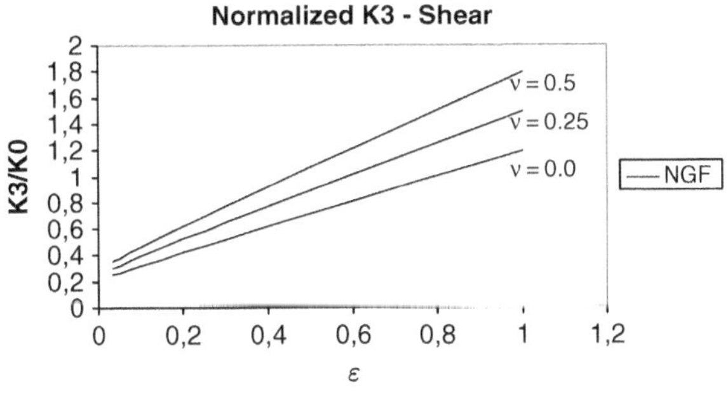

Fig. 6 Normalized shear force intensity factor K3/K0= $K_3/\frac{Q\sqrt{10a}}{h(1+\nu)}$ for shear problem, varying plate thickness ε and ν = 0.0, 0.25 and 0.50

Two examples will be discussed initially, i. e., the bending and torsion applied loads of Fig. 1. In these examples the dimensionless plate thickness, $\varepsilon = h/(a\sqrt{10})$, varies from 0.02 to 1.0, and the Poisson's ratio, ν is chosen to be 0.0, 0.25 and 0.5, like in Wang (1968, 1970). Each side of the finite plate, simulating infinite medium, is discretized with 4 quadratic elements and the crack line is subdivided into 10 segments. The results for the normalized force intensity factors (FIF), K_1, K_2 and K_3, plotted in Figs. 2, 3 and 4, compare well with those presented by Wang (1968, 1970).

The FIF results for the shear load example of Fig. 1 are plotted in Figs. 5 and 6 and left as a reference since not any available comparison was found in the literature.

6.2 Finite Square Plate Subjected to Bending Moment

This example consists of a square plate, with sides equal to 2L, with a centred embedded crack of length $2a$ and subjected to uniform bending moments along its horizontal boundaries, as depicted in Fig. 7. The external boundary is discretized into 32 quadratic elements, 8 being on the boundaries away from the crack, where the bending moment M is applied and 24 on the other two lateral boundaries. These 24 elements reduce in size as they approach the axis of the crack and the smallest one is equal to L/30. The crack line is subdivided into 12 segments, enough to model examples with ratio range $0.05 \leq h/a \leq 4.0$ (for L/a=∞, the range numerically evaluated is $0.005 \leq h/a \leq 4.0$). The NGF normalized moment intensity factors (MIF) are plotted in Fig. 7 for a variety of plate thickness to crack length ratio, h/a.

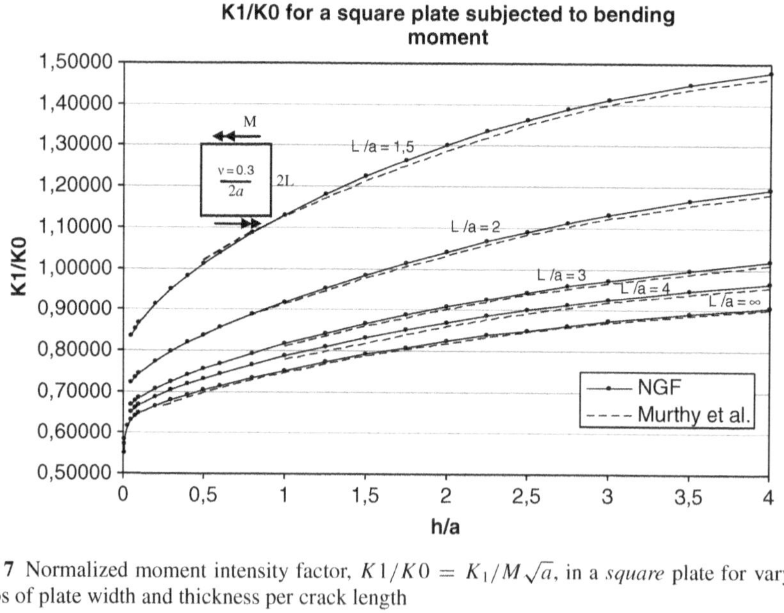

Fig. 7 Normalized moment intensity factor, $K1/K0 = K_1/M\sqrt{a}$, in a *square* plate for varying ratios of plate width and thickness per crack length

Aside from the good results observed in comparison to Murthy et al. (1981), Fig. 7 also shows that the NGF procedure allows for a consistent MIF evaluation, for small values of h/a, not achieved by the reference.

6.3 Two Collinear Cracks in a Finite Plate

A comparison between NGF and previous stress intensity factors evaluated by Boduroglu and Erdogan (1983) is presented in this example. It consists of two symmetrically located collinear cracks, embedded in a plate of finite width and subjected to uniform bending moments applied away from the crack region. Half the plate is modelled, split along its vertical axis of symmetry. The crack length is fixed and is equal to 2, $b = 10$, the thickness of the plate, $h = 1$ and Poisson's ratio, $\nu = 0.3$. The fundamental crack is subdivided into 16 segments to adjust the modelling to the whole varying range $0.12 \leq 2a/(c+d) \leq 0.95$. The external boundary is discretized using 60 quadratic elements. The elements get smaller near the crack axis to avoid integration errors when the crack is too near the external boundary. The smallest element size there is equal to $b/16$. The results for tips c (internal) and d (closer to boundary), shown in Figs. 8 and 9, present less than 1% difference if compared to Boduroglu and Erdogan (1983) for the range considered.

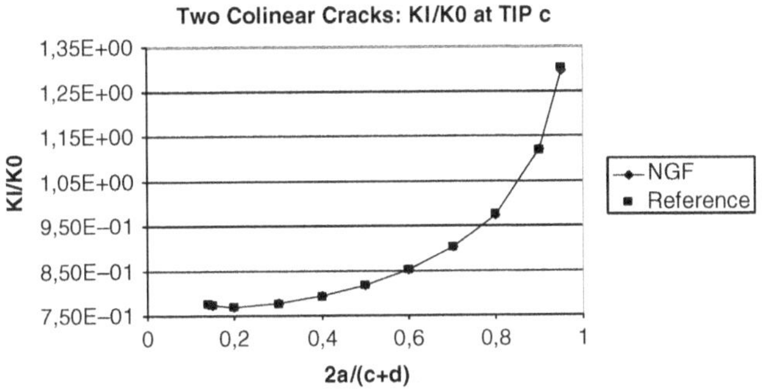

Fig. 8 Normalized stress intensity factor at tip c, $K_I\,/(\sigma_o\sqrt{a})$, $\sigma_o = 6M/h^2$, in a plate of finite width with two collinear cracks embedded and subjected to an uniform remote bending moment, M

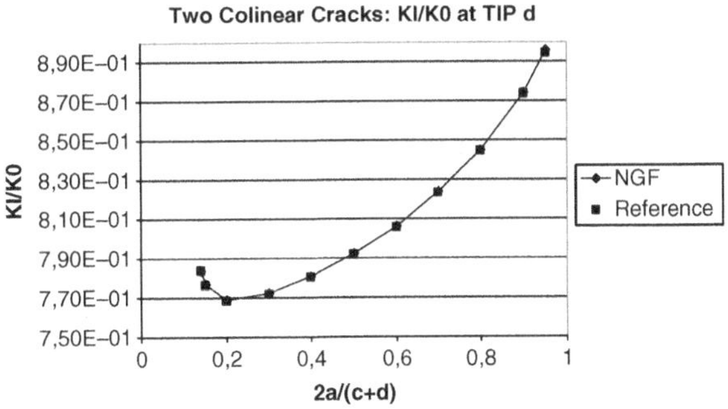

Fig. 9 Normalized stress intensity factor at tip d, $K_I/(\sigma_o\sqrt{a})$, $\sigma_o = 6M/h^2$, in a plate of finite width with two collinear cracks embedded and subjected to an uniform remote bending moment, M

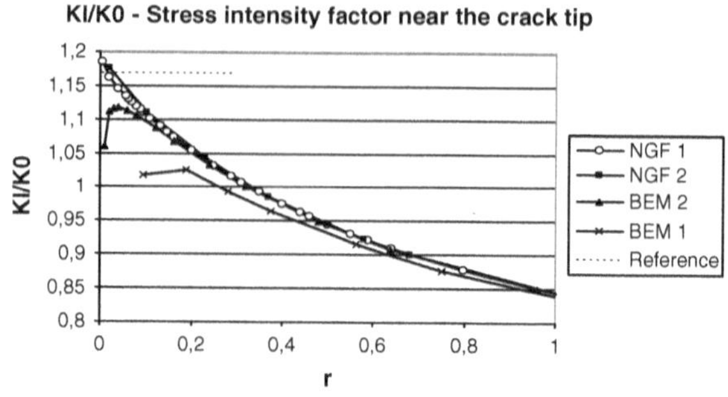

Fig. 10 Normalized stress intensity factors at Gauss or collocation points near the tip; 1 represents a more refined discretization

6.4 Edge Crack in a Finite Plate

A comparison between NGF and the standard boundary element formulation (Karam and Telles, 1988) evaluations of stress intensity factors is illustrated in this example. It consists of two symmetrically located collinear edge cracks of $a = 6$ placed in a plate of finite width, $b = 20$ and subjected to uniform bending moments applied away from the crack region. Poisson's ration is chosen to be 0.3. In this case the problem boundary will split the fundamental crack in two to simulate the edge crack.

Considering the NGF formulation, half the plate is modelled by separating it along the vertical axis of symmetry. The external boundary discretization is exactly the same used in the previous example. The only difference is the introduction of an extra node on the boundary to accommodate the presence of semi-discontinuous elements on the junction of the external boundary and the fundamental crack line to avoid possible singularities in the Green's function generation. 48 and 112 Gauss points are adopted over the crack line (notice that only half the crack and the respective Gauss points lie inside the domain; i.e., 24 and 56 respectivelly).

For the classical formulation, symmetry about the axis of the crack was additionally adopted to model the problem. Two discretization procedures for the crack line have been implemented with the intention of observing the results for the stress intensity factor at the collocating points. Here, 21 and 29 collocation points were adopted over the crack line, the smaller element near the tip is first equal to $0{,}03a$ and then $0{,}0033a$, respectively. Both formulations make use of quadratic elements.

The problem is solved using both techniques and the results are shown in Fig. 10. The stress intensity factor from Boduroglu and Erdogan (1983) is also present in Fig. 10 as a reference; the superior accuracy of the NGF procedure is clearly illustrated here.

7 Conclusion

The proposed NGF procedure uses a numerical solution for a fundamental cracked infinite medium, known as a Green's Function, as the kernel for the BEM technique to solve linear elastic fracture mechanics plate bending problems. Here the Reissner theory of bending of elastic plates is used to numerically obtain the Green's function, aiming at the evaluation of the stress intensity factors for this class of problems. The implementation was found accurate enough and not to need the application of regression formulas to compute the stress intensity factors, since the modelling allows for the positioning of Gauss points very near the tip, almost numerically simulating the limit when $r' \to 0$. The applications include bending moment, torsion and shear force loadings, confirming the precision of the current Reissner's plate NGF procedure proposed, in line with the previous different NGF applications presented by the authors in the recent past.

References

Boduroglu, H., Erdogan, F.: Internal and edge cracks in a plate of finite width under bending, Journal of Applied Mechanics, ASME, 50, 621–629, (1983).

Guimarães, S., Figueiredo, V.S., Telles, J.C.F.: Reissner's Plate Green's function for fracture mechanics (in Portuguese), XXI CILAMCE – Ibero-Latin-American Congress on Computational Methods in Engineering, (L. Vaz, ed.), Rio de Janeiro, November, (2000).

Karam, V.J., Telles, J.C.F.: On boundary elements for Reissner's plate bending, Engineering Analysis with Boundary Elements, 5, 1 21–27, (1988).

Murthy, M.V.V., Raju, K.N., Viswanath, S.: On the bending stress distribution at the tip of a stationary crack form Reissner's theory, International Journal of Fracture, 17, 537–552, (1981).

Reissner, E: On the bending of elastic plates, Quarterly of Applied Mechanics, 5, 55–68, (1947).

Silveira, N.P.P., Guimarães, S., Telles, J.C.F.: A numerical Green's function BEM formulation for crack growth simulation, Engineering Analysis with Boundary Elements, 29(11) 978–985, (2005).

Telles, J.C.F.. A Self-adaptive co-ordinate transformation for efficient numerical evaluation of general boundary element integrals, International Journal for Numerical Methods in Engineering, 24 959–973, (1987).

Telles, J.C.F., Castor, G.S., Guimarães, S.: A numerical Green's function approach for boundary elements applied to fracture mechanics, International Journal for Numerical Methods in Engineering, 38, 3259–3274, (1995).

Telles, J.C.F., Guimarães, S.: Green's function: a numerical generation for fracture mechanics problems via boundary elements, Computer Methods in Applied Mechanics and Engineering, 188 847–858, (2000).

Vander Weeën, F.: Application of the boundary integral equation method to Reissner's plate model, International Journal for Numerical Methods in Engineering, 18, 1–10, (1982).

Wang, N.M. Effects of plate thickness on the bending of an elastic plate containing a crack, Journal of Mathematics and Physics, 47, 371–390, (1968).

Wang, N.M. Twisting of an elastic plate containing a crack, International Journal of Fracture Mechanics, 6(4) 367–378, (1970).

Wearing, J.L., Ahmadi-Brooghami, S.Y.: Fracture analysis of plate bending problems using the boundary element method, In: Plate Bending Analysis With Boundary Elements, (M.H. Aliabadi, ed.), Computational Mechanics Publications, Southampton, (1998).

General Approaches on Formulating Weakly-Singular BIES for PDES

Z. D. Han and S. N. Atluri

Abstract Straight-forward systematic development of the weakly-singular boundary integral equations (BIEs) for general Partial Differential Equations (PDEs) are extended in the present study by utilizing the gradients of the fundamental solutions as the test function, following the original work reported by Okada, Rajiyah, and Atluri (J. Appl. Mech. April, 1988) and its extension by Han and Atluri (2002 CMES 3(6): 699–716) and Qian, Han, Ufimtsev, Atluri (2004 CMES 5(6): 541–562). The weak-forms and their algebraic combinations of the fundamental solutions are directly applied for the derivation of their "intrinsic proprieties", which are used to avoid the hyper-singularities and decomposing the singularities of the kernel functions of BIEs. The derivation of the weakly-singular BIEs based on the present method has been demonstrated by formulating BIEs for elasticity and acoustics with details, respectively. The corresponding formulation have been numerically implemented and verified by solving the numerical examples.

1 Introduction

In the past 25 years, much has been written about the integral equation formulations for the displacement and traction vectors in a solid body. Much of this work has been concentrated on linear elastic, homogenous, and isotropic solids. The focus in these derivations is on the "fundamental solution" in a linear elastic isotropic solid, viz., the Kelvin solution for a unit point load applied at an arbitrary location, in an arbitrary direction, in an infinite linear elastic solid. The Kelvin solution is well-understood, and is "singular". For clarity, we denote the various levels of singularity: If "r" is the distance between any arbitrary point (ξ) in the solid and ($\mathbf{x}$) (the point at which the unit load is applied in a 3-dimensional solid), we denote the $(1/r)$ type singularities as being "weakly-singular", the $(1/r^2)$ type singularities as being "strongly-singular", the $(1/r^3)$ type singularities as being "hyper-singular".

Z.D. Han (✉)
Center for Aerospace Research & Education University of California,
Irvine 5251 California Avenue, Suite 140 Irvine, CA, 92612, USA
e-mail: han.zhidong@gmail.com

G.D. Manolis, D. Polyzos (eds.), *Recent Advances in Boundary Element Methods*,
DOI 10.1007/978-1-4020-9710-2_12,

In the Kelvin solution, for a 3-D solid, it is well-known that the displacement-vector is "weakly-singular", and the stress-tensor is "strongly-singular". In the classical formulation, the integral equation for the displacement vector at any point [$\mathbf{x} \in \Omega$ or $\partial\Omega$] is "strongly singular". If the displacement integral equation at any point $\mathbf{x} \in \Omega$, is differentiated with respect to $\mathbf{x} \in \Omega$, one may obtain an integral equation for $\nabla \mathbf{x}[\nabla = \mathbf{e}_i \frac{\partial}{\partial x_i}]$ for any $\mathbf{x} \in \Omega$. From this equation for the displacement gradients, one may derive a traction boundary integral equation [tBIE] at any $\mathbf{x} \in \Omega$ or $\partial\Omega$. It is seen that this tBIE is "hyper-singular". Much has been written in the last 10~15 years on the "regularization" of the tBIE [i.e., render the "hyper-singular" tBIE into a "weakly-singular" tBIE], through what appears to be laborious mathematical exercises and "manipulations". This literature is too large to discuss here, but excellent summaries may be found in Cruse and Richardson (1996); Bonnet, Maier, and Polizzotto (1998); Li and Mear (1998).

In this paper, we revisit the Kelvin solution, and delineate certain "fundamental properties" of this solution. By using the global-weak-form [or the weighted-residual-equation"] of the momentum balance laws of linear elasticity, corresponding to a point load, on which the Kelvin solution is based, we derive an arbitrary number of these "fundamental properties", by simply using an arbitrary number of different types of "test functions" in writing the weak-forms. These fundamental properties of the Kelvin solution, which otherwise has a $(1/r^2)$ singularity for tractions, are shown to be the key in-gradients in any "regularization" of the tBIE, which is derived by differentiating the strongly-singular displacement integral equation.

On the other hand, as far back as 1988, Okada, Rajiyah, and Atluri (1988, 1989, 1990) have proposed a way to *directly derive* integral equations for $\nabla \mathbf{u}[\nabla = \mathbf{e}_i \frac{\partial}{\partial x_i}]$, rather than first derive the dBIE, and then differentiate it with respect to $\mathbf{x}$ as is most common is literature. Thus, one may also, from Okada, Rajiyah, and Atluri (1988, 1989), *directly derive* a tBIE. Thus, the directly derived tBIE is only "strongly-singular", as opposed to being "hyper-singular". It is shown in the present paper, that by using the "fundamental properties" of the Kelvin solution [which are also derived in the present paper], one may "regularize" the directly derived tBIE of Okada, Rajiyah, and Atluri (1988, 1989) in a very *straight-forward and simple manner*. In a like manner, it is shown here that the dBIE can also be "regularized" in a very straight-forward and simple manner.

It is also shown in this paper that the fundamental Kelvin solution for the stress-tensor can be naturally split into 2 parts, which we denote here as ϕ_{ij}^{*p} and ψ_{ij}^{*p}, respectively, and write $\sigma_{ij}^{*p} = \psi_{ij}^{*p} - \phi_{ij}^{*p}$, where σ_{ij}^{*p} is the Kelvin solution for stresses, ψ_{ij}^{*p} is divergence free, and the divergence of ϕ_{ij}^{*p} is the Dirac function.

We also discuss the numerical solution, by discretization, of the directly derived, tBIE of Okada, Rajiyah, and Atluri (1988, 1989), as well as of the regularized, weakly-singular dBIE. We write the general Petrov-Galerkin types of weak-forms of these integral equations at $\partial\Omega$. Thus, we introduce an arbitrary test function $\mathbf{w}(\mathbf{x})$, $\mathbf{x} \in \partial\Omega$. If $\mathbf{w}(\mathbf{x})$ us a Dirac function, and if the trial functions $\mathbf{t}(\mathbf{x})$ and $\mathbf{u}(\mathbf{x})$ are interpolated in terms of their nodal values over a contiguous (non-overlapping) set of elements at $\partial\Omega$ [boundary elements], one obtains the popular "Boundary Element Methods". On the other hand, if $\mathbf{w}(\mathbf{x})$ is chosen to be the same as the complementary [or energy-conjugate] trial function [i.e., in the tBIE, we use $\mathbf{u}(\mathbf{x})$ as the test function, and in the dBIE, we use $\mathbf{t}(\mathbf{x})$ as the test function], we obtain the so-called "Symmetric Galerkin Approaches" for dBIE and tBIE [Bonnet, Maier, and Polizzotto, 1998]. On the other hand, one may leave $\mathbf{w}(\mathbf{x})$ as arbitrary, and formulate a general Petrov-Galerkin Approach. It is further shown that, in the "Symmetric Galerkin" approach to solving the directly derived tBIE [Okada, Rajiyah, and Atluri, 1988, 1989], using the natural split of σ_{ij}^{*p}, and the use of the Stokes' theorem at $\partial\Omega$ when $\mathbf{w}(\mathbf{x})$ is a continuous function, the resulting discrete formulation results in certain further algebraic conveniences. These end results are found to be somewhat similar to the results in Li and Mear (1998), but the present results are different from those in Li and Mear (1998) in terms of the attendant kernel functions. However, the present formulations of the Patrov-Galerkin approaches to solving the directly derived tBIE [Okada, Rajiyah, and Atluri, 1988, 1989] are simple and straight-forward.

The present method is generic for developing the similar non-hyper-singular BIEs for other PDEs. Another examples demonstrated in the present paper is to derive the BIEs for the Helmholtz equations for acoustics. It is well known that the difficulties in dealing with hyper-singular integrals and the nonuniqueness are two of the major drawbacks of the boundary integral equation (BIE) methods for solving acoustic problems. The nonuniqueness of solutions, at certain frequencies of the associated acoustic problem in the interior of the solid body, is actually a purely mathematical issue arising from the boundary integral formulation, without any physical significance, as involved in the CHIEF method [Schenck, 1968]. Burton, and Miller (1971) developed a combination of the surface Helmholtz, integral equation for potential, and the integral equation for the normal derivative of potential at the surface, to circumvent the problem of nonuniqueness at characteristic frequencies. Their method was labeled as CHIE (Composite Helmholtz Integral Equation), or CONDOR (Composite Outward Normal Derivative Overlap Relation) by Reut (1985). The CHIE method introduces the hypersingular integrals, which are computationally costly. Moreover, in CHIE method, the accuracy of the integrations affects the results, and the conventional Gauss quadrature can not be used directly. Regularization techniques are commonly employed by the followers of the CHIE methodology, to improve the approach by reducing the problem to the one involving $O(r^{-1})$ singular integrals near the point of singularity. Chien, Rajiyah, and Atluri (1990) employed some known identities of the fundamental solution from the associated interior Laplace problem, to regularize the hypersingular integrals. This concept was used by many successive researchers: Hwang (1997) reduced the singularity of the Helmholtz integral equation also by using some identities from the associated Laplace equation. Yang (2000) also uses the identities of the fundamental solution of the Laplace problem, to efficiently solve the problem of acoustic scattering from a rigid body. Besides, Meyer, Bell, Zinn, and Stallybrass (1978) and Terai (1980) developed regularization techniques for planar elements. The regularized normal derivative equation [Wu, Seybert, and Wan, 1991] proposed by Wu et al. converged in the Cauchy principal value sense, rather than in the finite-part sense. The computation of tangential derivatives was required everywhere on the boundary. Another way commonly used in the literature is to develop the methods to directly evaluate the hypersingular integrals. To solve the intensive computation of double surface integral, Yan, Hung, and Zheng (2003) employed the concept of a discretized operator matrix to replace the evaluation of double surface integral with the evaluation of two discretized operator matrices. In summary, most of *regularization techniques* for evaluating the hyper-singular integrals in the acoustic BIEs arise from certain identities associated with the fundamental solution to the Laplace equations. In the present paper, novel non-hyper-singular boundary integral equations are derived directly, for the gradients of the velocity potential, in the same way as we derive the BIEs for the elastics. The acoustic potential gradients are related to the sound velocity in their physical meaning. The basic idea is to use the gradients of the fundamental solution to the Helmholtz differential equation for velocity potential, as *vector test-functions* to write the weak-form of the original Helmholtz differential equation for potential, and thereby directly derive *non-hyper-singular boundary integral equations for velocity potential gradients.* The current method can be shown to be fundamentally different from the regularized normal derivative equation by Wu, Seybert, and Wan (1991), which used tangential derivatives to reduce the singularity.

The structure of the paper contains two parts: we derive the BIEs for the elastics in Part I, and those for the acoustics in Part II. In Part I, we briefly discuss the well-known Galerkin vector potential for displacements in an elastic solid undergoing small displacements. We derive the displacement equations (dBIE), and *directly derive* the the traction BIE (tBIE) without differentiating the dBIE, and a large number of basic properties of σ_{ij}^{*p}, by using the weak-forms [with different test functions], of the balance laws for σ_{ij}^{*p}. The regularization of tBIE and dBIE is also demonstrated in Part I. In Part II, the boundary integral equations for the potential [labeled here as ϕ-BIE], and its gradient [labeled here as q-BIE], are directly derived and contain only strongly singular [$O(r^{-2})$]. The further regularization of these *strongly singular* ϕ-BIE, and q-BIE, to *only weakly singular* [$O(r^{-1})$] types, which are labeled here as R-ϕ-BIE, R-q-BIE, respectively, is achieved *by using certain basic identities of the fundamental solution of the Helmholtz differential equation for potential.* These basic identities, in their most general form, are also newly derived in this paper in systematic way. In addition, we formulate general Petrov-Galerkin methods to solve the R-ϕ-BIE, and R-q-BIE, in their weak senses. By using the test functions in these Petrov-Galerkin schemes to be the energy-conjugates of the respective trial functions, we develop Symmetric Galerkin Boundary Element Methods (SGBEM) for solving R-ϕ-BIE, and R-q-BIE, respectively. We label these SGBEM as SGBEM-R-ϕ -BIE and SGBEM-R-q-BIE, respectively. These SGBEM-R-ϕ -BIE and SGBEM-R-q-BIE are totally different from the ones in Chen, Hofstetter, and Mang (1997), Gray and Paulino (1997). In the present SGBEM-R-ϕ -BIE and SGBEM-R-q-BIE, C^0 continuity of ϕ and q over the boundary elements is sufficient for numerical implementation. Though the double surface integral may increase the numerical accuracy of the SGBEM, many fast SGBEM methods are proposed to speed up the computation of double surface integrals, including panel clustering methods, wavelet methods and so on Aimi, Diligenti, Lunardini, and Salvadori (2003), Breuer, Steinbach, and Wendland (2002).

2 Part I: Weakly-Singular BIEs for Elasticity

2.1 *Galerkin Vector Potential for Displacements in an Elastic Solid Undergoing Small Deformations*

Consider a linear elastic, homogeneous, isotropic body in a domain Ω, with boundary $\partial\Omega$. The Lame' constants of the linear elastic isotropic body are λ and μ; and the corresponding Young's modulus and Poisson's ratio are E and υ, respectively. We use Cartesian coordinates ξ_i, and the attendant base vectors $\mathbf{e}_i$, to describe the geometry in Ω. The solid is assumed to undergo infinitesimal deformations. The displacement vector, strain-tensor, and the stress-tensor in the elastic body are denoted as $\mathbf{u}$, ε and σ, respectively, with the corresponding dyadic representations, as follows:

$$\mathbf{u} = u_i \mathbf{e}_i; \tag{1}$$

$$\varepsilon = \varepsilon_{ij} \mathbf{e}_i \mathbf{e}_j; \tag{2}$$

$$\sigma = \sigma_{ij} \mathbf{e}_i \ \mathbf{e}_j \tag{3}$$

The equations of balance of linear and angular momentum can be written as:

$$\nabla \cdot \sigma + \mathbf{f} = \mathbf{0}; \quad \sigma = \sigma^{t}; \quad \nabla = \mathbf{e}_i \frac{\partial}{\partial \xi_i}; \quad \sigma_{ij,j} + f_i = 0; \quad \sigma_{ij} = \sigma_{ji} \tag{4}$$

The strain-displacement relations are:

$$\varepsilon = \frac{1}{2}(\nabla \mathbf{u} + \mathbf{u}\nabla); \quad \varepsilon_{ij} = \frac{1}{2}(u_{i,j} + u_{j,i}) \tag{5}$$

The constitutive relations of an isotropic linear elastic homogeneous solid are:

$$\begin{aligned} \sigma &= \lambda\, \mathbf{I}(\nabla \cdot \mathbf{u}) + 2\mu\varepsilon = \lambda\, \mathbf{I}(\nabla \cdot \mathbf{u}) + \mu(\nabla \mathbf{u} + \mathbf{u}\nabla) \\ &= \mu \left[\frac{1}{A}\, \mathbf{I}(\nabla \cdot \mathbf{u}) + \nabla \mathbf{u} + \mathbf{u}\nabla - 2\, \mathbf{I}(\nabla \cdot \mathbf{u}) \right] \end{aligned} \tag{6}$$

where

$$A = \frac{\mu}{\lambda + 2\mu} = \frac{1 - 2\upsilon}{2(1 - \upsilon)}$$

It is well known [Fung and Tong, 2001] that the displacement vector, which is a continuous function of ξ, can be derived, in general, from the Galerkin vector potential ϕ such that:

$$\mathbf{u} = \nabla^2\phi - \frac{1}{2(1-\upsilon)}\nabla(\nabla\cdot\phi) = A\nabla\Phi + \nabla\times\Psi = \mathbf{u}^{\Phi} + \mathbf{u}^{\Psi} \tag{7}$$

where, by definition,

$$\mathbf{u}^{\Phi} = A\nabla\Phi = A\nabla^2\phi;\ \Phi = \nabla\cdot\phi \tag{8}$$

$$\mathbf{u}^{\Psi} = \nabla\times\Psi = -\nabla\times\nabla\times\phi = \nabla^2\phi - \nabla(\nabla\cdot\phi);\ \Psi = -(\nabla\times\phi) \tag{9}$$

2.2 Fundamental Solutions in a Linear Elastic Isotropic Homogeneous Infinite Medium

Consider a point unit load applied in an arbitrary direction $\mathbf{e}^p$ at a generic location $\mathbf{x}$ in a linear elastic isotropic homogeneous infinite medium as shown in Fig. 1. It is well-known [Fung and Tong, 2001] that the displacement solution corresponding to this unit point load is given by the Galerkin vector displacement potential:

$$\phi^{*p} = (1-\upsilon)F^*\mathbf{e}^p \tag{10}$$

where

$$F^* = \frac{r}{8\pi\mu(1-\upsilon)} \quad \text{for 3D problems} \tag{11a}$$

and

$$F^* = \frac{-r^2\ln r}{8\pi\mu(1-\upsilon)} \quad \text{for 2D problems} \tag{11b}$$

where $r = \|\xi - \mathbf{x}\|$

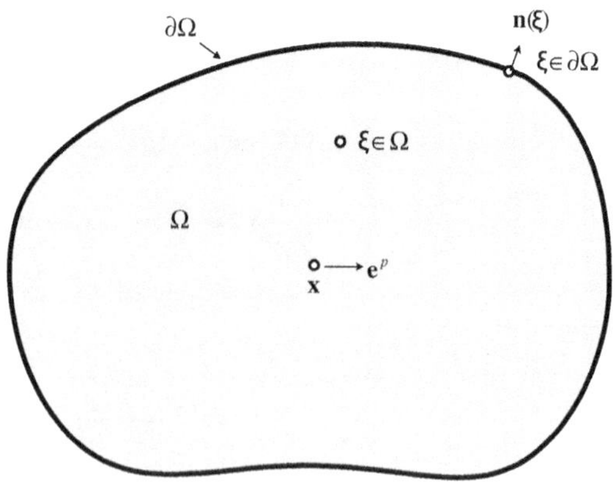

Fig. 1 A solution domain with source point **x** and target point ξ

Thus, ϕ^{*p} in (10) is the solution, in infinite space, to the differential equation (in the coordinates ξ),

$$\mu(1-\upsilon)\nabla^2\nabla^2 F^* + \delta(\mathbf{x},\xi) = 0 \tag{12}$$

The corresponding displacements are derived from the Galerkin vector displacement potential, using (7), as:

$$u_i^{*p}(\mathbf{x},\xi) = (1-\upsilon)\delta_{pi}F^*_{,kk} - \frac{1}{2}F^*_{,pi} \tag{13}$$

The gradients of the displacements in (13) are:

$$u_{i,j}^{*p}(\mathbf{x},\xi) = (1-\upsilon)\delta_{pi}F^*_{,kkj} - \frac{1}{2}F^*_{,pij} \tag{14}$$

The corresponding stresses in a linear elastic homogenous isotropic body are given by:

$$\sigma_{ij}^{*p}(\mathbf{x},\xi) = \mu[(1-\upsilon)\delta_{pi}F^*_{,kkj} + \upsilon\delta_{ij}F^*_{,pkk} - F^*_{,pij}] + \mu(1-\upsilon)\delta_{pj}F^*_{,kki} \tag{15}$$

These stresses are seen to satisfy the balance laws:

$$\sigma_{ij,i}^{*p}(\mathbf{x},\xi) = \mu(1-\upsilon)\delta_{pj}F^*_{,kkii} = -\mu\delta_{pj}\delta(\mathbf{x},\xi); \quad \sigma_{ij}^{*p} = \sigma_{ji}^{*p} \tag{16}$$

We define two functions ϕ_{ij}^{*p} and ψ_{ij}^{*p}, as

$$\phi_{ij}^{*p}(\mathbf{x},\xi) \equiv -\mu(1-\upsilon)\delta_{pj}F^*_{,kki} \tag{17a}$$

$$\psi_{ij}^{*p}(\mathbf{x},\xi) \equiv \sigma_{ij}^{*p}(\mathbf{x},\xi) + \phi_{ij}^{*p}(\mathbf{x},\xi) = \mu[(1-\upsilon)\delta_{pi}F^*_{,kkj} + \upsilon\delta_{ij}F^*_{,pkk} - F^*_{,pij}] \tag{17b}$$

Then, from (16) and (17), it can be seen that:

$$\phi_{ij,i}^{*p}(\mathbf{x},\xi) = -\sigma_{ij,i}^{*p}(\mathbf{x},\xi) = -\mu(1-\upsilon)\delta_{pj}F^*_{,kkii} = \mu\delta_{pj}\delta(\mathbf{x},\xi) \tag{18a}$$

$$\psi_{ij,i}^{*p}(\mathbf{x},\xi) = \mathbf{0}\ [\text{divergence of}\ \psi^*(\mathbf{x},\xi) = \mathbf{0}] \tag{18b}$$

Hence, as a divergence free tensor, $\psi^*(\mathbf{x},\xi)$ must be a curl of another divergence free tensor. We choose to rewrite it in term of F^* from Eq. (17b), as:

$$\begin{aligned}\psi_{ij}^{*p}(\mathbf{x},\xi) &= \mu[(1-\upsilon)\delta_{pi}F^*_{,kkj} + \upsilon\delta_{ij}F^*_{,pkk} - F^*_{,pij}] \\ &= \mu e_{tis}[(1-\upsilon)e_{tpj}F^*_{,kk} - e_{tkj}F^*_{,pk}]_{,s} \equiv e_{ist}G^{*p}_{tj,s}\end{aligned} \tag{19}$$

where, by definition,

$$G_{ij}^{*p}(\mathbf{x},\xi) = \mu[(1-\upsilon)e_{ipj}F^*_{,kk} - e_{ikj}F^*_{,pk}] \tag{20}$$

The derivatives of F^* for 3D and 2D problems are detailed in Han; Atluri (2003a).

2.3 Displacement & Traction BIE: Derivations From Unsymmetric Weak Forms of Balance Laws in Elasticity

For the present, we ignore the body forces f_i (but include them later, when necessary). The governing equations are reduced to:

$$\sigma_{ij,i} = 0 \text{ in} \Omega \tag{21}$$

For a homogeneous linear elastic isotropic homogeneous solid, the constitutive equation is

$$\sigma_{ij} = E_{ijkl}\varepsilon_{kl} = E_{ijkl}u_{k,l}, \tag{22}$$

and

$$\varepsilon_{kl} = \frac{1}{2}(u_{k,l} + u_{l,k}) \tag{23}$$

and

$$E_{ijkl} = \lambda\delta_{ij}\delta_{kl} + \mu(\delta_{ik}\delta_{jl} + \delta_{il}\delta_{jk}), \tag{24}$$

with λ and μ being the Lame's constants.

Let u_i be the trial functions for displacements, to satisfy Eq. (21), in terms of u_i, when Eqs. (22), (23) and (24) are used. Let $\bar{u}_j$ be the test functions to satisfy the momentum balance laws in terms of u_i, in a weak form. The weak form of the equilibrium Eq. (21) can then be written as,

$$\int_\Omega \sigma_{ij,i}\bar{u}_j d\Omega \equiv \int_\Omega (E_{ijmn}u_{m,n})_i \bar{u}_j d\Omega = 0 \tag{25}$$

Applying the divergence theorem two times* in Eq. (25), we obtain:

$$0 = \int_{\partial\Omega} n_i E_{ijmn}u_{m,n}\bar{u}_j \, dS - \int_{\partial\Omega} n_n E_{ijmn}u_m\bar{u}_{j,i} \, dS + \int_\Omega u_m(E_{ijmn}\bar{u}_{j,i})_{,n} \, d\Omega \tag{26}$$

---

* If we use the divergence theorem only once in Eq. 1)a)i)(1)(25), we obtain the "symmetric" weak form:

$$\int_\Omega \sigma_{ij}\bar{u}_{j,i} d\Omega - \int_{\partial\Omega} n_i\sigma_{ij}\bar{u}_j d\Omega = 0$$

Thus, in the symmetric weak form, both the trial functions u_i as well as the test functions $\bar{u}_j$ are only required to be once differentiable. However, in the "unsymmetric weak form" of Eq. 1)a)i)(1)(26), the test functions $\bar{u}_j$ in Ω are required be twice-differentiable, while there is no differentiability requirement on u_i in Ω.

Instead of the ***scalar*** weak form of Eq. (21), as in Eq. (25), we may also write a ***vector*** weak form of Eq. (21), by using the tensor test functions $\bar{u}_{i,k}$ [as originally proposed in Okada, Rajiyah, and Atluri, 1989] as:

$$\int_{\Omega} \sigma_{ij,i} \bar{u}_{j,k} d\Omega = 0 \quad k = 1, 2, 3 \tag{27}$$

By applying divergence theorem *three times* in (27), we may write:

$$0 = \int_{\Omega} E_{ijmn} u_{m,ni} \bar{u}_{j,k} \, d\Omega = \int_{\partial\Omega} n_i E_{ijmn} u_{m,n} \bar{u}_{j,k} \, dS - \int_{\partial\Omega} n_k E_{ijmn} u_{m,n} \bar{u}_{j,i} \, dS$$
$$+ \int_{\partial\Omega} n_n E_{ijmn} u_{m,k} \bar{u}_{j,i} \, dS - \int_{\Omega} u_{m,k} (E_{ijmn} \bar{u}_{j,i})_{,n} \, d\Omega \tag{28}$$

By taking the fundamental solution $u_i^{*p}(\mathbf{x}, \xi)$ as the test function $\bar{u}_i(\xi)$, we rewrite Eqs. (26) and (28), respectively, as,

$$u_p(\mathbf{x}) = \int_{\partial\Omega} t_j(\xi) u_j^{*p}(\mathbf{x}, \xi) \, dS - \int_{\partial\Omega} u_m(\xi) t_m^{*p}(\mathbf{x}, \xi) \, dS \tag{29}$$

$$-u_{p,k}(\mathbf{x}) = \int_{\partial\Omega} n_i(\xi) E_{ijmn} u_{m,n}(\xi) u_{j,k}^{*p}(\mathbf{x}, \xi) \, dS$$
$$- \int_{\partial\Omega} n_k(\xi) E_{ijmn} u_{m,n}(\xi) u_{j,i}^{*p}(\mathbf{x}, \xi) \, dS$$
$$+ \int_{\partial\Omega} n_n(\xi) E_{ijmn} u_{m,k}(\xi) u_{j,i}^{*p}(\mathbf{x}, \xi) \, dS \tag{30}$$

Equations (29) and (30) were originally given in Okada, Rajiyah, and Atluri (1988, 1989), and the notion of using unsymmetric weak-forms of the differential equations, to obtain integral representations for displacements, was presented in Atluri (1985). It should be noted that the integral equations for $u_p(\mathbf{x})$ and $u_{p,k}(\mathbf{x})$ as in (29) and (30) are derived independently of each other. On the other hand, if we derive the integral equation for displacement-gradients, by directly differentiating $u_p(\mathbf{x})$ in Eq. (29), i.e. by differentiating,

$$u_p(\mathbf{x}) = \int_{\partial\Omega} t_j(\xi) u_j^{*p}(\mathbf{x}, \xi) \, dS - \int_{\partial\Omega} u_m(\xi) t_m^{*p}(\mathbf{x}, \xi) \, dS$$

with respect to x_k, we obtain:

$$u_{p,k}(\mathbf{x}) = \int_{\partial\Omega} t_j(\xi) u_{j,k}^{*p}(\mathbf{x}, \xi) \, dS - \int_{\partial\Omega} u_m(\xi) t_{m,k}^{*p}(\mathbf{x}, \xi) \, dS \tag{31}$$

Thus, Eq. (31) is hypersingular, since $t_{m,k}^{*p}(\mathbf{x}, \xi)$ is of $O(r^{-3})$ for 3D problems. On the other hand, the directly derived integral equations for $u_{p,k}(\mathbf{x}, \xi)$ as in Eq. (30) contain only singularities of $O(r^{-2})$.

Equation (29) is the original displacement BIE (dBIE) in its strongly-singular form before any regularization. On the other hand, Eq. (30) are the non-hypersingular (strongly-singular) integral equations for displacement gradients in a homogeneous linear elastic solid, as originally derived in Okada, Rajiyah, and Atluri (1988, 1989). It is but a simple extension to derive a non-hypersingular integral equation for tractions in a linear elastic solid, from Eq. (30),

$$
\begin{aligned}
-E_{abpk}u_{p,k}(\mathbf{x}) = E_{abpk}\int_{\partial\Omega} t_j(\xi)u_{p,k}^{*j}(\mathbf{x}, \xi)\, dS \\
+E_{abpk}\int_{\partial\Omega} D_t u_m(\xi) e_{nkt}\sigma_{nm}^{*p}(\mathbf{x}, \xi)\, dS
\end{aligned} \tag{32}
$$

where the surface tangential operator D_t is defined as,

$$
D_t = n_r e_{rst}\frac{\partial}{\partial \xi_s} \tag{33}
$$

Then Eq. (32) can be re-written as,

$$
-\sigma_{ab}(\mathbf{x}) = \int_{\partial\Omega} t_q(\xi)\sigma_{ab}^{*q}(\mathbf{x}, \xi)\, dS + \int_{\partial\Omega} D_p u_q(\xi)\Sigma_{abpq}^{*}(\mathbf{x}, \xi)\, dS \tag{34}
$$

where by definition,

$$
\Sigma_{ijpq}^{*}(\mathbf{x}, \xi) = E_{ijkl}e_{nlp}\sigma_{nq}^{*k}(\mathbf{x}, \xi) \tag{35}
$$

Contracting Eq. (34) with $n_a(\mathbf{x})$, we have

$$
-t_b(\mathbf{x}) = \int_{\partial\Omega} t_q(\xi)n_a(\mathbf{x})\sigma_{ab}^{*q}(\mathbf{x}, \xi)\, dS + \int_{\partial\Omega} D_p u_q(\xi)n_a(\mathbf{x})\Sigma_{abpq}^{*}(\mathbf{x}, \xi)\, dS \tag{36}
$$

where the traction is defined as,

$$
t_b(\mathbf{x}) = n_a(\mathbf{x})\sigma_{ab}(\mathbf{x}) \tag{37}
$$

2.4 Some Basic Properties of the Fundamental Solution

Consider a body of an infinite extent, subject to a point force at a generic location $\mathbf{x}$ in the direction $\mathbf{e}^p$, as shown in Fig. 1. The fundamental solution, in infinite space, of the stress field, denoted by $\sigma^{*p}(\mathbf{x}, \xi)$, at any point ξ due to this point load at $\mathbf{x}$, is generated by the balance law:

$$\nabla \cdot \sigma^{*p}(\mathbf{x}, \xi) + \delta(\mathbf{x}, \xi)\mathbf{e}^p = \mathbf{0} \tag{38}$$

We write the weak form of Eq. (38) over the domain, using a constant c as a test function, as

$$\int_{\Omega} \nabla \cdot \sigma^{*p}(\mathbf{x}, \xi)\, c d\Omega + \mathbf{e}^p c = \mathbf{0} \tag{39a}$$

or

$$\int_{\partial\Omega} t_j^{*p}(\mathbf{x}, \xi) dS + \delta_{pj} = 0 \quad \mathbf{x} \in \Omega \tag{39b}$$

Equation (39) is a "basic identity" of the fundamental solution $\sigma^{*p}(\mathbf{x}, \xi)$. Equation (39) is simply an affirmation of the force balance law for Ω : if the point load is applied at a point $\mathbf{x} \in \Omega$ when Ω is entirely embedded in an infinite space, the tractions exerted by the surrounding infinite body on the finite-body Ω should be equilibrated with the applied point force at $\mathbf{x}$ inside Ω.

Secondly, if we consider an arbitrary function $\mathbf{u}(\mathbf{x})$ in Ω as the test function, we may write the corresponding weak-form, from Eq. (38), as:

$$\int_{\Omega} [\nabla \cdot \sigma^{*p}(\mathbf{x}, \xi) + \delta(\mathbf{x}, \xi)\mathbf{e}^p] \cdot \mathbf{u}(\mathbf{x}) d\Omega = \mathbf{0} \tag{40a}$$

or

$$\int_{\partial\Omega} t_j^{*p}(\mathbf{x}, \xi) u_j(\mathbf{x}) dS + u_p(\mathbf{x}) = 0 \quad \mathbf{x} \in \Omega \tag{40b}$$

Once the point $\mathbf{x}$ approaches a smooth boundary, i.e. $\mathbf{x} \in \partial\Omega$, the first term in Eq. (39b) can be written in a Cauchy Principal value (CPV) integral, denoted by $\int^{\mathrm{CPV}}$, as,

$$\lim_{\mathbf{x}\to\partial\Omega} \int_{\partial\Omega} t_j^{*p}(\mathbf{x}, \xi) dS = \int_{\partial\Omega}^{\mathrm{CPV}} t_j^{*p}(\mathbf{x}, \xi) dS - \frac{1}{2}\delta_{pj} \tag{41a}$$

and thus, one obtains:

$$\int_{\partial\Omega}^{\mathrm{CPV}} t_j^{*p}(\mathbf{x}, \xi) dS + \frac{1}{2}\delta_{pj} = 0 \quad \mathbf{x} \in \partial\Omega \tag{41b}$$

The second term on the right hand-side of Eq. (41a) results from the principal value of the singular integral involving t_j^{*p}, which has a $O(1/r^2)$ singularity. Eq. (41b) may also be physically explained as below. σ_{ij}^{*p} (and thus t_j^{*p}) are solutions due to a point load applied in an infinite space. In reality, the point load can

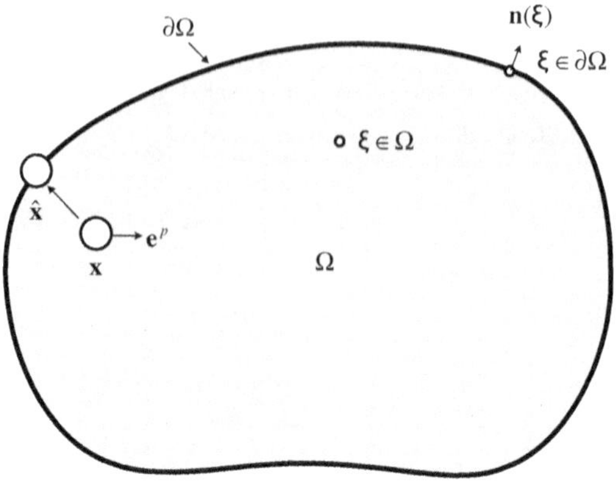

Fig. 2 A loading point **x** approaching the boundary

be assumed to be distributed over a small-sphere, of radius ε, in an infinite body. The tractions distributed over this sphere, that result in a point load, are of $O(1/\varepsilon^2)$; while the surface area of the sphere is $O(\varepsilon^2)$. As long as this sphere is inside Ω, and while Ω is a part of the infinite space, the load applied on Ω is still unity. Suppose $\mathbf{x} \to \hat{\mathbf{x}}$ at $\partial\Omega$ shown in Fig. 2, then the sphere of radius ε is centered at the boundary. As long as the boundary is smooth, only one-half of the sphere of radius ε is actually inside Ω, when $\mathbf{x} \to \hat{\mathbf{x}}$ at $\partial\Omega$. Thus while the load applied, in infinite space, on a sphere of radius ε at $\hat{\mathbf{x}} \in \partial\Omega$, is still unity, the actual load applied on Ω is only 1/2. Thus we obtain Eq. (41b).

We can write Eqs. (39) and (40) for $\mathbf{x} \in \partial\Omega$, with Eq. (41), as:

$$\int_{\partial\Omega}^{CPV} t_j^{*p}(\mathbf{x}, \xi)dS + \frac{1}{2}\delta_{pj} = 0 \quad \mathbf{x} \in \partial\Omega \tag{42}$$

$$\int_{\partial\Omega}^{CPV} t_j^{*p}(\mathbf{x}, \xi)u_j(\mathbf{x})dS + \frac{1}{2}\delta_{pj}u_j(\mathbf{x}) = 0 \quad \mathbf{x} \in \partial\Omega \tag{43}$$

From Eq. (18a), we also see that

$$-\int_{\partial\Omega}^{CPV} t_j^{*p}(\mathbf{x}, \xi)dS = \int_{\partial\Omega}^{CPV} n_i(\xi)\phi_{ij}^{*p}(\mathbf{x}, \xi)dS \tag{44}$$

for both $\mathbf{x} \in \Omega$ and $\mathbf{x} \in \partial\Omega$. Thus, we can write identities for ϕ_{ij}^{*p} which are similar to those in Eqs. (39d) and (40d) for σ_{ij}^{*p}, as:

$$\int_{\partial\Omega}^{CPV} n_i(\xi)\phi_{ij}^{*p}(\mathbf{x}, \xi)dS - C\delta_{pj} = 0 \tag{45}$$

and

$$\int_{\partial\Omega}^{CPV} n_i(\xi)\phi_{ij}^{*p}(\mathbf{x},\xi)u_j(\mathbf{x})dS - Cu_p(\mathbf{x}) = 0 \tag{46}$$

where $C = \frac{1}{2}$ for $\mathbf{x} \in \partial\Omega$ and $C = 1$ for $\mathbf{x} \in \Omega$.

The corresponding equations for Ψ_{ij}^{*p} can also be written as,

$$\int_{\partial\Omega} n_i(\xi)\Psi_{ij}^{*p}(\mathbf{x},\xi)dS = 0 \tag{47}$$

and

$$\int_{\partial\Omega} n_i(\xi)\Psi_{ij}^{*p}(\mathbf{x},\xi)\cdot u_j(\mathbf{x})dS = 0 \tag{48}$$

Third, we consider the weak form of Eq. (38), and consider the test functions to be the gradients of an arbitrary function $\mathbf{u}(\xi)$ in Ω. This function $\mathbf{u}(\xi)$ is so chosen that it has constant gradients, as:

$$u_{j,k}(\xi) = u_{j,k}(\mathbf{x}) \tag{49}$$

Then, the weak form of Eq. (38) may be written as:

$$\int_{\Omega} [\sigma_{ij,i}^{*p}(\mathbf{x},\xi) + \delta(\mathbf{x},\xi)e_j^p]u_{j,k}(\mathbf{x})d\Omega = 0 \tag{50a}$$

Applying the divergence theorem, we obtain,

$$\int_{\partial\Omega} t_m^{*p}(\mathbf{x},\xi)u_{m,k}(\mathbf{x})d\Omega + u_{p,k}(\mathbf{x}) = 0 \quad \mathbf{x} \in \Omega \tag{50b}$$

In addition, we may observe that the first two terms in Eq. (30) have the following identity, as:

$$\begin{aligned}\int_{\partial\Omega} n_i(\xi)E_{ijmn}u_{m,n}(\mathbf{x})u_{j,k}^{*p}(\mathbf{x},\xi)\,dS - \int_{\partial\Omega} n_k(\xi)E_{ijmn}u_{m,n}(\mathbf{x})u_{j,i}^{*p}(\mathbf{x},\xi)\,dS \\ = \int_{\Omega} E_{ijmn}u_{m,n}(\mathbf{x})u_{j,ki}^{*p}(\mathbf{x},\xi)\,dS - \int_{\Omega} E_{ijmn}u_{m,n}(\mathbf{x})u_{j,ik}^{*p}(\mathbf{x},\xi)\,dS = 0\end{aligned} \tag{51}$$

By adding Eq. (51) into Eq. (50b), we obtain,

$$\begin{aligned}\int_{\partial\Omega} n_i(\xi)E_{ijmn}u_{m,n}(\mathbf{x})u_{j,k}^{*p}(\mathbf{x},\xi)\,dS \\ + \int_{\partial\Omega} e_{nkt}D_t u_m(\mathbf{x})E_{ijmn}u_{j,i}^{*p}(\mathbf{x},\xi)\,dS + u_{p,k}(\mathbf{x}) = 0\end{aligned} \tag{52}$$

Multiplying Eq. (52) by E_{abpk}, we obtain the following identity for the corresponding stresses $\sigma_{ab}(\mathbf{x})$, as:

$$\int_{\partial\Omega} n_p(\xi)\sigma_{pq}(\mathbf{x})\sigma_{ab}^{*q}(\mathbf{x},\xi)dS + \int_{\partial\Omega} D_p u_q(\mathbf{x})\Sigma_{abpq}^{*}(\mathbf{x},\xi)\,dS + \sigma_{ab}(\mathbf{x}) = 0 \quad (53)$$

where Σ_{abpq}^{*} is defined in Eq. (35).

It is clear that "properties" similar to the above, can be derived for the fundamental solutions for any set of partial differential equations, such as those that arise in fluid mechanics, acoustics, electromagnetism, etc. These will be presented in subsequent papers by the present authors.

2.5 Regularization of tBIE

Contracting Eq. (53) with $n_a(\mathbf{x})$, as shown in Fig.3, and using the resulting equation in Eq. (36), we can obtain the fully regularized form of Eq. (36), as

$$\begin{aligned} 0 = &\int_{\partial\Omega} [t_q(\xi) - n_p(\xi)\sigma_{pq}(\mathbf{x})]n_a(\mathbf{x})\sigma_{ab}^{*q}(\mathbf{x},\xi)\,dS \\ &+ \int_{\partial\Omega} [D_p u_q(\xi) - (D_p u_q)(\mathbf{x})]n_a(\mathbf{x})\Sigma_{abpq}^{*}(\mathbf{x},\xi)\,dS \end{aligned} \quad (54)$$

Equation (54), may be satisfied in a *weak-form* at $\partial\Omega$, using a Petrov-Galerkin scheme, as:

$$\begin{aligned} 0 = &\int_{\partial\Omega} w_b(\mathbf{x})dS_x \int_{\partial\Omega} [t_q(\xi) - n_p(\xi)\sigma_{pq}(\mathbf{x})]n_a(\mathbf{x})\sigma_{ab}^{*q}(\mathbf{x},\xi)\,dS_\xi \\ &+ \int_{\partial\Omega} w_b(\mathbf{x})dS_x \int_{\partial\Omega} [D_p u_q(\xi) - (D_p u_q)(\mathbf{x})]n_a(\mathbf{x})\Sigma_{abpq}^{*}(\mathbf{x},\xi)\,dS_\xi \end{aligned} \quad (55)$$

where $w_b(\mathbf{x})$ is a test function. If $w_b(\mathbf{x})$ is chosen as a Dirac delta function, i.e. $w_b(\mathbf{x}) = \delta(\mathbf{x},\mathbf{x}_m)$ at $\partial\Omega$, we obtain the standard "collocation" traction boundary element method.

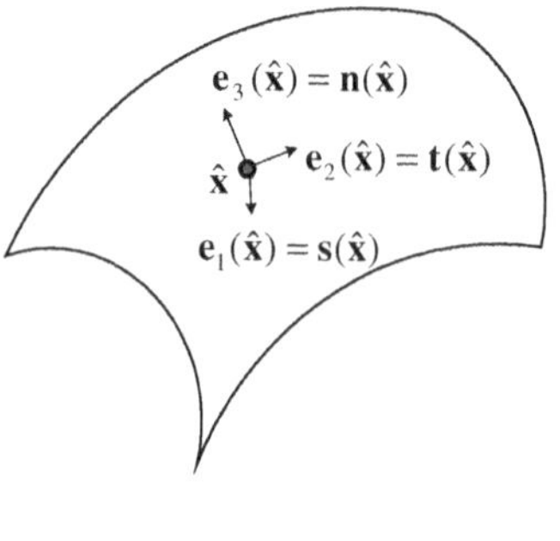

Fig. 3 The local coordinates at a boundary point x

From Eqs. (6) and (15), Σ^*_{ijpq} can be written in terms of F^* as:

$$\Sigma^*_{ijpq}(\mathbf{x}, \xi) = E_{ijkl} e_{nlp} \sigma^{*k}_{nq}(\mathbf{x}, \xi) = \mu^2[(e_{inp} F_{,jqn} - e_{inp}\delta_{jq} F_{,bbn} + e_{int} e_{tqk} e_{jpm} F_{,kmn}) + \upsilon(e_{inq}\delta_{jp} F_{,bbn} + e_{jnq}\delta_{ip} F_{,bbn})] \quad (56)$$

We also have the divergence of Σ^*_{ijpq} as:

$$\Sigma^*_{ijpq,i}(\mathbf{x}, \xi) = \mu^2 \upsilon e_{jnq} F_{,bbnp} \equiv \Lambda^*_{ijpq,i}(\mathbf{x}, \xi) \quad (57)$$

where, by definition,

$$\Lambda^*_{ijpq}(\mathbf{x}, \xi) = \mu^2 \upsilon e_{jiq} F_{,bbp} \quad (58)$$

We observe that:

$$\int_{\partial\Omega} D_p u_q(\xi) \Lambda^*_{ijpq}(\mathbf{x}, \xi) dS = - \int_{\partial\Omega} u_q(\xi) D_p \Lambda^*_{ijpq}(\mathbf{x}, \xi) dS = - \int_{\partial\Omega} u_q(\xi) D_p [\mu^2 \upsilon e_{jiq} F_{,bb}]_{,p} dS = 0 \quad (59)$$

Then we have

$$\int_{\partial\Omega} D_p u_q(\xi) \Sigma^*_{ijpq}(\mathbf{x}, \xi)\, dS = \int_{\partial\Omega} D_p u_q(\xi) [\Sigma^*_{ijpq}(\mathbf{x}, \xi) - \Lambda^*_{ijpq}(\mathbf{x}, \xi)]\, dS \equiv \int_{\partial\Omega} D_p u_q(\xi) \mathrm{K}^*_{ijpq}(\mathbf{x}, \xi)\, dS \quad (60)$$

where, by definition

$$\mathrm{K}^*_{ijpq}(\mathbf{x}, \xi) = \Sigma^*_{ijpq}(\mathbf{x}, \xi) - \Lambda^*_{ijpq}(\mathbf{x}, \xi) = \mu^2 e_{int}[(\delta_{tp} F_{,jq} - \delta_{tp}\delta_{jq} F_{,bb} + e_{tqk} e_{jpm} F_{,km}) + \upsilon(\delta_{tq}\delta_{jp} F_{,bb} + e_{tpm} e_{jmq} F_{,bb})]_{,n} \equiv e_{int} H^*_{tjpq,n}(\mathbf{x}, \xi) \quad (61)$$

We have H^*_{ijpq}, by definition, as

$$H^*_{ijpq}(\mathbf{x}, \xi) = \mu^2[-\delta_{ij} F_{,pq} + 2\delta_{ip} F_{,jq} + 2\delta_{jq} F_{,ip} - \delta_{pq} F_{,ij} - 2\delta_{ip}\delta_{jq} F_{,bb} + 2\upsilon\delta_{iq}\delta_{jp} F_{,bb} + (1-\upsilon)\delta_{ij}\delta_{pq} F_{,bb})] \quad (62)$$

With Eqs. (11)a, we can write H^*_{ijpq} for 3D problems as:

$$H^*_{ijpq}(\mathbf{x}, \xi) = \frac{\mu}{8\pi(1-\upsilon)r}[4\upsilon\delta_{iq}\delta_{jp} - \delta_{ip}\delta_{jq} - 2\upsilon\delta_{ij}\delta_{pq} + \delta_{ij} r_{,p} r_{,q} + \delta_{pq} r_{,i} r_{,j} - 2\delta_{ip} r_{,j} r_{,q} - \delta_{jq} r_{,i} r_{,p}] \quad (63)$$

for 2D plain strain problems as:

$$H^*_{ijpq}(\mathbf{x}, \xi) = \frac{1}{4\pi(1-\upsilon)}[-4\upsilon \ln r\delta_{iq}\delta_{jp} + \ln r\delta_{ip}\delta_{jq} + 2\upsilon \ln r\delta_{ij}\delta_{pq}$$
$$+\delta_{ij}r_{,p}r_{,q} + \delta_{pq}r_{,i}r_{,j} - 2\delta_{ip}r_{,j}r_{,q} - \delta_{jq}r_{,i}r_{,p}] \tag{64}$$

From Eqs. (57) and (61), some properties of the kernel functions can be found as following:

$$\nabla \cdot \Sigma^*(\mathbf{x}, \xi) = \nabla \cdot \Lambda^*(\mathbf{x}, \xi),\ \nabla \cdot \mathrm{K}^*(\mathbf{x}, \xi) = \mathbf{0},\ \mathrm{K}^*(\mathbf{x}, \xi) = \nabla \times H^*(\mathbf{x}, \xi) \tag{65}$$

We may take $w_b(\mathbf{x})$ to be any continuous function in Eq. (55), and derive a Petrov-Galerkin boundary element method. If $w_b(\mathbf{x})$ is continuous, one may use Stokes' theorem, and write:

$$-\int_{\partial\Omega} w_b(\mathbf{x})t_b(\mathbf{x})dS_x = \int_{\partial\Omega} w_b(\mathbf{x})dS_x \int_{\partial\Omega} t_q(\xi)n_a(\mathbf{x})[\Psi^{*q}_{ab}(\mathbf{x}, \xi) - \phi^{*q}_{ab}(\mathbf{x}, \xi)]dS_\xi$$
$$+\int_{\partial\Omega} w_b(\mathbf{x})dS_x \int_{\partial\Omega} D_p u_q(\xi)n_a(\mathbf{x})\mathrm{K}^*_{abpq}(\mathbf{x}, \xi)\, dS_\xi \tag{66}$$

in which, Eqs. (53) and (60) are applied. As $w_b(\mathbf{x})$ is continuous, one may use Stokes' theorem, and re-write Eq. (66) as:

$$-\frac{1}{2}\int_{\partial\Omega} t_b(\mathbf{x})w_b(\mathbf{x})dS_x = \int_{\partial\Omega} D_a w_b(\mathbf{x})dS_x \int_{\partial\Omega} t_q(\xi)G^{*q}_{ab}(\mathbf{x}, \xi)\, dS_\xi$$
$$-\int_{\partial\Omega} t_q(\xi)\, dS_\xi \int_{\partial\Omega}^{CPV} n_a(\mathbf{x})w_b(\mathbf{x})\phi^{*q}_{ab}(\mathbf{x}, \xi)dS_x \tag{67}$$
$$+\int_{\partial\Omega} D_a w_b(\mathbf{x})dS_x \int_{\partial\Omega} D_p u_q(\xi)H^*_{abpq}(\mathbf{x}, \xi)\, dS_\xi$$

If the test function $w_b(\mathbf{x})$ is chosen to be identical to a function that is energy-conjugate to t_b, namely, the trial function $\hat{u}_b(\mathbf{x})$, we generate the symmetric Galerkin BEM as:

$$-\frac{1}{2}\int_{\partial\Omega} t_b(\mathbf{x})\hat{u}_b(\mathbf{x})dS_x = \int_{\partial\Omega} D_a\hat{u}_b(\mathbf{x})dS_x \int_{\partial\Omega} t_q(\xi)G^{*q}_{ab}(\mathbf{x}, \xi)\, dS_\xi$$
$$-\int_{\partial\Omega} t_q(\xi)\, dS_\xi \int_{\partial\Omega}^{CPV} n_a(\mathbf{x})\hat{u}_b(\mathbf{x})\phi^{*q}_{ab}(\mathbf{x}, \xi)dS_x \tag{68}$$
$$+\int_{\partial\Omega} D_a\hat{u}_b(\mathbf{x})dS_x \int_{\partial\Omega} D_p u_q(\xi)H^*_{abpq}(\mathbf{x}, \xi)\, dS_\xi$$

The results in Eq. (67) are similar to those reported in Li and Mear, (1998) but are different from those in Li and Mear (1998) in the kernel functions appearing in Eq. (68). However, here, we obtain these results in a very straightforward and simple manner.

2.6 Regularization of dBIE

In this section, we consider the regularization of the displacement BIE (29), in order to render it tractable for numerical implementation. We also consider the possibility of satisfying the dBIE, at $\partial\Omega$, in *a weak form*, through a general Petrov-Galerkin scheme.

We subtract Eq. (40d) from Eq. (29), and obtain,

$$0 = \int_{\partial\Omega} t_j(\xi) u_j^{*p}(\mathbf{x}, \xi)\, dS - \int_{\partial\Omega} n_i(\xi)[u_j(\xi) - u_j(\mathbf{x})]\sigma_{ij}^{*p}(\mathbf{x}, \xi)\, dS \tag{69}$$

We can also use a Petrov-Galerkin scheme to write a weak-form for Eq. (69) as:

$$\begin{aligned} 0 = &\int_{\partial\Omega} w_p(\mathbf{x}) dS_x \int_{\partial\Omega} t_j(\xi) u_j^{*p}(\mathbf{x}, \xi)\, dS \\ &- \int_{\partial\Omega} w_p(\mathbf{x}) dS_x \int_{\partial\Omega} n_i(\xi)[u_j(\xi) - u_j(\mathbf{x})]\sigma_{ij}^{*p}(\mathbf{x}, \xi)\, dS \end{aligned} \tag{70}$$

where $w_p(\mathbf{x})$ is a test function. If $w_p(\mathbf{x})$ is chosen as a Dirac delta function, i.e. $w_p(\mathbf{x}) = \delta(\mathbf{x}, \mathbf{x}_m)$ at $\partial\Omega$, we obtain the standard "collocation" displacement boundary element method.

Using (17b), i.e., $\sigma_{ij}^{*p} = \psi_{ij}^{*p}(\mathbf{x}, \xi) - \phi_{ij}^{*p}(\mathbf{x}, \xi)$ [with Ψ_{ij}^{*p} is defined in Eq. (19). and ϕ_{ij}^{*p} is defined in Eq. (17a)], we re-write Eq. (70), with Eq. (46), as:

$$\begin{aligned} \frac{1}{2}\int_{\partial\Omega} w_p(\mathbf{x}) u_j(\mathbf{x}) dS_x = &\int_{\partial\Omega} w_p(\mathbf{x}) dS_x \int_{\partial\Omega} t_j(\xi) u_j^{*p}(\mathbf{x}, \xi)\, dS_\xi \\ &- \int_{\partial\Omega} w_p(\mathbf{x}) dS_x \int_{\partial\Omega} n_i(\xi) u_j(\xi) \psi_{ij}^{*p}(\mathbf{x}, \xi)\, dS_\xi \\ &+ \int_{\partial\Omega} w_p(\mathbf{x}) dS_x \int_{\partial\Omega}^{CPV} n_i(\xi) u_j(\xi) \phi_{ij}^{*p}(\mathbf{x}, \xi)\, dS_\xi \end{aligned} \tag{71}$$

Applying Stokes' theorem to Eq. (71), and if $w_p(\mathbf{x})$ is chosen to be identical to a function which is energy-conjugate to u_p, viz., the trial function $\hat{t}_p(\mathbf{x})$, we obtain the symmetric Galerkin dBEM, as

$$\frac{1}{2}\int_{\partial\Omega}\hat{t}_p(\mathbf{x})u_p(\mathbf{x})dS_x = \int_{\partial\Omega}\hat{t}_p(\mathbf{x})dS_x\int_{\partial\Omega}t_j(\xi)u_j^{*p}(\mathbf{x},\xi)\,dS_\xi$$
$$+\int_{\partial\Omega}\hat{t}_p(\mathbf{x})dS_x\int_{\partial\Omega}D_i(\xi)u_j(\xi)G_{ij}^{*p}(\mathbf{x},\xi)\,dS_\xi \quad (72)$$
$$+\int_{\partial\Omega}\hat{t}_p(\mathbf{x})dS_x\int_{\partial\Omega}^{CPV}n_i(\xi)u_j(\xi)\phi_{ij}^{*p}(\mathbf{x},\xi)\,dS_\xi$$

3 Part II: Weakly-Singular BIEs for Acoustics

3.1 The Governing Wave Equation, and its Fundamental Solution

The propagation of acoustic waves through an unbounded homogeneous medium is described by the wave equation:

$$\nabla^2\phi(r,t) - \frac{1}{c^2}\frac{\partial^2\phi(r,t)}{\partial t^2} = 0 \quad (73)$$

where ∇^2 denotes the Laplacian operator, ϕ is the acoustic velocity potential at a point r at time t, and c is the speed of sound in the medium at the equilibrium state. For time-harmonic waves with a time factor $e^{-i\omega t}$, the Helmholtz differential equation for ϕ can be written as follows:

$$\nabla^2\phi + k^2\phi = 0 \quad (74)$$

where i is the imaginary unit, ω is the angular frequency of the acoustic wave, and $k = \omega/c$ is the wave number.

The acoustic pressure p and the velocity $\mathbf{u}$ of the fluid particles induced by the sound waves are determined through the velocity potential ϕ, as:

$$p = -\rho_0\frac{\partial\phi}{\partial t}; \quad (75)$$

$$\mathbf{u} = \nabla\phi \quad (76)$$

where ∇ is the gradient operator, and ρ_0 denotes the density of the fluid at the equilibrium state. For time-harmonic waves, we have:

$$p = i\omega\rho_0\phi \quad (77)$$

The power-flux-density of sound waves is given by:

$$\mathbf{P} = p\mathbf{u} = p\nabla\phi \quad (78)$$

In general, the acoustic velocity potential can be represented as a sum of the incident potential and the scattered potential:

$$\phi = \phi^i + \phi^s \tag{79}$$

The fundamental solution of the Helmholtz differential equation (74) at any field point ξ due to a point sound source at $\mathbf{x}$, is well known as the free-space Green's function $G_k(\mathbf{x}, \xi)$, which is listed here for 2- and 3-D problems, respectively, as follows:For a 2D problem, the Green's function is:

$$G_k(\mathbf{x}, \xi) = \frac{i}{4} H_0^{(1)}(kr) \tag{80a}$$

$$\frac{\partial G_k}{\partial r}(\mathbf{x}, \xi) = -\frac{i}{4} k H_1^{(1)}(kr), \; G_{k,i}(\mathbf{x}, \xi) = -\frac{i}{4} k H_1^{(1)}(kr) r_{,i} \tag{80b}$$

where $H_0^{(1)}(kr)$ denotes the Hankel function of the first kind, and r represents the distance between the field point ξ and the source point $\mathbf{x}$. For a 3D problem,

$$G_k(\mathbf{x}, \xi) = \frac{1}{4\pi r} e^{-ikr} \tag{80c}$$

$$\frac{\partial G_k}{\partial r}(\mathbf{x}, \xi) = \frac{e^{-ikr}}{4\pi r^2}(-ikr - 1), \; G_{k,i}(\mathbf{x}, \xi) = \frac{e^{-ikr}}{4\pi r^2}(-ikr - 1) r_{,i} \tag{80d}$$

in which $r = |\mathbf{x} - \xi|$, and $r_{,i} \equiv \dfrac{\partial r}{\partial \xi_i} = -\dfrac{x_i - \xi_i}{r}$.

In the following sections, we use $\phi^*(\mathbf{x}, \xi)$ to denote $G_k(\mathbf{x}, \xi)$, without losing any generality. Consider a body of an infinite extent, subject to a point sound source at a generic location $\mathbf{x}$. The fundamental solution is the sound field, denoted by $\phi^*(\mathbf{x}, \xi)$, at any point ξ due to the point sound source, and is governed by the wave equation:

$$\phi^*_{,ii}(\mathbf{x}, \xi) + k^2 \phi^*(\mathbf{x}, \xi) + \delta(\mathbf{x}, \xi) = 0 \tag{81}$$

We also refer to Eq. (81) as the Helmholtz potential equation governing the fundamental solution.

3.2 Boundary Integral Equations for the Velocity Potential ϕ, and its Gradient $\phi_{,k}$

3.2.1 The Current State-of-Science for BIE in Acoustics: Strongly Singular ϕ and Hyper-Singular $\phi_{,k}$

Let $\bar{\phi}$ be the test function chosen to enforce the Helmholtz equation (74), in terms of the trial function ϕ, in a weak form. The weak form of Helmholtz equation can then be written as:

$$\int_\Omega \left(\nabla^2 \phi + k^2 \phi\right) \bar{\phi} \, d\Omega = 0 \tag{82}$$

If we apply the divergence theorem once in Eq. (82), we obtain a "symmetric" weak form:

$$\int_{\partial\Omega} n_i \phi_{,i} \bar{\phi} dS - \int_{\Omega} \phi_{,i} \bar{\phi}_{,i} d\Omega + \int_{\Omega} k^2 \phi \bar{\phi} d\Omega = 0 \tag{83}$$

Thus, in the "symmetric weak-form", both the trial function ϕ, as well as the test functions $\bar{\phi}$ are only required to be first-order differentiable. If we apply the divergence theorem twice in Eq. (82), we obtain:

$$\int_{\partial\Omega} n_i \phi_{,i} \bar{\phi} dS - \int_{\partial\Omega} n_i \phi \bar{\phi}_{,i} dS + \int_{\Omega} \phi \left(\bar{\phi}_{,ii} + k^2 \bar{\phi}\right) d\Omega = 0 \tag{84}$$

We label Eq. (84) as the "unsymmetric weak-form". Now, the test functions $\bar{\phi}$ are required to be second-order differentiable, while ϕ is not required to be differentiable [Han; Atluri, 2003a] in Ω.

If we take the fundamental solution $\phi^*(\mathbf{x}, \xi)$ to be the test function $\bar{\phi}$ in Eq. (84), and noting the property from Eq. (81), we have:

$$\begin{aligned} \phi(\mathbf{x}) &= \int_{\partial\Omega} n_i(\xi) \phi_{,i}(\xi) \phi^*(\mathbf{x}, \xi) dS - \int_{\partial\Omega} n_i(\xi) \phi(\xi) \phi^*_{,i}(\mathbf{x}, \xi) dS \\ &\equiv \int_{\partial\Omega} q(\xi) \phi^*(\mathbf{x}, \xi) dS - \int_{\partial\Omega} \phi(\xi) \Theta^*(\mathbf{x}, \xi) dS \end{aligned} \tag{85}$$

where by definition,

$$q(\xi) = \frac{\partial \phi(\xi)}{\partial n_\xi} = n_k(\xi) \phi_{,k}(\xi) \quad \xi \in \partial\Omega \tag{86}$$

and the kernel function,

$$\Theta^*(\mathbf{x}, \xi) = \frac{\partial \phi^*(\mathbf{x}, \xi)}{\partial n_\xi} = n_k(\xi) \phi^*_{,k}(\mathbf{x}, \xi) \quad \xi \in \partial\Omega \tag{87}$$

Thus, $q(\xi)$ is the potential gradient along the outward normal direction of the boundary surface.

Equation (85) is the "traditional" BIE for ϕ, that is widely used in literature. We refer to Eq. (85), hereafter, as ϕ-BIE. The nonuniqueness of the solution of Eq. (85) arises because, this homogeneous equation has nontrivial solutions at some characteristic frequencies [Chien, Rajiyah, and Atluri, 1990]. As noted in the introduction, many researchers have investigated and expended substantial efforts in solving this problem of nonuniqueness.

If we differentiate Eq. (85) directly with respect to $\mathbf{x}_k$, we obtain the integral equations for potential gradients $\phi_{,k}(\mathbf{x})$ as:

$$\frac{\partial\phi(\mathbf{x})}{\partial x_k} = \int_{\partial\Omega} q(\xi)\frac{\partial\phi^*(\mathbf{x},\xi)}{\partial x_k}dS - \int_{\partial\Omega}\phi(\xi)\frac{\partial\Theta^*(\mathbf{x},\xi)}{\partial x_k}dS \tag{88}$$

The second term on the right hand side of Eq. (88) is hyper-singular, since $\frac{\partial\Theta^*(\mathbf{x},\xi)}{\partial x_k}$ is of order $O\left(r^{-3}\right)$ for a 3D problem. Eq. (88) is also the integral equations for the gradients of $\phi(\mathbf{x})$ that are widely used in the literature; and hence a wide body of literature is devoted to deal with the hyper-singularity in this equation.

3.2.2 Presently Proposed Non-Hyper-Singular BIE for $\phi_{,k}$

On the other hand, the new method proposed in this paper starts *from writing a vector weak-form* [as opposed to a scalar weak-form] of the governing equation (82) by using the vector test functions $\bar{\phi}_{,k}$, as in Okada, Rajiyah, and Atluri (1988, 1989):

$$\int_{\Omega}\left(\phi_{,ii} + k^2\phi\right)\bar{\phi}_{,k}d\Omega = 0 \quad for\ k = 1,2,3 \tag{89}$$

After applying the divergence theorem three times in Eq. (89), we can write:

$$\begin{aligned}&\int_{\partial\Omega} n_i\phi_{,i}\bar{\phi}_{,k}dS - \int_{\partial\Omega} n_k\phi_{,i}\bar{\phi}_{,i}dS + \int_{\partial\Omega} n_i\phi_{,k}\bar{\phi}_{,i}dS \\ &\quad + \int_{\partial\Omega} k^2 n_k\phi\bar{\phi}dS - \int_{\Omega}\left(\bar{\phi}_{,ii} + k^2\bar{\phi}\right)\phi_{,k}d\Omega = 0\end{aligned} \tag{90}$$

Using the gradients of the fundamental solution, viz., $\phi^*_{,k}(\mathbf{x},\xi)$, as the test functions, and using the identity from Eq. (81), we obtain

$$\begin{aligned}-\phi_{,k}(\mathbf{x}) = &\int_{\partial\Omega} q(\xi)\phi^*_{,k}(\mathbf{x},\xi)dS - \int_{\partial\Omega} n_k(\xi)\phi_{,i}(\xi)\phi^*_{,i}(\mathbf{x},\xi)dS \\ &+ \int_{\partial\Omega}\phi_{,k}(\xi)\Theta^*(\mathbf{x},\xi)dS + \int_{\partial\Omega} k^2 n_k(\xi)\phi(\xi)\phi^*(\mathbf{x},\xi)dS\end{aligned} \tag{91}$$

It should be noted that the integral equations for $\phi(\mathbf{x})$ [Eq. (85)], and $\phi_{,k}(\mathbf{x})$ [Eq. (91)], *are derived independently of each other*. The most interesting feature of the "directly derived" integral equations (91), for $\phi_{,k}(\mathbf{x})$, *is that they are non-hyper-singular*, viz, *the highest order singularity in the kernels appearing in Eq. (91) is only* $O\left(r^{-2}\right)$ *in a 3D problem.*

We may rewrite (91) as:

$$-\phi_{,k}(\mathbf{x}) = \int_{\partial\Omega} q(\xi)\,\phi^*_{,k}(\mathbf{x},\xi)\,dS + \int_{\partial\Omega} D_t\phi(\xi)\,e_{ikt}\phi^*_{,i}(\mathbf{x},\xi)\,dS$$
$$+\int_{\partial\Omega} k^2 n_k(\xi)\,\phi(\xi)\,\phi^*(\mathbf{x},\xi)\,dS \tag{92}$$

3.2.3 Some Basic Physical & Mathematical Properties of the Fundamental Solution ϕ^*, Which Permit Simple and Direct Regularizations of the Strongly Singular BIEs for ϕ and $q(\phi_{,k})$

We write the weak form of Eq. (81), governing the fundamental solution, over the domain, using a constant c as the test function, and obtain

$$\int_{\partial\Omega} n_i(\xi)\,\phi^*_{,i}(\mathbf{x},\xi)\,dS + \int_{\Omega} k^2\phi^*(\mathbf{x},\xi)\,d\Omega + 1 = 0 \text{ at } \quad \mathbf{x}\in\Omega \tag{93}$$

Equation (93) is a "basic identity" of the fundamental solution $\phi^*(\mathbf{x},\xi)$. Now, consider an arbitrary function $\phi(\mathbf{x})$ in Ω as the test function, and once again write the weak form of (81), as

$$\int_{\partial\Omega} \Theta^*(\mathbf{x},\xi)\,\phi(\mathbf{x})\,dS + \int_{\Omega} k^2\phi^*(\mathbf{x},\xi)\,\phi(\mathbf{x})\,d\Omega + \phi(\mathbf{x}) = 0 \quad \text{at} \quad \mathbf{x}\in\Omega \tag{94}$$

Once the point $\mathbf{x}$ approaches a smooth boundary, i.e., $\mathbf{x}\in\partial\Omega$, the first term in Eq. (94) can be written as

$$\lim_{\mathbf{x}\to\partial\Omega}\int_{\partial\Omega} \Theta^*(\mathbf{x},\xi)\,\phi(\mathbf{x})\,dS = \int_{\partial\Omega}^{CPV} \Theta^*(\mathbf{x},\xi)\,\phi(\mathbf{x})\,dS - \frac{1}{2}\phi(\mathbf{x}) \tag{95}$$

in which we introduce the notion of a Cauchy Principal Value (CPV) integral. The physical meaning of Eq. (95) can be understood by rewriting Eq. (93) and (94), respectively as:

$$\int_{\partial\Omega}^{CPV} \Theta^*(\mathbf{x},\xi)\,dS + \int_{\Omega} k^2\phi^*(\mathbf{x},\xi)\,d\Omega + \frac{1}{2} = 0 \quad \text{at } \mathbf{x}\in\partial\Omega \tag{96a}$$

$$\int_{\partial\Omega}^{CPV} \Theta^*(\mathbf{x},\xi)\,\phi(\mathbf{x})\,dS + \int_{\Omega} k^2\phi^*(\mathbf{x},\xi)\,\phi(\mathbf{x})\,d\Omega + \frac{1}{2}\phi(\mathbf{x}) = 0 \quad \text{at } \mathbf{x}\in\partial\Omega \tag{96b}$$

Equation (96a) implies that only a half of the sound source at point $\mathbf{x}$ is applied to the domain Ω, when the point $\mathbf{x}$ approaches a smooth boundary, $\mathbf{x}\in\partial\Omega$. Eq. (96b) can be likewise interpreted physically.

We again consider another weak form of Eq. (81), by taking the vector test functions to be the gradients of an arbitrary function $\phi(\xi)$ in Ω, which are so chosen that they have constant values, as:

$$\phi_{,k}(\xi) = \phi_{,k}(\mathbf{x}) \tag{97}$$

Then the weak form of Eq. (81) may be written as:

$$\int_{\Omega} \left[\phi^{*}_{,ii}(\mathbf{x},\xi) + k^2\phi^{*}(\mathbf{x},\xi)\right]\phi_{,k}(\xi)\,d\Omega + \phi_{,k}(\mathbf{x}) = 0 \tag{98}$$

After applying the divergence theorem, we can obtain from (98):

$$\int_{\partial\Omega} \Theta^{*}(\mathbf{x},\xi)\,\phi_{,k}(\mathbf{x})\,dS + \int_{\Omega} k^2\phi^{*}(\mathbf{x},\xi)\,\phi_{,k}(\mathbf{x})\,d\Omega + \phi_{,k}(\mathbf{x}) = 0 \tag{99}$$

In addition, we may observe that the first two terms in Eq. (91) have the following property:

$$\begin{aligned}&\int_{\partial\Omega} n_i(\xi)\,\phi_{,i}(\mathbf{x})\,\phi^{*}_{,k}(\mathbf{x},\xi)\,dS - \int_{\partial\Omega} n_k(\xi)\,\phi_{,i}(\mathbf{x})\,\phi^{*}_{,i}(\mathbf{x},\xi)\,dS\\ &= \int_{\Omega} \phi_{,i}(\mathbf{x})\,\phi^{*}_{,ki}(\mathbf{x},\xi)\,d\Omega - \int_{\partial\Omega} \phi_{,i}(\mathbf{x})\,\phi^{*}_{,ik}(\mathbf{x},\xi)\,dS = 0\end{aligned} \tag{100}$$

By adding Eq. (100) and (99), we have:

$$\begin{aligned}&\int_{\partial\Omega} n_i(\xi)\,\phi_{,i}(\mathbf{x})\,\phi^{*}_{,k}(\mathbf{x},\xi)\,dS + \int_{\partial\Omega} e_{ikt}D_t\phi(\mathbf{x})\,\phi^{*}_{,i}(\mathbf{x},\xi)\,dS\\ &+ \int_{\Omega} k^2\phi^{*}(\mathbf{x},\xi)\,\phi_{,k}(\mathbf{x})\,d\Omega + \phi_{,k}(\mathbf{x}) = 0\end{aligned} \tag{101}$$

We now use the fundamental properties of ϕ^{*}, as enumerated in this section, to give simple, straightforward and elegant physical and mathematical regularizations of the strongly-singular BIEs for ϕ, and $\phi_{,k}$, as given in Eq. (85) and (91) respectively.

3.2.4 Regularization of ϕ-BIE, Eq. (85)

In this section, we consider the regularization of ϕ-BIE, as well as the possibility of satisfying the *ϕ-BIE itself in a weak form*, at $\partial\Omega$, through a general Petrov-Galerkin scheme. It is well known the Eq. (85) is numerically tractable if it is restricted only for boundary points, ie., $\mathbf{x} \in \partial\Omega$, because $n_k(\xi)\,\phi^{*}_{,k}(\mathbf{x},\xi)$ contains the weak singularity $[O(r^{-1})]$. Most researchers have implemented the ϕ-BIE based on this equation and solved the boundary problems. On another hand, considering a domain point which is approaching the boundary, one may encounter the higher singularity $[\mathrm{O}\left(r^{-2}\right)]$ with Eq. (85).

Subtracting Eq. (94) from Eq. (85), and obtain:

$$\begin{aligned}&\int_{\partial\Omega} q(\xi)\,\phi^*(\mathbf{x},\xi)\,dS - \int_{\partial\Omega}[\phi(\xi)-\phi(\mathbf{x})]\,\Theta^*(\mathbf{x},\xi)\,dS\\&+\int_{\Omega} k^2\phi^*(\mathbf{x},\xi)\,\phi(\mathbf{x})\,d\Omega = 0\end{aligned} \tag{102}$$

With Eq. (96b), Eq. (102) is applicable at point $\mathbf{x}$ on the boundary $\partial\Omega$, as:

$$\begin{aligned}&\int_{\partial\Omega} q(\xi)\,\phi^*(\mathbf{x},\xi)\,dS - \int_{\partial\Omega}[\phi(\xi)-\phi(\mathbf{x})]\,\Theta^*(\mathbf{x},\xi)\,dS\\&= \int_{\partial\Omega}^{CPV}\Theta^*(\mathbf{x},\xi)\,\phi(\mathbf{x})\,dS + \frac{1}{2}\phi(\mathbf{x})\end{aligned} \qquad \mathbf{x}\in\partial\Omega \tag{103}$$

One can see that $\phi(\xi)-\phi(\mathbf{x})$ becomes $\mathrm{O}(r)$ when $\xi\to\mathbf{x}$, and thus Eq. (103) becomes weakly-singular $[\mathrm{O}\left(r^{-1}\right)]$. For a point close to the boundary, a reference node on the boundary may be used for regularization [Han; Atluri, 2003a]. Hence, all the integrals in Eq. (103) can be evaluated numerically, for both the boundary points and the points close to the boundary. *We refer to Eq. (103) as the regularized ϕ-BIE or "R-ϕ-BIE".*

On the other hand, when $\partial\Omega$ has corners, ϕ may be expected to have a variation of $r^{+\lambda}$ $(\lambda<1)$ near the corners. In such cases, $\phi(\xi)-\phi(\mathbf{x})$ may become $O\left(r^{\lambda-1}\right)$ when $\xi\to\mathbf{x}$, and thus, in a theoretical sense, Eq. (103) is no longer weakly singular. However, in a numerical solution of R-ϕ-BIE (103) directly, through a collocation process, to derive a *ϕ Boundary Element Method (BEM-R- ϕ-BIE)*, we envision using only C^0 polynomial interpolations of ϕ and q. Thus, in the numerical implementation of the *BEM-R-ϕ -BIE by a collocation of (103)*, we encounter only weakly singular integrals. This method of BEM-R-ϕ-BIE, using a direct collocation of (103), is presented elsewhere Qian, Han, Ufimtsev, Atluri (2004). By using C^0 elements and employing an adaptive boundary-element refinement strategy near corners at the boundary, one may extract the value of $(\lambda<1)$ in the asymptotic solution for ϕ near such a corner.

We can also use a Petrov-Galerkin scheme to write the weak-form for Eq. (103). If $w(\mathbf{x})$ is chosen to be identical to a function which is energy-conjugate to $\phi(\mathbf{x})$, viz. the trial function $\hat{q}(\mathbf{x})$, we obtain the symmetric Galerkin ϕ-BIE form as [Han; Atluri, 2003a]

$$\begin{aligned}\frac{1}{2}\int_{\partial\Omega}\hat{q}(\mathbf{x})\,\phi(\mathbf{x})\,dS_x &= \int_{\partial\Omega}\hat{q}(\mathbf{x})\,dS_x\int_{\partial\Omega} q(\xi)\,\phi^*(\mathbf{x},\xi)\,dS_\xi\\&\quad-\int_{\partial\Omega}\hat{q}(\mathbf{x})\,dS_x\int_{\partial\Omega}^{CPV}\Theta^*(\mathbf{x},\xi)\,\phi(\xi)\,dS_\xi\end{aligned} \tag{104}$$

Equation (104) leads to the present novel formulation for a Symmetric Galerkin Boundary Element Method for the weakly singular BEM-R-ϕ-BIE. *We label this as SGBEM-R-ϕ-BIE for convenience.*

3.2.5 Regularization of q-BIE, Eq. (90)

We can obtain the fully regularized form of Eq. (92) as:

$$\begin{aligned}\int_{\partial\Omega} [q(\xi) - n_i(\xi)\psi_i(\mathbf{x})]\,\phi^*_{,k}(\mathbf{x},\xi)\,dS + \int_{\partial\Omega} e_{ikt}[D_t\phi(\xi) \\ - (D_t\phi)(\mathbf{x})]\phi^*_{,i}(\mathbf{x},\xi)\,dS + \int_{\partial\Omega} k^2 n_k(\xi)\,\phi(\xi)\,\phi^*(\mathbf{x},\xi)\,dS \\ + \int_{\partial\Omega}^{CPV} \Theta^*(\mathbf{x},\xi)\,q(\mathbf{x})\,dS + \frac{1}{2}q(\mathbf{x}) = 0\end{aligned} \tag{105}$$

We define a kernel function as:

$$\hat{\Theta}^*(\mathbf{x},\xi) = -\frac{\partial\phi^*(\mathbf{x},\xi)}{\partial n_x} = n_k(\mathbf{x})\frac{\partial\phi^*(\mathbf{x},\xi)}{\partial\xi_k} \tag{106}$$

If we contract with $n_k(\mathbf{x})$ on both sides of Eq. (105), we obtain

$$\begin{aligned}\int_{\partial\Omega} [q(\xi) - n_i(\xi)\psi_i(\mathbf{x})]\,\hat{\Theta}^*(\mathbf{x},\xi)\,dS + \int_{\partial\Omega} [D_t\phi(\xi) \\ - (D_t\phi)(\mathbf{x})]n_k(\mathbf{x})\,e_{ikt}\phi^*_{,i}(\mathbf{x},\xi)\,dS + \int_{\partial\Omega} k^2 n_k(\mathbf{x})\,n_k(\xi)\,\phi(\xi)\,\phi^*(\mathbf{x},\xi)\,dS \\ + \int_{\partial\Omega}^{CPV} \Theta^*(\mathbf{x},\xi)\,q(\mathbf{x})\,dS + \frac{1}{2}q(\mathbf{x}) = 0\end{aligned} \tag{107}$$

We label Eq. (107) as the regularized *q-BIE, or "R-q-BIE"*. When $\partial\Omega$ is smooth, one can see that $[q(\xi) - n_i(\xi)\psi_i(\mathbf{x})]$ and $[D_t\phi(\xi) - (D_t\phi)(\mathbf{x})]$ become $O(r)$ when $\xi \to \mathbf{x}$, and Eq. (105) becomes weakly singular $[O(r^{-1})]$ on a 3D problem. Thus, all the integrals in Eq. (107) can be evaluated numerically, and applicable to any point $\mathbf{x}$ on the boundary $\partial\Omega$. On the other hand, when $\partial\Omega$ has corners, $[q(\xi) - n_i(\xi)\psi_i(\mathbf{x})]$ and $[D_t\phi(\xi) - (D_t\phi)(\mathbf{x})]$ may become $O(r^{\lambda-1})$ when $\xi \to \mathbf{x}$, and thus, in a theoretical sense, Eq. (107) is no longer weakly singular. However, in a numerical implementation of the R-q-BIE, viz. Eq. (107), directly, *through a collocation process, to derive a qBoundary Element Method (BEM-R-q-BIE)*, we envision using only C^0 polynomial interpolations of ϕ and q. Thus, in the numerical implementation of the BEM-q-BIE by a collocation of (107), we encounter only weakly singular integrals. The method of BEM-R-q-BIE using a direct collocation of (107), is presented elsewhere [Qian, Han, and Atluri, 2003].

We can also use a Petrov-Galerkin scheme to write a weak form for Eq. (107). If $w(\mathbf{x})$ is chosen as a Dirac delta function, i.e., $w(\mathbf{x}) = \delta(\mathbf{x}, \mathbf{x}_m)$ at $\partial\Omega$, we obtain the standard "collocation" boundary element method. [BEM-R-q-BIE] Also, we can directly use a Petrov-Galerkin scheme to write a weak-form for Eq. (92) as:

$$
\begin{aligned}
-\int_{\partial\Omega} q(\mathbf{x})\, w(\mathbf{x})\, dS_x = &\int_{\partial\Omega} w(\mathbf{x})\, dS_x \int_{\partial\Omega} q(\xi)\,\hat{\Theta}^*(\mathbf{x},\xi)\, dS_\xi \\
&+ \int_{\partial\Omega} w(\mathbf{x})\, dS_x \int_{\partial\Omega} D_t\phi(\xi)\, e_{ikt} n_k(\mathbf{x})\, \phi^*_{,i}(\mathbf{x},\xi)\, dS_\xi \\
&+ \int_{\partial\Omega} w(\mathbf{x})\, dS_x \int_{\partial\Omega} k^2 n_k(\xi)\,\phi(\xi)\, n_k(\mathbf{x})\,\phi^*(\mathbf{x},\xi)\, dS_\xi
\end{aligned} \tag{108}
$$

The first integral at the right side of Eq. (108) can be written as:

$$
\begin{aligned}
\int_{\partial\Omega} w(\mathbf{x})\, dS_x \int_{\partial\Omega} q(\xi)\,\hat{\Theta}^*(\mathbf{x},\xi)\, dS_\xi = &\int_{\partial\Omega} q(\xi)\, dS_\xi \int_{\partial\Omega}^{CPV} w(\mathbf{x})\,\hat{\Theta}^*(\mathbf{x},\xi)\, dS_x \\
&- \int_{\partial\Omega} \frac{1}{2} q(\mathbf{x})\, w(\mathbf{x})\, dS_x
\end{aligned} \tag{109}
$$

in which Eq. (96b) is used, and $\dfrac{\partial}{\partial x_i} = -\dfrac{\partial}{\partial \xi_i}$.

The second integral on the right side of Eq. (108), can be simplified by using the Stokes theorem. We introduce a kernel function, Σ^*_{kt}, defined as

$$
\Sigma^*_{kt}(\mathbf{x},\xi) = e_{ikt}\phi^*_{,i}(\mathbf{x},\xi) = e_{ink}\left(-\delta_{nt}\phi^*_{,i}(\mathbf{x},\xi)\right) = e_{ink}\mathrm{H}^*_{nt,i}(\mathbf{x},\xi) \tag{110}
$$

Thus, by definition, $\mathrm{H}^*_{nt}(\mathbf{x},\xi) = -\delta_{nt}\phi^*(\mathbf{x},\xi)$, and this kernel function is seen to be:

$$
\mathrm{H}^*_{nt}(\mathbf{x},\xi) = -\frac{i}{4} H_0^{(1)}(kr)\,\delta_{nt} \quad \text{in 2D} \tag{111a}
$$

and

$$
\mathrm{H}^*_{nt}(\mathbf{x},\xi) = -\frac{e^{-ikr}}{4\pi r}\delta_{nt} \quad \text{in 3D} \tag{111b}
$$

The second integral on the right side of Eq. (108) is rewritten as

$$
\begin{aligned}
&\int_{\partial\Omega} w(\mathbf{x})\, dS_x \int_{\partial\Omega} D_t\phi(\xi)\, e_{ikt} n_k(\mathbf{x})\,\phi^*_{,i}(\mathbf{x},\xi)\, dS_\xi \\
&= \int_{\partial\Omega} D_k w(\mathbf{x})\, dS_x \int_{\partial\Omega} D_t\phi(\xi)\,\mathrm{H}_{kt}(\mathbf{x},\xi)\, dS_\xi
\end{aligned} \tag{112}
$$

Then, through combining Eqs. (108), (109) and (112), and the test function $w(\mathbf{x})$ is chosen to be identical to a function which is energy-conjugate to $q(\mathbf{x})$, viz. the trial function $\hat{\phi}(\mathbf{x})$, we obtain the symmetric Galerkin q-BIE form as

$$
\begin{aligned}
-\frac{1}{2}\int_{\partial\Omega} q(\mathbf{x})\hat{\phi}(\mathbf{x})\,dS_x &= \int_{\partial\Omega} q(\xi)\,dS_\xi \int_{\partial\Omega}^{CPV} \hat{\phi}(\mathbf{x})\,\hat{\Theta}^*(\mathbf{x},\xi)\,dS_x \\
&+ \int_{\partial\Omega} D_k\hat{\phi}(\mathbf{x})\,dS_x \int_{\partial\Omega} D_t\phi(\xi)\,\mathrm{H}_{kt}(\mathbf{x},\xi)\,dS_\xi \\
&+ \int_{\partial\Omega} n_k(\mathbf{x})\hat{\phi}(\mathbf{x})\,dS_x \int_{\partial\Omega} k^2 n_k(\xi)\phi(\xi)\,\phi^*(\mathbf{x},\xi)\,dS_\xi
\end{aligned} \tag{113}
$$

Equation (113) leads to the present novel formulation for a Symmetric Galerkin Boundary Element Method for the regularized R-q-BIE. *We label this as "SGBEM-R-q-BIE" for convenience, in this paper.*

3.2.6 Some Detailed Properties of Kernel Functions

(i) $\Theta^*(\mathbf{x},\xi)$ and $\hat{\Theta}^*(\mathbf{x},\xi)$
First of all, it is quite straight forward to see that

$$\Theta^*(\mathbf{x},\xi) = -\hat{\Theta}^*(\xi,\mathbf{x}) \tag{114}$$

from the definition of $\Theta^*(\mathbf{x},\xi)$ and $\hat{\Theta}^*(\mathbf{x},\xi)$. Eq. (114) results in the symmetry of the "SGBEM-R-ϕ-BIE" and "SGBEM-R-q-BIE", as seen in the next section.

(ii) $\mathrm{H}^*(\mathbf{x},\xi)$
From the definition of $\Sigma^*(\mathbf{x},\xi)$ in Eq. (110), we know that

$$\nabla\cdot\Sigma^*(\mathbf{x},\xi) = 0 \tag{115a}$$

$$\Sigma^*(\mathbf{x},\xi) = \nabla\times\mathrm{H}^*(\mathbf{x},\xi) \tag{115b}$$

which means that $\Sigma^*(\mathbf{x},\xi)$ spans a solenoidal field, and there exists a potential field, in this case, $\mathrm{H}^*(\mathbf{x},\xi)$, to construct the solenoidal field by using a curl operator. These properties ensure the application of the Stokes theorem in Eq. (112) to obtain simplified boundary integral equations.

Also, $\mathrm{H}^*(\mathbf{x},\xi)$ becomes $O\left(r^{-1}\right)$ in 3 dimensional problem, which is of the same order as $\phi^*(\mathbf{x},\xi)$, when $\xi\rightarrow\mathbf{x}$. Therefore, $\mathrm{H}^*(\mathbf{x},\xi)$ possesses the weak singularity, and it is convenient for the numerical implementation. Now, the ϕ-BIE and the q-BIE have been fully desingularized simply, and elegantly in the present work.

3.3 Closure

1. We have presented simple and straight-forward formulations for weakly-singular boundary integral equations for general PDEs. We have also presented a systematic way to decompose the corresponding kernel functions into divergence-free and companion curl-free parts, with the use of the properties of the fundamental solutions, for regularizing the BIEs. The fully regularized BIEs contain weakly singular integrals which are numerical tractable. The present approach has been demonstrated by developing the weakly-singular BIEs for elastics and acoustics in a simple and straight-forward manner. The present approach follows the methodologies presented in Okada, Rajiyah, and Atluri, (1989b) and extended in Han; Atluri (2003a). Clearly, these formulations can be extended to nonlinear problems, using the methodologies presented in Okada, Rajiyah, and Atluri, (1989b).
2. The traditional traction boundary element, or displacement boundary element methods can be derived the regularized tBIE and dBIE, respectively. In these methods, one uses a "mesh" at $\partial\Omega$, which consists of a set of contiguous (non-overlapping) "elements". In general, the trial functions u_b, t_b, and the test functions w_b at $\partial\Omega$ are interpolated in terms of their respective values at the nodes of the boundary elements. On the other hand, using the concepts of the meshless local Petrov-Galerkin methods (MLPG) developed in Atluri et al. (1998, 2002a,b), one may develop "meshless local Petrov-Galerkin boundary integral equation approaches". Also, tBIE and dBIE in general involve double integrals over $\partial\Omega$ on their right-hand sides, where $\partial\Omega$ is the entire global boundary. However, in using MLPG, the second integral over $\partial\Omega$ in the MLPG approach may be replaced by a local integral over a sub-region of $\partial\Omega$ only. In evaluating the first integral over the global $\partial\Omega$, one may use "shadow-elements", or alternatively, one may also develop a truly meshless MLPG method for integral equations. They are the subjects of our forthcoming papers [Atluri, Han, and Shen (2003); Han and Atluri (2003b)].

References

Atluri, S.N., Shen, S. (2002a). The meshless local Petrov-Galerkin (MLPG) method. Tech. Science Press, Beijng 440 pages.

Atluri, S.N., Shen, S. (2002b). The meshless local Petrov-Galerkin (MLPG) method: A simple & less-costly alternative to the finite element and boundary element method. *Comput. Model. Eng. Sci.*, vol. 3 no. 1, pp. 11–52.

Atluri, S.N., Zhu, T. (1998). A new meshless local Petrov-Galerkin (MLPG) approach in computational mechanics. *Comput. Mech.*, vol. 22, pp. 117–127.

Bonnet, M.; Maier, G.; Polizzotto, C. (1998): Symmetric Galerkin boundary element methods, *Appl Mech. Rev.*, vol. 51, pp. 669–704.

Chen, Z.S., Hofstetter, G., Mang, H.A. (1997). A symmetric Galerkin formulation of the boundary element method for acoustic radiation and scattering. *J. Comp. Acoust.*, vol. 5, no. 2, pp. 219–241.

Chien, C.C., Rajiyah, H., Atluri, S.N. (1990). An effective method for solving the hypersingular integral equations in 3-D acoustics. *J. Acoust. Soc. Am.* vol. 88, no. 2, pp. 918–937.

Cruse, T.A., Richardson, J.D. (1996). Non-singular Somigliana stress identities in elasticity, *Int. J. Numer. Meth. Eng.*, vol. 39, pp. 3273–3304.

Han. Z.D., Atluri, S. N. (2002). SGBEM (for Cracked Local Subdomain) – FEM (for uncracked global Structure) Alternating Method for Analyzing 3D Surface cracks and their fatigue-growth. *Comput. Model. Eng. Sci.*,vol. 3, no. 6, pp. 699–716.

Han, Z.D., Atluri, S.N. (2003a): On simple formulations of weakly-singular tBIE&dBIE, and Petrov-Galerkin approaches, *Comput. Model. Eng. Sci.*, vol. 4, no. 1, pp. 5–20.

Han, Z.D., Atluri, S.N. (2003b). Truly Meshless Local Petrov-Galerkin (MLPG) solutions of traction & displacement BIEs, *Comput. Model. Eng. Sci.*, vol. 4 no. 6 pp. 665–678.

Li, S., Mear, M.E. (1998). Singularity-reduced integral equations for displacement discontinuities in three-dimensional linear elastic media, *Int. J. Fract.*, vol. 93, pp. 87–114.

Meyer, W.L., Bell, W.A., Zinn, B.T., Stallybrass, M.P. (1978). Boundary integral solutions of three dimensional acoustic radiation problems, *J. Sound Vib.*, vol. 59, no. 2, pp. 245–262

Okada, H., Rajiyah, H., Atluri, S.N. (1988). A Novel Displacement Gradient Boundary Element Method for Elastic Stress Analysis with High Accuracy, *J. Applied Mech.*, vol. 55 no. 4 pp. 786–794.

Okada, H., Rajiyah, H., Atluri, S.N. (1989). Non-hyper-singular integral representations for velocity (displacement) gradients in elastic/plastic solids (small or finite deformations), *Comput. Mech.*, vol. 4, pp. 165–175.

Okada, H., Rajiyah, H., Atluri, S.N. (1990). A full tangent stiffness field-boundary-element formulation for geometric and material non-linear problems of solid mechanics, *Int. J. Numer. Meth. Eng.*, vol. 29, no. 1, pp. 15–35.

Qian, Z.Y., Han, Z.D., Atluri, S.N. (2004). Directly derived non-hyper-singular boundary integral equations for acoustic problems, and their solution through Petrov-Calerkin schemes, *Comput. Model. Eng. Sci.*, vol. 5, no. 6, pp. 541–562.

Qian, Z.Y., Han, Z.D., Ufimtsev, P., Atluri, S.N. (2004). Non-hyper-singular boundary integral equations for acoustic problems, implemented by the collocation-based boundary element method, *Comput. Model. Eng. Sci.*, vol. 6, no. 2, pp. 133–144.

Wu, T.W., Seybert, A.F., Wan, G.C. (1991): On the numerical implementation of a Cauchy principal value integral to insure a unique solution for acoustic radiation and scattering. *J. Acoust. Soc. Am.* vol.90, pp. 554–560.

Dynamic Inelastic Analysis with BEM: Results and Needs

George D. Hatzigeorgiou

Abstract This paper presents the techniques, advances and problems in analysis of dynamic inelastic problems with boundary element methods. The developed techniques for inelastic modeling within the BEM framework are reviewed. Adoptions of the appropriate fundamental solutions, corresponding solution techniques for the discretized system of equations, as well as the accuracy of the methods are addressed. Future improvements of the techniques and their applications are discussed.

1 Introduction

For many types of structures, the use of linear structural analysis is no longer considered adequate for achieving a realistic and safe design. To obtain more insight into the behavior, one must take into account nonlinearities due to inelasticity of the material. There is a growing interest in structures, which appear inelastic behavior under dynamic loading in civil and structural engineering (e.g. buildings, bridges, tunnels, tanks), and naval, aerospace and mechanical engineering (e.g. vehicles, ships, airplanes). It should be noted that for the aforementioned everyday engineering problems, dynamic inelastic analysis is carried out exclusively by numerical methods due to their vast complexity. In the last four decades, with the drastic evolution of digital computers, the finite element method (FEM) has assumed an important role in the solution of these complex engineering problems. The boundary element method (BEM) is about two decades younger for dynamic inelastic problems. This paper is explicitly concentrated to BEM's for inelastic problems under dynamic loading conditions. The main purpose of this paper has to do with the presentation and critical discussion for the application of the BEM's for these problems, presented between 1995 and 2007. Before 1995, the reader is strongly suggested to consult the review paper of Beskos (1995).

G.D. Hatzigeorgiou (✉)
Department of Environmental Engineering, Democritus University of Thrace,
GR-67100, Xanthi, Greece
e-mail: gchatzig@env.duth.gr

G.D. Manolis, D. Polyzos (eds.), *Recent Advances in Boundary Element Methods,*
DOI 10.1007/978-1-4020-9710-2_13, © Springer Science+Business Media B.V. 2009

2 Dynamic Inelastic Analysis with Boundary Elements

The presently available BEM's for inelastic analysis under dynamic loads can be divided into three major categories: 2-D solids and structures, 3-D solids and structures and analysis of other structural types as beams and plates.

2.1 Analysis of Two-Dimensional Solids and Structures

The first category is related to 2-D structures under plane strain or plane stress case. This category has reached a rather mature level and six methodologies have already been developed.

In the first one, the BEM in its direct conventional form and in conjunction with the elastostatic fundamental solution of the problem has been successfully used for the analysis of 2-D elastoplastic solids and structures under dynamic loading. This approach is also known as Domain/Boundary Element Method (D/BEM). For a two-dimensional body Ω which is bounded by its surface Γ (in two-dimensions this is a polyline or a curve), the Somigliana identity for the dynamic elastoplastic case is defined as

$$c_{ij}u_j(\xi,t) = \int_\Gamma u^*_{ij}(\xi,X)\,p_j(X,t)\,d\Gamma(X) - \int_\Gamma p^*_{ij}(\xi,X)\,u_j(X,t)\,d\Gamma(X) \\ - \rho\int_\Omega u^*_{ij}(\xi,X)\,\ddot{u}_j(X,t)\,d\Omega(X) + C_i \tag{1}$$

where

$$C_i = \int_\Omega \varepsilon^*_{jki}(\xi,X)\,\sigma^p_{jk}(X,t)\,d\Omega(X) \tag{2}$$

for formulation associated to the initial stress approach, and

$$C_i = \int_\Omega \sigma^*_{jki}(\xi,X)\,\varepsilon^p_{jk}(X,t)\,d\Omega(X) \tag{3}$$

associated to the initial strain approach. In the above, t is the time, ρ the constant mass density of the body and $u^*_{ij}(\xi,X)$, $p^*_{ij}(\xi,X)$, $\varepsilon^*_{jki}(\xi,X)$, and $\sigma^*_{jki}(\xi,X)$ are the fundamental solution components of the elastostatic problem representing the displacement, traction, strain and stress, respectively. Besides, u_j, $\ddot{u}_j$, p_j, σ^p_{jk} and ε^p_{jk} represent the displacements, accelerations, tractions, inelastic stresses and inelastic strains, respectively. Furthermore, c_{ij} is the usual free coefficient of elastostatic analysis where $c_{ij} = \delta_{ij}$ for any interior point and $c_{ij} = \delta_{ij}/2$ for any point on the smooth boundary where δ_{ij} is the Kronecker constant. If the tangent plane on the boundary is continuous but not smooth one can consult Hartmann (1983).

In Eq. (1), the Kelvin fundamental solution is adopted where Ω is assumed to be an infinite 2-D elastic medium and consequently its surface Γ is extended to infinity. The fundamental expressions for displacement, traction, strain and stress are given

$$u_{ij}^{*}(\xi, X) = \frac{(3-4\nu)\, ln\,(1/r)\,\delta_{ij} + r_{,i}r_{,j}}{8\pi\,(1-\nu)\,G} \tag{4}$$

$$p_{ij}^{*}(\xi, X) = \frac{\left[(1-2\nu)\,\delta_{ij} + 2r_{,i}r_{,j}\right]\frac{\partial r}{\partial n} + (1-2\nu)\left(r_{,j}n_i - r_{,i}n_j\right)}{4\pi\,(\nu-1)\,r} \tag{5}$$

$$\varepsilon_{jki}^{*}(\xi, X) = \frac{(1-2\nu)\left(r_{,k}\delta_{ij} + r_{,j}\delta_{ik}\right) - r_{,i}\delta_{jk} + 2r_{,i}r_{,j}r_{,k}}{8\pi\,(\nu-1)\,Gr} \tag{6}$$

and

$$\sigma_{jki}^{*}(\xi, X) = \frac{(1-2\nu)\left(r_{,k}\delta_{ij} + r_{,j}\delta_{ki} - r_{,i}\delta_{jk}\right) + 2r_{,i}r_{,j}r_{,k}}{4\pi\,(\nu-1)\,r^2} \tag{7}$$

In the above relations, n is the surface (boundary) normal vector and r and $r_{,i}$ is the distance and its derivative in i-axis, respectively, between field point X and collocation point ξ. Equation (1) represents the equation of motion of the body in integral form. For $\rho = 0$ this equation is reduced to the static case.

The boundary of the 2-D body is discretized into NB boundary elements and the domain is discretized into NV volume cells. Adopting the initial stress formulation, Eq. (1) becomes

$$\begin{aligned} c_{ij}u_j(\xi, t) = &\sum_{m=1}^{NB}\left\{\int_{\Gamma m} u_{ij}^{*}(\xi, X)\,\Phi d\Gamma\right\} p_j(X, t) \\ &-\sum_{m=1}^{NB}\left\{\int_{\Gamma m} p_{ij}^{*}(\xi, X)\,\Phi d\Gamma\right\} u_j(X, t) \\ &-\rho\sum_{n=1}^{NV}\left\{\int_{\Omega n} u_{ij}^{*}(\xi, X)\,\Phi d\Omega\right\} \ddot{u}_j(X, t) \\ &+\sum_{n=1}^{NV}\left\{\int_{\Omega n} \varepsilon_{jki}^{*}(\xi, X)\,\Phi d\Omega\right\} \sigma_{jk}^{p}(X, t) \end{aligned} \tag{8}$$

where Φ is the matrix of the shape functions. The boundary element implementation transforms the system of integral equations to an equivalent algebraic system, which in matrix notation reads

$$[G]\{p(t)\} - [H]\{u(t)\} - [M]\{\ddot{u}(t)\} + [Q]\left\{\sigma^{p}(t)\right\} = \{0\} \tag{9}$$

In the above equation matrices $[G]$ and $[H]$ correspond to the boundary integrals and $[M]$ and $[Q]$ to the inertial and initial stress domain integrals, respectively. One can mention here the works of Coda and Venturini (2000), Soares, Telles and Mansur (2006a, b) and Soares, Telles and Carrer (2007) using the domain (D)/BEM treat both plastic stresses and inertial forces by internal cells. Furthermore, Hatzigeorgiou and Beskos (2002a) apply this approach to compute the seismic inelastic response of masonry bridges using damage mechanics. In all these cases, the volume discretization is required for the whole structure.

The second presented approach has to do with the adoption of time-dependent fundamental solutions of the problem. The analysis also requires the determination of inelastic stresses and an internal discretization is needed. However, the internal discretization is applied only in those regions of the interior domain where the inelasticity is expected. Equation (1) can also express the integral equation of motion using initial stress or initial strain approach and the elastodynamic fundamental solution, under zero initial conditions and zero body forces. In this case, ξ and X corresponds to field and source point, respectively, while the symbol $(*)$ denotes convolution. Furthermore, $u^*_{ij}(\xi, X, T)$, $p^*_{ij}(\xi, X, T)$, $\varepsilon^*_{ikj}(\xi, X, T)$ and $\sigma^*_{ikj}(\xi, X, T)$ are the fundamental expressions for point X at time T due to a unit force vector applied at a point ξ at a preceding time τ, where

$$u^*_{ij} * p_j = \int_0^T u^*_{ij}(X, T; \xi, \tau)\, p_j(X, \tau)\, d\tau \tag{10}$$

$$p^*_{ij} * u_j = \int_0^T p^*_{ij}(X, T; \xi, \tau)\, u_j(X, \tau)\, d\tau \tag{11}$$

$$\varepsilon^*_{jki} * \sigma^p_{jk} = \int_0^T \varepsilon^*_{jki}(X, T; \xi, \tau)\, \sigma^p_{jk}(X, \tau)\, d\tau \tag{12}$$

$$\sigma^*_{jki} * \varepsilon^p_{jk} = \int_0^T \sigma^*_{jki}(X, T; \xi, \tau)\, \varepsilon^p_{jk}(X, \tau)\, d\tau \tag{13}$$

Stresses at interior points derived from displacements or from integral equations. According to the former approach, displacement state leads to strain state, where using appropriate constitutive relations the stress state is arisen. This procedure is computationally efficient. On the other hand, using the second approach, the integral equations for the stress state derived in incremental form by

$$\begin{aligned}\delta\sigma_{jk}(\xi, T) = &\int_\Gamma \bar{u}^*_{ijk}(\xi, X, T) * \delta p_i(X, T)\, d\Gamma(X) - \\ &- \int_\Gamma \bar{p}^*_{ijk}(\xi, X, T) * \delta u_i(X, T)\, d\Gamma(X) + \\ &+ \int_\Omega \bar{\varepsilon}^*_{imjk}(\xi, X) * \delta\sigma^p_{im}(X, T)\, d\Omega(X) + R^\delta_{imjk}\sigma^p_{im}(\xi, T)\end{aligned} \tag{14}$$

The knowledge of the response at a time T_N requires the discretization of the time axis into N equal time intervals, i.e.

$$T_N = \sum_{n=1}^{N} n\Delta T \tag{15}$$

Application of Eqs. (10), (11), (12) (13) and (15) into (1) gives the equation of motion of the structure in incremental form

$$\begin{aligned} c_{ij}(\xi)\,\Delta u_j(\xi, T_N) = &\int_{T_0}^{T_{N-1}} \int_\Gamma u_{ij}^* \Delta p_j d\Gamma d\tau + \int_{T_{N-1}}^{T_N} \int_\Gamma u_{ij}^* \Delta p_j d\Gamma d\tau - \\ &- \int_{T_0}^{T_{N-1}} \int_\Gamma p_{ij}^* \Delta u_j d\Gamma d\tau - \int_{T_{N-1}}^{T_N} \int_\Gamma p_{ij}^* \Delta u_j d\Gamma d\tau + \\ &+ \int_{T_0}^{T_{N-1}} \int_\Omega \varepsilon_{ikj}^* \Delta\sigma_{ik}^p d\Omega d\tau + \int_{T_{N-1}}^{T_N} \int_\Omega \varepsilon_{ikj}^* \Delta\sigma_{ik}^p d\Omega d\tau \end{aligned} \tag{16}$$

One must assume the variation of the field variables (i.e. displacements, tractions and stresses) during a time step. The simplest consideration assuming a constant variation of these variables during a time step, while linear variation gives

$$u_j(X,\tau) = \sum_{n=1}^{N} M_I^n u_j^{n-1}(X) + M_F^n u_j^n(X) \tag{17}$$

and

$$p_j(X,\tau) = \sum_{n=1}^{N} M_I^n p_j^{n-1}(X) + M_F^n p_j^n(X) \tag{18}$$

while,

$$\sigma_{ik}^p(X,\tau) = \sum_{n=1}^{N} M_I^n \left(\sigma_{ik}^p\right)^{n-1}(X) + M_F^n \left(\sigma_{ik}^p\right)^n(X) \tag{19}$$

Furthermore, *MI* and *MF* are temporal interpolation functions of the form

$$M_I^n = \frac{T_n - \tau}{\Delta T}\phi_n(\tau) \tag{20}$$

and

$$M_F^n = \frac{\tau - T_{n-1}}{\Delta T}\phi_n(\tau) \tag{21}$$

where

$$\phi_n(\tau) = H(\tau - (n-1)\Delta T) - H(\tau - n\Delta T) \quad (22)$$

and H being the Heaviside function. After the usual time and spatial discretization and integrations, the integral equation of motion (Eq. 16) is transformed into an equivalent system of matrix equations of the form

$$\begin{aligned}
&\left[A_F^1\right]\left\{\Delta X^N\right\} - \left[B_F^1\right]\left\{\Delta Y^N\right\} - \left[C_F^1\right]\left\{(\Delta\sigma^p)^N\right\} \\
&= -\sum_{n=2}^{N}\left[A_F^n + A_I^{n-1}\right]\left\{\Delta X^{N-n+1}\right\} + \sum_{n=2}^{N}\left[B_F^n + B_I^{n-1}\right]\left\{\Delta Y^{N-n+1}\right\} - \\
&\quad - \sum_{n=2}^{N}\left[C_F^n + C_I^{n-1}\right]\left\{\Delta(\Delta\sigma^p)^{N-n+1}\right\}
\end{aligned} \quad (23)$$

or

$$\left[A_F^1\right]\left\{\Delta X^N\right\} = \left[B_F^1\right]\left\{\Delta Y^N\right\} + \left[C_F^1\right]\left\{(\Delta\sigma^p)^N\right\} + \left\{R^N\right\} \quad (24)$$

Similarly, the integral equation for stresses can be written as

$$\left\{\Delta\sigma^N\right\} = \left[A_F^1\right]_\sigma\left\{\Delta X^N\right\} + \left[B_F^1\right]_\sigma\left\{\Delta Y^N\right\} + \left[C_F^1\right]_\sigma\left\{(\Delta\sigma^p)^N\right\} + \left\{R^N\right\}_\sigma \quad (25)$$

One can mention here the work of Telles, Carrer and Mansur (1999). This approach presents the advantage of eliminating the inertial volume integrals and thus the domain discretization is restricted to those parts of the domain where plastic stresses are expected to develop. Nevertheless, the method appears to be particularly complicated and time consuming because of the complex kernels involved and the need to satisfy causality at every time step (Telles, Carrer and Mansur 1999). Moreover, problems of stability may appear during the time integration process (Siebrits and Peirce 1997).

In the third one, the BEM in its direct conventional form and in conjunction with the elastostatic fundamental solution of the problem is formulated. However, the dual reciprocity technique (DR-BEM) is applied to transform the inertial volume integrals into surface integrals. More specifically, the inertial domain integral of Eq. (1) can be transformed into boundary integrals by approximating the accelerations $\ddot{u}_j$ within the domain. The displacements $u_i(X,t)$ can be expressed by a sum of m coordinate functions $f^k(X)$ multiplied by the unknown time dependent functions $a_i^k(t)$, i.e.

$$u_i(X,t) = a_i^k(t)\, f^k(X) \quad (26)$$

with summation on $k = 1$ to m implied. Similarly, the accelerations take the form

$$\ddot{u}_i(X,t) = \ddot{a}_i^k(t) f^k(X) \tag{27}$$

Thus, the interior discretization due to inelasticity is restricted only to regions expected to become inelastic. One can mention here the work of Czyz and Fedelinski (2006).

The fourth methodology has to do with the BEM in its symmetric Galerkin form and in conjunction with the elastostatic fundamental solution of the problem, as in the 1st case, which has also been successfully applied to 2-D dynamic elastoplastic problems. Generally, the symmetric Galerkin form is a non-traditional BEM formulation using the classical Betti's work reciprocity theorem with single-layer and double-layer sources, in such a way that the integral operator turns out to be symmetric with respect to a suitably defined bilinear form. The space discretization is achieved either on the basis of variational properties of the solution, or using a weighted-residual technique according to Galerkin's classical correlation between shape and weight functions. On can mention here the works of Frangi (1998) and Frangi and Maier (1999).

The fifth one consists with the hybrid BEM/FEM schemes in the time domain, which appropriately combine the advantages of both the FEM and the BEM. The finite element method, for instance, is well suited for materials with inelastic behavior. For this reason, the finite element discretization is applied to regions expected to become inelastic. On the other hand, for systems with infinite extension, the use of the boundary element method in conjunction with the elastodynamic fundamental solution of the problem is by far more beneficial. Thus, the FEM/BEM coupling in the time domain has been successfully used to solve 2-D nonlinear dynamic soil/structure interaction problems where the inelastic structure and the surrounding soil part expected to become inelastic is simulated by the FEM and the remaining soil assumed to behave linearly by means of the BEM. One can mention here the works related to underground structures by Adam (1997) and Takemiya and Adam (1998), earth dams by Abouseeda and Dakoulas (1998), concrete gravity dams by Yazdchi, Khalili and Valliappam (1999), wall structures by von Estorff and Firuziaan (2000). A similar FEM/BEM coupling in the time domain has been successfully used to solve 2-D nonlinear dynamic fluid/structure interaction problems (an inelastic structure with a fluid of finite or semi-infinite extension) by Czygan (2002), Czygan and von Estorff (2002), Soares (2004) and Soares, von Estorff and Mansur (2005). FEM/BEM schemes for 2-D general structures have been developed by Soares, von Estorff and Mansur (2004) and for reinforced media by Coda (2001).

Finally, the sixth methodology has to do with the combination of the first and the second BEM's. More specifically, Soares, von Estorff and Mansur (2005) appropriately coupling of the aforementioned boundary element formulations to obtain time-domain numerical solution of dynamic non-linear problems. Thus, the domain is divided into two sub-domains: the sub-domain that behaves elastically is modeled by the elastodynamic time-domain BEM formulation (2nd aforementioned method) while the sub-domain that presents non-linear behavior is modeled by the D/BEM formulation (1st abovementioned approach).

2.2 Analysis of Three-Dimensional Solids and Structures

The second category has to do with the dynamic inelastic analysis of 3-D solids and structures by means of boundary element methods. During the examined period, a sufficient amount of progress is realized in this topic. However, many evolutions must be taken place, mainly as extension of the mature 2-D formulations in the corresponding three-dimensional space. Two basic methodologies have already been developed.

The first presented approach has to do with BEM using the time-independent (static) fundamental solutions of the problem (D/BEM). As it is expected, the analysis requires the determination of inelastic stresses and inertia terms, which are internal quantities, and therefore, an internal discretization with 3-D volume cells is needed. The integral equation of motion using initial stress approach is also given by Eq. (1), where the 3-D fundamental solutions are given by

$$u_{ij}^{*}(\xi, X) = \frac{(3-4\nu)\,\delta_{ij} + r_{,i}r_{,j}}{16\pi\,(1-\nu)\,Gr} \tag{28}$$

$$p_{ij}^{*}(\xi, X) = \frac{\left[(1-2\nu)\,\delta_{ij} + 3r_{,i}r_{,j}\right]\frac{\partial r}{\partial n} + (1-2\nu)\left(r_{,j}n_i - r_{,i}n_j\right)}{8\pi\,(\nu-1)\,r^2} \tag{29}$$

and

$$\varepsilon_{jki}^{*}(\xi, X) = \frac{(1-2\nu)\left(r_{,k}\delta_{ij} + r_{,j}\delta_{ik}\right) - r_{,i}\delta_{jk} + 3r_{,i}r_{,j}r_{,k}}{16\pi\,(\nu-1)\,Gr^2} \tag{30}$$

One can mention here the works of Beskos (2003), Hatzigeorgiou and Beskos (2001a, 2002b) for the analysis of 3-D elastoplastic structures and Hatzigeorgiou and Beskos (2000, 2002c) for 3-D damaged structures. Furtermore, Hatzigeorgiou (2001), and Hatzigeorgiou and Beskos (2001b) developed appropriate D/BEM methodologies to solve dynamic inelastic three-dimensional soil/structure interaction problems for underground structures. It should be noted that the D/BEM presents the advantages of stability and low computational cost, which are essential for the trying 3-D dynamic inelastic analysis.

In the second one, the BEM in its direct conventional form and in conjunction with the elastodynamic fundamental solution of the problem has been successfully combined with the FEM for the dynamic analysis of 3-D elastoplastic problems.

This promising shceme seems to be immature in 3-D formulations and only the work of Firuziaan and von Estorff (2002) can be mentioned here.

2.3 Analysis of Structural Elements: Beams, Plates and Shells

This section examines the third category, which is related to dynamic inelastic boundary element analysis of structural elements. Two major subcategories are existed. The first one is related to dynamic inelastic analysis of plates. In this important research area, various D/BEM techniques are developed by Providakis and his co-workers to determine the dynamic response of elastoplastic plates. Various effects have been taken into account as the influence of internal support conditions and resting on Winkler-type foundation for thin flexural (Kirchhoff) plates by Providakis (1996, 1997, 2000a) and Providakis and Toungelidis (1998). These techniques are also applied for thick (Reissner–Mindlin) plates, where the shear deformations are taken into account by Providakis (2000b, 2007) and Providakis and Beskos (2000). Amongst others, explicit expressions for the above formulations can be found in the review paper of Providakis and Beskos (1999).

The second subcategory is related to dynamic elastoplastic analysis of beam structures. One can mention here the works of Adam (1998) and Adam and Ziegler (1997a, b). In these works Green's functions in conjunction with modal analysis are adopted to create special BEMs for these problems.

3 Numerical Examples

This section examines four representative numerical examples to illustrate the various BEM's described in this paper and demonstrate their capabilities.

3.1 A Circular Elastoplastic Cavity Under Dynamic Load

Soares Carrer and Mansur (2005) examine a circular cavity under plane strain conditions, which is subjected to a uniform internal pressure suddenly applied and kept constant in time. In this work, the domain is divided into two sub-domains: one that behaves elastically is modeled by the elastodynamic time-domain BEM (TD-BEM) while the one presents non-linear behavior is modeled by D/BEM.

The physical properties of the examined model are: Poisson's ratio $\nu = 0.2308$, modulus of elasticity $E = 6.5277 \cdot 10^8$ N/m^2 and mass density $\rho = 1.804 \cdot 10^3$ kg/m^3. A perfectly plastic material obeying the Mohr–Coulomb yield criterion is assumed with cohesion $c = 4.8263 \cdot 10^6$ N/m^2 and internal friction $\phi =$ angle30°. The geometry data (Fig. 1) are defined by $R = 3.048$ m and $d - 3.658$ m.

Figure. 2 shows the elastic and inelastic response of A and B points, for radial and circumferential stresses σ_R and σ_C, respectively.

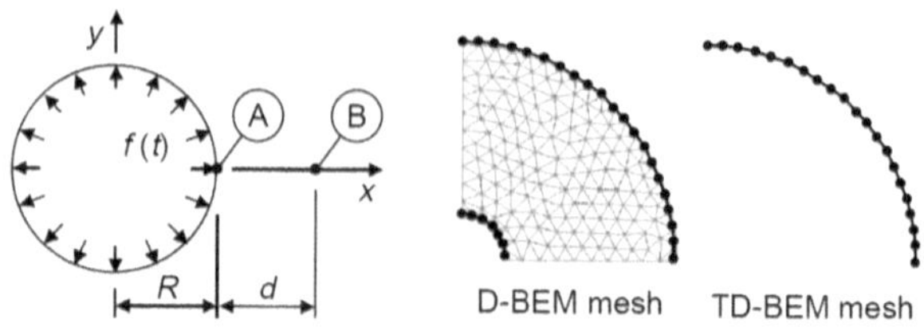

Fig. 1 Structure and boundary element/internal cell discretization

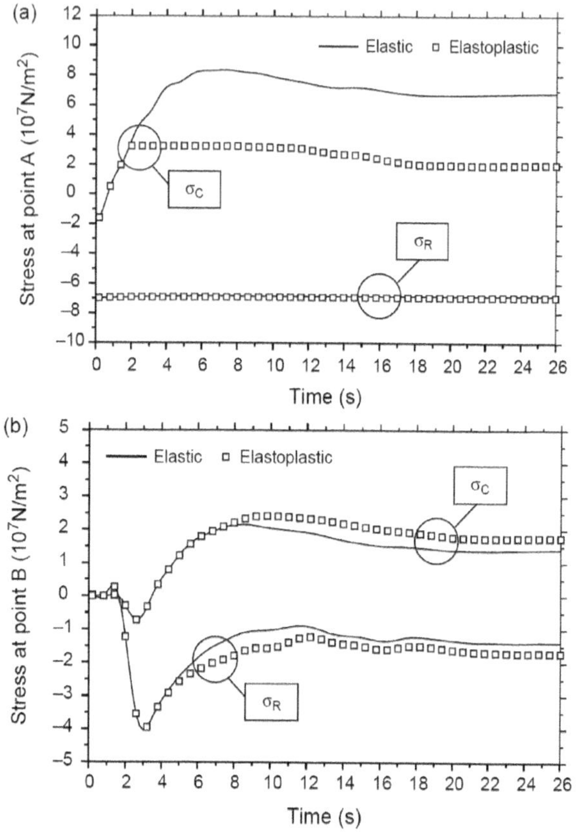

Fig. 2 Radial and circumferential stresses time history

3.2 Seismic Inelastic Analysis of a Masonry Bridge

The historic Arta Bridge made of masonry has been analyzed by Hatzigeorgiou and Beskos (2002a). In this work, a D/BEM under plane stress conditions is applied for the loading case of self-weight plus the first 5 s of the NS El-Centro (1940) earthquake with maximum horizontal and vertical acceleration of 0.16 and 0.10 g, respectively to match local conditions. The bridge consists of four main arches with spans of 23.95, 15.83, 15.43 and 16.16 m, while its width is 3.70 m. Its material parameters assumed as elasticity modulus $E = 3.0$ GPa, Poisson's ratio

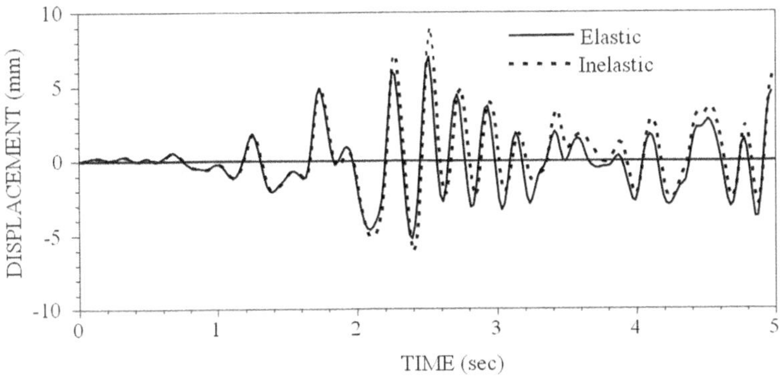

Fig. 3 Horizontal top displacement time history

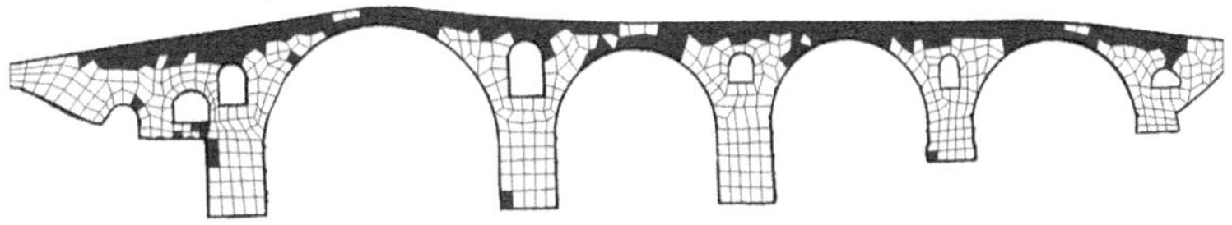

Fig. 4 Discretization and damaged region of the structure

$\nu = 0.22$, uniaxial compressive strength $f_c = 30.00$ MPa, biaxial compressive strength $f_{c-2D} = 34.8$ MPa, uniaxial tensile strength $f_t = 0.30$ MPa, specific fracture energy $G_f = 20$ N/m and density $\rho = 2700$ kg/m^3.

Figure 3 shows the time history of the horizontal displacement at the top of the largest arch as obtained under elastic and inelastic material behavior.

The discretization of the Arta Bridge is shown in Fig. 4. This figure also shows with grey color the damaged (inelastic) regions in the bridge, on the assumption that an element is considered failed (or fully damaged) when damage index d is equal to 1.00. It is evident that the damage concentrates at the upper part of the bridge, while the piers show no damage.

3.3 Dynamic Analysis of an Elastoplastic 3-D Beam

In this example, a cantilever steel beam subjected to an impact loading P=10 kN at its free end, as shown in Fig. 8a, is analyzed numerically by Hatzigeorgiou and Beskos (2002b) using the three-dimensional D/BEM approach. Figure 5a contains the geometry and the 3-D BEM discretization of the structure. The material parameters are: Young modulus E=210.0 GPa, inelastic modulus $E^T = 0.$, Poisson's ratio $\nu = 0.30$, mass density $\rho = 7850$ kg/m^3 and yield stress $\sigma_y = 400.0$ MPa. Figure 5b depicts the elastic and inelastic time history of the vertical displacement at the load point, as computed by the D/BEM and a commercial program based on the finite element method.

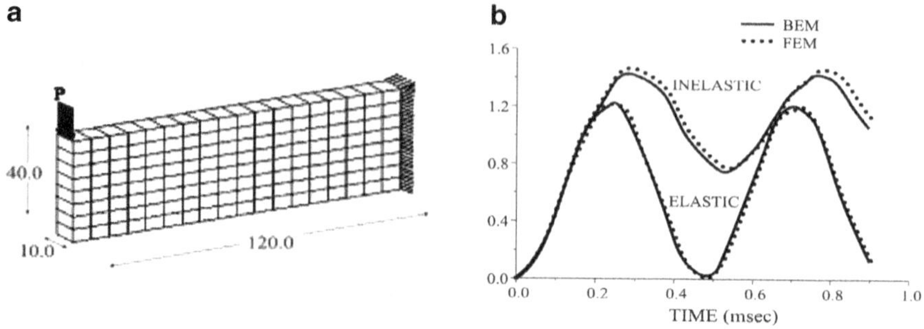

Fig. 5 (**a**) Geometry (in mm), (**b**) Vertical displacement history at load point

3.4 Dynamic Elastoplastic Analysis of a Thick Elliptical Plate

Providakis (2000b) examined a simply supported thick plate with elliptical boundary geometry under dynamic inelastic conditions. The two semi-axes of the elliptical shape are equal to 0.5 and 0.6 m while the plate thickness is $h = 0.15$ m. This structure subjected to a suddenly applied load uniformly distributed over the whole plate with intensity 100 N/m^2 and resting on a Winkler-type foundation. Two different foundation rigidities are examined: $k = 0$ and $k = 50 \cdot 10^8$ N/m. The material parameters for this example are: Young modulus $E = 200.0$ GPa, inelastic (hardening) modulus $E' = 0.6E$, Poisson's ratio $\nu = 0.3$, yield stress $\sigma_y = 488.0$ MPa and mass density $\rho = 76900$ kg/m^3. Figure 6 depicts the central

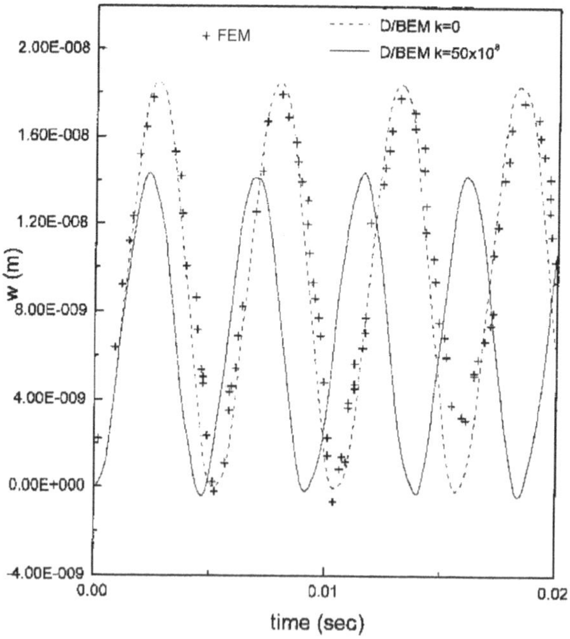

Fig. 6 Time history of vertical deflection of central point

deflection history of the elliptical plate as obtained by the D/BEM approach and using a FEM program.

4 Conclusions and General Criticism

This study leads to three conclusions for dynamic inelastic analysis with BEM:

a. Many BEM's have been developed to solve dynamic inelastic problems. Sections 2 and 3 present the formulations and their capabilities.
b. Among the existing BEM's the two most important approaches are those that employ the elastodynamic and those employ the elastostatic fundamental solution. The first one has the advantage of restricting the interior discretization to those parts of the structure expected to become inelastic. The latter approach is quite simpler and computationally effective than the former.
c. The BEM/D-BEM approach and the hybrid BEM/FEM approach seem to be very promising schemes. These techniques combine the advantages of concerned methods and are ideal for systems with limited inelastic regions as soil-structure interaction and fracture mechanics problems.

One major advantage of BEM for dynamic inelastic problem has to do with the determination at once and for all time or loading steps of the concerned matrices. On the other hand and due to inelasticity in FEM, stiffness matrix is continuously varied and should be computed in every time or loading step. However, the BEM matrices are full populated and non-symmetric (except for the symmetric Galerkin approach), while the FEM matrices are symmetric and sparse. It should be noticed that, especially for dynamic inelastic problems, there are three drawbacks for the BEM:

a. Due to inelasticity, and sometimes due to inertia terms (e.g. in D/BEM formulation), the analysis requires the determination of internal quantities as inelastic stresses. Therefore, an internal discretization is needed. In this way, the advantage of surface discretization is broken down.
b. No one commercial boundary element code has been developed for these problems. To the best of the author's knowledge, two general-purpose boundary element programs are existed, the BEASY (2007) and the GPBEST (2007) codes. However, these programs are irrelevant to inelastic material behavior problems, even for static loading conditions.
c. The theoretical development with BEM for many types of these problems is not existed and should be formulated and verified. In the followings, recommendations for future progress are presented:
    - The basic two- and three-dimensional D/BEM formulations with damage should be extended for beams, plates and shells.
    - The promising hybrid BEM/FEM and BEM/D-BEM schemes should be extended in three-dimensions.

- Appropriate extensions to dynamic problems are required both to 2- and 3-D problems for non-local plasticity or damage formulations.
- Much work is needed towards the developments of BEM's to fracture mechanics problems. The corresponding formulations are limited either to dynamic elastic or to static inelastic boundary element analysis.
- BEM's should be further extended to take into account large deformations or large strains/deformations in dynamic formulations.

References

Abouseeda H, Dakoulas P (1998) Non-linear dynamic earth dam-foundation interaction using a BE-FE method. Earthq. Eng. Struct. Dyn. 27: 917–936

Adam C (1998) Modal analysis of elastic-viscoplastic Timoshenko beam vibrations. Acta Mech. 126: 213–229

Adam C, Ziegler E (1997a) Moderately large forced oblique vibrations of elastic- Viscoplastic deteriorating slightly curved beams. Arch. Appl. Mech. 67: 375–392

Adam C, Ziegler E (1997b) Forced flexural vibrations of elastic-plastic composite beams with thick layers. Composites B 28B: 201–213

Adam M (1997) Nonlinear seismic analysis of irregular sites and underground structures by coupling BEM to FEM. PhD Thesis, Okayama University, Japan

BEASY Software (2007) http://www.beasy.com.

Beskos DE (1995) Dynamic inelastic structural analysis by boundary element method. Arch. Comput. Methods Eng. 2: 55–87

Beskos DE (2003) Dynamic analysis of structures and structural systems. In Beskos, D.E., Maier, G. (eds) Boundary Element Advances in Solid Mechanics, CISM International Centre for Mechanical Sciences 440: pp. 1–54, Springer, New York

Boundary Element Software Technology Corporation (2007) http://www.gpbest.com.

Coda HB (2001) Dynamic and static non-linear analysis of reinforced media: a BEM/FEM coupling approach. Comput. Struct. 79: 2751–2765

Coda HB, Venturini WS (2000) Dynamic non-linear stress analysis by the mass matrix BEM. Eng. Anal. Bound. Elem. 24: 323–332

Czygan O (2002) Fluid/structure coupling of 2D and axisymmetric systems taking into account a nonlinear structural behavior. PhD Thesis, TU Hamburg-Harburg, Germany

Czygan O, von Estorff O (2002) Fluid-structure interaction by coupling BEM and nonlinear FEM. Eng. Anal. Bound. Elem. 26: 773–779

Czyz T, Fedelinski P (2006) Boundary element formulation for dynamic analysis of inelastic structures. Comput. Assist. Mech. Eng. Sci. 13: 379–394

Firuziaan M, von Estorff O (2002) Transient 3D soil/structure interaction analyses including non-linear effects. In Grundmann H, Schuller GI (eds). Structural Dynamics EURODYN 2002, Lisse, Germany: 1291–1302

Frangi A (1998) Some developments in the symmetric Galerkin boundary element method. PhD Thesis. Politecnico di Milano, Milan, Italy

Frangi A, Maier G (1999) Dynamic elastic–plastic analysis by a symmetric Galerkin boundary element method with time independent kernels. Comput. Methods Appl. Mech. Eng. 171:281–308

Hartmann F (1983) Computing the C-matrix in non-smooth boundary points. In Brebbia CA (ed.) New Developments in Boundary Element Methods. pp. 367–379, CML, Southampton

Hatzigeorgiou GD (2001) Seismic inelastic analysis of underground structures by means of boundary and finite elements. PhD Thesis, Department of Civil Engineering, University of Patras, Patras, Greece

Hatzigeorgiou GD, Beskos DE (2000) 3-D boundary element analysis of damaged solids and structures. In Katsikadelis JT, Beskos DE, Gdoutos EE (eds) Recent Advances in Applied Mechanics: Honorary Volume for Professor A.N. Kounadis, NTUA, Athens

Hatzigeorgiou GD, Beskos DE (2001a) Transient dynamic response of 3-D elastoplastic structures by the D/BEM. In Beskos DE, Brebbia CA, Katsikadelis JT, Manolis GD (eds), BEM XXIII, Lemnos

Hatzigeorgiou GD, Beskos DE (2001b) Inelastic response of 3-D underground structures in rock under seismic loading. ERES 2001, Malaga

Hatzigeorgiou GD, Beskos DE (2002a) Dynamic analysis of 2-D and 3-D quasi-brittle solids and structures by D/BEM. Theor. Appl. Mech. 27: 39–48

Hatzigeorgiou GD, Beskos DE (2002b) Dynamic elastoplastic analysis of 3-D structures by the domain/boundary element method. Comput. Struct. 80: 339–347

Hatzigeorgiou GD, Beskos DE (2002c) Dynamic response of 3-D damaged solids and structures by BEM. Comput. Model Eng. Sci. 3: 791–802

Providakis CP (1996) A general and advanced boundary element transient analysis of elastoplastic plates. Eng. Anal. Bound. Elem. 17: 133–143

Providakis CP (1997) Transient boundary element analysis of elastoplastic plates on elastic foundation. Soil Dyn. Earthq. Eng. 16: 21–27

Providakis CP (2000a) Transient boundary element algorithm for elastoplastic building floor slab analysis. Int. J. Solids Struct. 37: 5839–5853

Providakis CP (2000b) Transient dynamic response of elastoplastic thick plates resting on Winkler-Type Foundation. Nonlinear Dyn. 23: 285–302

Providakis CP (2007) The effect of internal support conditions to the elastoplastic transient response of reissner-mindlin plates. Comput. Model. Eng. Sci. 18: 247–258

Providakis CP, Beskos DE (1999) Dynamic analysis of plates by boundary elements. Appl. Mech. Rev. 52: 213–236

Providakis CP, Beskos DE (2000) Inelastic transient dynamic analysis of Reissner-Mindlin plates by the D/BEM. Int. J. Numer. Methods Eng. 49: 383–397

Providakis CP, Toungelidis GA (1998) D/BEM approach to the transient response analysis of elastoplastic plates with internal supports. Eng. Comput. 15: 501–517

Siebrits E, Peirce AP (1997) Implementation and application of elastodynamic boundary element discretizations with improved stability properties. Eng. Comput. 14: 669–691

Soares Jr D (2004) Dynamic analysis of non-linear soil-fluid-structure coupled systems by the finite element method and the boundary element method. PhD Thesis. Federal University of Rio de Janeiro, Brazil

Soares Jr D, Carrer JAM, Mansur WJ (2005) Non-linear elastodynamic analysis by the BEM: an approach based on the iterative coupling of the D-BEM and TD-BEM formulations. Eng. Anal. Bound. Elem. 29: 761–774

Soares Jr D, von Estorff O, Mansur WJ (2004) Iterative coupling of BEM and FEM for nonlinear dynamic analysis. Comput. Mech. 34: 67–73

Soares Jr D, von Estorff O, Mansur WJ (2005) Efficient non-linear solid–fluid interaction analysis by an iterative BEM/FEM coupling. Int. J. Numer. Methods Eng. 64: 1416–1431

Soares Jr D, Telles JCF, Mansur WJ (2006a) Boundary elements with equilibrium satisfaction – a consistent formulation for dynamic problems considering non-linear effects. Int. J. Numer. Methods Eng. 65: 701–713

Soares Jr D, Telles JCF, Mansur WJ (2006b) A time-domain boundary element formulation for the dynamic analysis of non-linear porous media. Eng. Anal. Bound. Elem. 30: 363–370

Soares Jr D, Telles JCF, Carrer JAM (2007) A boundary element formulation with equilibrium satisfaction for thermo-mechanical problems considering transient and non-linear aspects. Eng. Anal. Bound. Elem. 31: 942–948

Takemiya H, Adam M (1998) 2D nonlinear seismic ground analysis by FEM-BEM: The case of Kobe in the Hyogo–Ken Nanbu earthquake. Struct. Eng./ Earthq. Eng. 15: 19–27

Telles JFC, Carrer JAM, Mansur WJ (1999) Transient dynamic elastoplastic analysis by the time-domain BEM formulation. Eng. Anal. Bound. Elem. 23: 479–86

Yazdchi M, Khalili N, Valliappan S (1999) Non-linear seismic behavior of concrete gravity dams using coupled finite element-boundary element technique. Int. J. Numer. Methods Eng. 44: 101–130

von Estorff O, Firuziaan M (2000) Coupled BEM/FEM approach for nonlinear soil/structure interaction. Eng. Anal. Bound. Elem. 24: 715–725

Quantifier-Free Formulae for Inequality Constraints Inside Boundary Elements

Nikolaos I. Ioakimidis

Abstract In this chapter we reconsider the problem of construction of quantifier-free formulae for the positivity of the unknown function in the BEM (boundary element method) along the whole boundary element. Such a situation appears in crack problems in fracture mechanics, where the crack opening displacement should be a positive function so that no contact between the crack edges be present. It also appears in contact problems where the pressure distribution should also be a positive quantity. Here our previous related results on the basis of classical Sturm sequences are improved: (i) by studying the unknown function on the interval $[0,\infty)$, (ii) by using *Mathematica* for all of the present computations and (iii) by directly using the nodal values of the originally unknown function also in the cubic polynomial. Both cases of the quadratic polynomial (three nodes on the boundary element) and the cubic polynomial (four nodes on the boundary element) are studied.

1 Introduction

The BEM (boundary element method) is a very well-known and efficient method for the numerical solution of boundary value problems on the basis of boundary integral equations with an extensive literature available during the last four decades. The problem is finally reduced to a system of linear algebraic equations and the unknown function is determined (or the unknown functions are determined) at the nodes of the boundary elements.

Generally, the obtained approximate solution is completely acceptable. But in few rare cases inequality constraints of physical importance should be satisfied as well. The two main such cases include crack problems in fracture mechanics and contact problems in elasticity.

N.I. Ioakimidis (✉)
Division of Applied Mathematics and Mechanics, Department of Engineering Sciences, School of Engineering, University of Patras, GR-26504, Rion, Patras, Greece
e-mail: n.ioakimidis@upatras.gr; n.ioakimidis@otenet.gr

A chapter in honour of Dimitri Beskos' 62nd birthday

G.D. Manolis, D. Polyzos (eds.), *Recent Advances in Boundary Element Methods*,
DOI 10.1007/978-1-4020-9710-2_14,

At first a straight crack problem (crack on $[a, b]$) in two-dimensional fracture mechanics is reduced to the following hypersingular (finite-part) integral equation:

$$\frac{1}{\pi}\int_a^b \left[\frac{1}{(t-x)^2} + k(t,x)\right] v(t)\,\mathrm{d}t = -c_1 p(x), \quad a < x < b \tag{1}$$

where $k(t, x)$ is a known kernel, $v(t)$ the unknown crack opening displacement, c_1 a known constant and $p(x)$ the known symmetrical compressive loading on the crack edges. Naturally, the crack opening displacement $v(x)$ should not be negative, it should be a continuously positive function on the crack $[a, b]$ so that no contact between the crack edges can be possible. Therefore, we must have

$$v(x) > 0 \text{ for all } x \text{ such that } a < x < b \tag{2}$$

Next, a contact problem in plane elasticity (for a half-plane under the indentation of a frictionless rigid punch of known profile) on the interval $[a, b]$ is reduced to the following weakly singular (with a logarithmic kernel) integral equation:

$$\frac{1}{\pi}\int_a^b [\log|t-x| + k(t,x)]\,p(t)\,\mathrm{d}t = -c_2 v(x), \quad a < x < b \tag{3}$$

where $k(t, x)$ is again a known kernel, $p(t)$ the unknown pressure distribution, $-c_2$ a known constant and $v(t)$ the known displacement on the contact interval. Naturally, the pressure distribution $p(t)$ should not be negative, it should be a continuously positive function so that an actual pressure on the half-plane be present. Therefore, we must have

$$p(x) > 0 \quad \text{for all } x \text{ such that } a < x < b \tag{4}$$

At this point we should mention that in this chapter we make the assumption that our unknown functions $v(t)$ and $p(t)$ take only positive values and we ignore the case of zero values. Therefore, we restrict our attention just to the avoidance of negative values. This assumption is reasonable if we take into account the approximate character of the BEM together with the fact of the practically zero probability of a zero value for the unknown function, here $v(t)$ or $p(t)$.

Under these circumstances we assume that we have already found the approximate numerical solution of the boundary integral equation (1) or (3) by using the BEM and we have available the nodal values of the unknown function $v(t)$ or $p(t)$. Now our wish is to check the continuous positivity of the unknown function evidently on each separate boundary element generally with $-1 < \xi < 1$ on it.

Therefore, our continuous positivity problem on each boundary element $[-1, 1]$ takes the form

$$v(\xi) \text{ or } p(\xi) > 0 \quad \text{for all } \xi \text{ such that } -1 < \xi < 1 \tag{5}$$

Two very common cases for the approximate unknown function on the basis of its numerical values at the nodes are that this function be a quadratic polynomial (three nodes on each boundary element) or a cubic polynomial (four nodes on each boundary element), i.e.

$$v(\xi) \text{ or } p(\xi) = a\xi^2 + b\xi + c \text{ or } v(\xi) \text{ or } p(\xi) = a\xi^3 + b\xi^2 + c\xi + d \quad (6)$$

with the coefficients in these two polynomials determined from the numerically computed nodal values of the unknown function by using the BEM.

The construction of a logical–algebraic formula assuring the continuous positivity of the unknown function (here a quadratic or a cubic polynomial) on each boundary element (here on the interval $-1 < \xi < 1$) constitutes a quantifier elimination problem. The resulting logical–algebraic formula constitutes a quantifier-free formula. This does not depend on the variable ξ on each boundary element $[-1, 1]$. Here we will determine the quantifier-free formulae for the above two cases of the quadratic polynomial (in Section 2) and of the cubic polynomial (in Section 3).

The approach of using quantifier elimination to construct quantifier-free formulae for the positivity of the unknown function in the BEM seems to have been suggested by the author (Ioakimidis 1995) by using Sturm sequences, but only for the cases of the linear and the quadratic polynomials and on the original and somewhat inconvenient interval $[-1, 1]$. This had as a result a rather complicated final quantifier-free formula for the quadratic polynomial. Next, the author (Ioakimidis 1996) generalized these results to the cubic polynomial as well and he used the more convenient interval $[1, \infty)$. Here the even more convenient interval $[0, \infty)$ will be used instead and this will have as a result an even simpler polynomial for the construction of the Sturm sequence. Next here the powerful and very popular computer algebra system *Mathematica* (Wolfram 1999) has been used in all of the present computations.

Moreover, in the present chapter exactly as in the first author's paper (Ioakimidis 1995) the nodal values of the originally unknown function will be directly used in the two quantifier-free formulae to be derived and this is done for the first time for the cubic polynomial by using Sturm sequences (Davenport et al. 1993). On the other hand, for the quadratic polynomial the quantifier-free formula to be derived here will be much simpler than the original formula (Ioakimidis 1995) for the same polynomial.

Naturally, the use of specialized computer algebra software for the derivation of quantifier-free formulae is also an interesting possibility and this possibility has been already used by the author (Ioakimidis 1999) with the help of the REDLOG Reduce package). Yet this possibility is not very convenient and it requires the additional use of the underlying computer algebra software: in the case of REDLOG of Reduce, which is presently less popular than *Mathematica*.

Concluding, the present use of the classical Sturm sequences (Davenport et al. 1993) seems to be a sufficiently general possibility and it does not require an expertise in quantifier elimination techniques and the related specialized packages. Moreover, it does not have any serious restriction on the degree of the polynomial used in the BEM although, naturally, the final quantifier-free formula becomes much more complicated as the degree of the polynomial in the BEM increases.

2 Quantifier Elimination for the Quadratic Polynomial

At first we consider the case where the boundary element $[-1, 1]$ has three nodes: $\xi_1 = -1$, $\xi_2 = 0$ and $\xi_3 = 1$. Then the originally unknown and next numerically determined function $u_\xi(\xi)$ along the boundary element $[-1, 1]$ will be a quadratic polynomial of the form

$$u_\xi(\xi) = a\xi^2 + b\xi + c \tag{7}$$

The values u_1, u_2 and u_3 of $u_\xi(\xi)$ at the above three nodes ξ_1, ξ_2 and ξ_3 respectively are known from the numerical solution of the problem by using the BEM. We have

$$u(-1) = u_1, \quad u(0) = u_2, \quad u(1) = u_3 \tag{8}$$

By using *Mathematica* (Wolfram 1999), we can easily determine the quadratic polynomial $u_\xi(\xi)$ in terms of its values u_1, u_2 and u_3 at the three aforementioned nodes $\xi_1 = -1$, $\xi_2 = 0$ and $\xi_3 = 1$ respectively. The related well-known result is

$$u_\xi(\xi) = (1/2)(u_1 - 2u_2 + u_3)\xi^2 + (1/2)(-u_1 + u_3)\xi + u_2 \tag{9}$$

Now we change the variable on the boundary element from ξ to η with $\xi = 2\eta - 1$, equivalently $\eta = (\xi + 1)/2$. Then the interval for η on the boundary element becomes $[0, 1]$. We easily obtain

$$u_\eta(\eta) = (2u_1 - 4u_2 + 2u_3)\eta^2 + (-3u_1 + 4u_2 - u_3)\eta + u_1 \tag{10}$$

with $u_\eta(\eta) = u_\xi(\xi) = u_\xi(2\eta - 1)$ and naturally $u_\eta(0) = u_1$, $u_\eta(1/2) = u_2$ and $u_\eta(1) = u_3$.

At this point surely we can proceed to our computations, here with the help of Sturm sequences (Davenport et al. 1993), by using the interval $[0, 1]$ on the boundary element. (Incidentally, we can add that this was made by Ioakimidis (1995).) Yet a more convenient approach is to transform this second interval $[0, 1]$ to the semi-infinite interval $[1, \infty]$. This will result in a computationally simpler Sturm sequence. We can also add that this is the approach having been used by Ioakimidis (1996). This transformation is easily achieved through the change of variable $\eta = 1/x$. The result for our polynomial (previously $u_\eta(\eta)$ and now $u_x(x)$) is

$$u_x(x) = u_1x^2 + (-3u_1 + 4u_2 - u_3)x + 2u_1 - 4u_2 + 2u_3 \tag{11}$$

For the derivation of this polynomial $u_x(x)$ we have also taken the liberty to multiply it by x^2 so that no denominator be present. Since x^2 is always a positive quantity on the interval $[1, \infty)$, this has naturally no consequence on the quantifier elimination problem that we try to solve here.

Our final step is one more variable transformation: from x to $z = x - 1$. This has as a result the further change of the present boundary element interval from $[1, \infty)$ with respect to x to the final interval $[0, \infty)$ with respect to z. In this way, we have achieved to find our final polynomial $u(z) := u_z(z) = u_x(x)$ as

$$u(z) = u_1 z^2 + (-u_1 + 4u_2 - u_3)z + u_3 \tag{12}$$

For convenience we have also used the simpler symbol $u(z)$ instead of $u_z(z)$. This interval $[0, \infty)$ is the interval that was used by Collins (1994, Private communication on quantifier elimination) for the derivation of the quantifier-free formula for the positivity of the general quadratic polynomial $p_2(z) = az^2+bz+c$ by his interesting approach yet a quite different approach from the present one.

Finally we can add that by comparing the right-hand sides of Eqs. (11) and (12) we remark that our final polynomial $u(z)$ is sufficiently simpler than the previous polynomial $u_x(x)$. Therefore we selected $u(z)$ as our final polynomial in the present quantifier-elimination-related symbolic computations. For this polynomial $u(z)$ we wish that it should be positive along the whole boundary element now on $[0, \infty)$.

We have already mentioned that in this chapter we will use the classical method of Sturm sequences (Davenport et al. 1993) as the approach for the derivation of the positivity-related quantifier-free formula that we will derive. The Sturm sequence related to the present quadratic polynomial $u(z)$ is sufficiently simple to derive (here with the help of *Mathematica*). It consists of the following three polynomials:

$$h_0(z) = u(z) = u_1 z^2 + (-u_1 + 4u_2 - u_3)z + u_3 \tag{13}$$

$$h_1(z) = \mathrm{d}h_0(z)/\mathrm{d}z = \mathrm{d}u(z)/\mathrm{d}z = 2u_1 z - u_1 + 4u_2 - u_3 \tag{14}$$

$$h_2(z) = -\mathrm{rem}[h_0(z), h_1(z)] = [u_1^2 + (-4u_2 + u_3)^2 - 2u_1(4u_2 + u_3)]/(4u_1) \tag{15}$$

where rem denotes the remainder of the division of two polynomials here of the first two polynomials $h_0(z)$ and $h_1(z)$. But since the quantity u_1 is positive (it is the nodal value at the left tip $\xi = -1$ of the boundary element) and we are interested only in the signs of the polynomials $h_0(z)$, $h_1(z)$ and $h_2(z)$, it is reasonable to use instead of $h_2(z)$ the modified and simpler polynomial

$$h_{2a}(z) = 4u_1 h_2(z) = u_1^2 + (-4u_2 + u_3)^2 - 2u_1(4u_2 + u_3) \tag{16}$$

having been derived through a multiplication of $h_2(z)$ by the positive quantity $4u_1$.

It is easily observed that this polynomial (naturally here simply a constant) $h_{2a}(z)$ can also be written in the somewhat simpler form

$$h_{2a}(z) = (-u_1 + 4u_2 - u_3)^2 - 4u_1 u_3 = \Delta \tag{17}$$

where Δ is the well-known discriminant of the polynomial $h_0(z) = u(z)$, i.e.

$$\Delta = (-u_1 + 4u_2 - u_3)^2 - 4u_1u_3 \tag{18}$$

This situation is an expected result in Sturm sequences (Davenport et al. 1993), that is its last term is proportional to the discriminant of the original polynomial. (Here the equality resulted after a multiplication of $h_2(z)$ by $4u_1$.)

Our next task in the present Sturm-sequences-related computations is to compute the values of the three polynomials $h_0(z)$, $h_1(z)$ and $h_2(z)$ at the left end $z_1 = 0$ and at the right end $z_2 = \infty$ of the present interval $[0, \infty)$. We denote these values at the left end $z_1 = 0$ by $h_{0,1}$, $h_{1,1}$ and $h_{2,1}$. Because of Eqs. (13), (14) and (17) we have

$$h_{0,1} = u_3, \quad h_{1,1} = -u_1+4u_2-u_3, \quad h_{2,1} = \Delta = (-u_1+4u_2-u_3)^2-4u_1u_3 \tag{19}$$

On the other hand, we denote the corresponding values at the right end $z_2 = \infty$ by $h_{0,2}$, $h_{1,2}$ and $h_{2,2}$. Since $z_2 = \infty$, we have to take the values $h_{0,2}$ and $h_{1,2}$ as the coefficients of z^2 and z of $h_0(z)$ and $h_1(z)$ respectively. Again because of Eqs. (13), (14) and (17) we have

$$h_{0,2} = u_1, \quad h_{1,2} = 2u_1, \quad h_{2,2} = \Delta = (-u_1 + 4u_2 - u_3)^2 - 4u_1u_3 \tag{20}$$

In agreement with Sturm's theorem (Davenport et al. 1993) we are now ready to evaluate the number of sign variations both at the left end $z_1 = 0$ and at the right end $z_2 = \infty$ of our final interval $[0, \infty)$. At first, we remember that all three quantities u_1, u_2 and u_3 should be positive so that originally $u_\xi(\xi)$ (on the original interval $[-1,1]$ of the boundary element) and finally $u(z)$ (on the final interval $[0, \infty)$ of the boundary element) be continuously positive. Naturally, this is not a sufficient condition for this continuous positivity, but it is surely a necessary condition. Therefore, three of the above six quantities, the quantities

$$h_{0,1} = u_3, \quad h_{0,2} = u_1 \text{ and } h_{1,2} = 2u_1 \tag{21}$$

have already been assumed to be positive. On the other hand, two more quantities, the quantities $h_{2,1}$ and $h_{2,2}$, are equal to the discriminant Δ of the quadratic polynomial $u(z)$, whereas the last of these six quantities, the quantity $h_{1,1}$, is given by the second of Eqs. (19), i.e.

$$h_{2,1} = h_{2,2} = \Delta = (-u_1 + 4u_2 - u_3)^2 - 4u_1u_3, \quad h_{1,1} = -u_1 + 4u_2 - u_3 \tag{22}$$

Under these circumstances the numbers of sign variations var(1) and var(2) in the above Sturm sequence at the ends $z_1 = 0$ and $z_2 = \infty$ of the present semi-infinite interval $[0, \infty)$ for the variable z on the boundary element are exclusively determined by the discriminant Δ and the quantity $h_{1,1} = -u_1 + 4u_2 - u_3$. From the logic point of view we denote the positivity of Δ simply by D = true and the positivity of $h_{1,1} = -u_1 + 4u_2 - u_3$ by A = true, i.e.

$$\text{if } \Delta > 0, \text{ then } D = \text{true, whereas if } -u_1 + 4u_2 - u_3 > 0, \text{ then } A = \text{true} \quad (23)$$

Now we directly conclude that under the original assumption that $u_3 > 0$ and Eqs. (19) the number of sign variations var(1) at the left end $z_1 = 0$ of the final interval $[0, \infty)$ will be given by

$$\text{var}(1) = 0 \quad \text{if} \quad A \text{ and } D = \text{true} \quad (24)$$

$$\text{var}(1) = 1 \quad \text{if} \quad (\text{not } D) = \text{true} \quad (25)$$

$$\text{var}(1) = 2 \quad \text{if} \quad (\text{not } A) \text{ and } D = \text{true} \quad (26)$$

Similarly, we can also easily conclude that under the original assumption that $u_1 > 0$ and Eqs. (20) the number of sign variations var(2) at the right end $z_2 = \infty$ of the same interval $[0, \infty)$ will be determined by

$$\text{var}(2) = 0 \quad \text{if} \quad D = \text{true} \quad (27)$$

$$\text{var}(2) = 1 \quad \text{if} \quad (\text{not } D) = \text{true} \quad (28)$$

$$\text{var}(2) = 2 \quad \text{never} \quad (29)$$

Therefore, since $u_1 > 0$ and $u_3 > 0$, the continuous positivity of $u(z)$ on the whole boundary element (finally on the semi-infinite interval $[0, \infty)$) is assured if the difference var(1) − var(2) is equal to zero, i.e.

$$\text{var}(1) - \text{var}(2) = 0 \quad (30)$$

Because of Eqs. (24), (25), (26), (27), (28) and (29), it is clear that this happens if and only if

$$\text{var}(1) = \text{var}(2) = 0 \text{ or } \text{var}(1) = \text{var}(2) = 1 \quad (31)$$

(since var(2) = 2 never happens) and further if and only if

$$((A \text{ and } D) \text{ and } D) \text{ or } ((\text{not } D) \text{ and } (\text{not } D)) = \text{true} \quad (32)$$

This logical expression finally reduces (again with the help of *Mathematica*) to

$$A \text{ or } (\text{not } D) = \text{true} \quad (33)$$

In other words, in order that the positivity of the unknown function $u_\xi(\xi)$ or $u(z)$ holds true on the whole boundary element its nodal values should be positive and, moreover, the logical equation (33) should simultaneously hold true. Therefore our quantifier-free formula is

$$\text{QFF} = u_1 > 0 \text{ and } u_3 > 0 \text{ and (either } 4u_2 > u_1 + u_3 \text{ or } \Delta < 0) \qquad (34)$$

If this formula is valid, then $u_\xi(\xi)$ or equivalently $u(z)$ are continuously positive on the boundary element under consideration. Otherwise they are not.

Now we can proceed to the more difficult case of the cubic polynomial.

3 Quantifier Elimination for the Cubic Polynomial

Now we consider the more difficult case where the boundary element $[-1, 1]$ has four nodes: $\xi_1 = -1$, $\xi_2 = -1/3$, $\xi_3 = 1/3$ and $\xi_4 = 1$. Then we will have a cubic polynomial for the originally unknown and next numerically determined function $u_\xi(\xi)$ along this boundary element $[-1, 1]$. Therefore, this function will have the form

$$u_\xi(\xi) = a\xi^3 + b\xi^2 + c\xi + d \qquad (35)$$

Again the nodal values of $u_\xi(\xi)$ at the aforementioned four nodes are known from the numerical solution of the problem by using the BEM. We have

$$u(-1) = u_1, \quad u(-1/3) = u_2, \quad u(1/3) = u_3, \quad u(1) = u_4 \qquad (36)$$

Exactly as in the previous section, by using *Mathematica* we can easily determine the cubic polynomial $u_\xi(\xi)$ in terms of its values u_1, u_2, u_3 and u_4 at the above four nodes $\xi_1 = -1$, $\xi_2 = -1/3$, $\xi_3 = 1/3$ and $\xi_4 = 1$ respectively. The result is now

$$\begin{aligned} u_\xi(\xi) = {} & [-9(u_1 - 3u_2 + 3u_3 - u_4)\xi^3 + 9(u_1 - u_2 - u_3 + u_4)\xi^2 \\ & + (u_1 - 27u_2 + 27u_3 - u_4)\xi - u_1 + 9u_2 + 9u_3 - u_4]/16 \end{aligned} \qquad (37)$$

By changing the variable on the boundary element from ξ (originally) to η (now) with $\xi = 2\eta - 1$, we get the interval $[0, 1]$ for η on the boundary element. In this way, we easily obtain

$$\begin{aligned} u_\eta(\eta) = {} & -9(u_1 - 3u_2 + 3u_3 - u_4)\eta^3 + 9(2u_1 - 5u_2 + 4u_3 - u_4)\eta^2 \\ & + (-11u_1 + 18u_2 - 9u_3 + 2u_4)\eta + 2u_1 \end{aligned} \qquad (38)$$

with $u_\eta(\eta) = 2u_\xi(\xi) = 2u_\xi(2\eta - 1)$ (for convenience we have multiplied $u_\xi(\xi)$ by 2) and also obviously now $u_\eta(0) = 2u_1$, $u_\eta(1/3) = 2u_2$, $u_\eta(2/3) = 2u_3$ and $u_\eta(1) = 2u_4$.

At this point we can proceed again to our computations (finally with Sturm sequences) by using the computationally more convenient interval $[1, \infty)$. This will result in a simpler Sturm sequence. (We can add again that this is the approach having been used by Ioakimidis 1996.) This transformation is easily achieved through a second change of variable: $\eta = 1/x$. The result for our new polynomial (our third polynomial) $u_x(x)$ is

$$\begin{aligned} u_x(x) &= 2u_1x^3 + (-11u_1 + 18u_2 - 9u_3 + 2u_4)x^2 \\ &\quad + 9(2u_1 - 5u_2 + 4u_3 - u_4)x - 9(u_1 - 3u_2 + 3u_3 - u_4) \end{aligned} \tag{39}$$

For the derivation of this polynomial $u_x(x)$ we took also the liberty to multiply it by x^3 so that no denominator be present. (Since x is always a positive quantity in the interval $[1, \infty)$, this has no consequence on the present positivity problem.)

Exactly as in the previous section, our final step is one more variable transformation: from x to $z = x - 1$. This has again as a result the change of the boundary element interval from $[1, \infty)$ with respect to x to $[0, \infty)$ with respect to z. In this way, we have found our final polynomial $u(z) := u_z(z) = u_x(x)$ as

$$\begin{aligned} u(z) &= 2u_1z^3 + (-5u_1 + 18u_2 - 9u_3 + 2u_4)z^2 \\ &\quad + (2u_1 - 9u_2 + 18u_3 - 5u_4)z + 2u_4 \end{aligned} \tag{40}$$

For convenience we have also used the simpler symbol $u(z)$ instead of $u_z(z)$. Again this semi-infinite interval $[0, \infty)$ is the interval that has been used by Collins (1994, Private communication on quantifier elimination) for the derivation of the quantifier-free formula for the general cubic polynomial $p_3(z) = az^3 + bz^2 + cz + d$ by his interesting approach yet quite different from the present approach.

Finally we can add that by comparing the right-hand sides of Eqs. (39) and (40), it is clear that our final polynomial $u(z)$ is sufficiently simpler than the previous polynomial $u_x(x)$. Therefore, exactly as in the previous section we selected $u(z)$ as our final polynomial in the present quantifier-elimination-related symbolic computations. For this polynomial $u(z)$ we simply wish that it should be positive along the whole boundary element finally (after three variable transformations!) on the semi-infinite interval $[0, \infty)$.

Exactly as in the previous section we are now ready to construct (again by using *Mathematica*) the Sturm sequence for our final cubic polynomial $u(z)$ in Eq. (40).

The present Sturm sequence consists of the following four polynomials with $h_0(z) = u(z)$:

$$h_0(z) = 2u_1z^3 + (-5u_1 + 18u_2 - 9u_3 + 2u_4)z^2 + (2u_1 - 9u_2 + 18u_3 - 5u_4)z + 2u_4 \tag{41}$$

$$h_1(z) = \mathrm{d}h_0(z)/\mathrm{d}z = 6u_1z^2 + 2(-5u_1 + 18u_2 - 9u_3 + 2u_4)z + 2u_1 - 9u_2 + 18u_3 - 5u_4 \tag{42}$$

$$h_2(z) = -\mathrm{rem}[h_0(z), h_1(z)] = \{2[13u_1^2 - 2(63u_2 + 9u_3 - 5u_4)u_1 + (18u_2 - 9u_3 + 2u_4)^2]z - [10u_1^2 + (-81u_2 + 108u_3 + 7u_4)u_1 + (18u_2 - 9u_3 + 2u_4)(9u_2 - 18u_3 + 5u_4)]\}/(18u_1) \tag{43}$$

$$h_3(z) = -\mathrm{rem}[h_1(z), h_2(z)] = 81u_1\Delta_1/\{2[13u_1^2 - 2(63u_2 + 9u_3 - 5u_4)u_1 + (18u_2 - 9u_3 + 2u_4)^2]^2\} \tag{44}$$

with the symbol Δ_1 in the numerator of Eq. (44) defined by

$$\Delta_1 = 4u_1^4 + (-84u_2 + 48u_3 + 20u_4)u_1^3 + 3[219u_2^2 - 6(62u_3 + 15u_4)u_2 - 24u_3^2 + 11u_4^2 - 60u_3u_4]u_1^2 + 2[10u_4^3 - 45(2u_2 + 3u_3)u_4^2 + 18(31u_2^2 + 35u_2u_3 + 31u_3^2)u_4 - 81(u_2 - 2u_3)(14u_2^2 - 17u_2u_3 - 4u_3^2)]u_1 + (18u_2 - 9u_3 + 2u_4)^2[9u_2^2 - 6(6u_3 + u_4)u_2 + (u_4 - 6u_3)^2] \tag{45}$$

We can add that in Eqs. (43) and (44) rem has been already defined as the remainder of two polynomials. We also mention that the quantity u_1 in the denominator of Eq. (43) and in the numerator of Eq. (44) has been assumed to be positive.

On the other hand, we can directly compute the discriminant Δ of the cubic polynomial $u(z)$ in Eq. (40). This gives

$$\Delta = 2u_4[-8u_1(2u_1 - 9u_2 + 18u_3 - 5u_4)^3 + (-5u_1 + 18u_2 - 9u_3 + 2u_4)^2 (2u_1 - 9u_2 + 18u_3 - 5u_4)^2 + 72u_1u_4(-5u_1 + 18u_2 - 9u_3 + 2u_4) (2u_1 - 9u_2 + 18u_3 - 5u_4) - 8u_4(-5u_1 + 18u_2 - 9u_3 + 2u_4)^3 - 432u_1^2u_4^2] \tag{46}$$

It is verified (always by using *Mathematica*) that the quantity Δ_1 in Eq. (45) and the discriminant Δ in Eq. (46) of the cubic polynomial in Eq. (40) are related by

$$\Delta = 18u_4\Delta_1 \tag{47}$$

We can repeat that this situation is again an expected result in Sturm sequences. (Here the equality resulted after a multiplication of Δ_1 by $18u_4$.)

Therefore, since the nodal values u_1 and u_4 are positive, it is clear that the discriminant Δ of $u(z)$ can be used instead of the polynomial $h_3(z)$ in Eq. (44) (which is simply a constant) as far as its sign in concerned. Thus we can simply put

$$h_{3a}(z) = \Delta \tag{48}$$

analogously to Eq. (17) in the previous section for the quadratic polynomial there. Of course, now we must admit that the discriminant Δ in Eq. (46) is much more complicated than the corresponding discriminant Δ in Eq. (18) for the related quadratic polynomial.

As in the previous section, our next task in the present Sturm-sequences-related computations is again to compute the values of the polynomials $h_0(z)$, $h_1(z)$ $h_{2a}(z) = 18u_1h_2(z)$ (for convenience since u_1 is a positive quantity) and $h_{3a}(z)$ (now four polynomials instead of three) at the left end $z_1 = 0$ and at the right end $z_2 = \infty$ of the present semi-infinite interval $[0, \infty)$. We denote these values at the left end $z_1 = 0$ by $h_{0,1}$, $h_{1,1}$, $h_{2,1}$ and $h_{3,1}$. Because of Eqs. (41), (42), (43) and (48), we have

$$\begin{aligned} h_{0,1} &= 2u_4, \quad h_{1,1} = 2u_1 - 9u_2 + 18u_3 - 5u_4, \\ h_{2,1} &= -[10u_1^2 + (-81u_2 + 108u_3 + 7u_4)u_1 \\ &\quad + (18u_2 - 9u_3 + 2u_4)(9u_2 - 18u_3 + 5u_4)], \\ h_{3,1} &= \Delta \end{aligned} \tag{49}$$

Analogously, we denote the corresponding values at the right end $z_2 = \infty$ by $h_{0,2}$, $h_{1,2}$, $h_{2,2}$ and $h_{3,2}$. Since $z_2 = \infty$, we compute the values $h_{0,2}$ and $h_{1,2}$ simply as the coefficients of z^3, z^2 and z of $h_0(z)$, $h_1(z)$ and $h_{2a}(z)$ respectively. Again because of Eqs. (41), (42) and (43) (here the latter multiplied by $9u_1 > 0$), we get

$$\begin{aligned} &h_{0,2} = 2u_1, \quad h_{1,2} = 6u_1, \\ &h_{2,2} = 13u_1^2 - 2(63u_2 + 9u_3 - 5u_4)u_1 + (18u_2 - 9u_3 + 2u_4)^2, \quad h_{3,2} = \Delta \end{aligned} \tag{50}$$

And now we are completely ready to evaluate the numbers of sign variations of the present Sturm sequence at the left end $z_1 = 0$ and at the right end $z_2 = \infty$ of the semi-infinite interval $[0, \infty)$ analogously to our computations for the related quadratic polynomial in the previous section. Of course, we do not forget that all four nodal values u_1, u_2, u_3 and u_4 were already assumed to be positive so that originally $u_\xi(\xi)$ (on the original interval $[-1,1]$ of the boundary element) and finally $u(z)$ (on the final interval $[0, \infty)$ of the same element) be continuously positive. Evidently, this is a necessary (but not sufficient) positivity condition. Therefore, three of the above eight quantities, the quantities

$$h_{0,1} = 2u_4, \quad h_{0,2} = 2u_1 \quad \text{and} \quad h_{1,2} = 6u_1 \tag{51}$$

are already positive. On the other hand, two more quantities, the quantities $h_{3,1}$ and $h_{3,2}$, are equal to the discriminant Δ of the cubic polynomial $u(z)$, i.e.

$$h_{3,1} = h_{3,2} = \Delta \tag{52}$$

Finally, the remaining three of these eight quantities, the quantities $h_{1,1}$, $h_{2,1}$ and $h_{2,2}$, are given by the second and the third of Eqs. (49) and by the third of Eqs. (50) respectively, i.e.

$$\begin{aligned} h_{1,1} &= 2u_1 - 9u_2 + 18u_3 - 5u_4 \\ h_{2,1} &= -[10u_1^2 + (-81u_2 + 108u_3 + 7u_4)u_1 \\ &\qquad + (18u_2 - 9u_3 + 2u_4)(9u_2 - 18u_3 + 5u_4)] \\ h_{2,2} &= 13u_1^2 - 2(63u_2 + 9u_3 - 5u_4)u_1 + (18u_2 - 9u_3 + 2u_4)^2 \end{aligned} \tag{53}$$

Therefore the numbers of sign variations var(1) and var(2) at the ends $z_1 = 0$ and $z_2 = \infty$ of the present interval $[0, \infty)$ for the variable z on the boundary element are exclusively determined by the discriminant Δ and the above three quantities $h_{1,1}$, $h_{2,1}$ and $h_{2,2}$. From the logic point of view it is convenient to denote the positivity of Δ by $D =$ true (exactly as in the previous section), the positivity of $h_{1,1}$ by $A =$ true, the positivity of $h_{2,1}$ by $B =$ true and the positivity of $h_{2,2}$ by $C =$ true, i.e.

$$\begin{aligned} &\text{if } \Delta > 0, \text{ then } D = \text{true}, \text{ if } h_{1,1} > 0, \text{ then } A = \text{true}, \\ &\text{if } h_{2,1} > 0, \text{ then } B = \text{true}, \text{ if } h_{2,2} > 0, \text{ then } C = \text{true} \end{aligned} \tag{54}$$

Now we directly conclude that under the original assumption that $u_4 > 0$ the number of sign variations var(1) at the left end $z_1 = 0$ of the interval $[0, \infty)$, which evidently may take one of the values 0, 1, 2 and 3, will be given by

$$\text{var}(1) = 0 \quad \text{if} \quad A \text{ and } B \text{ and } D \tag{55}$$

$$\begin{aligned} \text{var}(1) = 1 \quad &\text{if } (A \text{ and } B \text{ and } (\text{not } D)) \text{ or } (A \text{ and } (\text{not } B) \\ &\quad \text{and } (\text{not } D)) \text{ or } ((\text{not } A) \text{ and } (\text{not } B) \text{ and } (\text{not } D)) \\ &= (A \text{ and } (\text{not } D)) \text{ or } ((\text{not } B) \text{ and } (\text{not } D)) \end{aligned} \tag{56}$$

$$\text{var}(1) = 2 \text{ if } ((\text{not } A) \text{ or } (\text{not } B)) \text{ and } D \tag{57}$$

$$\text{var}(1) = 3 \text{ if } (\text{not } A) \text{ and } B \text{ and } (\text{not } D)) \tag{58}$$

Similarly, we can also easily conclude that under the original assumption that $u_1 > 0$ the number of sign variations var(2) at the right end $z_2 = \infty$ of the semi-infinite interval $[0, \infty)$ which theoretically may take one of the values 0, 1, 2 and 3, will be determined by

$$\mathrm{var}(2) = 0 \quad \text{if} \quad C \text{ and } D \tag{59}$$

$$\mathrm{var}(2) = 1 \quad \text{if} \quad (\text{not D}) \tag{60}$$

$$\mathrm{var}(2) = 2 \quad \text{if} \quad (\text{not } C) \text{ and } D \tag{61}$$

$$\mathrm{var}(2) = 3 \quad \text{never} \tag{62}$$

Therefore, since $u_1 > 0$ and $u_4 > 0$, the continuous positivity of the final function $u(z)$ on the whole boundary element (in our case finally with the variable z on the semi-infinite interval $[0, \infty)$) is assured if the difference var(1) – var(2) is equal to zero. Therefore, we must have

$$\mathrm{var}(1) - \mathrm{var}(2) = 0 \tag{63}$$

Because of Eqs. (55), (56), (57), (58), (59), (60), (61) and (62), it is clear that this happens if and only if

$$\mathrm{var}(1) = \mathrm{var}(2) = 0 \ \text{ or } \ \mathrm{var}(1) = \mathrm{var}(2) = 1 \ \text{ or } \ \mathrm{var}(1) = \mathrm{var}(2) = 2 \tag{64}$$

(since var(2) = 3 can never happen) and further if and only if

$$\begin{aligned}&[(A \text{ and } B \text{ and } D) \text{ and } (C \text{ and } D)] \text{ or } [((\text{not } D) \text{ and } A) \text{ or } ((\text{not } D)\\ &\quad \text{and } (\text{not } B)) \text{ and } (\text{not } D)] \text{ or } [(((\text{not } A) \text{ or } (\text{not } B)) \text{ and } D)\\ &\quad \text{and } ((\text{not } C) \text{ and } D)] = \text{true}\end{aligned} \tag{65}$$

After a simplification (always by using *Mathematica*, Wolfram 1999) this logical expression takes the form of our final quantifier-free formula

$$\begin{aligned}\mathrm{QFF} = {} & u_1 > 0 \text{ and } u_4 > 0 \text{ and } (A \text{ and } (\text{not } D)) \text{ or } ((\text{not } B)\\ & \text{and } (\text{not } C)) \text{ or } ((\text{not } B) \text{ and } (\text{not } D)) \text{ or } (A \text{ and } B \text{ and } C) \text{ or } ((\text{not } A)\\ & \text{and } (\text{not } C) \text{ and } D) = \text{true}\end{aligned} \tag{66}$$

In other words, in order that the positivity of the unknown function $u(\xi)$ or finally $u(z)$ holds true on the whole boundary element (its nodal values assumed positive in advance, i.e. from the numerical solution of the problem by using the BEM) the above quantifier-free-formula (66) should hold true. Otherwise this positivity is false. In this case, the numerical solution of the problem by using the BEM will not be completely acceptable from the physical point of view. But, evidently, this is a rare case.

4 Conclusions

From the above results it is concluded that the method of Sturm sequences is easily applicable (with the help of a computer algebra system such as the popular system *Mathematica*, Wolfram 1999) to the construction of quantifier-free formulae for the continuous positivity of the originally unknown function in the BEM along a whole boundary element. Originally this possibility has been used by the author (Ioakimidis 1995) for the quadratic polynomial and extended to the case of the cubic polynomial (Ioakimidis 1996). As was already seen, such a quantifier-free formula can be constructed to contain only the nodal values of the unknown function, which are numerically obtained by using the BEM.

This possibility is useful in the BEM in special cases (such as in crack and contact problems) where the positivity of the unknown function is really required from the physical point of view. The above method provides us with general closed-form algebraic–logical formulae such as those derived here for the usual cases of three nodes (quadratic polynomial) and of four nodes (cubic polynomial) on the boundary element.

Additional similar cases such as the case of a boundary element in a crack problem with a node at a crack tip or the case of a boundary element with two end-point nodes, but with both values of the unknown function and its derivative numerically determined by using the BEM, can also be studied in the same way.

Of course, the method of Sturm sequences is not the only approach to the solution of the present positivity problem and additional approaches are also of interest and actually used (see, e.g., Ioakimidis 1999). But these approaches are generally based on special computer algebra packages and mainly concern simple cases. On the contrary the method of Sturm sequences is a general method applicable without difficulties to the polynomial of any degree. Evidently, the final logical result should be simplified as much as possible and such a simplification command is offered by *Mathematica* as was seen in the present results of Sections 2 and 3.

References

Davenport JH, Siret Y, Tournier E (1993) Computer algebra: systems and algorithms for algebraic computation, 2nd edn. Academic Press, London, pp. 124–125

Ioakimidis NI (1995) Elementary quantifier-free formulae in boundary elements. Bound Elem Comm 6:107–112

Ioakimidis NI (1996) Inequality constraints in rectangular finite/boundary elements. Comput Struct 60:415–431

Ioakimidis NI (1999) Automatic derivation of positivity conditions inside boundary elements with the help of the REDLOG computer logic package. Eng Anal Bound Elem 23:847–856

Wolfram S (1999) The *Mathematica* book, 4th edn. Wolfram Media, Champaign, IL, and Cambridge University Press, Cambridge, UK

Matrix Decomposition Algorithms Related to the MFS for Axisymmetric Problems

Andreas Karageorghis and Yiorgos–Sokratis Smyrlis

Abstract In this paper we review some applications of the Method of Fundamental Solutions (MFS) to certain elliptic boundary value problems in rotationally symmetric domains. In particular, we show how efficient matrix decomposition MFS algorithms can be developed for such problems. The efficiency of these algorithms is optimized by using Fast Fourier Transforms.

1 Introduction

The Method of Fundamental Solutions (MFS) is a meshless Trefftz-type (Kita and Kamiya, 1995) boundary method which has become very popular in recent years primarily because of the simplicity with which it can be implemented and, unlike the boundary element method, it does not require an elaborate discretization of the boundary. Also, it can be applied even in the case of domains with irregular boundaries. The advantages and disadvantages of the MFS compared to other numerical methods, implementational details as well as a wide range of applications can be found in the survey papers (Cho et al. 2004; Fairweather 2007; Fairweather and Karageorghis 1998; Fairweather et al. 2003; Golberg and Chen 1999). One interesting application of the MFS is to elliptic boundary value problems in rotationally symmetric domains. In particular, in Karageorghis and Fairweather (1998, 1999, 2000) the MFS was applied to the solution of axisymmetric acoustics, potential and elasticity and problems, respectively. In these studies, the MFS was used to solve the axisymmetric version of the governing equations which, despite reducing the dimension of the problem by one, led to certain difficulties. In particular, the fundamental solutions of these equations involved the potentially troublesome evaluation of complete elliptic integrals.

Moreover, when the boundary conditions of the problem under consideration were not axisymmetric, this approach required the solution of a sequence of

---

A. Karageorghis (✉)
Department of Mathematics and Statistics, University of Cyprus, P.O.Box 20537
1678 Nicosia, Cyprus
e-mail: andreask@ucy.ac.cy

G.D. Manolis, D. Polyzos (eds.), *Recent Advances in Boundary Element Methods*,
DOI 10.1007/978-1-4020-9710-2_15, © Springer Science+Business Media B.V. 2009

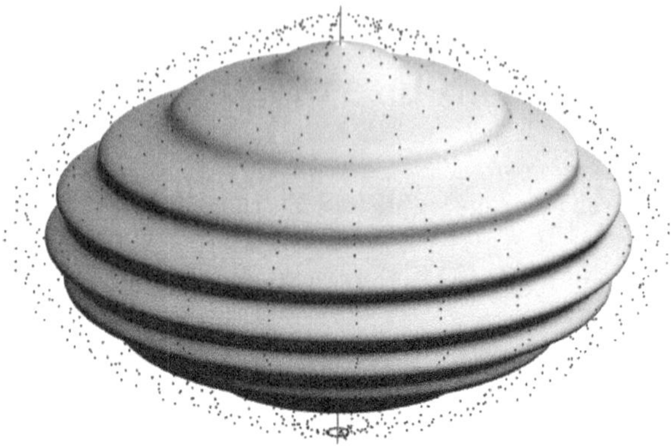

Fig. 1 Typical distribution of singularities on a pseudo–boundary of an axisymmetric domain

problems in order to approximate a finite Fourier sum. Alternatively, the full three-dimensional version of the governing equations may be considered. In such cases, matrix decomposition algorithms (Bialecki and Fairweather 1993) may be developed for the efficient solution of the resulting systems. The solution of three-dimensional harmonic problems in axisymmetric domains was considered in Smyrlis and Karageorghis (2004b) and the corresponding biharmonic case in Fairweather et al. (2005). The solution of such problems in hollow axisymmetric domains is described in Tsangaris et al. (2006a), while the extension of these algorithms to stationary heat conduction problems is presented in Smyrlis and Karageorghis (2006). Finally, elasticity and thermo-elasticity problems are considered in Karageorghis and Smyrlis (2007). All matrix decomposition algorithms algorithms make use of Fast Fourier Transforms (FFTs) and can be seen as generalizations of the basic ideas used for the solution of the corresponding two-dimensional problems in a disk (Smyrlis and Karageorghis 2001, 2004a; Tsangaris et al. 2004, 2006b).

In this paper, we first describe in detail the matrix decomposition algorithm for the solution of three-dimensional harmonic problems and then describe how it can be modified for the solution of three-dimensional elasticity problems. We also show the simplicity of the approach by including a `MATLAB` code implementing the algorithm for three-dimensional harmonic problems. Some numerical results are also included.

2 Axisymmetric Potential Problems

We consider the three-dimensional boundary value problem

$$\begin{cases} \Delta u = 0 \text{ in } \Omega, \\ u = f \quad \text{on } \partial\Omega, \end{cases} \tag{1}$$

where, Δ denotes the Laplace operator and f is a given function. The region $\Omega \subset \mathbb{R}^3$ is axisymmetric, which means that it is formed by rotating a region

$\Omega' \subset \mathbb{R}^2$ about the z-axis. The boundaries of Ω and Ω' are denoted by $\partial\Omega$ and $\partial\Omega'$, respectively. The solution u is approximated by

$$u_{MN}(\boldsymbol{c}, \boldsymbol{Q}; P) = \sum_{m=1}^{M} \sum_{n=1}^{N} c_{m,n} \, \mathcal{K}_3(P, Q_{m,n}), \qquad P \in \overline{\Omega},$$

where $\boldsymbol{c} \in \mathbb{R}^{MN}$ and $\boldsymbol{Q}$ is a $3MN$-vector containing the coordinates of the sources $Q_{m,n}$, $m = 1, \ldots, M,\ n = 1, \ldots, N$, which lie outside $\overline{\Omega}$. The function $\mathcal{K}_3(P, Q)$ is a fundamental solution of the Laplace equation in $\mathbb{R}^3$ given by

$$\mathcal{K}_3(P, Q) = \frac{1}{4\pi |P - Q|},$$

with $|P - Q|$ denoting the distance between the points P and Q. The singularities $Q_{m,n}$ are fixed on the boundary $\partial\tilde{\Omega}$ of a solid $\tilde{\Omega}$ surrounding Ω. The solid $\tilde{\Omega}$ is generated by the rotation of the planar domain $\tilde{\Omega}'$ which is similar to Ω'. A set of MN collocation points $\{P_{i,j}\}_{i=1,j=1}^{M,N}$ is chosen on $\partial\Omega$ in the following way. We first choose N points on the boundary $\partial\Omega'$ of Ω'. These can be described by their polar coordinates (r_{P_j}, z_{P_j}), $j = 1, \cdots, N$, where r_{P_j} denotes the vertical distance of the point P_j from the z-axis and z_{P_j} denotes the z- coordinate of the point P_j. The points on $\partial\Omega$ are taken to be

$$x_{P_{i,j}} = r_{P_j} \cos\phi_i, \quad y_{P_{i,j}} = r_{P_j} \sin\phi_i, \quad z_{P_{i,j}} = z_{P_j},$$

where $\phi_i = 2(i-1)\pi/M,\ i = 1, \ldots, M$. Similarly, we choose a set of MN singularities $\{Q_{m,n}\}_{i=m,n=1}^{M,N}$ on $\partial\tilde{\Omega}$ by taking $Q_{m,n} = (x_{Q_{m,n}}, y_{Q_{m,n}}, z_{Q_{m,n}})$, and

$$x_{Q_{i,j}} = r_{Q_j} \cos\theta_i, \quad y_{Q_{i,j}} = r_{Q_j} \sin\theta_i, \quad z_{Q_{i,j}} = z_{Q_j},$$

where $\theta_i = 2(\alpha + i - 1)\pi/M,\ i = 1, \ldots, M$. The angular parameter α $(0 \le \alpha < 1/2)$ indicates that the sources are rotated by an angle $2\pi\alpha/M$ in the angular direction. The coefficients $\boldsymbol{c}$ are determined so that the boundary condition is satisfied at the boundary points

$$u_{MN}(\boldsymbol{c}, \boldsymbol{Q}; P_{i,j}) = f(P_{i,j}), \quad i = 1, \ldots, M,\ j = 1, \ldots, N.$$

By ordering both the collocation points $P_{i,j}$ and $Q_{i,j}$ the sources in the following way:

$$k^{\text{th}} \text{ point} = (j-1)M + i,\ i = 1, \ldots, M,\ j = 1, \ldots, N \tag{2}$$

this yields an $MN \times MN$ linear system of the form

$$Gc = f, \tag{3}$$

for the coefficients $\boldsymbol{c}$, where

$$\boldsymbol{f} = (f_{11}, f_{21}, \ldots, f_{M1}, \ldots, f_{1N}, \ldots, f_{MN})^T,$$
$$\boldsymbol{c} = (c_{11}, c_{21}, \ldots, c_{M1}, \ldots, c_{1N}, \ldots, c_{MN})^T,$$

and the elements of the matrix G are given by

$$G_{(j-1)M+i,(n-1)M+m} = \frac{1}{4\pi |P_{i,j} - Q_{m,n}|},$$

$i,\ m = 1, \ldots, M,\ \ j,\ n = 1, \ldots, N$. The matrix G has the following block structure

$$G = \begin{pmatrix} A_{11} & A_{12} & \cdots & A_{1N} \\ A_{21} & A_{22} & \cdots & A_{2N} \\ \vdots & \vdots & & \vdots \\ A_{N1} & A_{N2} & \cdots & A_{NN} \end{pmatrix},$$

where the matrices $A_{j,n},\ \ j, n = 1, \cdots, N$, are $M \times M$ circulant matrices each defined by the row

$$a_m^{j,n} = \left(A_{1,m}\right)_{j,n} = \frac{1}{4\pi |P_{1,j} - Q_{m,n}|}, \qquad m = 1, \ldots, M \quad j,\ n = 1, \ldots, N.$$

We also write the vectors $\boldsymbol{c}$ and $\boldsymbol{f}$ as

$$\boldsymbol{c} = \begin{pmatrix} \boldsymbol{c}_1 \\ \boldsymbol{c}_2 \\ \vdots \\ \boldsymbol{c}_N \end{pmatrix}, \quad \boldsymbol{f} = \begin{pmatrix} \boldsymbol{f}_1 \\ \boldsymbol{f}_2 \\ \vdots \\ \boldsymbol{f}_N \end{pmatrix}, \tag{4}$$

where $\boldsymbol{c}_n = (c_{1n}, c_{2n}, \ldots, c_{Mn})^T$ and $\boldsymbol{f}_n = (f_{1n}, f_{2n}, \ldots, f_{Mn})^T,\ \ n = 1, \ldots, N$. System (3) can thus be written as

$$\begin{pmatrix} A_{11} & A_{12} & \cdots & A_{1N} \\ A_{21} & A_{22} & \cdots & A_{2N} \\ \vdots & \vdots & & \vdots \\ A_{N1} & A_{N2} & \cdots & A_{NN} \end{pmatrix} \begin{pmatrix} \boldsymbol{c}_1 \\ \boldsymbol{c}_2 \\ \vdots \\ \boldsymbol{c}_N \end{pmatrix} = \begin{pmatrix} \boldsymbol{f}_1 \\ \boldsymbol{f}_2 \\ \vdots \\ \boldsymbol{f}_N \end{pmatrix}.$$

If we define U_M to be the unitary $M \times M$ Fourier matrix which is the conjugate of the matrix

$$U_M^* = \frac{1}{\sqrt{M}} \begin{pmatrix} 1 & 1 & 1 & \cdots & 1 \\ 1 & \omega & \omega^2 & \cdots & \omega^{M-1} \\ 1 & \omega^2 & \omega^4 & \cdots & \omega^{2(M-1)} \\ \vdots & \vdots & \vdots & & \vdots \\ 1 & \omega^{M-1} & \omega^{2(M-1)} & \cdots & \omega^{(M-1)(M-1)} \end{pmatrix}, \quad \text{where} \quad \omega = \mathrm{e}^{2\pi \mathrm{i}/M},$$

and I_N to be the $N \times N$ identity matrix, pre–multiplication of (3) by $I_N \otimes U_M$ yields

$$(I_N \otimes U_M)\, G \left(I_N \otimes U_M^*\right)(I_N \otimes U_M)\, \boldsymbol{c} = (I_N \otimes U_M)\, \boldsymbol{f},$$

or

$$\tilde{G}\tilde{\boldsymbol{c}} = \tilde{\boldsymbol{f}}, \tag{5}$$

where

$$\tilde{G} = \begin{pmatrix} D_{11} & D_{12} & \cdots & D_{1N} \\ D_{21} & D_{22} & \cdots & D_{2N} \\ \vdots & \vdots & & \vdots \\ D_{N1} & D_{N2} & \cdots & D_{NN} \end{pmatrix}, \tag{6}$$

and

$$\tilde{\boldsymbol{c}} = \begin{pmatrix} \tilde{\boldsymbol{c}}_1 \\ \tilde{\boldsymbol{c}}_2 \\ \vdots \\ \tilde{\boldsymbol{c}}_N \end{pmatrix}, \quad \tilde{\boldsymbol{f}} = \begin{pmatrix} \tilde{\boldsymbol{f}}_1 \\ \tilde{\boldsymbol{f}}_2 \\ \vdots \\ \tilde{\boldsymbol{f}}_N \end{pmatrix}. \tag{7}$$

In (6), each of the $M \times M$ matrices $D_{j,n}$, $j, n = 1, \cdots, N$ is diagonal which is a result of the properties of circulant matrices and

$$D_{j,n} = U_M A_{j,n} U_M^*, \; j, n = 1, \cdots, N.$$

Further, if

$$D_{j,n} = \operatorname{diag}\left(d_1^{j,n}, d_2^{j,n}, \ldots, d_M^{j,n}\right),$$

we have, for $j, n = 1, \cdots, N$,

$$d_\ell^{j,n} = \sum_{k=1}^{M} a_k^{j,n} \omega^{(k-1)(\ell-1)}, \ \ell = 1, \cdots, M. \tag{8}$$

In (5),

$$\tilde{\boldsymbol{c}} = (I_N \otimes U_M)\boldsymbol{c} \quad \text{and} \quad \tilde{\boldsymbol{f}} = (I_N \otimes U_M)\boldsymbol{f},$$

and, equivalently, in (7),

$$\tilde{\boldsymbol{c}}_\ell = U_M \boldsymbol{c}_\ell \quad \text{and} \quad \tilde{\boldsymbol{f}}_\ell = U_M \boldsymbol{f}_\ell, \quad \ell = 1, \cdots, N,$$

where the vectors $\boldsymbol{c}_\ell$ and $\boldsymbol{f}_\ell$ are defined in (4). Because of the structure of matrix $\tilde{G}$, the solution of system (5) is equivalent to solving the M systems of order N

$$E_m \boldsymbol{x}_m = \boldsymbol{y}_m, \quad m = 1, \cdots, M, \tag{9}$$

where

$$(E_m)_{j,n} = d_m^{j,n}, \ j, n = 1, \cdots, N$$

and

$$(\boldsymbol{x}_m)_j = \left(\tilde{\boldsymbol{c}}_j\right)_m, \ (\boldsymbol{y}_m)_j = \left(\tilde{\boldsymbol{f}}_j\right)_m, \ j = 1, \cdots, N. \tag{10}$$

The solution of the M systems (9) yields the vectors $\boldsymbol{x}_m, \ m = 1, \cdots, M$ from which we can readily recover the vectors $\tilde{\boldsymbol{c}}_n, \ n = 1, \cdots, N$ from (10). Finally, the vectors $\boldsymbol{c}_n, \ n = 1, \cdots, N$ may be calculated from

$$\boldsymbol{c}_n = U_M^* \tilde{\boldsymbol{c}}_n. \tag{11}$$

The algorithm described in the section may thus be summarized as follows:

Step 1. Compute $\tilde{\boldsymbol{f}}_\ell = U_M \boldsymbol{f}_\ell, \quad \ell = 1, \cdots, N$.
Step 2. Construct the diagonal matrices $D_{j,n}$ from (8).
Step 3. Solve the M, $N \times N$ systems (9) to obtain the $\{\boldsymbol{x}_m\}_{m=1}^M$, and subsequently the $\{\tilde{\boldsymbol{c}}_n\}_{n=1}^N$.
Step 4. Recover the vector of coefficients $\boldsymbol{c}$ from (11).

Cost. In Step 1 and Step 4, the operations can be carried out via Fast Fourier Transforms (FFTs) at a cost of order $\mathcal{O}(NM \log M)$ operations. FFTs can also be used for the evaluation of the N^2 matrices $D_{j,n}$ in Step 2 at a cost of $\mathcal{O}(N^2 M \log M)$

operations. The FFTs are performed using the MATLAB commands `fft` and `ifft`. Finally, in Step 3, the cost of solving M complex linear systems of order N via LU–factorization with partial pivoting is $\mathcal{O}(M\,N^3)$ operations.

Remark. The algorithm described in this section is different from the one described in Smyrlis and Karageorghis (2004b) the sense that the system (3) is set up differently than the corresponding system in Smyrlis and Karageorghis (2004b). The current ordering leads to a block matrix, where each block is a circulant matrix in contrast to the ordering in Smyrlis and Karageorghis (2004b) which leads to a block circulant matrix. Clearly, the two formulations are equivalent.

3 Axisymmetric Elasticity Problems

We consider the boundary value problem in $\mathbb{R}^3$ governed by the Cauchy–Navier equations of elasticity

$$\begin{cases} (\lambda+\mu)\,u_{k,ki}+\mu\,u_{i,kk}=0 \text{ in } & \Omega, \\ u_i=f_i & \text{on } \partial\Omega. \end{cases}$$

The region $\Omega\subset\mathbb{R}^3$ is, as in Section 2, axisymmetric. The displacements $\boldsymbol{u}=(u_1,u_2,u_3)$ at the point $P\in\mathbb{R}^3$ are approximated by

$$u_i^{M,N}(\boldsymbol{c},\boldsymbol{Q};P)=\sum_{m=1}^{M}\sum_{n=1}^{N}\sum_{j=1}^{3}c_{m,n}^{j}\,g_{ij}(P-Q_{m,n}),\ i=1,2,3,$$

where $\boldsymbol{Q}=(Q_{m,n})_{m=1,\dots,M}^{n=1,\dots,N}$ with $Q_{m,n}=\left(x_{m,n}^{Q},y_{m,n}^{Q},z_{m,n}^{Q}\right)\in\mathbb{R}^3$ are the coordinates of the sources. The fundamental solution in this case is a 3×3 matrix defined by

$$g_{ij}(\boldsymbol{x})=-\frac{3\mu+\lambda}{8\pi\mu\,(2\mu+\lambda)}\cdot\frac{\delta_{ij}}{|\boldsymbol{x}|}-\frac{\mu+\lambda}{8\pi\mu\,(2\mu+\lambda)}\cdot\frac{x_i x_j}{|\boldsymbol{x}|^3}.$$

The discretization of the axisymmetric domain is carried out as in Section 2. The satisfaction of the boundary conditions at the boundary points $\{P_{k,\ell}\}_{k=1,\ell=1}^{M,N}$ yields

$$u_1^{M,N}(\boldsymbol{c},\boldsymbol{Q};P_{k,\ell})=f_1(P_{k,\ell}),$$

$$u_2^{M,N}(\boldsymbol{c},\boldsymbol{Q};P_{k,\ell})=f_2(P_{k,\ell}),$$

$$u_3^{M,N}(\boldsymbol{c},\boldsymbol{Q};P_{k,\ell})=f_3(P_{k,\ell}),$$

$k = 1, \ldots, M$, $\ell = 1, \ldots, N$. By ordering both the boundary points and the singularities as in (2), this gives a $3MN \times 3MN$ linear system of the form

$$G\,\boldsymbol{c} = \begin{pmatrix} A_{1,1} & A_{1,2} & \cdots & A_{1,N} \\ A_{2,1} & A_{2,2} & \cdots & A_{2,N} \\ \vdots & \vdots & \ddots & \vdots \\ A_{N,1} & A_{N,2} & \cdots & A_{N,N} \end{pmatrix} \begin{pmatrix} \boldsymbol{c}^1 \\ \boldsymbol{c}^2 \\ \vdots \\ \boldsymbol{c}^N \end{pmatrix} = \begin{pmatrix} \boldsymbol{f}^1 \\ \boldsymbol{f}^2 \\ \vdots \\ \boldsymbol{f}^N \end{pmatrix} = \boldsymbol{f}, \tag{12}$$

where

$$A_{\ell,\nu} = \begin{pmatrix} A^{1,1}_{\ell,\nu} & A^{1,2}_{\ell,\nu} & \cdots & A^{1M,}_{\ell,\nu} \\ A^{2,1}_{\ell,\nu} & A^{2,2}_{\ell,\nu} & \cdots & A^{2,M}_{\ell,\nu} \\ \vdots & \vdots & \ddots & \vdots \\ A^{M,1}_{\ell,\nu} & A^{M,2}_{\ell,\nu} & \cdots & A^{M,M}_{\ell,\nu} \end{pmatrix},$$

where $A^{k\mu}_{\ell,\nu} = g(P_{k,\ell} - Q_{\mu,\nu}) \in \mathbb{R}^{3\times 3}$,

$$\boldsymbol{c}^{\nu} = \left(\boldsymbol{c}^{1,\nu}, \ldots, \boldsymbol{c}^{M,\nu}\right) = \left(c^1_{1,\nu}, c^2_{1,\nu}, c^3_{1,\nu}, c^1_{2,\nu}, c^2_{2,\nu}, c^3_{2,\nu}, \ldots, c^1_{M,\nu}, c^2_{M,\nu}, c^3_{M,\nu}\right),$$

$$\boldsymbol{f}^{\ell} = \left(\boldsymbol{f}^{1,\ell}, \ldots, \boldsymbol{f}^{M,\ell}\right) = \left(f^1_{1,\ell}, f^2_{1,\ell}, f^3_{1,\ell}, f^1_{2,\ell}, f^2_{2,\ell}, f^3_{2,\ell}, \ldots, f^1_{M,\ell}, f^2_{M,\ell}, f^3_{M,\ell}\right),$$

and $f^i_{k,\ell} = f_i(P_{k,\ell})$, $i = 1, 2, 3$, $k, \mu = 1, \ldots, M$ and $\ell, \nu = 1, \ldots, N$. We pre–multiply system (12) by the matrix

$$\mathcal{R} = I_N \otimes R$$

where

$$R = \begin{pmatrix} R_{\vartheta_1} & 0 & 0 & \cdots & 0 & 0 \\ 0 & R_{\vartheta_2} & 0 & \cdots & 0 & 0 \\ \vdots & \vdots & \ddots & \vdots & \vdots & \vdots \\ 0 & 0 & 0 & \cdots & R_{\vartheta_{M-1}} & 0 \\ 0 & 0 & 0 & \cdots & 0 & R_{\vartheta_M} \end{pmatrix},$$

with

$$R_{\vartheta_k} = \begin{pmatrix} \cos\vartheta_k & \sin\vartheta_k & 0 \\ \sin\vartheta_k & -\cos\vartheta_k & 0 \\ 0 & & 0\;1 \end{pmatrix}$$

and $\vartheta_k = \frac{2\pi(k-1)}{M}$, to obtain, using the properties of $\mathcal{R}$,

$$\mathcal{R}G\boldsymbol{c} = \mathcal{R}G\mathcal{R}\mathcal{R}\boldsymbol{c} = \mathcal{R}\boldsymbol{f}, \quad \text{or} \quad \tilde{G}\tilde{\boldsymbol{c}} = \tilde{\boldsymbol{f}}, \tag{13}$$

where $\tilde{G} = \mathcal{R}G\mathcal{R}$, $\tilde{\boldsymbol{c}} = \mathcal{R}\boldsymbol{c}$ and $\tilde{\boldsymbol{f}} = \mathcal{R}\boldsymbol{f}$. The matrix $\tilde{G}$ has the form

$$\tilde{G} = \begin{pmatrix} \tilde{A}_{1,1} & \tilde{A}_{1,2} & \cdots & \tilde{A}_{1,N} \\ \tilde{A}_{2,1} & \tilde{A}_{2,2} & \cdots & \tilde{A}_{2,N} \\ \vdots & \vdots & \ddots & \vdots \\ \tilde{A}_{N,1} & \tilde{A}_{N,2} & \cdots & \tilde{A}_{N,N} \end{pmatrix},$$

where

$$\tilde{A}_{\ell,\nu} = RA_{\ell,\nu}R = \begin{pmatrix} \tilde{A}_{\ell,\nu}^{1,1} & \tilde{A}_{\ell,\nu}^{1,2} & \cdots & \tilde{A}_{\ell,\nu}^{1,M} \\ \tilde{A}_{\ell,\nu}^{2,1} & \tilde{A}_{\ell,\nu}^{2,2} & \cdots & \tilde{A}_{\ell,\nu}^{2,M} \\ \vdots & \vdots & \ddots & \vdots \\ \tilde{A}_{\ell,\nu}^{M,1} & \tilde{A}_{\ell,\nu}^{M,2} & \cdots & \tilde{A}_{\ell,\nu}^{M,M} \end{pmatrix}$$

and

$$\tilde{A}_{\ell,\nu}^{k,\mu} = R_{\vartheta_k} A_{\ell,\nu}^{k,\mu} R_{\vartheta_\mu}.$$

At this point we re-order the vectors $\tilde{\boldsymbol{c}}$ and $\tilde{\boldsymbol{f}}$ as $\hat{\boldsymbol{c}}$ and $\hat{\boldsymbol{f}}$, respectively, where

$$\hat{\boldsymbol{c}} = \begin{pmatrix} \hat{\boldsymbol{c}}^1 \\ \hat{\boldsymbol{c}}^2 \\ \vdots \\ \hat{\boldsymbol{c}}^N \end{pmatrix}, \quad \hat{\boldsymbol{f}} = \begin{pmatrix} \hat{\boldsymbol{f}}^1 \\ \hat{\boldsymbol{f}}^2 \\ \vdots \\ \hat{\boldsymbol{f}}^N \end{pmatrix},$$

where

$$\hat{\boldsymbol{c}}^\nu = \left(\tilde{c}_{1,\nu}^1, \ldots \tilde{c}_{M,\nu}^1, \tilde{c}_{1,\nu}^2, \ldots \tilde{c}_{M,\nu}^2, \tilde{c}_{1,\nu}^3, \ldots \tilde{c}_{M,\nu}^3\right) \in \mathbb{R}^{3M},$$

$$\hat{\boldsymbol{f}}^\nu = \left(\tilde{f}_{1,\nu}^1, \ldots \tilde{f}_{M,\nu}^1, \tilde{f}_{1,\nu}^2, \ldots \tilde{f}_{M,\nu}^2, \tilde{f}_{1,\nu}^3, \ldots \tilde{f}_{M,\nu}^3\right) \in \mathbb{R}^{3M}.$$

With this re-ordering, system (13) becomes

$$\hat{G}\hat{\boldsymbol{c}} = \hat{\boldsymbol{f}}, \tag{14}$$

where

$$\hat{G} = \begin{pmatrix} \hat{A}_{1,1} & \hat{A}_{1,2} & \cdots & \hat{A}_{1,N} \\ \hat{A}_{2,1} & \hat{A}_{2,2} & \cdots & \hat{A}_{2,N} \\ \vdots & \vdots & \ddots & \vdots \\ \hat{A}_{N,1} & \hat{A}_{N,2} & \cdots & \hat{A}_{N,N} \end{pmatrix}$$

where the matrices

$$\hat{A}_{\ell,\nu} = \begin{pmatrix} \hat{A}^{1,1}_{\ell,\nu} & \hat{A}^{1,2}_{\ell,\nu} & \hat{A}^{1,3}_{\ell,\nu} \\ \hat{A}^{2,1}_{\ell,\nu} & \hat{A}^{2,2}_{\ell,\nu} & \hat{A}^{2,3}_{\ell,\nu} \\ \hat{A}^{3,1}_{\ell,\nu} & \hat{A}^{3,2}_{\ell,\nu} & \hat{A}^{3,3}_{\ell,\nu} \end{pmatrix}$$

are $M \times M$, and

$$\left(\hat{A}^{k,\mu}_{\ell,\nu}\right)_{i,j} = \left(\tilde{A}^{i,j}_{\ell,\nu}\right)_{k,\mu}. \tag{15}$$

Each of the $M \times M$ matrices $\hat{A}^{k,\mu}_{\ell,\nu}$ is now circulant (Karageorghis and Smyrlis 2007) and thus premultiplication of (14) by $I_N \otimes I_3 \otimes U_M$ yields

$$(I_N \otimes I_3 \otimes U_M)\,\hat{G}\left(I_N \otimes I_3 \otimes U_M^*\right)(I_N \otimes I_3 \otimes U_M)\,\hat{\boldsymbol{c}} = (I_N \otimes I_3 \otimes U_M)\,\hat{\boldsymbol{f}}, \tag{16}$$

or

$$D\boldsymbol{\gamma} = \boldsymbol{h}, \tag{17}$$

where

$$D = \begin{pmatrix} \hat{D}_{11} & \hat{D}_{12} & \cdots & \hat{D}_{1N} \\ \hat{D}_{21} & \hat{D}_{22} & \cdots & \hat{D}_{2N} \\ \vdots & \vdots & & \vdots \\ \hat{D}_{N1} & \hat{D}_{N2} & \cdots & \hat{D}_{NN} \end{pmatrix},$$

with

$$\hat{D}_{\ell,\nu} = \begin{pmatrix} \hat{D}^{1,1}_{\ell,\nu} & \hat{D}^{1,2}_{\ell,\nu} & \hat{D}^{1,3}_{\ell,\nu} \\ \hat{D}^{2,1}_{\ell,\nu} & \hat{D}^{2,2}_{\ell,\nu} & \hat{D}^{2,3}_{\ell,\nu} \\ \hat{D}^{3,1}_{\ell,\nu} & \hat{D}^{3,2}_{\ell,\nu} & \hat{D}^{3,3}_{\ell,\nu}, \end{pmatrix}, \tag{18}$$

while $\boldsymbol{\gamma} = (I_N \otimes I_3 \otimes U_M)\,\hat{\boldsymbol{c}}$ and $\boldsymbol{h} = (I_N \otimes I_3 \otimes U_M)\,\hat{\boldsymbol{f}}$. In (18), each of the $M \times M$ matrices $\hat{D}^{k,\mu}_{\ell,\nu} = \operatorname{diag}\left(d^{k,\mu^1}_{\ell,\nu}, d^{k,\mu^2}_{\ell,\nu}, \cdots, d^{k,\mu^M}_{\ell,\nu}\right)$ is diagonal and its elements may be calculated as in (8). Solving (17) is thus equivalent to solving M systems of order $3N$ of the form

$$E_m\,\boldsymbol{x}_m = \boldsymbol{y}_m, \quad m = 1, \cdots, M \tag{19}$$

where for $m = 1, \cdots, M$

$$(E_m)_{i,j} = \left(d^{k,\mu^m}_{\ell,\nu}\right), \quad i = 3(\ell-1)+k, \quad j = 3(\nu-1)+\mu,$$

$\ell, \nu = 1, \cdots, N,\ k, \mu = 1, 2, 3$. The vectors $\boldsymbol{x}_m$ and $\boldsymbol{y}_m$ are defined accordingly. Having obtained $\boldsymbol{\gamma}$ one may recover $\hat{\boldsymbol{c}} = \left(I_N \otimes I_3 \otimes U_M^*\right)\boldsymbol{\gamma}$ and subsequently $\boldsymbol{c}$.

The algorithm described in the section may thus be summarized as follows:

Step 1. Compute $\tilde{\boldsymbol{f}} = \mathcal{R}\boldsymbol{f}$.
Step 2. Construct the first rows of the $M \times M$ submatrices $\hat{A}^{k,\mu}_{\ell,\nu}$, $k, \mu = 1, 2, 3$, $\ell, \nu = 1, \ldots, N$ from (15).
Step 3. Compute $\boldsymbol{h} = (I_N \otimes I_3 \otimes U_M)\,\hat{\boldsymbol{f}}$.
Step 4. Construct the matrices $\hat{D}^{k,\mu}_{\ell,\nu}$ using (8).
Step 5. Solve the M systems of order $3N$ in (19)
Step 6. Recover the vector of coefficients $\hat{\boldsymbol{c}} = \left(I_N \otimes I_3 \otimes U_M^*\right)\boldsymbol{\gamma}$ and subsequently $\boldsymbol{c}$.

Cost. In Step 3, Step 4 and Step 6, the operations can be carried out via FFTs at a cost of order $\mathcal{O}(NM \log M)$ operations. In Step 5, the cost of solving M complex linear systems of order $3N$ via $LU-$factorization with partial pivoting is $\mathcal{O}(M\,N^3)$ operations.

Remark. The algorithm described in this section is different from the one described in Karageorghis and Smyrlis (2007) the sense that the system is set up differently than the corresponding system in Karageorghis and Smyrlis (2007). As in Section 2 the two formulations are equivalent.

4 Numerical Results

We considered problem (1) with f corresponding to the exact solution

$$u(x, y, z) = e^{ax} \cos by \sin cz$$

when Ω is the unit sphere. The absolute value of the maximum error was calculated for various values of $M = N$ on a uniformly distributed set of points (different from the collocation points) on the unit sphere. Plots of the maximum error versus the radius R of the pseudo–boundary are presented in Fig. 2 for $\alpha = 0$ and

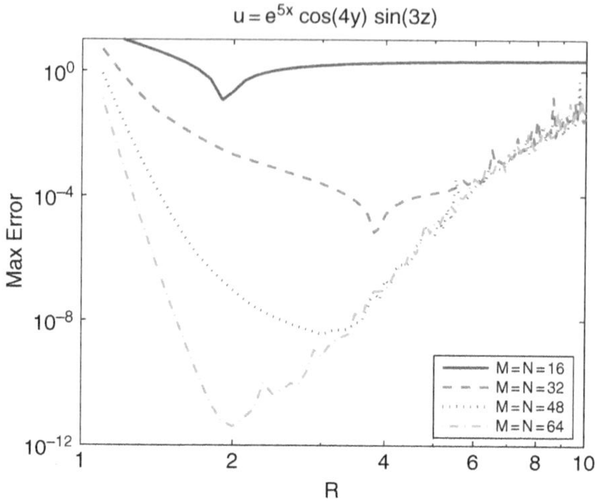

Fig. 2 Maximum error versus R

$a = 5$, $b = 4$, $c = 3$. In all cases we observe that the accuracy of the approximation improves as we increase M. Also, because of ill-conditioning, for larger values of R the accuracy deteriorates. In Fig. 3 we present plots of the maximum error versus the angular parameter α for a fixed pseudo-boundary of radius $R = 1.01$ for $a = 0.5$, $b = 0.4$, $c = 0.3$. These confirm the observations reported in previous studies, i.e. that when the pseudo-boundary is close to the boundary, the variation of the angular parameter does improve the accuracy of the MFS approximation, with a minimum reached for $\alpha \approx 0.25$. In order to show the simplicity of the implementation of the algorithm presented in this work, in the Appendix we present a `MATLAB` code performing the calculations presented.

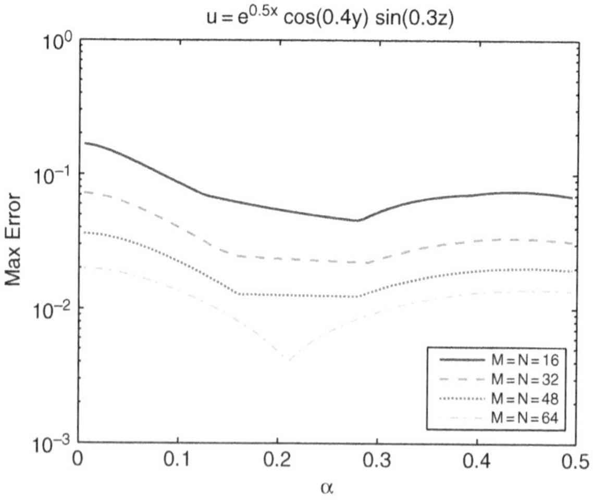

Fig. 3 Maximum error versus α for $R = 1.01$

5 Concluding Remarks

In this paper we describe how the MFS can be efficiently implemented for the solution of problems in rotationally symmetric domains using FFT-based domain decomposition algorithms. We present the algorithm for the cases of the Laplace equation and the Cauchy-Navier equations of elasticity. Numerical results for the former are presented as well as the MATLAB code used for the numerical tests.

Appendix

```
function mfs_mda(f,mp,iter,rp,ds,alfa,n,m)
fi=2*pi/m;th=pi/(n+1);rs=rp+ds;om=ones(1,m);
onm=ones(n,m);omp=ones(mp,mp);at=th*(1:n);
for ii=1:iter
   rs=rp+ii*ds;alfa=.5*ii/(iter+1);af=fi*(0:m-1);
   cat=cos(at);sat=sin(at);ca=cos(af);sa=sin(af);
   ca1=cos(af+fi*alfa);sa1=sin(af+fi*alfa);
   xp=rp*sat'*ca;yp=rp*sat'*sa;zp=rp*cat'*om;
   b=feval(f,xp,yp,zp);xs=rs*sat'*ca1;
   ys=rs*sat'*sa1;zs=rs*cat'*om;
   for in=1:n
      x=xp(in,1)*onm-xs;y=yp(in,1)*onm-ys;
      z=zp(in,1)*onm-zs;r=sqrt(x.^2+y.^2+z.^2);
      aaa(in,:,:)=(1/(4*pi))/r;
   end
   for k=1:n
      fff(k,:)=fft(b(k,:)')/sqrt(m);
   end
   for im=1:m
      for in=1:n
         for jn=1:n
            d(in,jn,:)=ifft(aaa(in,jn,:))*m;
            matr(in,jn)=d(in,jn,im);
         end
      end
      sol=matr\fff(:,im)'; c(:,im)=sol;
   end
   for k=1:n
      ct(k,:)=real(ifft(c(k,:)')*sqrt(m));
   end
   angf=th*(0:mp-1);angt=fi*(1:mp);car=cos(angt);
   sar=sin(angt); caf=cos(angf);saf=sin(angf);
   xpm=rp*sar'*caf;ypm=rp*sar'*saf;
   zpm=rp*car'*ones(1,mp);exa=feval(f,xpm,ypm,zpm);
```

```
    for jn=1:n
        for jm=1:m
            x=xpm-xs(jn,jm)*omp;y=ypm-ys(jn,jm)*omp;
            z=zpm-zs(jn,jm)*omp;r=sqrt(x.^2+y.^2+z.^2);
            approx=approx+((1/(4*pi))*ct(jn,jm)./r);
        end
    end
    error=max(max(approx-exa));
end
```

References

Bialecki, B., Fairweather, G.: Matrix decomposition algorithms for separable elliptic boundary value problems in two space dimensions. J. Comput. Appl. Math. **46**(3), 369–386 (1993)

Cho, H.A., Golberg, M.A., Muleshkov, A.S., Li, X.: Trefftz methods for time dependent partial differential equations. Comput. Mat. Cont. **1**(1), 1–37 (2004)

Fairweather, G.: The method of fundamental solutions – a personal perspective. Plenary talk at the First International Workshop on the Method of Fundamental Solutions (MFS 2007), Ayia Napa, Cyprus, June 11–13, 2007

Fairweather, G., Karageorghis, A.: The method of fundamental solutions for elliptic boundary value problems. Numerical treatment of boundary integral equations. Adv. Comput. Math. **9**(1–2), 69–95 (1998)

Fairweather, G., Karageorghis, A., Martin, P.A.: The method of fundamental solutions for scattering and radiation problems. Eng. Anal. Bound. Elem. **27**, 759–769 (2003)

Fairweather, G., Karageorghis, A., Smyrlis, Y.S.: A matrix decomposition MFS algorithm for axisymmetric biharmonic problems. Adv. Comput. Math. **23**(1-2), 55–71 (2005)

Golberg, M.A., Chen, C.S.: The method of fundamental solutions for potential, Helmholtz and diffusion problems. In: Boundary integral methods: numerical and mathematical aspects, *Comput. Eng.*, vol. 1, pp. 103–176. WIT Press/Comput. Mech. Publ., Boston, MA (1999)

Karageorghis, A., Fairweather, G.: The method of fundamental solutions for axisymmetric acoustic scattering and radiation problems. J. Acoust. Soc. Amer. **104**, 3212–3218 (1998)

Karageorghis, A., Fairweather, G.: The method of fundamental solutions for axisymmetric potential problems. Int. J. Numer. Meth. Engng. **44**, 1653–1669 (1999)

Karageorghis, A., Fairweather, G.: The method of fundamental solutions for axisymmetric elasticity problems. Comput. Mech. **25**(6), 524–532 (2000)

Karageorghis, A., Smyrlis, Y.S.: Matrix decomposition mfs algorithms for elasticity and thermo-elasticity problems in axisymmetric domains. J. Comput. Appl. Math. **206**(2), 774–795 (2007)

Kita, E., Kamiya, N.: Trefftz method: an overview. Adv. Eng. Softw. **24**, 3–12 (1995)

Smyrlis, Y.S., Karageorghis, A.: Some aspects of the method of fundamental solutions for certain harmonic problems. J. Sci. Comput. **16**(3), 341–371 (2001)

Smyrlis, Y.S., Karageorghis, A.: A linear least-squares MFS for certain elliptic problems. Numer. Algorithms **35**(1), 29–44 (2004a)

Smyrlis, Y.S., Karageorghis, A.: A matrix decomposition MFS algorithm for axisymmetric potential problems. Eng. Anal. Bound. Elem. **28**, 463–474 (2004b)

Smyrlis, Y.S., Karageorghis, A.: The method of fundamental solutions for stationary heat conduction problems in rotationally symmetric domains. SIAM J. Sci. Comput. **27**(4), 1493–1512 (electronic) (2006)

Tsangaris, T., Smyrlis, Y.S., Karageorghis, A.: A matrix decomposition MFS algorithm for biharmonic problems in annular domains. Comput. Materi. Continua **1**(3), 245–258 (2004)

Tsangaris, T., Smyrlis, Y.S., Karageorghis, A.: A matrix decomposition MFS algorithm for problems in hollow axisymmetric domains. J. Sc. Comput. **28**, 31–50 (2006a)

Tsangaris, T., Smyrlis, Y.S., Karageorghis, A.: Numerical analysis of the method of fundamental solutions for harmonic problems in annular domains. Numer. Methods Partial Differential Equations **22**(3), 507–539 (2006b)

Boundary Element Analysis of Gradient Elastic Problems

G.F. Karlis, S.V. Tsinopoulos and D. Polyzos

Abstract A boundary element method (BEM), suitable for solving two dimensional (2D) and three dimensional (3D) gradient elastic problems under static loading, is presented. The simplified Form-II gradient elastic theory (a simple version of Mindlin's Form II general gradient elastic theory) is employed and the corresponding fundamental solution is exploited for the formulation of the integral representation of the problem. Three noded quadratic line and eight noded quadratic quadrilateral boundary elements are utilized and the discretization is restricted only to the boundary. The boundary element methodology is explained and presented. The importance of satisfying the correct boundary conditions, being compatible with Mindlin's theory is demonstrated with a simple example. Three numerical examples are reported to illustrate the method and exhibit its merits.

1 Introduction

Experimental observations have shown that macroscopically many materials are significantly affected by their microstructure and exhibit a mechanical behavior which is different than that classically expected. Polycrystals (Smyshlyaev and Fleck 1996; Dillard et al. 2006), polymers (Lakes 1983; Chen and Lakes 1989; Lam et al. 2003), metallic foams (Chen and Fleck 2002), granular materials (Vardoulakis and Sulem 1995), concrete (Van Vliet and Van Mier 1999), porous media (Lakes 1986), bones (Lakes 1995) and particle reinforced composites (Lloyd 1994; Nan and Clarke 1996) are some examples of such materials. These microstructural effects become more pronounced especially when the size or a dimension of a tested specimen becomes small as well as in cases where generated wavelengths have the same order of magnitude with the microstructure of the considered materials.

Due to the lack of internal length scale parameters, which would correlate the microstructure with the macrostructure, classical theory of linear elasticity fails

D. Polyzos (✉)
Department of Mechanical and Aeronautical Engineering, University of Patras, GR-26500 Patras, Greece
e-mail: polyzos@mech.upatras.gr

G.D. Manolis, D. Polyzos (eds.), *Recent Advances in Boundary Element Methods*,
DOI 10.1007/978-1-4020-9710-2_16,

to describe such behaviour. On the other hand, as it is stated in Exadaktylos and Vardoulakis (2001), in fracture mechanics and dislocation problems where near the tip of the crack and dislocation lines abrupt changes of strains and stresses are observed, enhanced elastic theories should be applied since classical elasticity works only under the assumption that all the material and field parameters vary linearly throughout the representative volume element of the considered material or structure.

Thus, resort should be made to other enhanced elastic theories where internal length scale constants, correlating the microstructure with the macrostructure, are involved in the constitutive equations of the continuum. Such theories are the non-local elasticity (Eringen 1992), the higher order gradient elasticity (Mindlin 1964) the couple stresses elastic theories (Tiersten and Bleustein 1974) and the micropolar elasticity (Eringen 1999). Since the present paper deals with a Boundary Element Method (BEM) for solving gradient elastostatic problems the non-local and couple stresses-micropolar theories are not discussed here.

Among those who have developed higher order gradient elastic theories one can mention Mindlin (1964, 1965), Mindlin and Eshel (1968), Green and Rivlin (1964), Aifantis (1992), Ru and Aifantis (1993), Vardoulakis and Sulem (1995), Exadaktylos and Vardoulakis (1998), and Fleck and Hutchinson (1997, 2001). In the regime of isotropic linear elastic behaviour, the most general and comprehensive gradient elastic theory is the one due to Mindlin (1964). However, in order to balance the dimensions of strains and higher order gradients of strains as well as to correlate the micro-strains with macro-strains, Mindlin utilized eighteen new constants rendering thus his initial general theory very complicated from physical and mathematical point of view. In the sequel, considering long wave-lengths and the same deformation for macro and micro structure Mindlin proposed three new simplified versions of his theory, known as Form I, II and III, where beyond the two Lame constants other five ones are introduced instead of sixteen employed in his initial model. In Form-I, the strain energy density function is assumed to be a quadratic form of the classical strains and the second gradient of displacement; in Form-II the second displacement gradient is replaced by the gradient of strains and in Form-III the strain energy function is written in terms of the strain, the gradient of rotation, and the fully symmetric part of the gradient of strain. The most important difference among the aforementioned three simplified versions of the general Mindlin's theory is the fact that the Form-II leads to a total stress tensor, which is symmetric as in the case of classical elasticity. This symmetry avoids the problems introduced by the non-symmetric stress tensors in Cosserat and couple stress theories. Aifantis (1992) and Ru and Aifantis (1993) proposed a very simple gradient elastic model requiring only one new gradient elastic constant plus the standard Lame ones. As it is mentioned in Vardoulakis et al. (1996) and explained in the next section, this gradient elastic model can be considered as the simplest possible special case of Form-II version of Mindlin's theory. The main problem with Aifantis' model is that due to the complete lack of a variational formulation, the considered boundary conditions are not compatible with the corresponding correct ones provided by Mindlin. The correction on the boundary conditions is made later in the paper of

Vardoulakis et al. (1996). The gradient elastic with surface energy theory proposed by Vardoulakis and Sulem (1995) is slightly more complicated than that of Aifantis and co-workers but it is a direct consequence of the continuum model proposed by Casal (1972) and not a special case of Mindlin's general theory. Finally, Fleck and Hutchinson (1997, 2001) decomposed the second gradient of displacement into the stretch gradient and the rotation gradient tensors proposed thus an alternative version of Mindlin's Form I gradient elastic theory.

As in the case of classical elasticity, the solution of gradient elastic problems with complicated geometry and boundary conditions requires the use of numerical methods such as the finite element method (FEM) and the boundary element method (BEM). Shu et al. (1999) and Tsepoura et al. (2002) were the first to use FEM and BEM, respectively, for solving elastostatic problems in the framework of the gradient elasticity theories of Mindlin. Since then many papers dealing with numerical solutions of gradient elastic problems have been appeared to the literature. Here one can mention the FEM formulations of Amanatidou and Aravas (2002), Amanatidou et al. (2005), Engel et al. (2002), Tenek and Aifantis (2002), Peerlings and Fleck (2004), Soh and Wanji (2004), Imatani et al. (2005), Askes and Gutierrez (2006), Dessouky et al. (2003, 2006), Akarapu and Zbib (2006) and Markolefas et al. (2007), the BEM formulations of Tsepoura and Polyzos (2003), Polyzos et al. (2003, 2005), Tsepoura et al. (2003), Polyzos (2005), Karlis et al. (2007, 2008) and the Meshless Local Petrov-Galerkin formulation of Tang et al. (2003).

In the present paper the BEM in its direct form is employed for the solution of two-dimensional (2D) and three-dimensional (3D) elastostatic problems in the framework of the simplified Form-II strain-gradient theory of Mindlin. The paper consists of the following five sections: Section 2 deals with the integral representation of a gradient elastic boundary value problem in the context of Mindlin's simplified Form-II gradient elastic theory. Section 3 presents the corresponding BEM for solving 2D and 3D gradient elastostatic problems. For the sake of simplicity the method is presented in two dimensions. Three numerical examples are presented in Section 4 to illustrate the accuracy of the method and demonstrate its merits. Finally, Section 5 consists of the conclusions pertaining to this work.

2 Simplified Gradient Elastic Model, Integral Representation of the Problem

Mindlin in Form II version of his gradient elastic theory (Mindlin 1964) considering the potential energy density to be a quadratic form of the strains and the gradient of strains and taking its variation, obtained an equilibrium equation of the form

$$\partial_j \sigma_{jk} = \partial_j (\hat{\tau}_{jk} - \partial_i \hat{\mu}_{ijk}) = 0 \tag{1}$$

$$\hat{\tau}_{ij} = 2\mu e_{ij} + \lambda e_{kk} \delta_{ij} \tag{2}$$

where $\sigma_{jk}, \hat{\tau}_{jk}, -\partial_i \hat{\mu}_{ijk}$ represent the symmetric total, Cauchy and relative stress tensors, respectively, with the third order double stress tensor $\hat{\mu}_{ijk}$ given by

$$\begin{aligned}\mu_{ijk} &= \frac{1}{2}\hat{\alpha}_1 \left(\partial_l \partial_l u_i \delta_{jk} + \partial_k \partial_l u_l \delta_{ij} + \partial_i \partial_l u_l \delta_{jk} + \partial_j \partial_l u_l \delta_{ik}\right) + 2\hat{\alpha}_2 \partial_i \partial_l u_l \delta_{jk} \\ &+ \frac{1}{2}\hat{\alpha}_3 \left(\partial_l \partial_l u_k \delta_{ij} + \partial_k \partial_l u_l \delta_{ij} + \partial_l \partial_l u_j \delta_{ik} + \partial_j \partial_l u_l \delta_{ik}\right) \\ &+ \hat{\alpha}_4 \left(\partial_i \partial_j u_k + \partial_l \partial_k u_j\right) + \frac{1}{2}\hat{\alpha}_5 \left(2\partial_j \partial_k u_i + \partial_l \partial_l u_k + \partial_l \partial_k u_j\right)\end{aligned} \tag{3}$$

The seven constants λ, μ and $\hat{\alpha}_1 \div \hat{\alpha}_5$ appearing in eq. (3) are illustrated in Mindlin (1964). Equation (1) is accompanied by the classical boundary conditions

$$u_k = \bar{u} \text{ and } \hat{p}_k = \bar{p} \tag{4}$$

and the non-classical ones

$$n_l \partial_l u_k = \bar{q},\ \hat{R}_k = \bar{R} \tag{5}$$

$$\text{and } \hat{E}_k = \bar{E} \tag{6}$$

where $\bar{u},\ \bar{p},\ \bar{q},\ \bar{R},\ \bar{E}$ represent prescribed vectors, $\hat{\mathbf{n}}$ is the unit vector normal to the boundary and $\hat{p}_k,\ \hat{R}_k,\ \hat{E}_k$ are the traction, double traction and jump traction vectors, respectively, defined as

$$\hat{p}_k = n_j \hat{\tau}_{jk} - n_i n_j D \hat{\mu}_{ijk} - (n_j D_i + n_i D_j)\hat{\mu}_{ijk} + (n_i n_j D_l n_l - D_j n_i)\hat{\mu}_{ijk} \tag{7}$$

$$\hat{R}_k = n_i n_j \hat{\mu}_{ijk} \tag{8}$$

$$\hat{E}_k = \left\| n_i m_j \hat{\mu}_{ijk} \right\| \tag{9}$$

The non-classical boundary condition (9) exists only when non-smooth boundaries are considered. Double brackets indicate that the enclosed quantity is the difference between its values taken on the two sides of a corner while $\hat{\mathbf{m}}$ is a vector being tangential to the corner line.

Although very elegant, the use of the Form II gradient elastic theory for the solution of real problems is discouraging since totally seven new material constants, i.e. $\lambda, \mu, \hat{\alpha}_1 \div \hat{\alpha}_5$ have to be determined. During the decade of 1990s, Aifantis and co-workers proposed a simpler gradient elastic theory according to which the constitutive eq. (2) remains the same with λ, μ being the classical Lamè constants, while the double stresses, given by eq. (3), obtain the much simpler form

$$\mu_{ijk} = g^2 \partial_i \mu_{jk} \tag{10}$$

where g^2 is the volumetric strain gradient energy coefficient or simply the gradient coefficient, which is introduced to balance the dimensions of strains and strain gradients as well as to relate the microstructure with the macrostructure representing a characteristic length of the material. However, it is easy to see of that eq. (10) can be obtained from (3) by replacing $\hat{\alpha}_1 \div \hat{\alpha}_5$ with

$$\hat{\alpha}_1 = \hat{\alpha}_3 = \hat{\alpha}_5 = 0, \quad \hat{\alpha}_2 = \frac{\lambda}{2} g^2, \quad \hat{\alpha}_4 = \mu g^2 \tag{11}$$

Thus, Aifantis' gradient elastic theory can be considered as the simplest possible special case of Mindlin's theory. As it has already been mentioned in the introduction, the problem with Aifantis' theory, as it was initially proposed, is the complete lack of a variational formulation, which should provide the correct boundary conditions, compatible with the corresponding ones provided by Mindlin, i.e. eqs. (4), (5) and (6). The importance of considering the correct boundary conditions in this kind of problems is demonstrated with a simple example in the next section.

Adopting the above simplified Form-II theory of Mindlin, Polyzos et al. (2003) derived analytically the fundamental solution of the equilibrium eq. (1) and proposed a new reciprocity identity for gradient elastic solids. In the sequel, considering a gradient elastic material of volume V surrounded by a smooth surface S and characterized by two Lame constants λ, μ and a gradient coefficient g^2, they showed that for any static gradient elastic boundary value problem the following integral representation is satisfied

$$\begin{aligned} &\tilde{\mathbf{c}}(\mathbf{x}) \cdot \mathbf{u}(\mathbf{x}) + \int_S \tilde{\mathbf{P}}^*(\mathbf{x}, \mathbf{y}) \cdot \mathbf{u}(\mathbf{y}) dS_\mathbf{y} + \int_S \tilde{\mathbf{R}}^*(\mathbf{x}, \mathbf{y}) \cdot \mathbf{q}(\mathbf{y}) dS_\mathbf{y} \\ &= \int_S \tilde{\mathbf{U}}^*(\mathbf{x}, \mathbf{y}) \cdot \mathbf{P}(\mathbf{y}) dS_\mathbf{y} + \int_S \tilde{\mathbf{Q}}^*(\mathbf{x}, \mathbf{y}) \cdot \mathbf{R}(\mathbf{y}) dS_\mathbf{y} \end{aligned} \tag{12}$$

for gradient elastic problems with smooth boundaries and

$$\begin{aligned} &\tilde{\mathbf{c}}(\mathbf{x}) \cdot \mathbf{u}(\mathbf{x}) + \int_S \left(\tilde{\mathbf{P}}^*(\mathbf{x}, \mathbf{y}) \cdot \mathbf{u}(\mathbf{y})\right) dS_\mathbf{y} \\ &+ \int_S \left(\tilde{\mathbf{R}}^*(\mathbf{x}, \mathbf{y}) \cdot \mathbf{q}(\mathbf{y})\right) dS_\mathbf{y} + \sum_a \oint_{C_a} \tilde{\mathbf{E}}^*(\mathbf{x}, \mathbf{y}) \cdot \mathbf{u}(\mathbf{y}) dc \\ &= \int_S \tilde{\mathbf{U}}^*(\mathbf{x}, \mathbf{y}) \cdot \mathbf{p}(\mathbf{y}) dS_\mathbf{y} + \int_S \tilde{\mathbf{Q}}^*(\mathbf{x}, \mathbf{y}) \cdot \mathbf{R}(\mathbf{y}) dS_\mathbf{y} + \sum_a \oint_{C_a} \tilde{\mathbf{U}}^*(\mathbf{x}, \mathbf{y}) \cdot \mathbf{E}(\mathbf{y}) dc \end{aligned} \tag{13}$$

for problems with non-smooth boundaries containing α corner lines C_a. $\tilde{\mathbf{U}}^*(\mathbf{x}, \mathbf{y})$, $\tilde{\mathbf{Q}}^*(\mathbf{x}, \mathbf{y})$ are the fundamental solution of eq. (1) and its normal derivative, respectively, $\tilde{\mathbf{P}}^*(\mathbf{x}, \mathbf{y})$, $\tilde{\mathbf{R}}^*(\mathbf{x}, \mathbf{y})$ and $\tilde{\mathbf{E}}^*(\mathbf{x}, \mathbf{y})$ are the fundamental traction, double stress traction and jump traction tensors, respectively and $\tilde{\mathbf{c}}(\mathbf{x})$ is the well-known jump tensor being equal to $(1/2)\tilde{\mathbf{I}}$ for $\mathbf{x} \in S$ and equal to $\tilde{\mathbf{I}}$ when $\mathbf{x} \in V \cap S$.

Observing eqs. (12) and (13), one realizes that these equations contain four unknown vector fields, $\mathbf{u}(\mathbf{x})$, $\mathbf{P}(\mathbf{x})$, $\mathbf{R}(\mathbf{x})$ and $\mathbf{q}(\mathbf{x}) = \partial\mathbf{u}/\partial n$. Thus, the evaluation of the unknown fields requires the existence of one more integral equation. This integral equation is obtained by applying the operator $\partial/\partial n_x$ on eqs. (12) and (13), i.e., respectively

$$\begin{aligned}\tilde{\mathbf{c}}(\mathbf{x})\cdot\mathbf{q}(\mathbf{x}) + \int_S \frac{\partial\tilde{\mathbf{P}}^*(\mathbf{x},\mathbf{y})}{\partial n_{\mathbf{x}}}\cdot\mathbf{u}(\mathbf{y})dS_{\mathbf{y}} + \int_S \frac{\partial\tilde{\mathbf{R}}^*(\mathbf{x},\mathbf{y})}{\partial n_{\mathbf{x}}}\cdot\mathbf{q}(\mathbf{y})dS_{\mathbf{y}} \\ = \int_S \frac{\partial\tilde{\mathbf{U}}^*(\mathbf{x},\mathbf{y})}{\partial n_{\mathbf{x}}}\cdot\mathbf{P}(\mathbf{y})dS_{\mathbf{y}} + \int_S \frac{\partial\tilde{\mathbf{Q}}^*(\mathbf{x},\mathbf{y})}{\partial n_{\mathbf{x}}}\cdot\mathbf{R}(\mathbf{y})dS_{\mathbf{y}}\end{aligned} \tag{14}$$

$$\begin{aligned}\tilde{\mathbf{c}}(\mathbf{x})\cdot\mathbf{q}(\mathbf{x}) + \int_S \frac{\partial\tilde{\mathbf{P}}^*(\mathbf{x},\mathbf{y})}{\partial n_{\mathbf{x}}}\cdot\mathbf{u}(\mathbf{y})dS_{\mathbf{y}} + \int_S \frac{\partial\tilde{\mathbf{R}}^*(\mathbf{x},\mathbf{y})}{\partial n_{\mathbf{x}}}\cdot\mathbf{q}(\mathbf{y})dS_{\mathbf{y}} \\ + \sum_a \oint_{C_a} \frac{\partial\tilde{\mathbf{E}}^*(\mathbf{x},\mathbf{y})}{\partial n_{\mathbf{x}}}\cdot\mathbf{u}(\mathbf{y})dc = \int_S \frac{\partial\tilde{\mathbf{U}}^*(\mathbf{x},\mathbf{y})}{\partial n_{\mathbf{x}}}\cdot\mathbf{P}(\mathbf{y})dS_{\mathbf{y}} \\ + \int_S \frac{\partial\tilde{\mathbf{Q}}^*(\mathbf{x},\mathbf{y})}{\partial n_{\mathbf{x}}}\cdot\mathbf{R}(\mathbf{y})dS_{\mathbf{y}} + \sum_a \oint_{C_a} \frac{\partial\tilde{\mathbf{U}}^*(\mathbf{x},\mathbf{y})}{\partial n_{\mathbf{x}}}\cdot\mathbf{E}(\mathbf{y})dc\end{aligned} \tag{15}$$

All the kernels appearing in the integral eqs. (12), (13), (14) and (15) are given explicitly in Polyzos et al. (2003). Integral equations (12), (13), (14) and (15) accompanied by the classical and non-classical boundary conditions (4), (5) and (6) form the integral representation of a well-posed gradient elastostatic boundary value problem.

3 BEM Procedure

The goal of a BEM is to solve numerically the integral eqs. (12), (13), (14) and (15) of the just described gradient elastic problem. For the sake of simplicity only 2D gradient elastic bodies with smooth boundaries are considered. The boundary S is discretized into three-noded quadratic line isoparametric boundary elements and in cases of fracture mechanics problems special elements at either side of the crack tip are employed (Karlis et al. 2007, 2008). All the nodal parameters of the corresponding fields are expressed to the local coordinate system with the aid of the shape functions. Adopting a global numbering for all the considered nodes the integral equations (12) and (14) can be written in a discrete form as

$$\frac{1}{2}\mathbf{u}^k + \sum_{\beta=1}^{L} \tilde{\mathbf{H}}_{\beta}^{k} \cdot \mathbf{u}^{\beta} + \sum_{\beta=1}^{L} \tilde{\mathbf{K}}_{\beta}^{k} \cdot \mathbf{q}^{\beta} = \sum_{\beta=1}^{L} \tilde{\mathbf{G}}_{\beta}^{k} \cdot \mathbf{p}^{\beta} + \sum_{\beta=1}^{L} \tilde{\mathbf{L}}_{\beta}^{k} \cdot \mathbf{R}^{\beta} \tag{16}$$

$$\frac{1}{2}\mathbf{q}^k + \sum_{\beta=1}^{L} \tilde{\mathbf{S}}_{\beta}^{k} \cdot \mathbf{u}^{\beta} + \sum_{\beta=1}^{L} \tilde{\mathbf{T}}_{\beta}^{k} \cdot \mathbf{q}^{\beta} = \sum_{\beta=1}^{L} \tilde{\mathbf{V}}_{\beta}^{k} \cdot \mathbf{p}^{\beta} + \sum_{\beta=1}^{L} \tilde{\mathbf{W}}_{\beta}^{k} \cdot \mathbf{R}^{\beta} \tag{17}$$

where L is the total number of nodes.

The above discretized boundary integral equations are collocated at all nodal points and the following linear system of algebraic equations is obtained

$$\begin{bmatrix} \frac{1}{2}\tilde{\mathbf{I}} + \tilde{\mathbf{H}} & \tilde{\mathbf{K}} \\ \tilde{\mathbf{S}} & \frac{1}{2}\tilde{\mathbf{I}} + \tilde{\mathbf{T}} \end{bmatrix} \cdot \begin{Bmatrix} \mathbf{u} \\ \mathbf{q} \end{Bmatrix} = \begin{bmatrix} \tilde{\mathbf{G}} & \tilde{\mathbf{L}} \\ \tilde{\mathbf{V}} & \tilde{\mathbf{W}} \end{bmatrix} \cdot \begin{Bmatrix} \mathbf{P} \\ \mathbf{R} \end{Bmatrix} \tag{18}$$

where the influence matrices $\tilde{\mathbf{H}}$, $\tilde{\mathbf{K}}$, $\tilde{\mathbf{S}}$, $\tilde{\mathbf{T}}$, $\tilde{\mathbf{G}}$, $\tilde{\mathbf{L}}$, $\tilde{\mathbf{V}}$, and $\tilde{\mathbf{W}}$ contain integrals over the boundary elements involving products of fundamental kernels times shape functions and Jacobians (Polyzos et al. 2003; Tsepoura et al. 2003). Applying the boundary conditions (4) and (5) and rearranging eq. (18), one produces the final linear system of algebraic equations of the form

$$\tilde{\mathbf{A}} \cdot \mathbf{X} = \mathbf{B} \tag{19}$$

where the vectors **X** and **B** contain all the unknown and known nodal components of the boundary fields respectively. At the non-special elements, the singular and hyper-singular integrals involved, are evaluated with high accuracy applying the methodology for direct treatment of Cauchy principal value and hyper-singular integrals described in Tsepoura et al. (2003). The integrations over the special boundary elements utilized for 2D and 3D cracks are described in Karlis et al. (2007, 2008). Finally, the linear system is solved via a typical LU-decomposition algorithm and the vector **X**, comprising all the unknown nodal values of **u**, **q**, **R** and **P**, is evaluated.

It is apparent that for non-smooth boundaries the integral eqs. (13) and (15) instead of (12) and (14) should be employed as well as the extra non-classical boundary condition (6) should be taken into account. In order to demonstrate the importance of considering the correct integral equations and satisfying the correct boundary conditions in a gradient elastic problem, the following simple problem is numerically solved:

Consider a hollow gradient elastic cylinder with a uniform internal pressure. Exploiting the above described BEM for smooth surfaces, the problem is solved first considering an artificial interface in the circumferential direction (Fig. 1(a)) and next an interface separating the cylinder into two equal pieces (Fig. 1(b)). The corresponding deformation profiles of the cylinder in both cases are depicted in Figs. 1(c) and 1(d), respectively. It is apparent that the obtained solutions are not the

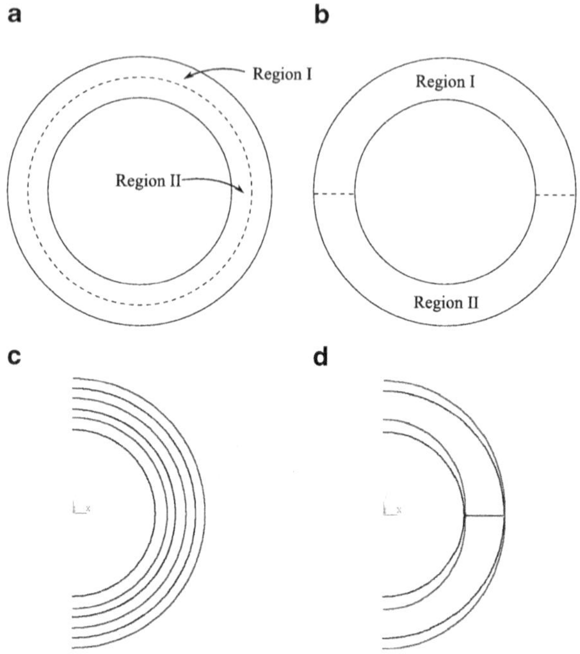

Fig. 1 (**a**) A hollow cylinder divided into two sub-regions by a circular interface (**b**) A hollow cylinder divided by a straight interface into two sub-regions of the same shape and size. (**c**) The deformed hollow cylinder cross-section with the circular interface. (**d**) The deformed hollow cylinder cross-section with the straight interface

same. The displacement profile of Fig. 1(c) is the same with the corresponding one taken when no interfaces are considered in the solution of the problem. On the other hand the displacement profile of the cylinder in Fig. 1(d) violates the radial symmetry of the problem and the solution exhibits significant errors near to the interface. The obvious explanation for these two different solutions of the same problem is that in the first case the two considered subregions have smooth boundaries, while in the second case they do not. Thus, for the second case the integral eqs. (13) and (15) instead of (12) and (14) should be employed in the BEM solution of the problem.

4 Numerical Results

The first benchmark concerns a spherical cavity of radius a embedded into an infinite gradient elastic 3-D space and subjected to an internal uniform pressure P_0. The classical and non-classical boundary conditions of the problem are

$$\mathbf{P}(r)|_{r=a} = P_0\hat{\mathbf{r}} \tag{20}$$

$$\mathbf{R}(r)|_{r=a} = \mathbf{0} \tag{21}$$

and its analytical solution (Tsepoura et al. 2003)

$$\mathbf{u} = \left\{ \mathrm{B}\frac{1}{\mathrm{r}^2} + D\,G\,(r) \right\} \hat{\mathbf{r}}, \text{ with } G(r) = \sqrt{\frac{\pi}{2\,(r\,/g)}}\mathrm{K}_{\frac{3}{2}}\,(\mathrm{r}\,/g) \tag{22}$$

With

$$B = \frac{-P_0\,(1-2\nu)\,(1+\nu)\,a^3\left(6g^3+6g^2a+3ga^2+a^3\right)}{2E\left[3g^3\,(-3+4\nu)+3g^2\,(-3+4\nu)\,a+3g\,(-1+2\nu)\,a^2+(-1+2\nu)\,a^3\right]} \tag{23}$$

$$D = -\frac{6e^{a/g}\,P_0\,(-1+2\nu)\,(1+\nu)\,\sqrt{g\,/r}\,a^4}{E\pi\sqrt{r\,/g}\left[3g^3\,(-3+4\nu)+3g^2\,(-3+4\nu)\,a+3g\,(-1+2\nu)\,a^2+(-1+2\nu)\,a^3\right]} \tag{24}$$

Assuming $a = 1$, $P_0\,/E = 1$ and $\nu = 0$ the radial displacements and the radial strains for three values of g are evaluated and depicted in Figs. 2(a) and (b), respectively. The calculated displacements and strains are compared to the corresponding analytical ones and as it is evident from Figs. 2(a and b) the agreement between the solutions is excellent.

The second example is concerned with the numerical solution of Boussinesque's problem for a gradient elastic half space. For the case of classical elasticity the corresponding static problem has an analytical solution, which in terms of vertical displacements reads

$$u_z = \frac{\left(1-\nu^2\right)P}{\pi\,Er} \tag{25}$$

where r is the radial distance form the force application point, P is the applied force and ν, E are the Poisson's ratio and Young's modulus, respectively. Observing

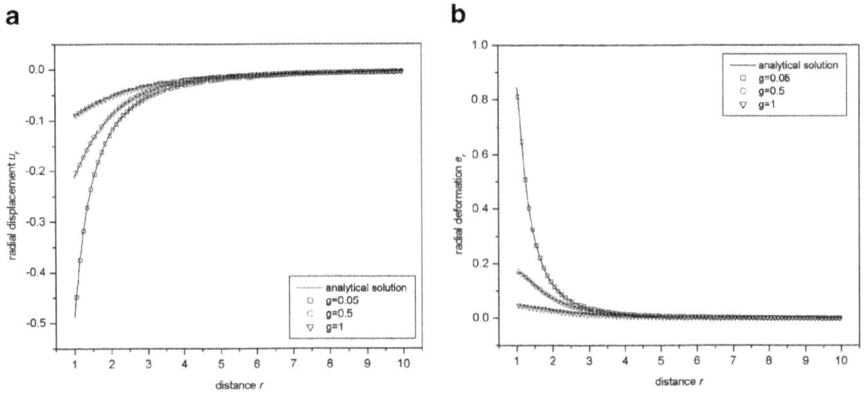

Fig. 2 (**a**) Radial displacements versus the radial distance r from the surface of the spherical cavity. (**b**) Radial strains versus the radial distance r from the surface of the spherical cavity

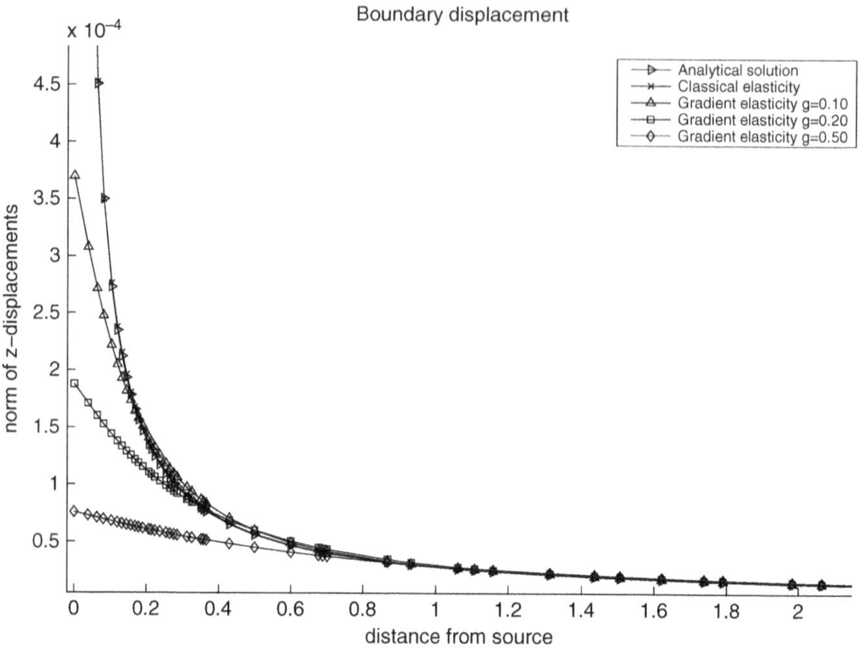

Fig. 3 Boundary vertical displacements for gradient elastic Boussinesque problem

eq. (25) one can easily notice that at the point where the force acts, vertical displacements become infinite. This is one of the well known drawbacks of classical elasticity theory. In order to examine the behavior of vertical displacements in a gradient elastic material a boundary surface of the half-space ($z = 0$) was discretized with 273 eight-noded quadrilateral elements and the elastic material constants was chosen to be E =0.178 GPa and ν=0.33. The applied force was considered to be $17.777E{+}03Nt$ and its application was facilitated by the use of a 9-noded quadratic quadrilateral element. The problem was solved for different values of the volumetric gradient elastic material constant g and the obtained results are depicted in Fig. 3.

Figure 3 reveals that approaching the force application point, the z-displacement corresponding to classical elasticity solution tends to infinity. On the other hand, the gradient elastic solutions demonstrate finite values, while the microstructural effects, represented by g, are apparent. This result is in agreement with the analytical solution provided for the corresponding two dimensional gradient elastic half-space problem by Georgiadis and Anagnostou (2008) and Li et al. (2004).

The last example deals with a Mode I fracture mechanics problem for a gradient elastic plate. More precisely, a square gradient elastic plate in a state of plane strain is considered. The plate contains a central horizontal line crack and is subjected to a uniform tensile traction $P_0 = 100$ MPa applied normal to its top and bottom sides, as shown in Fig. 4. The crack length is chosen to be equal to $2a = 1$m and the side of the square plate is $L = 16a$. The Young modulus and the Poisson ratio of the gradient

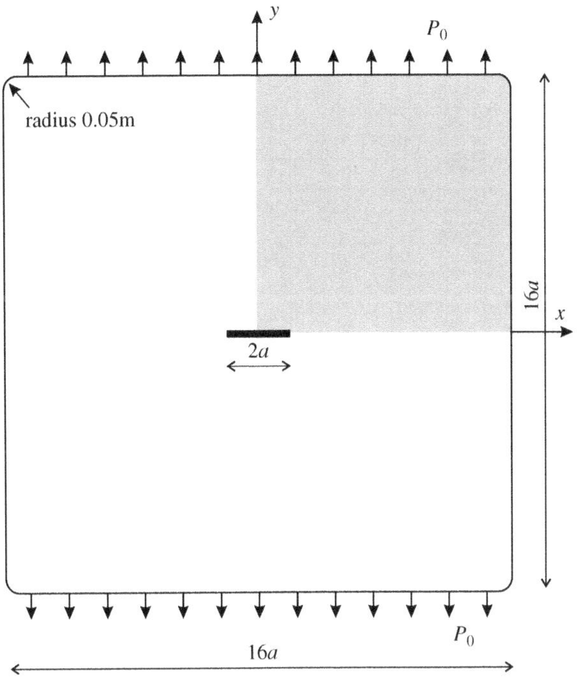

Fig. 4 Gradient elastic plate with a horizontal line crack

elastic plate are E = 210 GPa and ν = 0.2, respectively. Due to the symmetry of the problem, only one quarter of the plate is discretized, with the following classical and non-classical boundary conditions: $\mathbf{P}(x, 0) = \mathbf{0}$ and $\mathbf{R}(x, 0) = \mathbf{0}$ for $0 \leq x < a$, $u_y(x, 0) = 0$ and $\mathbf{R}(x, 0) = \mathbf{0}$ for $a \leq x \leq 8a$, $\mathbf{P}(x, 8a) = P_0\hat{\mathbf{y}}$ and $\mathbf{R}(x, 8a) = \mathbf{0}$, $u_x(0, y) = 0$ and $\mathbf{R}(0, y) = \mathbf{0}$ and $\mathbf{P}(8a, y) = \mathbf{R}(8a, y) = \mathbf{0}$.

Figure 5 displays the upper-right-quarter of the crack opening displacement profile obtained by the present BEM for four different values of the material volumetric

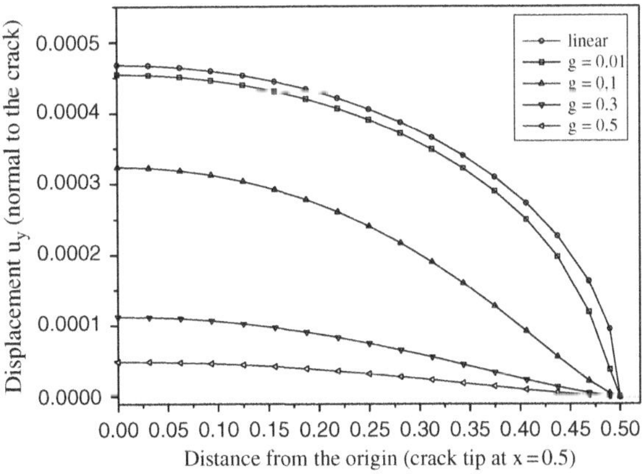

Fig. 5 Upper right quarter of the crack opening displacement profile

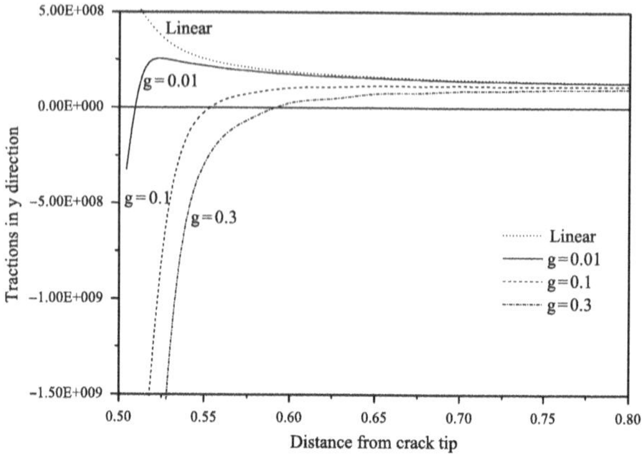

Fig. 6 Traction values near the crack tip

strain energy coefficient g (0.01, 0.1, 0.3, 0.5). In the same figure, the crack profile taken in the classical elastic case (g=0) is also shown. The main conclusion here is that the crack profile of the gradient elastic case remains sharp at the crack tip and is not blunted as in the classical case. This cusp like crack profile agrees with the one coming out of Barenblatt's cohesive zone theory (Barenblatt 1962). Also, it should be noticed that as the volumetric strain energy gradient coefficient g increases, the crack becomes stiffer.

Finally Fig. 6 depicts the traction field near the crack tip, for various values of the gradient coefficient g. As it is apparent near the crack tip tractions are compressive and note tensile as in classical elasticity. More precisely, as it is noted in Karlis et al. (2007), stresses near the crack tip go to infinity with different order ($r^{-3/2}$) than those of classical elasticity ($r^{-1/2}$). This is one more result that agrees with Barenblatt's theory.

5 Conclusions

On the basis of the material presented in the previous sections, the following conclusions can be drawn:

A boundary element method has been presented for the static analysis of 2D and 3D bodies characterized by a linear elastic material behavior, taking into account microstructural effects with the aid of a simple strain gradient elastic theory.

The possible error that can be made in a gradient elastic problem when the non-classical boundary conditions proposed by Mindlin's theory are not satisfied is illustrated with a simple example.

Three numerical examples have been used to demonstrate the presented here BEM and its high accuracy, as well as to illustrate some of the advantages of exploiting higher order gradient elastic theories in the solution of engineering problems.

References

Aifantis EC (1992): On the role of gradients in the localization of deformation and fracture, Int. J. Eng. Sci., Vol. 30, pp. 1279–1299.

Akarapu S, Zbib HM (2006): Numerical analysis of plane cracks in strain-gradient elastic materials, Int. J. Fract., Vol. 141, pp. 403–430.

Amanatidou E, Aravas N (2002): Mixed finite element formulations of strain-gradient elasticity problems, Comput. Methods Appl. Mech. Engng., Vol. 191, pp. 1723–1751.

Amanatidou E, Giannakopoulos A, Aravas N (2005): Finite element models of strain-gradient elasticity: accuracy and error estimates. In: G. Georgiou, P. Papanastasiou, M. Papadrakakis (Eds), *Proceedings of 5th GRACM International Congress on Computational Mechanics*, University of Cyprus, Nicosia, pp. 797–804.

Askes H, Gutierrez MA (2006): Implicit gradient elasticity, Int. J. Numer. Meth. Eng., Vol. 67, pp. 400–416.

Barenblatt GI (1962): Mathematical theory of equilibrium cracks in brittle fracture, Adv. Appl. Mech., Vol. 7, pp. 55–129.

Casal P (1972): La theorie du second gradient et la capillarite, C.R. Acad. Sc. A247, pp. 1571–1574.

Chen CP, Fleck NA (2002) Size effects in the constrained deformation of metallic foams, J. Mech. Phys. Solids., Vol. 50, pp. 955–977.

Chen CP, Lakes RS (1989) Dynamic wave dispersion and loss properties of conventional and negative Poisson's ratio polymeric cellular polymers. Cell. Polym., Vol. 8, pp. 343–369

Dessouky S, Masad E, Little D, Zbib H (2006): Finite-element analysis of hot mix asphalt microstructure using effective local material properties and strain gradient elasticity, J. Eng. Mech., Vol. 132, pp. 158–171.

Dessouky S, Masad E, Zbib H, Little D (2003): Gradient elasticity finite element model for the microstructure analysis of asphaltic materials, In: K. J. Bathe (Ed.), *Computational Fluid and Solid Mechanics*, Elsevier, London, pp. 228–233.

Dillard T, Forest S, Ienny P (2006) Micromorphic continuum modelling of the deformation and fracture behaviour of nickel foams, Eur. J. Mech. A/Solids, Vol. 25, pp. 526–549.

Engel G, Garikipati K, Hughes TJR, Larson MG, Mazzei L, Taylor RL (2002): Continuous/discontinuous finite element approximations of fourth-order elliptic problems in structural and continuum mechanics with applications to thin beams and plates, and strain gradient elasticity, Comput. Meth. Appl. Mech. Eng., Vol. 191, pp. 3669–3750.

Eringen AC (1992): Vistas of nonlocal continuum physics, Int. J. Eng. Sci., Vol. 30, No. 10, pp. 1551–1565.

Eringen C (1999): *Microcontinuum Field Theories I: Foundations and Solids*, Springer-Verlag, New York.

Exadaktylos GE, Vardoulakis I (1998): Surface instability in gradient elasticity with surface energy, Int. J. Solids Struct., Vol. 35, No. 18, pp. 2251–2281.

Exadaktylos GE, Vardoulakis I (2001): Microstructure in linear elasticity and scale effects: a reconsideration of basic rock mechanics and rock fracture mechanics. Tectonophysics, Vol. 335, pp. 81–109.

Fleck NA, Hutchinson JW (1997): Strian gradient plasticity. Hutchinson, J.W., Wu, T.Y. (Eds.), Advances in Applied Mechanics, Academic Press, New York, Vol. 33, pp. 295–361.

Fleck NA, Hutchinson JW (2001): A reformulation of strain gradient plasticity, J. Mech. Phys. Solids, Vol. 49, pp. 2245–2271.

Georgiadis HG, Anagnostou DS (2008): Problems of the Flamant–Boussinesq and Kelvin type in dipolar gradient elasticity, J. Elasticity, Vol. 90, pp. 71–98.

Green AE, Rivlin RS (1964): Multipolar continuum mechanics, Arch. Ration. Mech. Anal., Vol. 17, pp. 113–147.

Imatani S, Hataday K, Maugin GA (2005): Finite element analysis of crack problems for strain gradient material model, Philos. Mag.., Vol. 85, No. 33–35, pp. 4245–4256.

Karlis GF, Tsinopoulos SV, Polyzos D, Beskos DE (2007): Boundary element analysis of mode I and mixed mode (I and II) crack problems of 2-D gradient elasticity, Comput. Meth. Appl. Mech. Eng., Vol. 196, pp. 5092–5103.

Karlis GF, Tsinopoulos SV, Polyzos D, Beskos DE (2008): 2D and 3D Boundary Element Analysis of Mode-I Cracks in Gradiend Elasticity, CMES-Computer Modeling in Engineering & Sciences, Vol. 26, No.3, pp. 189–208.

Lakes RS (1983): Size effects and micromechanics of a porous solid, J. Mater. Sci. Vol. 18, pp. 2572–2580.

Lakes RS (1986): Experimental microelasticity of two porous solids, Int. J. Solids Struct., Vol. 22, pp. 55–63.

Lakes RS (1995): Experimental methods for study of Cosserat elastic solids and other generalized elastic continua. In: H. Muhlhaus (Ed) *Continuum Models for Materials with Microstructure*, J. Wiley, New York, pp. 1–22.

Lam DCC, Yang F, Chong ACM, Wang J, Tong P (2003): Experiments and theory in strain gradient elasticity, J. Mech. Phys. Solids, Vol. 51, pp. 1477–1508.

Li S, Miskioglu I, Altan BS (2004): Solution to line loading of a semi-infinite solid in gradient elasticity, Int. J. Solids Struct., Vol. 41, pp. 3395–3410.

Lloyd DJ (1994): Particle-reinforced aluminum and magnesium matrix composites, Int. Mater. Rev., Vol. 39, pp. 1–23.

Markolefas SI, Tsouvalas DA, Tsamasphyros GI (2007): Theoretical analysis of a class of mixed, C^0 continuity formulations for general dipolar gradient elasticity boundary value problems, Int. J. Solids Struct., Vol. 44, pp. 546–572.

Mindlin RD (1964): Micro-structure in linear elasticity, Arch. Rat. Mech. Anal., Vol. 16, pp. 51–78.

Mindlin RD (1965): Second gradient of strain and surface-tension in linear elasticity, Int. J. Solids Struct., Vol. 1, pp. 417–438.

Mindlin RD, Eshel NN (1968): On first strain-gradient theories in linear elasticity, Int. J. Solids Struct.,Vol. 4, pp. 109–124.

Nan CW, Clarke DR (1996) The influence of particle size and particle fracture on the elastic/plastic deformation of metal matrix composites, Acta Mater., Vol. 44, pp. 3801–3811.

Peerlings RHJ, Fleck NA (2004): Computational evaluation of strain gradient elasticity constants, Int. J. Multiscale Comput. Eng., Vol. 2, No. 4, pp. 599–619.

Polyzos D (2005): 3D frequency domain BEM for solving dipolar gradient elastic problems, Comput. Mech., Vol. 35, pp. 292–304.

Polyzos D, Tsepoura KG, Beskos DE (2005): Transient dynamic analysis of 3-D gradient elastic solids by BEM, Comput. Struct., Vol. 83, pp. 783–792.

Polyzos D, Tsepoura KG, Tsinopoulos SV, Beskos DE (2003): A boundary element method for solving 2-D and 3-D static gradient elastic problems. Part I: Integral formulation, Comput. Meth. Appl. Mech. Eng., Vol. 192, pp. 2845–2873.

Ru CQ, Aifantis EC (1993): A simple approach to solve boundary value problems in gradient elasticity, Acta Mechanica, Vol. 101, pp. 59–68.

Shu JY, King WE, Fleck NA (1999): Finite elements for materials with strain gradient effects, Int. J. Numer. Meth. Eng. Vol. 44, pp. 373–391.

Smyshlyaev VP, Fleck NA (1996): The role of strain gradients in grain size effect for polycrystals, J. Mech. Phys. Solids, Vol. 44, pp. 465–495.

Soh AK, Wanji C (2004): Finite element formulations of strain gradient theory for microstructures and the C^{0-1} patch test, Int. J. Numer. Meth. Eng., Vol. 61, pp. 433–454.

Tang Z, Shen S, Atluri SN (2003): Analysis of materials with strain-gradient effects: A meshless local Petrov-Galerkin (MLPG) approach, with nodal displacements only, Comput. Model. Eng. Sci., Vol. 4, pp. 177–196.

Tenek LT, Aifantis EC (2002): A two-dimensional finite element implementation of a special form of gradient elasticity, Comput. Model. Eng., Sci., Vol. 3, pp. 731–741.

Tiersten HF, Bleustein JL (1974): Generalized elastic continua. In: G. Hermann (Ed), *Mindlin RD and Applied Mechanics*. Pergamon Press, New York, pp. 67–103.

Tsepoura KG, Papargyri-Beskou S, Polyzos D (2002): A boundary element method for solving 3-D static gradient elastic problems with surface energy, Comput. Mech., Vol. 29, pp. 361–381.

Tsepoura KG, Polyzos D (2003): Static and harmonic BEM solutions of gradient elasticity problems with axisymmetry, Comput. Mech., Vol. 32, pp. 89–103.

Tsepoura KG, Tsinopoulos SV, Polyzos D, Beskos DE (2003): A boundary element method for solving 2-D and 3-D static gradient elastic problems. Part II: Numerical implementation, Comput. Meth. Appl. Mech. Eng., Vol. 192, pp. 2875–2907.

Van Vliet MRA, Van Mier JGM (1999): Effect of strain gradients on the size effect of concrete in uniaxial tension, Int. J. Fracture, Vol. 95, pp. 195–219.

Vardoulakis I, Exadaktylos G, Aifantis E (1996): Gradient elasticity with surface energy: Mode-III crack problem, Int. J. Solids Struct., Vol. 33, pp. 4531–4559.

Vardoulakis I, Sulem J (1995): *Bifurcation Analysis in Geomechanics*. Blackie/Chapman and Hall, London.

The Fractional Diffusion-Wave Equation in Bounded Inhomogeneous Anisotropic Media. An AEM Solution

John T. Katsikadelis

Abstract In this chapter a numerical solution for the general linear fractional diffusion-wave equation in bounded inhomogeneous anisotropic bodies is developed. The response of the body is described by the complete second order partial differential equation with respect to the spatial derivative, while the time derivatives is of fractional order varying between zero and two. The employed solution method is based on the principle of the analog equation of Katsikadelis. Using this principle the fractional partial differential equation is converted into a system of ordinary fractional differential equations. The latter are solved using a numerical method, which has been recently developed by Katsikadelis for the solution of systems of multi-term fractional differential equations. Example problems are studied which illustrate the method and demonstrate its efficiency and accuracy, while they reveal the physical response of systems described by the fractional diffusion-wave equation.

1 Introduction

The fractional diffusion-wave equation is obtained from the classical diffusion or wave equation by replacing the first- or the second order time derivative term by a fractional derivative of order between zero and two. It has been shown that many of the universal electromagnetic, acoustic and mechanical responses can be modeled accurately using fractional diffusion-wave equations. Relaxation phenomena in complex viscoelastic materials are described by the fractional diffusion equation (Ginoa et al. 1992). Propagation of stress waves in viscoelastic solids are successfully described by the fractional wave equation (Mainardi 1997; Mainardi and Paradisi, 1997). More information on the physical phenomena that can be described by the fractional diffusion-wave equation and in general on the fractional

J.T. Katsikadelis (✉)
Office of theoretical and Applied Mechanics, Academy of Athens & School of Civil Engineering, National Technical University, Athens, GR-15773, Greece
e-mail: jkats@central.ntua.gr

A chapter in honor of Dimitri Beskos' 65th birthday

G.D. Manolis, D. Polyzos (eds.), *Recent Advances in Boundary Element Methods*,
DOI 10.1007/978-1-4020-9710-2_17,

differential equations can be found in the relevant literature, e.g. Agrawal (2002), Mainardi (1997), Oldham and Spanier (1974), Miller and Ross (1993), Podlubny (1999), Gorenflo and Mainardi (1997) and references cited there. Due to the mathematical complexity, the developed analytical solutions are very few and are restricted to the solution of simple FDEs. Therefore the development of effective and easy-to-use numerical schemes for solving such equations has acquired an increasing interest in the recent years. Regarding the initial boundary value problems for partial differential equations with fractional time derivatives, the analytical solutions are limited to the time diffusion-wave equation in one dimensional bounded domains, (e.g. Atanackovic, 2007). These solutions are given by infinite series, whose evaluation exhibits computational complexity. On the other hand numerical solutions of partial FDEs in two-dimensional domains of arbitrary shape, to the author's knowledge, have not been reported as yet.

In this paper a numerical solution for the diffusion-wave equation in inhomogeneous anisotropic two dimensional bodies is presented. The governing equation is the complete second order PDE with respect to the spatial derivatives, while the time derivatives are of fractional order greater than zero and not greater than two. The numerical solution was achieved using: (a) the AEM (Katsikadelis 1994, 2002a, 2005) to convert the partial FDE into a system of ordinary FDEs with relatively small number of degrees of freedom and (b) an efficient numerical method for solving systems of multi-term ordinary FDEs (Katsikadelis, 2008a). Example problems are presented, which illustrate the method and demonstrate its efficiency and accuracy, while they reveal the physical response of systems described by the fractional diffusion-wave equation.

2 The Initial Boundary Value Problem

The fractional diffusion-wave equation for an inhomogeneous anisotropic plane body occupying domain Ω bounded by the curve Γ (Fig. 1) reads

$$\rho \frac{\partial^\beta u}{\partial t^\beta} + \eta \frac{\partial^\alpha u}{\partial t^\alpha} = A \frac{\partial^2 u}{\partial x^2} + 2B \frac{\partial^2 u}{\partial x \partial y} + C \frac{\partial^2 u}{\partial y^2} + D \frac{\partial u}{\partial x} + E \frac{\partial u}{\partial y} + Fu + g(\mathbf{x}, t) \quad (1)$$

where $\mathbf{x}(x, y) \in \Omega$, $t > 0$, $0 < \alpha < \beta \leq 2$ and $\rho = \rho(\mathbf{x})$, $\eta = \eta(\mathbf{x})$ and $g(\mathbf{x}, t)$ are specified functions, whose physical meaning depends on that of the field function $u(\mathbf{x}, t)$; $\partial^\beta u/\partial t^\beta$ and $\partial^\alpha u/\partial t^\alpha$ are the fractional derivatives of order β and a. Herein, the Caputo fractional derivative is employed which is designated by D_c^α and is defined as (Podlubny, 1999)

$$D_c^\alpha u(t) = \frac{\partial^\alpha u}{\partial t^\alpha} = \begin{cases} \left[\frac{1}{\Gamma(m-\alpha)} \int_0^t \frac{d^m u(\tau)/dt^m}{(t-\tau)^{\alpha+1-m}} d\tau \right], & m-1 < \alpha < m, \\ d^m u(t)/dt^m & m = \alpha \end{cases} \quad (2)$$

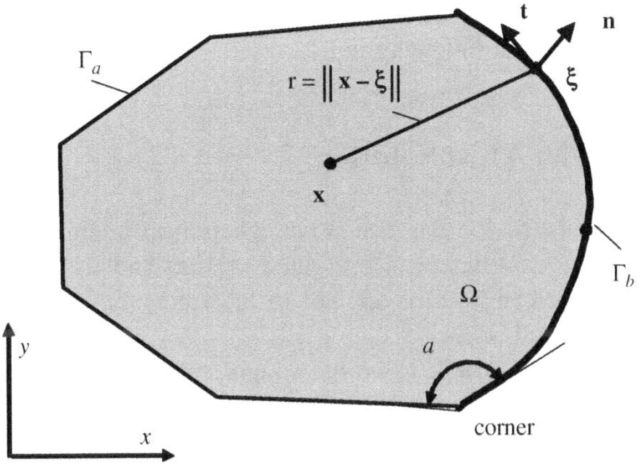

Fig. 1 Domain Ω occupied by the body and notation; $\Gamma = \Gamma_a \cup \Gamma_b$

where m is a positive integer. The advantage of this definition is that it permits the assignment of initial conditions which have direct physical significance. Apparently, we shall be able to recover: the classical diffusion equation for $\rho = 0, \alpha = 1$ and the classical wave equation in presence of damping for $\beta = 2, \alpha = 1$.

The position dependent coefficients $A = A(\mathbf{x}), B = B(\mathbf{x}), \ldots, F(\mathbf{x})$ satisfy the relations

$$A_{,x} + B_{,y} = D, \;\; B_{,x} + C_{,y} = E \tag{3}$$

Equation (1) is subjected to the boundary conditions

$$u = \alpha(\mathbf{x}, t), \;\; \mathbf{x} \in \Gamma_a, \quad k(\mathbf{x})u + \nabla u \cdot \mathbf{m} = \gamma(\mathbf{x}, t), \;\; \mathbf{x} \in \Gamma_b \tag{4a,b}$$

and to the initial conditions in Ω

$$u(\mathbf{x}, 0) = f_1(\mathbf{x}) \quad \text{if} \quad \beta \leq 1 \quad \text{or} \tag{5a}$$

$$u(\mathbf{x}, 0) = f_1(\mathbf{x}), \quad \dot{u}(\mathbf{x}, 0) = f_2(\mathbf{x}) \quad \text{if} \quad \beta > 1 \tag{5b}$$

where $\Gamma = \Gamma_a \cup \Gamma_b$ and $\mathbf{m} = (An_x + Bn_y)\mathbf{i} + (Bn_x + Cn_y)\mathbf{j}$ is a vector in the direction of the connormal on the boundary; $\mathbf{n}$ and $\mathbf{t}$ are the normal and the tangential unit vectors on the boundary. Finally, $\alpha(\mathbf{x}, t), \gamma(\mathbf{x}, t)$ are functions specified on Γ_a and Γ_b, respectively.

Using the identity $\nabla u \cdot \mathbf{m} = (\mathbf{m} \cdot \mathbf{n})u_{,n} + (\mathbf{m} \cdot \mathbf{n})u_{,t}$, the boundary conditions (4a,b) can be combined and written as (Katsikadelis, 2007)

$$\beta_1 u + \beta_2 q = \beta_3 \quad \text{on} \quad \Gamma \tag{6}$$

where $q = u_{,n}$.

The IBV problem (1), (4), (5) is solved using the AEM (Analog Equation Method) as following.

3 The AEM Solution

The AEM for classical wave equation in inhomogeneous bodies (free and forced vibrations) has been presented in Katsikadelis and Nerantzaki (2000). This procedure can be also applied to the fractional diffusion equation as well and it is described below.

Let $u = u(\mathbf{x}, t)$ be the sought solution to Eq. (1). If the Laplace operator is applied to it, we have

$$\nabla^2 u = b(\mathbf{x}, t) \tag{7}$$

where $b(\mathbf{x}, t)$ represents an unknown time dependent fictitious source. Equation (7) indicates that the solution of Eq. (1) could be established by solving the latter equation under the boundary conditions (4) and the initial conditions (5) provided that source $b(\mathbf{x}, t)$ is first established. For this purpose we write the solution of Eq. (7) in integral form (Katsikadelis, 2002b).

$$\varepsilon u(\mathbf{x}, t) = \int_\Omega u^* b d\Omega - \int_\Gamma (u^* q - q^* u) ds \quad \mathbf{x} \in \Omega \cup \Gamma \tag{8}$$

in which $q = u,_n$; $u^* = \ell nr/2\pi$ is the fundamental solution of Eq. (7) and $q^* = u,_n^*$ its derivative normal to the boundary with $r = \|\xi - \mathbf{x}\|$, $\mathbf{x} \in \Omega \cup \Gamma$ and $\xi \in \Gamma$; ε is the free term coefficient ($\varepsilon = 1$ if $\mathbf{x} \in \Omega$, $\varepsilon = a/2\pi$ if$\mathbf{x} \in \Gamma$ and $\varepsilon = 0$ if $\mathbf{x} \notin \Omega \cup \Gamma$; a is the interior angle between the tangents of boundary at point $\mathbf{x}$). Note that it is $\varepsilon = 1/2$ for points where the boundary is smooth.

The domain integral in Eq. (8) is converted to a boundary line integral by setting

$$b = \sum_{j=1}^{M} \alpha_j(t) f_j \tag{9}$$

where $f_j = f_j(r)$, $r = \|\mathbf{x} - \mathbf{x}_j\|$, is a set of radial basis approximating functions; $\mathbf{x}_j$ are collocation points in Ω and $a_j(t)$ time dependent coefficients to be determined. Using the Green's reciprocal identity (Katsikadelis, 2002b) we have

$$\int_\Omega u^* b d\Omega = \sum_{j=1}^{M} a_j \int_\Omega u^* f_j d\Omega$$

$$= \sum_{j=1}^{M} a_j \left\{ \varepsilon \hat{u}_j + \int_\Gamma \left[u^* \hat{q}_j - q^* \hat{u}_j \right] ds \right\} \tag{10}$$

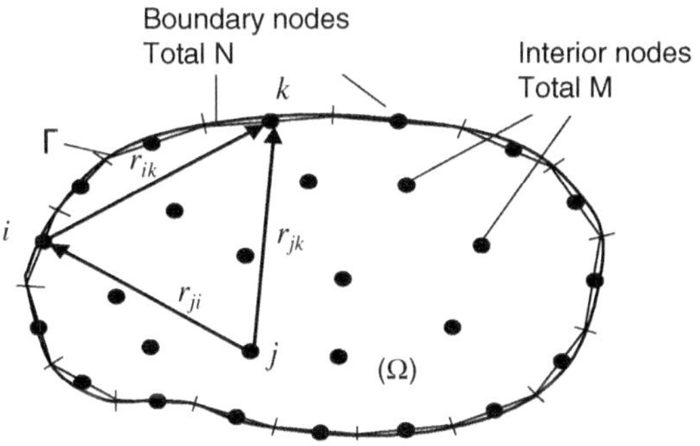

Fig. 2 Boundary discretization and domain nodal points

in which $\hat{u}_j(\mathbf{x})$ is a particular solution of the equation

$$\nabla^2 \hat{u}_j = f_j \quad j = 1, 2, \ldots, M \tag{11}$$

The AEM is now implemented numerically. To this end, the BEM with constant elements is used to approximate the boundary integrals in Eqs. (8) and (10). In this case it is $a = \pi$, hence $\varepsilon = 1/2$ on the elements. If N is the number of the boundary nodal points (see Fig. 2), we obtain (Katsikadelis 2005)

$$\mathbf{Hu} - \mathbf{Gq} + \mathbf{Ka} = \mathbf{0} \tag{12}$$

where $\mathbf{a}$ is the vector of the M coefficients $a_j(t)$; $\mathbf{u}$, $\mathbf{q}$ are the vectors of the N boundary nodal values of u and q, respectively, and

$$K_{ij} = \frac{1}{2}\hat{u}_j^i - \sum_{k=1}^{N} \tilde{H}_{ik}\hat{u}_j^k + \sum_{k=1}^{N} G_{ik}\hat{q}_j^k \tag{13a}$$

$$G_{ik} = \int_k u^*(r_{ik})ds \tag{13b}$$

$$H_{ik} = \int_k q^*(r_{ik})ds \tag{13c}$$

$$H_{ik} = \int_k q^*(r_{ik})ds - \delta_{ik}, \tag{13d}$$

in which $\int_k$ indicates integration on the k element and $\hat{u}_j^i = \hat{u}(r_{ji})$, $\hat{u}_j^k = \hat{u}(r_{jk})$. Moreover, the boundary condition (6), when applied to the N boundary nodal points, yields

$$\beta_1 \mathbf{u} + \beta_2 \mathbf{q} = \beta_3 \tag{14}$$

where β_1, β_2 are known $N \times N$ diagonal matrices and β_3 a vector including the values of β_i, $(i = 1, 2, 3)$ at the N boundary nodal points. For points $\mathbf{x}$ inside Ω, the discretised counterpart of Eq. (8) becomes

$$u(\mathbf{x}, t) = \sum_{j=1}^{M} K_{\mathbf{x}j} a^j(t) + \sum_{k=1}^{N} \tilde{H}_{\mathbf{x}k} u^k - \sum_{k=1}^{N} G_{\mathbf{x}k} q^k \tag{15}$$

Equations (12) and (14) can be used to eliminate the boundary quantities $\mathbf{u}, \mathbf{q}$ from Eq. (15). Thus we obtain

$$u(\mathbf{x}, t) = \sum_{j=1}^{M} a_j(t) S_j(\mathbf{x}) + s(\mathbf{x}, t) \tag{16}$$

For homogeneous boundary conditions it is $s(\mathbf{x}, t) = 0$.

The spatial derivatives of the solution inside Ω are obtained by direct differentiation of Eq. (8). Thus we have

$$u,_{mn}(\mathbf{x}, t) = \int_{\Omega} u,_{mn}^{*} \, b d\Omega - \int_{\Gamma} (u,_{mn}^{*} q - q,_{mn}^{*} u) ds, \quad m, n = 0, x, y \tag{17}$$

Note that notation implies $u,_{00} = u$, $u,_{x0} = u,_x$, etc. Using the same discretization we obtain after elimination of the boundary quantities

$$u(\mathbf{x}, t),_{mn} = \sum_{j=1}^{M} a_j(t) S_j(\mathbf{x}),_{mn} + s(\mathbf{x}, t),_{mn} \quad m, n = 0, x, y \tag{18}$$

Applying Eqs. (16) and (18) at the M collocation points inside Ω yields

$$\mathbf{u},_{mn} = \mathbf{S},_{mn} \mathbf{a} + \mathbf{s},_{mn}, \quad m, n = 0, x, y \tag{19}$$

where $\mathbf{S},_{mn}$, $\mathbf{s},_{mn}$ are known matrices and vectors, respectively; $\mathbf{u},_{mn}$ are the vectors of the nodal values of the solution and its derivatives and $\mathbf{a}$ is the vector of the unknown time dependent expansion coefficients.

Finally, Eq. (1) is applied at the M collocation points and gives

$$\rho D_c^{\beta} \mathbf{u} + \eta \, D_c^{\alpha} \mathbf{u} = \mathbf{A}\mathbf{u},_{xx} + 2\mathbf{B}\mathbf{u},_{xy} + \mathbf{C}\mathbf{u},_{yy} + \mathbf{D}\mathbf{u},_x + \mathbf{E}\mathbf{u},_y + \mathbf{F}\mathbf{u} + \mathbf{g} \tag{20}$$

where ρ, η, $\mathbf{A}, \mathbf{B}, \mathbf{C}, \mathbf{D}, \mathbf{E}, \mathbf{F}$ are known diagonal $M \times M$ matrices and $\mathbf{g}$ a known vector.

Substituting $\mathbf{u},_{mn}$ from Eq. (19) into Eq. (20) we obtain

$$\mathbf{M} D_c^{\beta} \mathbf{a} + \mathbf{H} \, D_c^{\alpha} \mathbf{a} + \mathbf{K}\mathbf{a} = \mathbf{p} \tag{21}$$

where the matrices in the above equations are defined as

$$\mathbf{M} = \rho\mathbf{S} \tag{22a}$$

$$\mathbf{H} = \eta\mathbf{S} \tag{22b}$$

$$\mathbf{K} = -(\mathbf{AS},_{xx} + 2\mathbf{BS},_{xy} + \mathbf{CS},_{yy} + \mathbf{DS},_{x} + \mathbf{ES},_{y} + \mathbf{FS}) \tag{22c}$$

$$\mathbf{p} = -(\rho D_c^{\beta}\mathbf{s} + \eta\, D_c^{\alpha}\mathbf{s}) + (\mathbf{As},_{xx} + 2\mathbf{Bs},_{xy} + \mathbf{Cs},_{yy} + \mathbf{Ds},_{x} + \mathbf{Es},_{y} + \mathbf{Fs}) + \mathbf{g} \tag{22d}$$

Note that for time independent boundary conditions (no support excitations) we have

$$\mathbf{p} = (\mathbf{As},_{xx} + 2\mathbf{Bs},_{xy} + \mathbf{Cs},_{yy} + \mathbf{Ds},_{x} + \mathbf{Es},_{y} + \mathbf{Fs}) + \mathbf{g} \tag{23}$$

while for homogeneous boundary conditions

$$\mathbf{p} = \mathbf{g} \tag{24}$$

Equation (21) represents the set of semi-discretised ordinary fractional diffusion-wave equations which are solved using the numerical solution developed by Katsikadelis (2008). This solution procedure is described in the following section.

4 Solution of the Semi-Discretised Equations

4.1 The Three-Term Fractional Differential Equation

We consider the system of the M three-term fractional ODEs

$$\mathbf{a}_1 D_c^{\beta}\mathbf{u} + \mathbf{a}_2 D_c^{\alpha}\mathbf{u} + \mathbf{a}_3\mathbf{u} = \mathbf{p}(t) \tag{25}$$

$$\text{with } 0 < \alpha < \beta \leq 2,\ t > 0,\ \mathbf{a}_i \in \mathbb{R},\ \det(\mathbf{a}_1) \neq 0$$

under the initial conditions

$$\mathbf{u}(0) = \mathbf{u}_0, \quad \text{if} \quad \beta \leq 1 \tag{26a}$$

or

$$\mathbf{u}(0) = \mathbf{u}_0,\ \dot{\mathbf{u}}(0) = \dot{\mathbf{u}}_0, \quad \text{if} \quad 1 < \beta \leq 2 \tag{26b}$$

Let $\mathbf{u} = \mathbf{u}(t)$ be the sought solution of Eq. (25). Then, if the operator D_c^β is applied to $\mathbf{u}$ we have

$$D_c^\beta \mathbf{u} = \mathbf{q}(t), \quad 0 < \beta \leq 2, \quad t > 0 \tag{27}$$

where $\mathbf{q}(t)$ is a vector of unknown fictitious sources. Equation (27) is the analog equation of (25). It indicates that the solution of Eq. (25) can be obtained by solving Eq. (27) with the initial conditions (26), if the $\mathbf{q}(t)$ is first established. This is achieved by working as following.

Using the Laplace transform method we obtain the solution of Eq. (27) as

$$\mathbf{u}(t) = \mathbf{u}_0 + [\text{ceil}(\beta) - 1]\dot{\mathbf{u}}_0 t + \frac{1}{\Gamma(\beta)}\int_0^t \mathbf{q}(\tau)(t-\tau)^{\beta-1}d\tau \tag{28}$$

where ceil( ) represents the *ceiling* function, e.g. ceil(β) yields the integer greater or equal to β. The use of this function permits to realize computationally the proper initial conditions prescribed by Eqs. (26).

Equation (28) is an integral equation for $\mathbf{q}(t)$, which can be solved numerically within a time interval $[0, T]$ as following. The interval $[0, T]$ is divided into N equal intervals $\Delta t = h$, $h = T/N$ (Fig. 3), in which $\mathbf{q}(t)$ is assumed to vary according to a certain law, e.g. constant, linear etc. In this analysis $\mathbf{q}(t)$ is assumed to be constant and equal to the mean value in each interval h. Hence, Eq. (28) at instant $t = nh$ can be written as

$$\begin{aligned}\mathbf{u}_n &= \mathbf{u}_0 + [\text{ceil}(\beta) - 1]nh\dot{\mathbf{u}}_0 + \tfrac{1}{\Gamma(\beta)}\Big[\mathbf{q}_1^m \textstyle\int_0^h (nh-\tau)^{\beta-1}d\tau \\ &+\mathbf{q}_2^m \textstyle\int_h^{2h} (nh-\tau)^{\beta-1}d\tau + \cdots + \mathbf{q}_n^m \int_{(n-1)h}^{nh} (nh-\tau)^{\beta-1}d\tau\Big]\end{aligned} \tag{29}$$

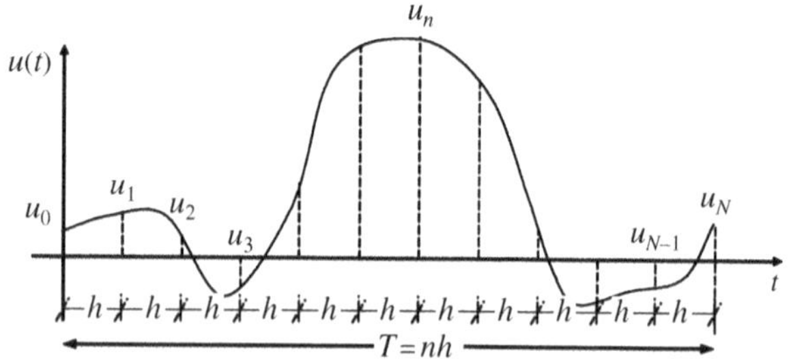

Fig. 3 Discretization of the interval $[0, T]$ into N equal intervals $h = T/N$

which after evaluation of the integrals yields

$$\mathbf{u}_n = \mathbf{u}_0 + [\text{ceil}(\beta) - 1]nh\dot{\mathbf{u}}_0 + c\sum_{r=1}^{n-1}\left[(n+1-r)^\beta - (n-r)^\beta\right]\mathbf{q}_r^m + \frac{c}{2}(\mathbf{q}_{n-1} + \mathbf{q}_n) \tag{30}$$

where

$$c = \frac{h^\beta}{\beta\Gamma(\beta)}, \quad \mathbf{q}_r^m = \frac{1}{2}(\mathbf{q}_{r-1} + \mathbf{q}_r) \tag{31}$$

Equation (30) can be further written as

$$-\frac{c}{2}\mathbf{q}_n + \mathbf{u}_n = \mathbf{u}_0 + [\text{ceil}(\beta) - 1]nh\dot{\mathbf{u}}_0 + c\sum_{r=1}^{n-1}\left[(n+1-r)^\beta - (n-r)^\beta\right]\mathbf{q}_r^m + \frac{c}{2}\mathbf{q}_{n-1} \tag{32}$$

We now set

$$D_c^\alpha \mathbf{u} = \bar{\mathbf{q}}(t) \tag{33}$$

where $\bar{\mathbf{q}}(t)$ is another unknown vector. We can establish a relation between $\mathbf{q}(t)$ and $\bar{\mathbf{q}}(t)$ by considering the Laplace transform of Eqs. (27) and (33). Thus, we can write

$$\mathbf{U}(s) = \mathbf{u}_0\frac{1}{s} + [\text{ceil}(\beta) - 1]\dot{\mathbf{u}}_0\frac{1}{s^2} + \frac{1}{s^\beta}\mathbf{Q}(s) \tag{34a}$$

$$\mathbf{U}(s) = \mathbf{u}_0\frac{1}{s} + [\text{ceil}(\alpha) - 1]\dot{\mathbf{u}}_0\frac{1}{s^2} + \frac{1}{s^\alpha}\bar{\mathbf{Q}}(s) \tag{34b}$$

Equating the right-hand sides of the above equations we have

$$\bar{\mathbf{Q}}(s) = ([\text{ceil}(\beta) - \text{ceil}(\alpha)])\,\dot{\mathbf{u}}_0\frac{1}{s^{2-\alpha}} + \frac{1}{s^{\beta-\alpha}}\mathbf{Q}(s), \quad \alpha \neq \beta \tag{35}$$

Taking the inverse Laplace transform of Eq. (35) we obtain

$$\bar{\mathbf{q}} = ([\text{ceil}(\beta) - \text{ceil}(\alpha)])\,\dot{\mathbf{u}}_0\frac{t^{1-\alpha}}{\Gamma(2-\alpha)} + \frac{1}{\Gamma(\beta-\alpha)}\int_0^t \mathbf{q}(\tau)(t-\tau)^{\beta-\alpha-1}d\tau \tag{36}$$

Using the same disctetization of the interval $[0, T]$ to approximate the integral in Eq. (36), we obtain

$$\bar{\mathbf{q}}_n = ([\mathrm{ceil}(\beta) - \mathrm{ceil}(\alpha)])\, n^{1-\alpha} d\dot{\mathbf{u}}_0 + \bar{c} \sum_{r=1}^{n} \left[(n+1-r)^{\beta-\alpha} - (n-r)^{\beta-\alpha}\right] \mathbf{q}_r^m \tag{37}$$

where

$$\bar{c} = \frac{h^{\beta-\alpha}}{(\beta-\alpha)\Gamma(\beta-\alpha)}, \quad d = \frac{h^{1-\alpha}}{\Gamma(2-\alpha)}, \quad \mathbf{q}_r^m = \frac{1}{2}(\mathbf{q}_{r-1} + \mathbf{q}_r) \tag{38}$$

Equation (37) can be further written as

$$-\frac{\bar{c}}{2}\bar{\mathbf{q}}_n + \mathbf{q}_n = [\mathrm{ceil}(\beta) - \mathrm{ceil}(\alpha)] n d\dot{\mathbf{u}}_0 + \bar{c} \sum_{r=1}^{n-1} \left[(n+1-r)^{\beta-\alpha} - (n-r)^{\beta-\alpha}\right] \mathbf{q}_r^m + \frac{\bar{c}}{2}\mathbf{q}_{n-1} \tag{39}$$

Equations (25), (32) and (39) can be combined as

$$\mathbf{A}\mathbf{x} = \mathbf{b} \tag{40}$$

where

$$\mathbf{A} = \begin{bmatrix} \mathbf{a}_1 & \mathbf{a}_2 & \mathbf{a}_3 \\ -\frac{c}{2}\mathbf{I}_{M\times M} & 0 & \mathbf{I}_{M\times M} \\ -\frac{\bar{c}}{2}\mathbf{I}_{M\times M} & \mathbf{I}_{M\times M} & 0 \end{bmatrix}, \mathbf{x} = \begin{Bmatrix} \mathbf{q}_n \\ \bar{\mathbf{q}}_n \\ \mathbf{u}_n \end{Bmatrix} \tag{41a,b}$$

$$\mathbf{b} = \begin{Bmatrix} \mathbf{p}_n \\ \mathbf{u}_0 + [\mathrm{ceil}(\beta) - 1] nh\dot{\mathbf{u}}_0 + c\sum_{r=1}^{n-1}\left[(n+1-r)^{\beta} - (n-r)^{\beta}\right]\mathbf{q}_r^m + \frac{c}{2}\mathbf{q}_{n-1} \\ [\mathrm{ceil}(\beta) - \mathrm{ceil}(\alpha)] nd\dot{\mathbf{u}}_0 + \bar{c}\sum_{r=1}^{n-1}\left[(n+1-r)^{\beta-\alpha} - (n-r)^{\beta-\alpha}\right]\mathbf{q}_r^m + \frac{\bar{c}}{2}\mathbf{q}_{n-1} \end{Bmatrix} \tag{41c}$$

Since $\det(\mathbf{a}_1) \neq 0$, the coefficient matrix $\mathbf{A}$ in Eq. (40) is not singular and it can be solved successively for $n = 1, 2, \ldots$ to yield the solution $\mathbf{u}_n$ and the fractional derivatives $\mathbf{q}_n, \bar{\mathbf{q}}_n$ at instant $t = nh \leq T$. For $n = 1$, the value $\mathbf{q}_0$ appears in the right hand side of Eq. (40). This value can be expressed in terms of $\bar{\mathbf{q}}_0$ as following.

Equation (25) for $t = 0$ gives

$$\mathbf{a}_1\mathbf{q}_0 + \mathbf{a}_2\bar{\mathbf{q}}_0 + \mathbf{a}_3\mathbf{u}_0 = \mathbf{p}_0 \tag{42}$$

The above equation includes two unknowns, $\mathbf{q}_0, \bar{\mathbf{q}}_0$. These values can be evaluated using the relations derived in Appendix.

$$\left.\begin{aligned} \bar{\mathbf{q}}_0 &= \left[\mathbf{a}_1 \frac{\Gamma(2-\alpha)\left(2-2^{1-\beta}\right)}{\Gamma(2-\beta)\left(2-2^{1-\alpha}\right)} h^{\alpha-\beta} + \mathbf{a}_2\right]^{-1} (\mathbf{p}_0 - \mathbf{a}_3\mathbf{u}_0) \\ \mathbf{q}_0 &\cong \frac{\Gamma(2-\alpha)}{\Gamma(2-\beta)} \frac{\left(2-2^{1-\beta}\right)}{\left(2-2^{1-\alpha}\right)} h^{\alpha-\beta} \bar{\mathbf{q}}_0 \end{aligned}\right\} \quad \text{if} \quad 0 < \alpha < \beta \le 1 \tag{43a}$$

$$\left.\begin{aligned} \bar{\mathbf{q}}_0 &= \left[\mathbf{a}_1 \frac{\Gamma(3-\alpha)\left(2-2^{2-\beta}\right)}{\Gamma(3-\beta)\left(2-2^{2-\alpha}\right)} h^{\alpha-\beta} + \mathbf{a}_2\right]^{-1} (\mathbf{p}_0 - \mathbf{a}_3\mathbf{u}_0) \\ \mathbf{q}_0 &\cong \frac{\Gamma(3-\alpha)}{\Gamma(3-\beta)} \frac{\left(2-2^{2-\beta}\right)}{\left(2-2^{2-\alpha}\right)} h^{\alpha-\beta} \bar{\mathbf{q}}_0 \end{aligned}\right\} \quad \text{if} \quad 1 < \alpha < \beta \le 2 \tag{43b}$$

$$\left.\begin{aligned} \bar{\mathbf{q}}_0 &= \frac{1}{\Gamma(2-a)} h^{1-a} \left(2-2^{1-\alpha}\right) \dot{\mathbf{u}}_0 \\ \mathbf{q}_0 &= \mathbf{a}_1^{-1} (\mathbf{p}_0 - \mathbf{a}_3\mathbf{u}_0 - \mathbf{a}_2\bar{\mathbf{q}}_0) \end{aligned}\right\} \quad \text{if} \quad 0 < \alpha \le 1 \text{ and } 1 < \beta \le 2 \tag{43c}$$

4.2 The Solution Algorithm

On the basis of the above analysis we write the following algorithm

ALGORITHM (Three-term simultaneous equations $0 < \alpha < \beta \le 2$)

Input Variables

M = the number of the equations to be solved
$\mathbf{a}$ = the real-valued $M \times 3M$ array that contains the coefficient matrices $\mathbf{a}_1, \mathbf{a}_2, \mathbf{a}_3$ ($\det(\mathbf{a}_1) \neq 0$) of Eq. (25)
$\mathbf{p}(t)$ = the $M \times 1$ array of the real-valued functions that defines the right-hand side of Eq. (25)
α = the lower order of the differential equation ($0 < \alpha < 2$)
β = the higher order of the differential equation ($0 < \beta \neq \alpha \le 2$)
$\bar{\mathbf{u}}_0$ = an $M \times \text{ceil}(\beta)$ array of real numbers that contains the elements of the vectors $\mathbf{u}_0$, $\dot{\mathbf{u}}_0$ of initial values
T = the upper bound of the interval where the solution is to be approximated
N = the number of the time steps that the solution is to be approximated

Output Variables

sol = the $3M \times N$ array $\{\mathbf{q}\ \bar{\mathbf{q}}\ \mathbf{u}\}^T$ of real numbers that contains the arrays of the approximate values of the fractional derivatives of order β and α and the solution u

Internal Variables

h = the step size of the algorithm (a positive real number)
c = the constant defined in Eq. (31)
$\bar{c}$ = the constant defined in Eq. (38)
d = the constant defined in Eq. (38)
$\bar{\mathbf{q}}_0 = D_c^\alpha \mathbf{u}(0)$
$\mathbf{q}_0 = D_c^\beta u(0)$
$\mathbf{A}$ = the $3M \times 3M$ co-efficient matrix defined resulting from Eq. (41a)
$\mathbf{b}$ = the right-hand side $3M \times 1$ array defined resulting from Eq. (41c)

Procedure

$$h := T/N$$

$$c := \frac{h^\beta}{\beta\Gamma(\beta)}$$

$$\bar{c} := \frac{h^{\beta-\alpha}}{(\beta-\alpha)\Gamma(\beta-\alpha)}$$

$$d := \frac{h^{1-\alpha}}{\Gamma(2-\alpha)}$$

$$\left.\begin{array}{l} \bar{\mathbf{q}}_0 = \left[\mathbf{a}_1 \dfrac{\Gamma(2-\alpha)\left(2-2^{1-\beta}\right)}{\Gamma(2-\beta)\left(2-2^{1-\alpha}\right)} h^{\alpha-\beta} + \mathbf{a}_2\right]^{-1} (\mathbf{p}_0 - \mathbf{a}_3\mathbf{u}_0) \\ \mathbf{q}_0 \cong \dfrac{\Gamma(2-\alpha)}{\Gamma(2-\beta)} \dfrac{\left(2-2^{1-\beta}\right)}{\left(2-2^{1-\alpha}\right)} h^{\alpha-\beta} \bar{\mathbf{q}}_0 \end{array}\right\} \text{if } 0 < \alpha < \beta \leq 1$$

$$\left.\begin{array}{l} \bar{\mathbf{q}}_0 = \left[\mathbf{a}_1 \dfrac{\Gamma(3-\alpha)\left(2-2^{2-\beta}\right)}{\Gamma(3-\beta)\left(2-2^{2-\alpha}\right)} h^{\alpha-\beta} + \mathbf{a}_2\right]^{-1} (\mathbf{p}_0 - \mathbf{a}_3\mathbf{u}_0) \\ \mathbf{q}_0 \cong \dfrac{\Gamma(3-\alpha)}{\Gamma(3-\beta)} \dfrac{\left(2-2^{2-\beta}\right)}{\left(2-2^{2-\alpha}\right)} h^{\alpha-\beta} \bar{\mathbf{q}}_0 \end{array}\right\} \text{if } 1 < \alpha < \beta \leq 2$$

$$\left.\begin{array}{l} \bar{\mathbf{q}}_0 = \dfrac{1}{\Gamma(2-a)} h^{1-a} \left(2-2^{1-\alpha}\right) \dot{\mathbf{u}}_0 \\ \mathbf{q}_0 = \mathbf{a}_1^{-1}(\mathbf{p}_0 - \mathbf{a}_3\mathbf{u}_0 - \mathbf{a}_2\bar{\mathbf{q}}_0) \end{array}\right\} \text{if } 0 < \alpha \leq 1 \text{ and } 1 < \beta \leq 2$$

$$\mathbf{A} := \begin{bmatrix} \mathbf{a}_1 & \mathbf{a}_2 & \mathbf{a}_3 \\ -\dfrac{c}{2}\mathbf{I}_{M\times M} & 0 & \mathbf{I}_{M\times M} \\ -\dfrac{\bar{c}}{2}\mathbf{I}_{M\times M} & \mathbf{I}_{M\times M} & 0 \end{bmatrix}$$

for $n := 1$ to N

for $r := 1$ to $n-1$

$$\mathbf{q}_r^m := (\mathbf{q}_{r-1} + \mathbf{q}_r)/2$$

$$\mathbf{sum1}_{n-1} := \sum_{r=1}^{n-1} \left[(n+1-r)^{\beta} - (n-r)^{\beta} \right] \mathbf{q}_r^m$$

$$\mathbf{sum2}_{n-1} := \sum_{r=1}^{n-1} \left[(n+1-r)^{\beta-\alpha} - (n-r)^{\beta-a} \right] \mathbf{q}_r^m$$

end

$$\mathbf{b} := \left\{ \begin{array}{c} \mathbf{p}_n \\ \mathbf{u}_0 + [\mathrm{ceil}(\beta) - 1] nh\dot{\mathbf{u}}_0 + c\mathbf{sum1}_{n-1} + \frac{c}{2}\mathbf{q}_{n-1} \\ [\mathrm{ceil}(\beta) - \mathrm{ceil}(\alpha)] n^{1-\alpha} d\dot{\mathbf{u}}_0 + \bar{c}\mathbf{sum2}_{n-1} + \frac{\bar{c}}{2}\mathbf{q}_{n-1} \end{array} \right\}$$

$$\mathbf{sol} := \mathbf{A}^{-1}\mathbf{b}$$

end

5 Examples

5.1 Example 1. Rectangular Homogeneous Body

The diffusion-wave equation in the rectangular plane body ($-a \leq x \leq a,\ -b \leq y \leq b$) has been studied using the following data: $\rho = 5$, $A = C = 20$, $B = D = E = F = 0$, $a = 6$, $b = 4.25$; $u(\mathbf{x}, t) = 0$, $\mathbf{x} \in \Gamma$. The time history of the solution at certain points obtained with $N = 82$, $M = 99$, $h = 0.01$ is shown in Fig. 4 through Fig. 8 as compared with the exact solution. Moreover, for case(ii), the variation of the error $e = u(0,0,t) - u_{exact}(0,0,t)$ versus time for two values

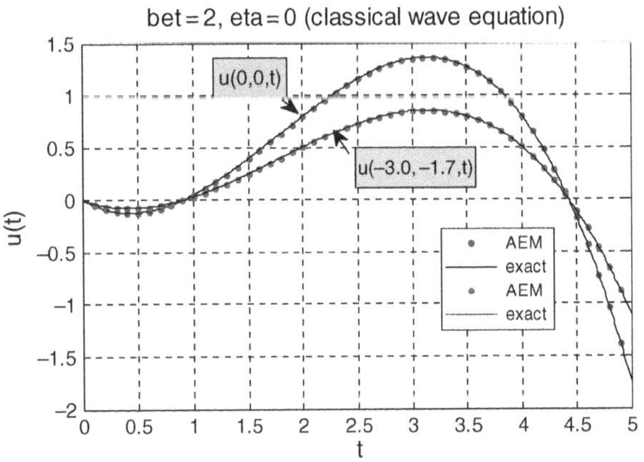

Fig. 4 Example 1(i)a; $u_{exact} = [(-.15t^3 + .8t^2 - .6t)(x^2 - a^2)(y^2 - b^2)]/a^2b^2$

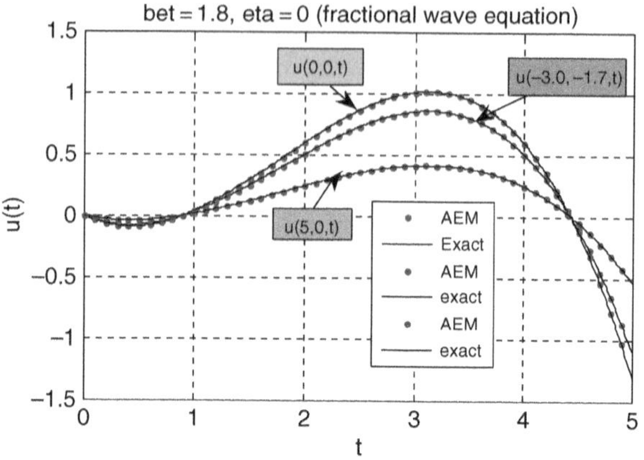

Fig. 5 Example 1(i)b; $u_{exact} = [(-.15t^3+.8t^2-.6t)(x^2-a^2)(y^2-b^2)]/a^2b^2$

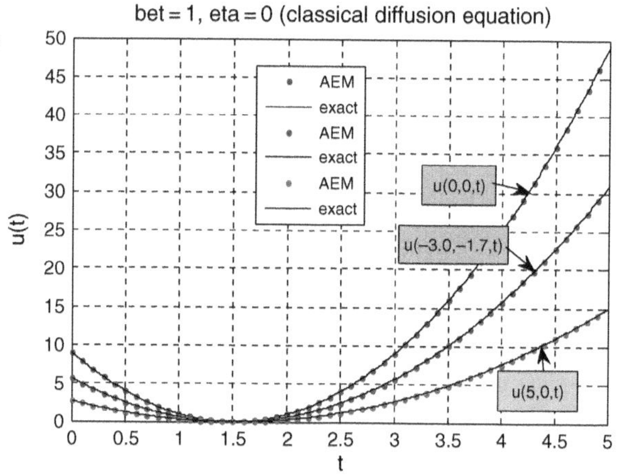

Fig. 6 Example 1(i)c; $u_{exact} = \left[(3-2t)^2(x^2-a^2) * (y^2-b^2)\right]/a^2b^2$

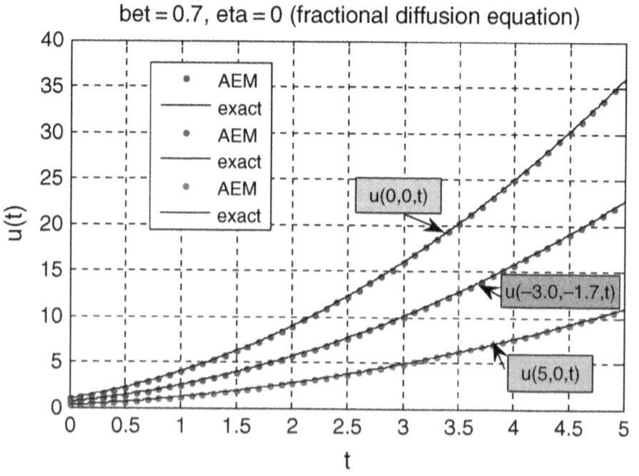

Fig. 7 Example 1(i)d; $u_{exact} = [(1+t)^2(x^2-a^2)(y^2-b^2)]/a^2b^2$

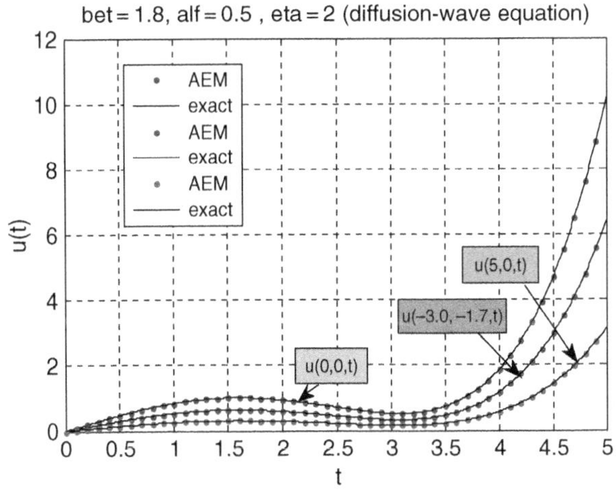

Fig. 8 Example 1(ii); $u_{exact} = [(t - t^3/6 + t^5/120)(x^2 - a^2)(y^2 - b^2)]/a^2b^2$

of the time step is presented in Fig. 9. These results demonstrate the accuracy of the method. The examined cases are:

(i) $\eta = 0$

(a) $\beta = 2$ (classical wave equation, no damping)

$$g = \big[(-4.5t + 8.)(x^2 - a^2)^*(y^2 - b^2)$$

$$-40[(y^2 - b^2) + (x^2 - a^2)](-.15t^3 + .8t^2 - .6t)]/a^2b^2\big]$$

$$\text{ICs } u(\mathbf{x}, 0) = 0, \ \dot{u}(\mathbf{x}, 0) = -0.6(x^2 - a^2)(y^2 - b^2)/a^2b^2$$

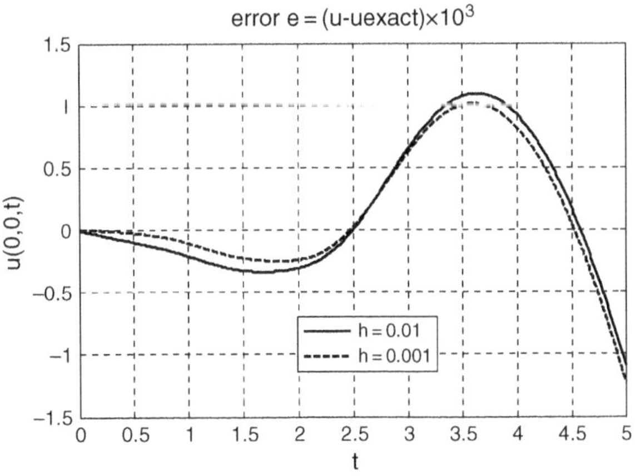

Fig. 9 Example 1(ii); error variation at the center of the domain

(b) $\beta = 1.8$ (fractional wave equation, no damping)

$$g = \left[(8.7130t^{1/5} - 4.0842t^{6/5})(x^2 - a^2)(y^2 - b^2)\right.$$
$$\left.-40[(y^2 - b^2) + (x^2 - a^2)](-0.15t^3 + 0.8t^2 - 0.6t)\right]$$
$$\text{ICs } u(\mathbf{x}, 0) = 0, \ \dot{u}(\mathbf{x}, 0) = -0.6(x^2 - a^2)(y^2 - b^2)/a^2b^2$$

(c) $\beta = 1$ (classical diffusion equation)

$$g = \left[20(-3 + 2t)(x^2 - a^2)(y^2 - b^2)\right.$$
$$\left.-40[(y^2 - b^2) + (x^2 - a^2)](3 - 2t)^2\right]/a^2b^2$$
$$\text{ICs } u(\mathbf{x}, 0) = 3(x^2 - a^2)(y^2 - b^2)/a^2b^2,$$
$$\dot{u}(\mathbf{x}, 0) = -12(x^2 - a^2)(y^2 - b^2)/a^2b^2$$

(d) $\beta = 0.7$ (fractional diffusion equation)

$$g = \left[(11.142t^{3/10} + 8.5711t^{13/10})(x^2 - a^2)(y^2 - b^2)\right.$$
$$\left.-40\left[(y^2 - b^2) + (x^2 - a^2)\right](1 + t)^2\right]/a^2b^2$$
$$\text{ICs } u(\mathbf{x}, 0) = (x^2 - a^2)(y^2 - b^2)/a^2b^2,$$
$$\dot{u}(\mathbf{x}, 0) = 2(x^2 - a^2)(y^2 - b^2)/a^2b^2$$

(ii) $\eta = 2, \beta = 1.8, \alpha = 0.5$

$$g = (-4.538t^{6/5} + 0.6446t^{16/5} + 2.2568t^{1/2} - .60180t^{5/2} + .03821t^{9/2})$$
$$(x^2 - a^2)(y^2 - b^2) \quad - 40[(y^2 - b^2) + (x^2 - a^2)] * (t - t^3/6 + t^5/120)/a^2b^2$$
$$\text{ICs } u(\mathbf{x}, 0) = 0, \ \dot{u}(\mathbf{x}, 0) = (x^2 - a^2)(y^2 - b^2)/a^2b^2$$

5.2 Example 2. Inhomogeneous Body

The diffusion-wave equation in the plane body shown in Fig. 10 has been studied. The boundary of the domain is defined by the curve $r = (ab)^{1/2}/[(\cos\theta/a)^2 + (\sin\theta/b)]^{(1/4)}[(\cos\theta/b)^2 + (\sin\theta/a)^2]^{(1/4)}$, $0 \leq \theta \leq 2\pi$ using the following data: $a = 3$, $b = 1.3$, $A = (y^2 - x^2 + 50)/50$, $B = 2xy/50$, $C = (x^2 - y^2 + 50)/50$, $D = E = 0$, $F = 0$; $\rho = 5\exp(-0.1\{|x| + |y|\})$, $\eta = 0.4(x^2 + y^2)^{1/2}$ $\beta = 1.7$, $\alpha = 0.8$:

Bcs: $u(\mathbf{x}, t) = 0, \ \mathbf{x} \in \Gamma$
ICs: $u(\mathbf{x}, 0) = 0, \dot{u}(\mathbf{x}, 0) = U(x, y)$

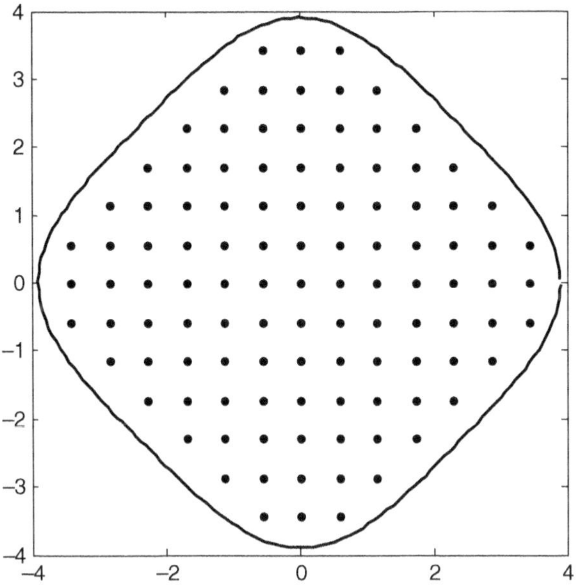

Fig. 10 Geometry of the plane body with distribution of the nodal points in Example 2

The external source: $g = \rho D_c^{1.7}T + \eta D_c^{0.8}T - (AU_{xx} + 2BU_{xy} + CU_{yy})$.

where

$$U(x, y) = a^2b^2 - \left[(x/a)^2 + (y/b)^2\right]\left[(x/b)^2 + (y/a)^2\right]$$

$$U_{xx} = -.78895x^2 - .72495y^2,\ U_{xy} = 0,\ U_{yy} = -.72495x^2 - .78895y^2$$

$$T(t) = t - t^3/6 + t^5/200,\ D_c^{1.7} = -4.2855t^{13/10} + .33878t^{33/10}$$

$$D_c^{0.8} = 5.4456t^{1/5} - 2.0627t^{11/5} + .92086e - t^{21/5}$$

The problem admits an exact solution $u_{exact} = T(t)U(x, y)$. The time history of the solution at certain points obtained with $N = 210$, $M = 109$, $h = 0.002$ is shown in Fig. 11 as compared with the exact solution.

5.3 Example 3. Fractional Oscillations of Inhomogeneous Membrane

The fractional oscillations of the inhomogeneous membrane of arbitrary shape, shown in Fig. 12, have been studied. Its boundary is defined by the curve $r = (5 + \sin\theta)(1.2\sin^4\theta + \cos^2\theta)$, $0 \le \theta \le 2\pi$. The membrane is under prestress due to imposed displacements $u_n = 0.005m$ along the boundary and in the direction normal to it, while in the tangential direction is $u_t = 0$. The response of the

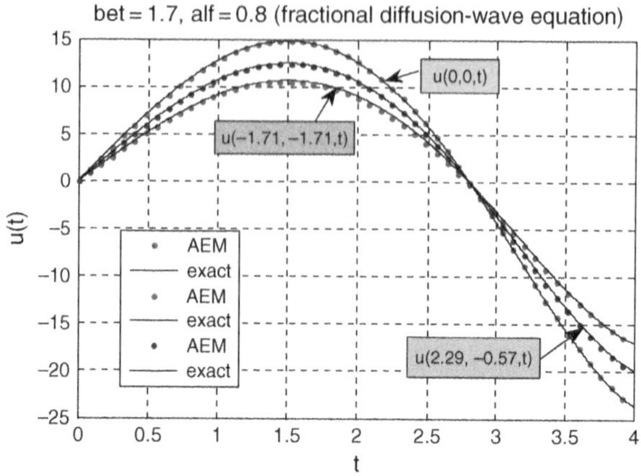

Fig. 11 Example 2; $u_{exact} = (t - t^3/6 + t^5/200)\{a^2b^2 - [(x/a)^2 + (y/b)^2][(x/b)^2 + (y/a)^2]\}$

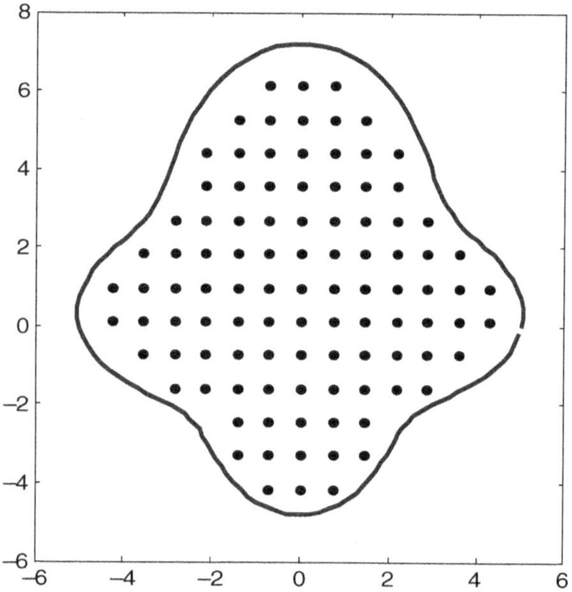

Fig. 12 Geometry of the membrane with distribution of the nodal points in Example 3

membrane is governed by the initial boundary value problem (Katsikadelis, 2008b)

$$\rho \frac{\partial^\beta w}{\partial t^\beta} = N_x \frac{\partial^2 w}{\partial x^2} + 2N_{xy} \frac{\partial^2 w}{\partial x \partial y} + N_y \frac{\partial^2 w}{\partial y^2} + g(\mathbf{x}, t), \quad 1 < \beta \leq 2 \tag{44a}$$

$$w(\mathbf{x}, t) = 0, \quad \mathbf{x} \in \Gamma \tag{44b}$$

$$w(\mathbf{x}, 0) = \tilde{w}(\mathbf{x}) \quad \dot{w}(\mathbf{x}, 0) = \dot{\tilde{w}}(\mathbf{x}) \tag{44c}$$

where $N_x(x, y)$, $N_{xy}(x, y)$, $N_y(x, y)$ are the stretching forces of the membrane.

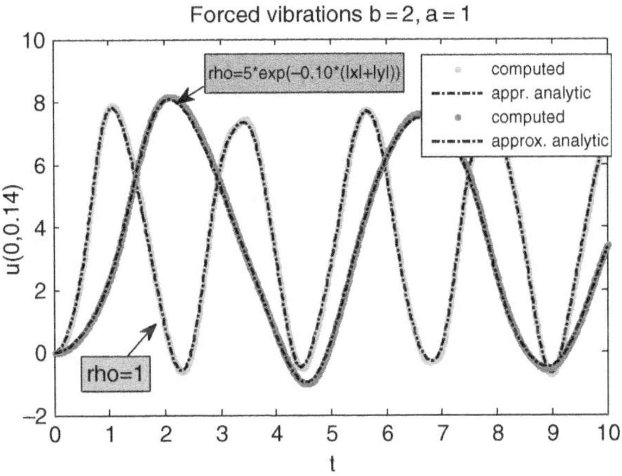

Fig. 13 Time history of the deflection $w(0, 0.14)$ in Example 3, case (a)

They are computed from the solution of the plane stress problem by integrating the Navier equations

$$N_{x,x} + N_{xy,y} = 0, \ N_{xy,x} + N_{y,y} = 0 \quad \text{in } \Omega \tag{44d}$$

$$u_n = 0.005m, \quad u_t = 0 \quad \text{on} \quad \Gamma \tag{44e}$$

This problem is solved using the BEM (Katsikadelis, 2002b). The resulting forces in the principal directions are tensile, i.d. $N_1, N_2 > 0$. Therefore, no wrinkling occurs.

The material of the membrane is homogeneous isotropic and linearly elastic with $E = 20000kN/m^2$ and $\nu = 0.20$. The forced vibrations of the membrane have been studied for the following cases:

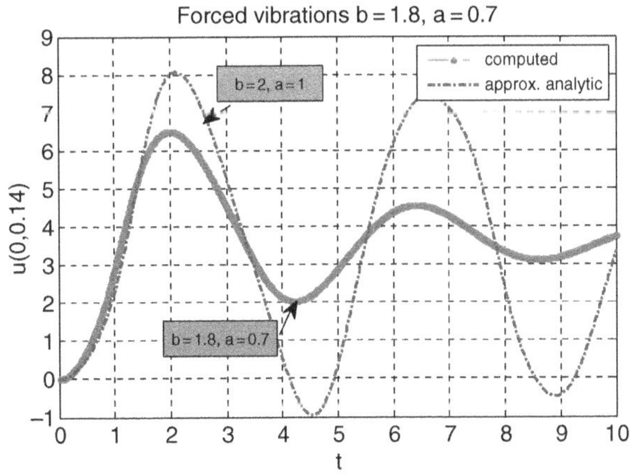

Fig. 14 Time history of the deflection $w(0, 0.14)$ in Example 3, case (b)

(a) $\beta = 2, \alpha = 1$ with $\rho = 1$ and $\rho = 5\exp(-0.1\{|x| + |y|\})$ (b) $\beta = 1.8, \alpha = 0.7$ with $\rho = 5\exp(-0.1\{|x| + |y|\})$. The employed initial conditions are $w(\mathbf{x}, 0) = 0$ $\dot{w}(\mathbf{x}, 0) = 0$ and the transverse load $g = 20\,\mathrm{H}(t)$.

The results are compared with an approximate solution using the Ritz method with 50 mode vectors. The time history of the deflection u(0,0.14) with $N = 300$, $M = 101$, $h = 0.01$ is shown in Figs. 13 and 14.

6 Conclusions

A numerical solution has been presented for the complete diffusion-wave equation in bounded inhomogeneous anisotropic bodies of arbitrary geometry. The numerical examples give insight in the response of systems described by the diffusion-wave equation, a problem that has attracted a lot of interest in recent years within the scientific community. The developed method is based on the concept of the analog equation. The basic characteristic of the method, besides its efficiency and accuracy, is its simple implementation.

Appendix Approximation of $D_c^\alpha u(0)$

The function $u(t)$ is expanded in Taylor series

$$u(t) = u_0 + \dot{u}_0 t + \frac{1}{2!}\ddot{u}_0 t^2 + \frac{1}{3!}u_0 t^3 \cdots \tag{45}$$

(i) $0 < \alpha \leq 1$

We have

$$\left.\begin{aligned} D_c^\alpha(1) &= 0 \\ D_c^\alpha(t) &= \frac{\Gamma(2)}{\Gamma(2-\alpha)} t^{1-\alpha} \\ D_c^\alpha(t^2) &= \frac{\Gamma(3)}{\Gamma(3-\alpha)} t^{2-\alpha} \\ &\vdots \\ D_c^\alpha(t^n) &= \frac{\Gamma(n+1)}{\Gamma(n+1-\alpha)} t^{n-\alpha} \end{aligned}\right\} \tag{46}$$

Then

$$D_c^\alpha u(t) = \dot{u}_0 \frac{\Gamma(2)}{\Gamma(2-\alpha)} t^{1-\alpha} + \frac{1}{2}\ddot{u}_0 \frac{\Gamma(3)}{\Gamma(3-\alpha)} t^{2-\alpha} + \cdots \tag{47}$$

From which we obtain

$$D_c^{\alpha} u(h) = \dot{u}_0 \frac{\Gamma(2)}{\Gamma(2-\alpha)} h^{1-\alpha} + \frac{1}{2} \ddot{u}_0 \frac{\Gamma(3)}{\Gamma(3-\alpha)} h^{2-\alpha} + \cdots \tag{48}$$

$$D_c^{\alpha} u(2h) = \dot{u}_0 \frac{1}{\Gamma(2-\alpha)} (2h)^{1-\alpha} + \ddot{u}_0 \frac{1}{\Gamma(3-\alpha)} (2h)^{2-\alpha} + \cdots \tag{49}$$

Linear extrapolation yields

$$D_c^{\alpha} u(0) = 2 D_c^{\alpha} u(h) - D_c^{\alpha} u(2h) \tag{50}$$

or after neglecting terms of $O(h^{2-\alpha})$

$$D_c^{\alpha} u(0) \simeq \frac{1}{\Gamma(2-\alpha)} h^{1-\alpha} \left(2 - 2^{1-\alpha}\right) \dot{u}_0 \tag{51}$$

(ii) $1 < \alpha \leq 2$

We have

$$\left.\begin{aligned} D_c^{\alpha}(1) &= 0 \\ D_c^{\alpha}(t) &= 0 \\ D_c^{\alpha}(t^2) &= \frac{\Gamma(3)}{\Gamma(3-\alpha)} t^{2-\alpha} \\ &\vdots \\ D_c^{\alpha}(t^n) &= \frac{\Gamma(n+1)}{\Gamma(n+1-\alpha)} t^{n-\alpha} \end{aligned}\right\} \tag{52}$$

Then

$$u(t) = u_0 + \dot{u}_0 t + \frac{1}{2!} \ddot{u}_0 t^2 + \frac{1}{3!} u_0 t^3 \cdots \tag{53}$$

$$D_c^{\alpha} u(t) = \ddot{u}_0 \frac{1}{\Gamma(3-\alpha)} t^{2-\alpha} + u_0 \frac{1}{\Gamma(4-\alpha)} t^{3-\alpha} + \cdots \tag{54}$$

Using Eq. (48) and neglecting $O(h^{3-\alpha})$ we take

$$D_c^{\alpha} u(0) \simeq \frac{1}{\Gamma(3-\alpha)} h^{2-\alpha} \left(2 - 2^{2-\alpha}\right) \ddot{u}_0 \tag{55}$$

Note that Eqs. (51) and (55) yield

$$D_c^{1} u(0) = \dot{u}_0, \qquad D_c^{2} u(0) = \ddot{u}_0 \tag{56}$$

References

Agrawal OP (2002), Solution for a fractional diffusion-wave equation defined in a bounded domain, Nonlinear Dynamics 29: 145–155.

Atanackovic TM, Pilipovic S, Zorica D (2007) A diffusion wave equation with two fractional derivatives, Journal of Physics A: Mathematical and Theoretical, 40: 5319–5333.

Ginoa M, Cerbelli S, Roman HE (1992) Fractional diffusion equation and relaxation in complex viscoelastic materials, Physica A 191: 449–453.

Gorenflo R, Mainardi F (1997) Integral and differential equations of fractional order, CISM Courses and Lectures No. 378, pp. 223–276, Springer Wien, New York.

Katsikadelis JT (1994) The analog equation method. A powerful BEM-based solution technique for solving linear and nonlinear engineering problems. In: Brebbia C.A. (ed.), Boundary Element Method XVI: 167–182, Computational Mechanics Publications, Southampton.

Katsikadelis JT (2002a) The analog boundary integral equation method for nonlinear static and dynamic problems in continuum mathematics, Journal of Theoretical and Applied Mechanics 40: 961–984.

Katsikadelis JT (2002b) Boundary Elements. Theory and Applications, Elsevier, London.

Katsikadelis JT (2005) The BEM for inhomogeneous problems, Archive of Applied Mechanics 74: 780–789.

Katsikadelis JT (2007) A generalized Ritz method for partial differential equations in domains of arbitrary shape using global shape functions, Engineering Analysis with Boundary Elements, doi: 10.1016/j.enganabound.2007.001

Katsikadelis JT (2008a) Numerical solution of multi-term fractional differential equations, (to be published)

Katsikadelis JT (2008b) Fractional vibrations of inhomogeneous membranes, Proceedings, 6th GRACM International Congress on Computational Mechanics, Thessaloniki, 19–21 June 2008

Katsikadelis JT, Nerantzaki MS (2000) A boundary-only solution to dynamic analysis of non-homogeneous elastic membranes, Computer Modeling in Engineering & Sciences 1(3): 1–9.

Mainardi F (1997) Fractional calculus: some basic problems in continuum and statistical mechanics. In: Carpinteri, A. and Mainardi, F. (eds), Fractals and Fractional Calculus in Continuum Mechanics, Springer, New York, pp. 291–348.

Mainardi F, Paradisi P (1997) Model of diffusive waves in viscoelasticity based on fractional calculus, Proceedings of the IEEE Conference on Decision and Control, Vol. 5, O. R. Gonzales, IEEE, New York, pp. 4961–4966.

Miller KS, Ross B (1993) An Introduction Fractional Calculus and Fractional Differential Equations, Wiley, New York.

Oldham KB, Spanier J (1974) The Fractional Calculus, Academic Press, New York.

Podlubny I (1999) Fractional Differential Equations, Academic Press, New York.

Efficient Solution for Composites Reinforced by Particles

Vladimír Kompiš, Mário Štiavnický and Qing-Hua Qin

Abstract A very effective method for some kind of problems is the Method of Fundamental Solutions (MFS). It is a boundary meshless method which does not need any mesh and in linear problems only nodes (collocation points) on the domain boundaries and a set of source functions (fundamental solutions, i.e. Kelvin functions) in points outside the domain are necessary to satisfy the boundary conditions. Another kind of source functions can be obtained from derivatives of the Kelvin source functions. Dipoles are the derivatives of Kelvin function in direction of acting force.

Putting the dipoles into the particle, i.e. outside the domain of the matrix of composite material, they can very efficiently simulate the interaction of the particle with the matrix and with the other particles. A dipole satisfies both the force and moment equilibrium and so the models do not require any additional condition for the simulation of interactions. If the particles of reinforcing material are in the form of spheres, or ellipsoids, a triple dipole (i.e. dipoles in the direction of main axes of the particle) can give satisfactory accuracy for practical problems. It is given how the models can be used for homogenization of composite material.

A novel method is used for evaluation of material properties of homogenized material in order to increase the numerical efficiency of the composite. The numerical results are compared with Mori-Tanaka analytic models with all rigid inclusions as well as for problems with elastic inclusions when the differences in stiffness of matrices and particle are small.

1 Introduction

Efficiency of a numerical method depends on how accurately it approximates the real conditions. It is closely related to the number of interpolation functions and

---

V. Kompiš (✉)
Department of Mechanical Engineering, Academy of Armed Forces of General M. R. Štefánik, Demänovská 393, Liptovský Mikuláš, 03119, Slovakia
e-mail: kompis@aoslm.sk

A chapter in honor of Dimitri Beskos' 65th birthday

G.D. Manolis, D. Polyzos (eds.), *Recent Advances in Boundary Element Methods*,
DOI 10.1007/978-1-4020-9710-2_18,

complexity of evaluated expressions necessary to satisfy the governing equations and boundary conditions of a given problem. Usually geometric concentrators, holes, inclusions, material inhomogeneities, contact with other bodies and other irregularities lead to local effects and large gradients in all mechanical, thermal, electromagnetic and coupled fields. In mechanics they lead to the effects known as stress concentration. Moreover, if such peculiarities are close to each other, or to the boundary, they can very strongly interact and further influence the fields.

Accurate approximation of the local effects is necessary to correct evaluation of such parameters as local and global behaviour of the structure, local and global damage conditions and other important properties.

The approximate function used to simulate the real structure will be more efficient, which will best approximate both governing equations and boundary conditions. The unknown functions are mostly approximated by polynomials. Correct approximation of large gradients requires very fine subregions, submeshing, remeshing during the solution when solution errors are controlled, etc. This reduces the efficiency of the model. If a structure can be defined as a continuum, then the local effects decay with the distance from the concentrator. Boundary type formulations, which keep the decaying effect, need much less parameters to obtain similar accuracy than the domain methods.

A very effective method for some kind of problems is the Method of Fundamental Solutions (MFS) (Golberg and Chen 1998; Karageorghis and Fairweather 1989). It is a boundary meshless method which does not need any mesh and in linear problems only nodes (collocation points) on the domain boundaries and a set of source functions (fundamental solutions) in points outside the domain are necessary to satisfy the boundary conditions. The MFS uses discrete source functions, the fundamental solutions, to approximate the fields in the domain. In domains with smooth boundary, if the boundary conditions are smooth it is necessary to use very many source points acting in some distances from the boundary in order to simulate smooth functions along the boundaries. Using polynomial Trefftz (T-)functions (Jirousek and Wroblewski 1997; Kompiš and Štiavnický 2006) in combination to MFS can improve both accuracy and numerical stability of the model. Oppositely, if the boundary conditions are discontinuous like contact problems of bodies with curved boundaries of different curvature, it is necessary to use continuous distribution of source functions placed along the boundary with corresponding discontinuity (Kompiš and Dekýš 2003) for the best approximation.

Problems with inhomogeneous micro/nano-structure and especially composite materials with stiff or weak inclusions are especially important (Shonaike and Advani 2003; Atieh et al. 2005; Schwarz et al. 2004). In structures like these, the boundary conditions are usually smooth, if the inhomogeneity surface is smooth, but the inhomogeneities introduce large gradients in local fields and many inhomogeneities would require to solve millions or billions equations after discretization by classical methods like FEM or BEM. It will be shown here how the use of discrete source functions, dipoles, in problems with small aspect ratio can lead to very effective solution for homogenization of Representative Volume Elements (RVE). The method has all attractive features of MFS, moreover, the inclusions in composite material do not contain any load for static isothermic problems and the loads by dipoles simulating the interaction are in force and moment equilibrium and do not need any additional conditions. Also the rigid body modes of particles are correctly introduced. The presented method can use all features of the Fast Multipole Solution (Greengard and Rokhlin 2002; Nishimura 2002; Nishimura et al. 1999) and of the Boundary Point Method (Ma and Qin, 2007; Ma et al. 2008; Zhang et al. 2004), i.e. reduced computations for far field interaction, moreover, also the near field interaction is considerably reduced in our formulations. Comparing to MFS usually only few collocation points are necessary to obtain results with required accuracy as the source points do not introduce large gradients along the interdomain (particle-matrix) surface.

A cube which contains specified distribution of reinforcing particles is chosen for RVE and auxiliary particles placed outside the RVE are used in order to increase the efficiency of the model for evaluation of homogenized material constants of the composite. The numerical results are compared with Mori-Tanaka analytic models with all rigid inclusions as well as for problems with elastic inclusions when the differences in stiffness of matrices and particle are small.

2 Description of the Model

Under discrete source functions we will understand all unit forces (the Kelvin solution (Beskos 1991; Cheng and Cheng 2005)), point dislocations, dipoles, couples, or some other type of Radial Basis Functions (RBF) satisfying homogeneous governing equations in infinite space or half space except for the source point force itself, i.e. this type of source functions are Trefftz functions, if the source points are located outside the discretized domain. A very attractive concept is dipoles having mechanical meaning of two collinear forces acting in the same point in opposite direction. Mathematically it is a derivative of the Kelvin function in direction of acting force. The displacement field of a dipole is

$$U_{pi}^{(D)} = U_{pi,p}^{(F)} = -\frac{1}{16\pi G\,(1-\nu)}\frac{1}{r^2}\left[3r_{,i}r_{,p}^2 - r_{,i} + 2\,(1-\nu)\,r_{,p}\delta_{ip}\right] \tag{1}$$

The upper index (D) denotes corresponding dipole field and (F) a force (Kelvin) field. The lower indices belong to components of the field; the first index corresponds to the component of the source quantity and the other indices the components of the field quantity. The index after comma denotes partial derivative, i.e.

$$r_{,i} = \partial r \big/ \partial x_i\,(t) = r_i/r \tag{2}$$

where r is the distance between the source point s, where the dipole or force are acting and the field point t, where the displacement (field variable) is expressed, i.e.

$$r = \sqrt{r_i r_i},\ r_i = x_i\,(t) - x_i\,(s) \tag{3}$$

and x_i are coordinates of the points. G, υ and δ_{ij} are material shear modulus, Poisson ratio and Kronecker delta function, respectively. Summation convection over repeated indices will be used, but it will not act over the repeated index p.

Corresponding strain and stress fields (see Kompiš et al. (2008) for more details) are

$$E_{pij}^{(D)} = \frac{1}{2}\left(U_{pi,j}^{(D)} + U_{pj,i}^{(D)}\right) = -\frac{1}{16\pi G(1-\nu)}\frac{1}{r^3}\left[-15r_{,i}r_{,j}r_{,p} + 3r_{,i}r_{,j} + \right.$$
$$\left. + 2(1-2\nu)\delta_{ip}\delta_{jp} + 6\nu\left(\delta_{ip}r_{,j}r_{,p} + \delta_{jp}r_{,i}r_{,p}\right) + \delta_{ij}\left(3r_{,p}^2 - 1\right)\right] \quad (4)$$

$$S_{pij}^{(D)} = 2GE_{pij}^{(D)} + \frac{2G\nu}{1-2\nu}\delta_{ij}E_{pkk}^{(D)}$$
$$= -\frac{1}{8\pi(1-\nu)}\frac{1}{r^3}\left[(1-2\nu)\left(2\delta_{ip}\delta_{jp} + 3r_{,p}^2\delta_{ij} - \delta_{ij}\right) + \right.$$
$$\left. + 6\nu r_{,p}\left(r_{,i}\delta_{jp} + r_{,j}\delta_{ip}\right) + 3\left(1 - 5r_{,p}^2\right)r_{,i}r_{,j}\right] \quad (5)$$

If unit forces acting in source points are located in discrete points outside the solution domain for computational models and also the collocation points (i.e. the points in which the boundary conditions have to be satisfied) are chosen in some discrete points of the domain boundary, then the method of solution is known as the Method of Fundamental Solutions (MFS) (Golberg and Chen 1998; Karageorghis and Fairweather 1989). The method is very simple one, it does not need any elements and any integration and thus, it is a fully meshless method. These functions are Trefftz functions and they serve as interpolators in the whole domain. Also any other Trefftz functions can be used for this purpose (Kompiš and Štiavnický 2006). They are very convenient to modeling of inhomogeneous materials with spherical, ellipsoidal, or other smooth inclusions or holes, especially when the density of particles is small. Note that the domain which is approximated by source functions is the matrix and thus, the domains of particles are outside of the domain. A dipole located inside the particle, i.e. outside of the domain represented by the matrix, gives both zero resulting force and moment along any closed surface and thus the global equilibrium is not destroyed by local errors as it can be by using MFS (Kompiš and Štiavnický 2006), however, the location of the source points is important for the best simulation of continuity and equilibrium along the interdomain boundaries.

In micromechanics (MM) the material properties are homogenized over the RVE. Corresponding average homogeneous material properties are obtained by integrals over the RVE (Qu and Cherkaoui 2006). The boundary conditions can influence the results and the inhomogeneities closed to boundaries give large gradients in integration surface.

In the present model the strains and stresses are split into a constant part (very small RVE is considered for MM) and a local part (corresponding to so called eigenstrain (Ma et al. 2008; Qu and Cherkaoui 2006)).

$$
\begin{aligned}
\varepsilon_{ij}^{tot} &= \varepsilon_{ij}^{\infty} + \varepsilon_{ij}^{loc} \\
\sigma_{ij}^{tot} &= \sigma_{ij}^{\infty} + \sigma_{ij}^{loc} \\
u_{ij}^{tot} &= u_{ij}^{\infty} + u_{ij}^{loc}
\end{aligned} \tag{6}
$$

where the upper indices *tot*, ∞ and *loc* denote the total component, component corresponding to constant strains and stresses acting in infinity on the continuous matrix material without any particles in it and local components acting in the matrix with eigenstrains from dipoles (including the eigenstrains in auxiliary points outside the RVE) with zero loads in infinity, respectively.

Corresponding displacements vary linearly in the RVE. In order to reduce the gradients we located auxiliary source points outside the RVE so that the local tractions are equal to zero (see Fig. 1) in the control points on the surface of the RVE. The problem in Fig. 1 is presented in the plane x_1x_3 for better understanding the model.

On each interface of matrix-inclusion six pairs of control points are chosen to prescribe the local displacements as shown in Fig. 2. A triple dipole (i.e. dipole in each coordinate direction) is placed into the centre of the inclusion and into auxiliary source points and there intensities are computed so that boundary displacements on inclusions and tractions on the RVE surface are satisfied. Note that we do not work with displacements in the model but with the differences of displacements in direction of vectors connecting corresponding points on opposite boundary of the particle.

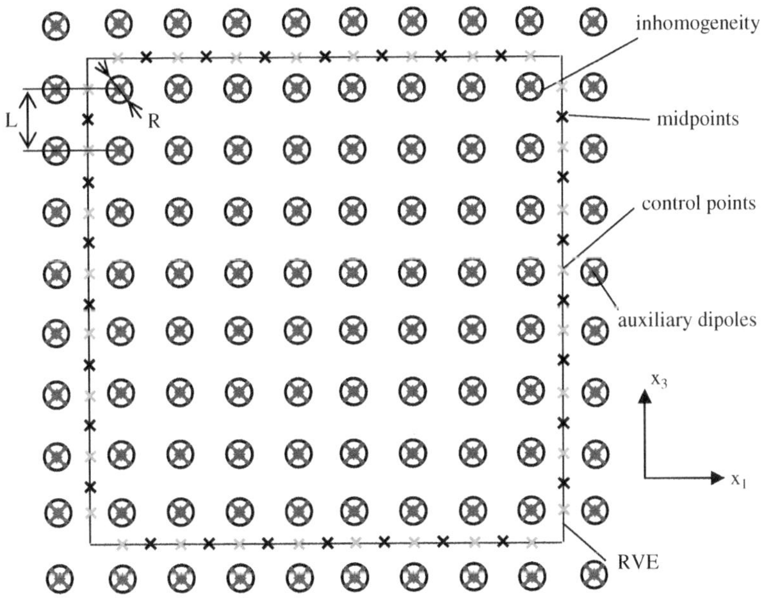

Fig. 1 RVE computational model

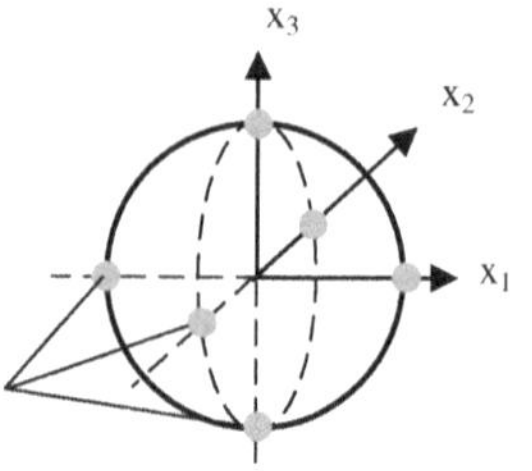

Fig. 2 Spherical inclusion and control points

In order to reduce the computations an elimination-iterative procedure is developed. In the first step the intensities of dipoles inside the inclusions (internal dipoles) are chosen to be equal those without influence the interaction and intensities in the auxiliary dipoles outside the RVE are computed. In the second step the intensities in internal dipoles are calculated from interaction of all dipoles and averaged multiplier is chosen for internal dipoles. Then intensities in auxiliary dipoles are recalculated. In the next steps the intensities of auxiliary dipoles are computed from intensities of internal dipoles by solving system of linear equations and intensity of each internal dipole is found from interaction with all dipoles (i.e. no averaging is used again).

The elastic constants of homogenized material are evaluated from the elastic energy. For this purpose both average strains and stresses are to be evaluated in the RVE. The total strains and stresses consist of the constant components (as given for material without inhomogeneities) and local strains (eigenstrains as defined in Qu and Cherkaoui (2006)) and the local stresses. The local averaged strains are evaluated from the total deformation of RVE by action of the dipoles. For this purpose the displacements of control points on the boundaries of the RVE are used and difference between the displacements on opposite boundaries are evaluated. In this way the strains are evaluated as the averaged deformation of the RVE. This deformation has to introduce the integral value of the whole RVE. As the displacement function of the RVE surface has complicated form given by the interaction of all dipoles (see Fig. 3), two procedures were used as follows. The control points were regularly distributed along the RVE surface. The midpoint and trapezoidal rules were used in the procedures. In regularly distributed particles the number of control points for midpoint rule integration was identical to the number of particles closest to corresponding side of the RVE. For trapezoidal rule points between these points and along the edges of the RVE were included. In following examples only axial reinforcement was evaluated from axial strains as

$$\varepsilon_{ii}^{loc} = \frac{1}{F_i} \int u_i dF_i \tag{7}$$

where F_i is crossectional area of the RVE with normal in direction of x_i coordinate axis. The integral is evaluated numerically.

The local averaged surface tractions, identical to corresponding normal stress component, are evaluated from the sum of dipole intensities inside the RVE in each

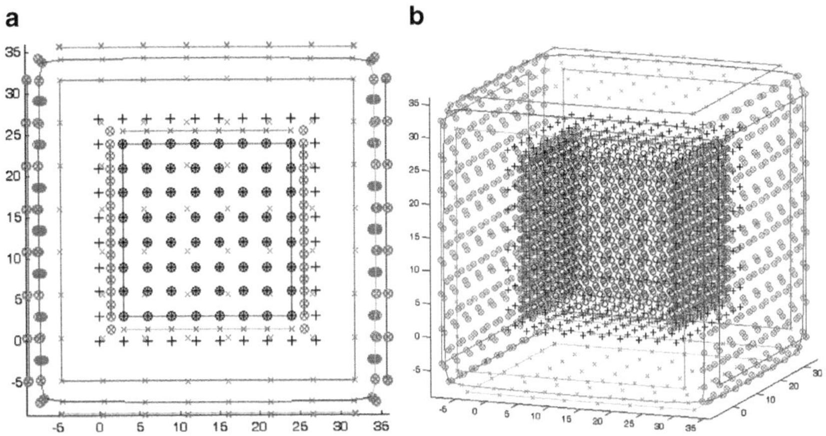

Fig. 3 Deformation of the matrix reinforced with a patch of particles

coordinate direction. The component sum is substituted by a "finite dipole" defined as constant traction acting in normal direction to corresponding side of the RVE and the distance between opposite sides. This is identical to division of the sum of corresponding component of the dipoles by the total volume of the RVE.

$$\sigma_{ii}^{loc} = \sum D_i^D / V_{RVE} \tag{8}$$

where D_i is intensity of a dipole in direction of x_i coordinate axis.

Elastic constants for the homogenized material are evaluated from the elastic energy

$$E_e = \frac{1}{2}\sigma_{ij}^{tot}\varepsilon_{ij}^{tot} = \frac{1}{2}\left(\sigma_{ij}^{\infty}\varepsilon_{ij}^{\infty} + \sigma_{ij}^{\infty}\varepsilon_{ij}^{loc} + \sigma_{ij}^{loc}\varepsilon_{ij}^{\infty} + \sigma_{ij}^{loc}\varepsilon_{ij}^{loc}\right) \tag{9}$$

and the stiffening is defined by relative increase of corresponding component of the elastic modulus.

The procedure will be explained on following examples.

3 Examples and Discussion

Let the matrix is an elastic material with modulus of elasticity $E = 1000$ and Poisson's ratio $\nu = 0.3$. Further let rigid spherical inclusions with radius R are regularly distributed in the RVE (Fig. 1) and there are $8 \times 8 \times 8$ inclusions inside the RVE. The hydrostatic stress state is chosen in this example with hydrostatic tension $p = \sigma_{11} = \sigma_{22} = \sigma_{33} = 25$, which corresponds to constant strain of the RVE without inclusion $\varepsilon_{11} = \varepsilon_{22} = \varepsilon_{33} = 0.01$. There are 512 triple-dipoles inside the RVE and 384 auxiliary triple-dipoles outside the RVE. If the problem would be solved taking into account all interactions, it has to be solved a system of 5760×5760 ($= 3.32 \times 10^7$ coefficients!) linear equations. Instead 1152×1152 equations was solved in 3–4 iterations (relative errors less than 0.5 %) for different percentage of the inclusions in the RVE, solution time in MATLAB in a notebook was less than 5 min. The stiffening effect was evaluated as an increase of stiffness in direction of coordinate axis x_1.

Note that if the elastic inclusions are considered then the distance of the control points on the interface changes according to corresponding tractions. For the stress state in inclusion it is sufficient to use constant terms. They correspond to linear terms in displacement fields. Recall that if Trefftz polynomials are used for approximation of homogeneous equilibrium equations then all constant and linear terms are Trefftz polynomials, i.e. they satisfy the homogeneous equilibrium equations (Kompiš and Štiavnický 2006) and there are 12 independent displacement modes corresponding to the constant and linear terms. Six modes correspond to rigid body displacements and another 6 terms to independent constant stress states. One need not keep for rigid body terms of the inclusions in the present formulation and they are correctly simulated in all inclusions contained in the RVE model and only 3 d.o.f. (3 dipoles) for simulation of each inclusion are necessary for approximate solution of the problem in this model.

Table 1 gives the stiffening of the RVE for different relative volume of stiffeners (1st column), L/R (where is the radius of particles and L is their distance) and present results are compared with Mori-Tanaka (Qu and Cherkaoui 2006) models as well. The last two columns give the results obtained by M.-T. and numerical models when the ratio of modules of elasticity of particle to the one of matrix is 2 : 1. As the Mori-Tanaka's (M.-T.) model is also an approximate model obtained analytically by simplified assumptions. The M.-T. model does not enable to study all fields by different configurations (probabilistic distribution of particles, probabilistic dimensions of particles, influence of the domain boundaries, etc. as it is in the present model.

Figure 3 shows deformation of the matrix reinforced with a patch of particles (only regularly distributed spherical particles with constant radius are shown in this paper). It is possible to see that the displacements are not constant between the particles because of the boundary effect of the patch. Recall that we have chosen the boundary conditions prescribed by constant (zero) local tractions in control points. Instead constant displacements can be chosen in such points.

Table 1 Stiffening effect

% Stiffener	L/R	M.-T., rigid	Numerical, rigid	M.-T., elastic	Numerical, elastic
0.0155	30	1.000316	1.000340	1.000104	1.000128
1	7.4882	1.0205	1.0226	1.0067	1.0083
3.35	5	1.0684	1.0744	1.0227	1.0279
15.5	3	1.3218	1.3746	1.1095	1.1341

4 Conclusions

We have shown that discrete source functions, which can be defined as Trefftz radial basis functions, are convenient approximation or interpolation functions for computational simulation (Oden et al. 2006) of material behaviour with local fields contain large gradients. They can very accurately simulate the decaying of the large gradients and thus, they are very useful tool for multiscale modeling and for simulation of large subdomains containing many large gradients inside. Problems with material inhomogeneities are important examples for practical applications. The hard inclusions are only one problem of the application. Also weak or mixed hard and weak inhomogeneities or microvoids can be solved very effectively by this technique. Very important applications come for materials with micro- or nano-structure, where interaction of micro/nano-particles with considerably larger particles comes into effect (Perez 2005). Important simulation of materials reinforced with hard particles is also the investigation of the interaction such material with free boundaries and the behavior of thin surface layers reinforced with the particles. Simulation of such problems is considered in the next future.

Presented model can be used for reinforcing particles with the aspect ratio close to one. If the aspect ratio is large, continuous source functions are to be used for better accuracy of the model as described in Kompiš, et al. (2008).

The interaction of fibre like inclusions is very complicated and very large gradients can rise in different part of the fibre. Discrete source functions are not able to simulate correctly the continuity conditions between fibre and matrix and continuous distribution of source functions is much more effective for this purpose than other methods.

Much research has to be done in all these simulations, as many effects not included into present models are important for correct simulation. We can mention the inter-layers in very small (nano-structures), curved fibers, nonlinear material behavior, interaction between micro/nanoparticles and polymeric and other matrix materials with fiber like structure, etc.

Acknowledgments The support of NATO (grant 001-AVT-SVK) and Slovak Agency APVV (grant APVT-20-035404) is gratefully acknowledged.

References

M. A. Atieh, et al., Multi-wall carbon nanotubes/natural rubber nanocomposites, AzoNano – Online Journal of Nanotechnology, 1, (2005), pp. 1–11.

D. E. Beskos, ed. Boundary Element Analysis of Plates and Shells. Springer-Verlag, Berlin (1991).

A. H. D. Cheng, D. T. Cheng, Heritage and early history of the boundary element method, Engineering Analysis with Boundary Elements, 29, (2005), pp. 268–302.

M. A. Golberg, C. S. Chen, The method of fundamental solutions for potential, Helmholtz and diffusion problems, in M. A. Golberg ed. Boundary Integral Methods – Numerical and Mathematical Aspect, Computational Mechanics Publications, Southampton (1998), pp. 103–176.

F. L. Greengard, V. Rokhlin, A fast algorithm for particle simulations, Journal of Computational Physics, 73, (1987), pp. 325–348.

J. Jirousek, A. Wroblewski, T-elements: State of the art and future trends, Archive of Applied Mechanics, 3, (1997), pp. 323–434.
A. Karageorghis, G. Fairweather, The method of fundamental solutions for the solution of nonlinear plane potential problems, IMA Journal Numerical Analysis, 9, (1989), pp. 231–242.
V. Kompiš, M. Štiavnický, M. Kompiš, Z. Murčinková, Q.H. Qin, Method of continuous source functions for modelling of matrix reinforced with finite fibres, in V. Kompiš, ed. Composites with Micro- and Nano-Structure, Chapter 3, (2008), pp. 27–46. Springer, New York.
V. Kompiš, V. Dekýš, Effective evaluation of local contact fields, in Proceedings of the 4th International Congress of Croatian Society of Mechanics, Bizovac, 2003, Croatia, (2003), CD ROM.
V. Kompiš, M. Štiavnický, Trefftz functions in FEM, BEM and meshless methods, Computer Assisted mechanics and Engineering Sciences, 13, (2006), pp. 417–426.
H. Ma, Q.-H. Qin, Solving potential problems by a boundary-type meshless method–the boundary point method based on BIE, Engineering Analysis with Boundary Elements, 31, (2007), pp. 749–761.
H. Ma, Q.-H. Qin, V. Kompiš, Computational models and solution procedure for inhomogeneous materials with eigen-strain formulation of boundary integral equations, in V. Kompiš, ed., Composites with Micro- and Nano-Structure, Chapter 13, (2008), pp. 239–256. Springer, New York.
N. Nishimura, Fast multipole accelerated boundary integral equations, Applied Mechanics Reviews, 55, (2002), pp. 299–324.
N. Nishimura, K. Yoshida, S. Kobayashi, A fast multipole boundary integral equation method for crack problems in 3D, Engineering Analysis with Boundary Elements, 23, (1999), pp. 97–105.
J. T. Oden et al. Simulation – Based Engineering Science: Revolutionazing Engineering Science through Simulation. Report NSF Blue Ribbon Panel, (2006).
I. Perez, Polymer Nanotube Composites, RTO-MP-AVT-122 (2005), pp. 13-1–13-11.
J. Qu, M. Cherkaoui, Fundamentals of Micromechanics of Solids, John Wiley &Sons, Hoboken, New Jersey (2006).
J. A. Schwarz, C. I. Contescu, K. Putyera, eds., Nanoscience and Nanotechnology, Vols. 1–5, Marcel Dekker, New York (2004).
G. O. Shonaike, S.G. Advani, eds., Advanced Polymeric Materials, Structure Property Relationships, CRC Press, Boca Raton (2003).
J. M. Zhang, M. Tanaka, T. Matsumoto, Meshless analysis of potential problems in threee dimensions with the hybrid boundary node method, International Journal for Numerical Methods in Engineering, 59, (2004). pp. 1147–1160.

Development of the Fast Multipole Boundary Element Method for Acoustic Wave Problems

Yijun Liu, Liang Shen and Milind Bapat

Abstract In this chapter, we review some recent development of the fast multipole boundary element method (BEM) for solving large-scale acoustic wave problems in both 2-D and 3-D domains. First, we review the boundary integral equation (BIE) formulations for acoustic wave problems. The Burton-Miller BIE formulation is emphasized, which uses a linear combination of the conventional BIE and hypersingular BIE. Next, the fast multipole formulations for solving the BEM equations are provided for both 2-D and 3-D problems. Several numerical examples are presented to demonstrate the effectiveness and efficiency of the developed fast multipole BEM for solving large-scale acoustic wave problems, including scattering and radiation problems, and half-space problems.

1 Introduction

Solving acoustic wave problems is one of the most important applications of the BEM, which can be used in analyzing sound fields for noise controls in automobiles, airplanes, and many other consumer products. Acoustic waves often exist in an infinite medium outside a structure which is in vibration (a radiation problem) or impinged upon by an incident wave (a scattering problem). With the BEM, only the boundary of the structure needs to be discretized. In addition, the boundary conditions at infinity can be taken into account analytically in the boundary integral equation formulations and thus these conditions can be satisfied exactly. The governing equation for acoustic wave problems is the Helmholtz equation, which has been solved using the BIE/BEM for more than four decades (see, e.g., some of the early work in Schenck 1968; Burton and Miller 1971; Ursell 1973; Kleinman and Roach 1974; Jones 1974; Meyer et al. 1978; Seybert et al. 1985; Kress 1985; Seybert and Rengarajan 1987; Cunefare and Koopmann 1989, 1991; Everstine and Henderson 1990; Martinez 1991; Cunefare and Koopmann 1991)). Especially, the

Y. Liu (✉)
Department of Mechanical Engineering, University of Cincinnati, Cincinnati
Ohio 45221-0072, USA
e-mail: Yijun.Liu@uc.edu

G.D. Manolis, D. Polyzos (eds.), *Recent Advances in Boundary Element Methods*,
DOI 10.1007/978-1-4020-9710-2_19, © Springer Science+Business Media B.V. 2009

work by Burton and Miller (1971) has been regarded as a classical one which provides a very elegant way to overcome the so called fictitious frequency difficulties existing in the conventional BIE for exterior acoustic wave problems. The Burton and Miller's BIE formulation has been used by many others in their research on the BEM for acoustic problems (e.g., Krishnasamy et al. 1990; Amini 1990; Wu et al. 1991; Liu and Rizzo 1992; Liu 1992; Yang 1997; Liu and Chen 1999).

The fast multipole method (FMM) developed by Rokhlin and Greengard (Rokhlin 1985; Greengard and Rokhlin 1987; Greengard 1988) has been extended to solving Helmholtz equation for quite some time (see, e.g., Rokhlin 1990; Rokhlin 1993; Coifman et al. 1993; Engheta et al. 1992; Lu and Chew 1993; Wagner and Chew 1994; Epton and Dembart 1995; Koc and Chew 1998; Gyure and Stalzer 1998; Greengard *et al.* 1998; Tournour and Atalla 1999; Gumerov and Duraiswami 2003; Darve and Havé 2004; Fischer et al. 2004; Chen and Chen 2004; Shen and Liu 2007). Most of these works are good for solving acoustic wave problems at either low frequencies or high frequencies. For example, Greengard *et al.*, (1998) suggested a diagonal translation in the FMM for low frequency range. Rokhlin (1993) and Lu and Chew (1993) proposed diagonal form of the translation matrices for high frequency range for the Helmholtz equation. Wagner and Chew (1994) used ray propagation approach to further accelerate the FMM for high frequency range. A new adaptive fast multipole BEM for 3-D acoustic wave problems was given in Shen and Liu (2007) and large acoustic models with degrees of freedom (in complex variables) above 200,000 have been solved successfully on laptop PCs (Shen and Liu 2007).

2 Basic Equations for Acoustic Wave Problems

Consider the Helmholtz equation governing time-harmonic acoustic wave fields:

$$\nabla^2\phi + k^2\phi = 0 \qquad \forall \mathbf{x} \in E, \tag{1}$$

where $\phi = \phi(\mathbf{x}, \omega)$ is the complex acoustic pressure, $k = \omega/c$ the wavenumber, ω the circular frequency, c the speed of sound, and $\nabla^2(\,) = \partial^2(\,)/\partial x_k \partial x_k = (\,)_{,kk}$. The acoustic domain E can be an *infinite* domain exterior to a body V (Fig. 1) or a *finite* domain interior to a closed surface.

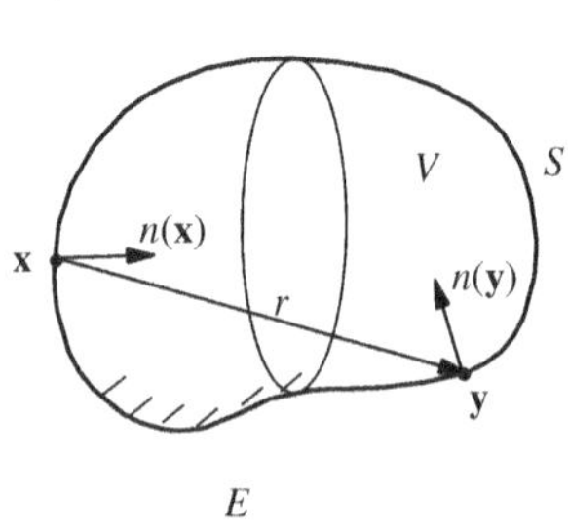

Fig. 1 The acoustic medium E, body V and boundary S

The boundary conditions for acoustic wave problems can be classified as follows:

$$(a)\ Pressure\ is\ given: \quad \phi = \overline{\phi}, \qquad \forall \mathbf{x} \in S; \tag{2}$$

$$(b)\ Velocity\ is\ given: q \equiv \frac{\partial \phi}{\partial n} = \overline{q} = i\omega\rho v_n, \quad \forall \mathbf{x} \in S; \tag{3}$$

$$(c)\ Impedance\ is\ given: \phi = Z v_n, \quad \forall \mathbf{x} \in S; \tag{4}$$

in which ρ is the mass density, v_n the normal velocity, Z the specific impedance, and the barred quantities indicate given values. For exterior (infinite domain) acoustic wave problems, the field at infinity must also satisfy the Sommerfeld radiation condition.

For 2-D problems, the fundamental solution is given by:

$$G(\mathbf{x}, \mathbf{y}, \omega) = \frac{i}{4} H_0^{(1)}(kr), \tag{5}$$

$$F(\mathbf{x}, \mathbf{y}, \omega) \equiv \frac{\partial G(\mathbf{x}, \mathbf{y}, \omega)}{\partial n(\mathbf{y})} = -\frac{ik}{4} H_1^{(1)}(kr) r,_l\, n_l(\mathbf{y}), \tag{6}$$

where r is the distance between $\mathbf{x}$ and $\mathbf{y}$, and $H_n^{(1)}()$ denotes the Hankel function of the first kind (Abramowitz and Stegun 1972). For 3-D problems, the fundamental solution is given by:

$$G(\mathbf{x}, \mathbf{y}, \omega) = \frac{1}{4\pi r} e^{ikr}, \tag{7}$$

$$F(\mathbf{x}, \mathbf{y}, \omega) \equiv \frac{\partial G(\mathbf{x}, \mathbf{y}, \omega)}{\partial n(\mathbf{y})} = \frac{1}{4\pi r^2}(ikr - 1) r,_j\, n_j(\mathbf{y}) e^{ikr}. \tag{8}$$

3 BIE Formulations

The solution of the Helmholtz equation is given by the representation integral:

$$\phi(\mathbf{x}) = \int_S [G(\mathbf{x}, \mathbf{y}, \omega) q(\mathbf{y}) - F(\mathbf{x}, \mathbf{y}, \omega)\phi(\mathbf{y})]\, dS(\mathbf{y}) + \phi^I(\mathbf{x}), \quad \forall \mathbf{x} \in E, \tag{9}$$

where $q = \partial\phi/\partial n$ and $\phi^I(\mathbf{x})$ is an incident wave. Equation (9) is the representation integral of the solution ϕ inside the domain E for Helmholtz equation (1) for both exterior and interior domain problems. Once the values of both ϕ and q are known on S, Eq. (9) can be applied to calculate ϕ everywhere in E, if needed.

Let the source point **x** approach the boundary S. We obtain the following conventional boundary integral equation (CBIE) for acoustic wave problems:

$$c(\mathbf{x})\phi(\mathbf{x}) = \int_S [G(\mathbf{x},\mathbf{y},\omega)q(\mathbf{y}) - F(\mathbf{x},\mathbf{y},\omega)\phi(\mathbf{y})]\, dS(\mathbf{y}) + \phi^I(\mathbf{x}), \quad \forall \mathbf{x} \in S, \tag{10}$$

where the constant $c(\mathbf{x}) = 1/2$, if S is smooth around **x**. The integral with the G kernel is a weakly-singular integral, while the one with the F kernel is a strongly-singular (CPV) integral. It is well known that this CBIE has a major defect for exterior domain problems, that is, it has nonunique solutions at a set of fictitious eigenfrequencies associated with the resonate frequencies of the corresponding interior problems (Burton and Miller 1971). This difficulty is referred to as the *fictitious eigenfrequency difficulty*. A remedy to this problem is to use the normal derivative BIE in conjunction with this CBIE. Taking the derivative of integral representation (9) with respect to the normal at the point **x** and letting **x** approach S, we obtain the following hypersingular boundary integral equation (HBIE):

$$\tilde{c}(\mathbf{x})q(\mathbf{x}) = \int_S [K(\mathbf{x},\mathbf{y},\omega)q(\mathbf{y}) - H(\mathbf{x},\mathbf{y},\omega)\phi(\mathbf{y})]\, dS(\mathbf{y}) + q^I(\mathbf{x}), \quad \forall \mathbf{x} \in S \tag{11}$$

where $\tilde{c}(\mathbf{x}) = 1/2$ if S is smooth. For 2-D problems, the two new kernels are:

$$K(\mathbf{x},\mathbf{y},\omega) \equiv \frac{\partial G(\mathbf{x},\mathbf{y},\omega)}{\partial n(\mathbf{x})} = \frac{ik}{4} H_1^{(1)}(kr) r_{,j}\, n_j(\mathbf{x}), \tag{12}$$

$$\begin{aligned} H(\mathbf{x},\mathbf{y},\omega) \equiv \frac{\partial F(\mathbf{x},\mathbf{y},\omega)}{\partial n(\mathbf{x})} &= \frac{ik}{4r} H_1^{(1)}(kr) n_j(\mathbf{x}) n_j(\mathbf{y}) \\ &\quad - \frac{ik^2}{4} H_2^{(1)}(kr) r_{,j}\, n_j(\mathbf{x}) r_{,l}\, n_l(\mathbf{y}). \end{aligned} \tag{13}$$

For 3-D problems, the two new kernels are:

$$K(\mathbf{x},\mathbf{y},\omega) \equiv \frac{\partial G(\mathbf{x},\mathbf{y},\omega)}{\partial n(\mathbf{x})} = -\frac{1}{4\pi r^2}(ikr - 1) r_{,j}\, n_j(\mathbf{x}) e^{ikr}, \tag{14}$$

$$\begin{aligned} H(\mathbf{x},\mathbf{y},\omega) \equiv \frac{\partial F(\mathbf{x},\mathbf{y},\omega)}{\partial n(\mathbf{x})} &= \frac{1}{4\pi r^3} \{(1 - ikr) n_j(\mathbf{y}) \\ &\quad + \left[k^2 r^2 - 3(1 - ikr)\right] r_{,j}\, r_{,l}\, n_l(\mathbf{y})\} \, n_j(\mathbf{x}) e^{ikr}, \end{aligned} \tag{15}$$

In HBIE (11), the integral with the kernel K is a strongly-singular (CPV) integral, while the one with the H kernel is a hypersingular (HFP) integral. For exterior acoustic wave problems, a dual BIE (CHBIE, or composite BIE (Liu and Rizzo 1992)) formulation using a linear combination of the CBIE (10) and HBIE (11) can be written as:

$$\text{CBIE} + \beta\text{HBIE} = 0, \tag{16}$$

where β is the coupling constant. This dual BIE formulation is called the Burton-Miller formulation (Burton and Miller 1971) for acoustic wave problems and has been shown by Burton and Miller to yield unique solutions at all frequencies, if β is a complex number (which, for example, can be chosen as $\beta = i/k$ (Kress 1985)).

CBIE (10) and HBIE (11) contain singular integrals that are difficult to evaluate analytically even on constant elements. Numerical integration can be employed to compute all the singular integrals with proper care, but it has been found not very efficient computationally with higher-order elements. As in all the other problems using the BIE/BEM, the best approach in such cases is to use the weakly-singular forms of these BIEs, which are obtained analytically and do not introduce any approximations. The weakly-ssingular forms of the BIEs for acoustic wave problems can be found in Liu and Rizzo (1992) and Liu and Chen (1999).

The discretized equations of the CBIE, HBIE, or the Burton-Miller's BIE formulation, in either singular or weakly-singular forms, can be written as:

$$\begin{bmatrix} a_{11} & a_{12} & \dots & a_{1N} \\ a_{21} & a_{22} & \cdots & a_{2N} \\ \vdots & \vdots & \ddots & \vdots \\ a_{N1} & a_{N2} & \cdots & a_{NN} \end{bmatrix} \begin{Bmatrix} \lambda_1 \\ \lambda_2 \\ \vdots \\ \lambda_N \end{Bmatrix} = \begin{Bmatrix} b_1 \\ b_2 \\ \vdots \\ b_N \end{Bmatrix}, \quad \text{or} \quad \mathbf{A}\boldsymbol{\lambda} = \mathbf{b}, \tag{17}$$

where $\mathbf{A}$ is the system matrix, $\boldsymbol{\lambda}$ the vector of unknown boundary variables at the nodes, $\mathbf{b}$ the known vector, and N the number of nodes on the boundary. For acoustic wave problems, this system of equations is in complex numbers, that is, all the coefficients and variables are complex numbers and thus the memory requirement is four times as large as its counterpart in potential problems. As a result of this, only small models have been solved using the conventional BEM.

4 Fast Multipole Formulation for 2-D Acoustic Wave Problems

We first discuss the fast multipole BEM formulation for 2-D acoustic wave problems (Nishimura 2002). Iterative solver GMRES will be used to solve the system of equations (17) in which the far field contributions will be evaluated using the fast multipole method.

The 2-D formulation is based on Graf's equation (Abramowitz and Stegun 1972) (page 363, equation (9.1.79)) for the kernel, that is, the far field expansion for the G

kernel can be represented as the following:

$$G(\mathbf{x},\mathbf{y},\omega)=\frac{i}{4}\sum_{n=-\infty}^{\infty}O_n(\mathbf{y}_c,\mathbf{x})I_{-n}(\mathbf{y}_c,\mathbf{y}),\quad |\mathbf{x}-\mathbf{y}_c|>|\mathbf{y}-\mathbf{y}_c|, \tag{18}$$

where $\mathbf{y}_c$ is the expansion point close to $\mathbf{y}$ and the two auxiliary functions O and I are given by:

$$O_n(\mathbf{x},\mathbf{y})=i^n H_n^{(1)}(kr)e^{in\alpha}, \tag{19}$$

$$I_n(\mathbf{x},\mathbf{y})=(-i)^n J_n(kr)e^{in\alpha}. \tag{20}$$

In the above two expressions, $J_n()$ denotes the Bessel-J function (Abramowitz and Stegun 1972) and α is the polar angle of the vector $\vec{r}$ from $\mathbf{x}$ to $\mathbf{y}$. Using Eq. (18), the far field expansion for the F kernel is given by:

$$F(\mathbf{x},\mathbf{y},\omega)=\frac{i}{4}\sum_{n=-\infty}^{\infty}O_n(\mathbf{y}_c,\mathbf{x})\frac{\partial I_{-n}(\mathbf{y}_c,\mathbf{y})}{\partial n(\mathbf{y})},\quad |\mathbf{x}-\mathbf{y}_c|>|\mathbf{y}-\mathbf{y}_c|, \tag{21}$$

in which the derivative is obtained by the formula:

$$\frac{\partial I_n(\mathbf{x},\mathbf{y})}{\partial n(\mathbf{y})}=\frac{(-i)^n k}{2}[J_{n+1}(kr)e^{i\delta}-J_{n-1}(kr)e^{-i\delta}]e^{in\alpha}, \tag{22}$$

with δ being the angle between the vector $\vec{r}$ from $\mathbf{x}$ to $\mathbf{y}$ and the outward normal.

Applying expansions in Eqs. (18) and (21), one can evaluate the G and Fintegrals in CBIE (10) on S_c (a subset of S that is away from the source point $\mathbf{x}$) with the following *multipole expansions*:

$$\int_{S_c}G(\mathbf{x},\mathbf{y},\omega)q(\mathbf{y})dS(\mathbf{y})=\sum_{n=-\infty}^{\infty}O_n(\mathbf{y}_c,\mathbf{x})M_n(\mathbf{y}_c),\ |\mathbf{x}-\mathbf{y}_c|>|\mathbf{y}-\mathbf{y}_c|, \tag{23}$$

$$\int_{S_c}F(\mathbf{x},\mathbf{y},\omega)\phi(\mathbf{y})dS(\mathbf{y})=\sum_{n=-\infty}^{\infty}O_n(\mathbf{y}_c,\mathbf{x})\widetilde{M}_n(\mathbf{y}_c),\ |\mathbf{x}-\mathbf{y}_c|>|\mathbf{y}-\mathbf{y}_c|, \tag{24}$$

where M_n and $\widetilde{M}_n$ are the *multipole moments* centered at $\mathbf{y}_c$ and given by:

$$M_n(\mathbf{y}_c)=\frac{i}{4}\int_{S_c}I_{-n}(\mathbf{y}_c,\mathbf{y})q(\mathbf{y})dS(\mathbf{y}), \tag{25}$$

$$\widetilde{M}_n(\mathbf{y}_c)=\frac{i}{4}\int_{S_c}\frac{\partial I_{-n}(\mathbf{y}_c,\mathbf{y})}{\partial n(\mathbf{y})}\phi(\mathbf{y})dS(\mathbf{y}). \tag{26}$$

When the multipole expansion center is moved from $\mathbf{y}_c$ to $\mathbf{y}_{c'}$, we have the following *M2M translations* for both M_n and $\widetilde{M}_n$:

$$M_n(\mathbf{y}_{c'}) = \sum_{m=-\infty}^{\infty} I_{n-m}(\mathbf{y}_{c'}, \mathbf{y}_c) M_m(\mathbf{y}_c), \tag{27}$$

which is derived using the following identity:

$$I_n(\mathbf{y}_{c'}, \mathbf{y}) = \sum_{m=-\infty}^{\infty} I_{n-m}(\mathbf{y}_{c'}, \mathbf{y}_c) I_m(\mathbf{y}_c, \mathbf{y}). \tag{28}$$

The *local expansion* for the G kernel integral in CBIE (10) is given as follows:

$$\int_{S_c} G(\mathbf{x}, \mathbf{y}, \omega) q(\mathbf{y}) dS = \sum_{n=-\infty}^{\infty} I_{-n}(\mathbf{x}_L, \mathbf{x}) L_n(\mathbf{x}_L), \tag{29}$$

where $\mathbf{x}_L$ is the local expansion point close to $\mathbf{x}$ ($|\mathbf{x} - \mathbf{x}_L| < |\mathbf{y} - \mathbf{x}_L|$) and the expansion coefficients are given by the following *M2L translation*:

$$L_n(\mathbf{x}_L) = \sum_{m=-\infty}^{\infty} (-1)^m O_{n-m}(\mathbf{x}_L, \mathbf{y}_c) M_m(\mathbf{y}_c). \tag{30}$$

This result, which is different from that given in (Nishimura 2002), is derived based on the following identity:

$$O_n(\mathbf{x}_L, \mathbf{y}) = \sum_{m=-\infty}^{\infty} (-1)^m O_{n-m}(\mathbf{x}_L, \mathbf{y}_c) I_m(\mathbf{y}_c, \mathbf{y}). \tag{31}$$

Similarly, the local expansion for the F kernel integral in CBIE (10) is given by:

$$\int_{S_c} F(\mathbf{x}, \mathbf{y}, \omega) \phi(\mathbf{y}) dS = \sum_{n=-\infty}^{\infty} I_{-n}(\mathbf{x}_L, \mathbf{x}) L_n(\mathbf{x}_L), \tag{32}$$

with $\widetilde{M}_n$ replacing M_n in the M2L translation (30).

The local expansion center in expansion (29) can be shifted from $\mathbf{x}_L$ to $\mathbf{x}_{L'}$ using the following *L2L translations*:

$$L_n(\mathbf{x}_{L'}) = \sum_{m=-\infty}^{\infty} I_m(\mathbf{x}_L, \mathbf{x}_{L'}) L_{n-m}(\mathbf{x}_L), \tag{33}$$

which is derived using the following identity:

$$O_n(\mathbf{x}_{L'}, \mathbf{y}) = \sum_{m=-\infty}^{\infty} I_m(\mathbf{x}_L, \mathbf{x}_{L'}) O_{n-m}(\mathbf{x}_L, \mathbf{y}). \tag{34}$$

For the HBIE (11), the local expansion of the K kernel integral can be written as:

$$\int_{S_c} K(\mathbf{x}, \mathbf{y}, \omega) q(\mathbf{y}) dS = \sum_{n=-\infty}^{\infty} \frac{\partial I_{-n}(\mathbf{x}_L, \mathbf{x})}{\partial n(\mathbf{x})} L_n(\mathbf{x}_L), \tag{35}$$

with the same local expansion coefficient $L_n(\mathbf{x}_L)$ given by Eq. (30). Similarly, the local expansion for the H kernel integral is given by:

$$\int_{S_c} H(\mathbf{x}, \mathbf{y}, \omega) \phi(\mathbf{y}) dS = \sum_{n=-\infty}^{\infty} \frac{\partial I_{-n}(\mathbf{x}_L, \mathbf{x})}{\partial n(\mathbf{x})} L_n(\mathbf{x}_L), \tag{36}$$

with $\widetilde{M}_n$ replacing M_n in Eq. (30) for evaluating $L_n(\mathbf{x}_L)$. Therefore, the same moments, M2M, M2L and L2L translations as used for the G and F integrals in the CBIE are used for the K and H integrals in the HBIE, respectively.

5 Fast Multipole Formulation for 3-D Acoustic Wave Problems

The fast multipole method for solving the Burton-Miller's BIE (16) is discussed in this section for the 3-D cases (Shen and Liu 2007). The fundamental solution $G(\mathbf{x}, \mathbf{y}, \omega)$ for Helmholtz equations in 3-D can be expanded as (see, e.g., Epton and Dembart 1995; Yoshida 2001):

$$G(\mathbf{x}, \mathbf{y}, \omega) = \frac{ik}{4\pi} \sum_{n=0}^{\infty} (2n+1) \sum_{m=-n}^{n} O_n^m(k, \mathbf{x} - \mathbf{y}_c) \bar{I}_n^m(k, \mathbf{y} - \mathbf{y}_c), \quad |\mathbf{x} - \mathbf{y}_c| > |\mathbf{y} - \mathbf{y}_c|, \tag{37}$$

where k is the wavenumber, $\mathbf{y}_c$ an expansion point near $\mathbf{y}$, O_n^m the outer function and I_n^m the inner function. Similarly, the kernel $F(\mathbf{x}, \mathbf{y}, \omega)$ can be expanded as:

$$F(\mathbf{x}, \mathbf{y}, \omega) = \frac{ik}{4\pi} \sum_{n=0}^{\infty} \sum_{m=-n}^{n} (2n+1) O_n^m(k, \mathbf{x} - \mathbf{y}_c) \frac{\partial \bar{I}_n^m(k, \mathbf{y} - \mathbf{y}_c)}{\partial n(\mathbf{y})}, \quad |\mathbf{x} - \mathbf{y}_c| > |\mathbf{y} - \mathbf{y}_c|. \tag{38}$$

Using Eqs. (37) and (38), we can evaluate the G and F integrals in CBIE (10) on S_c with the following *multipole expansions*:

$$\int_{S_c} G(\mathbf{x},\mathbf{y},\omega)q(\mathbf{y})dS(\mathbf{y}) = \frac{ik}{4\pi}\sum_{n=0}^{\infty}\sum_{m=-n}^{n}(2n+1)O_n^m(k,\mathbf{x}-\mathbf{y}_c)M_{n,m}(k,\mathbf{y}_c),$$
$$|\mathbf{x}-\mathbf{y}_c| > |\mathbf{y}-\mathbf{y}_c|, \tag{39}$$

$$\int_{S_c} F(\mathbf{x},\mathbf{y},\omega)\phi(\mathbf{y})dS(\mathbf{y}) = \frac{ik}{4\pi}\sum_{n=0}^{\infty}\sum_{m=-n}^{n}(2n+1)O_n^m(k,\mathbf{x}-\mathbf{y}_c)\widetilde{M}_{n,m}(k,\mathbf{y}_c),$$
$$|\mathbf{x}-\mathbf{y}_c| > |\mathbf{y}-\mathbf{y}_c|, \tag{40}$$

where $M_{n,m}$ and $\widetilde{M}_{n,m}$ are the *multipole moments* centered at $\mathbf{y}_c$ and given by:

$$M_{n,m}(k,\mathbf{y}_c) = \int_{S_c} \bar{I}_n^m(k,\mathbf{y}-\mathbf{y}_c)q(\mathbf{y})dS(\mathbf{y}), \tag{41}$$

$$\widetilde{M}_{n,m}(k,\mathbf{y}_c) = \int_{S_c} \frac{\partial \bar{I}_n^m(k,\mathbf{y}-\mathbf{y}_c)}{\partial n(\mathbf{y})}\phi(\mathbf{y})dS(\mathbf{y}). \tag{42}$$

The *M2M, M2L and L2L translations* for 3-D Helmholtz BIEs can be found in Shen and Liu (2007) and Yoshida (2001). Adaptive fast multipole algorithms (Shen and Liu 2007) have also been employed to further accelerate the solutions of the fast multipole BEM.

6 Numerical Examples

Several 2-D and 3-D examples of acoustic wave problems are presented in this section. Constant triangular elements are used in all these examples, for which one can use singularity subtraction approach to analytically evaluate the singular and hypersingular integrals involving the static kernels. In all the 3-D examples, the maximum number of elements in a leaf is set to 100, the number of multipole and local expansion terms set to 10 and the tolerance to 10^{-3}. All the computations for the 3-D examples were done on a laptop PC with an Intel 1.6 GHz Centrino processor and 512 MB memory.

6.1 Scattering from Cylinders in 2-D Medium

A 2-D scattering problem with a rigid cylinder and the incident wave coming from the right is considered first (Fig. 2). The cylinder has a radius $a = 1$ and is discretized with line elements. A relative error of 0.01% is achieved with 1,000 elements for $ka = 1$. Figure 2 shows the magnitude of the scattered pressure field outside the cylinder in a square region. Figure 3 shows the computed scattered field by an array of multiple cylinders with $ka = 0.1$.

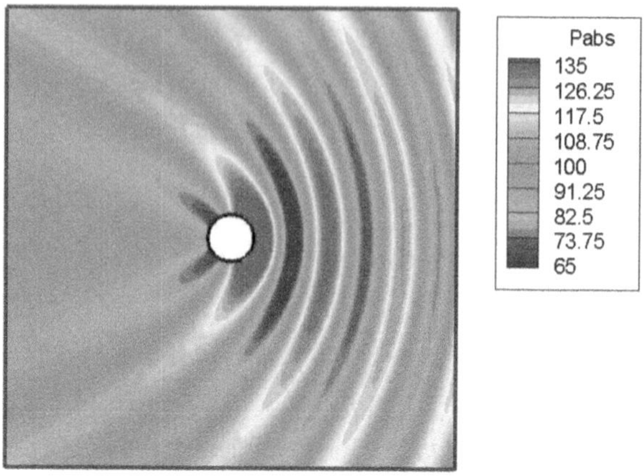

Fig. 2 Scattering from a single cylinder

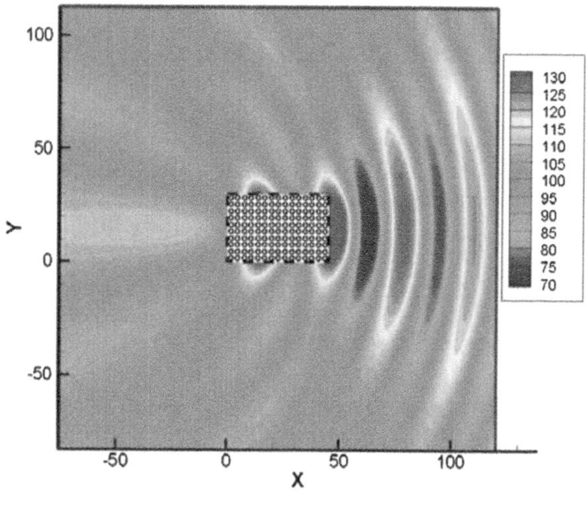

Fig. 3 Scattering from multiple cylinders

6.2 Radiation from a Pulsating Sphere

A pulsating sphere with radius $a = 1$ m is used to verify the fast multipole BEM code for 3-D radiation problems. The normalized wave number ka varies from 1 to 10. The total number of elements is 1,200. The velocity potentials at $(5a, 0, 0)$ are plotted in Fig. 4, which shows that the conventional BEM with the CBIE fails to predict the surface velocity potential at the fictitious frequencies ($ka = \pi, 2\pi, \ldots,$ for this case). The results using the conventional BEM with the Burton-Miller's (CHBIE) formulation agree well with the analytical solution at all wavenumbers. The fast multipole BEM with the CHBIE also yields very close results to those of the conventional BEM with the CHBIE, which suggests that the truncation error introduced in multipole expansions is very small for problems with ka ranging from 1 to 10.

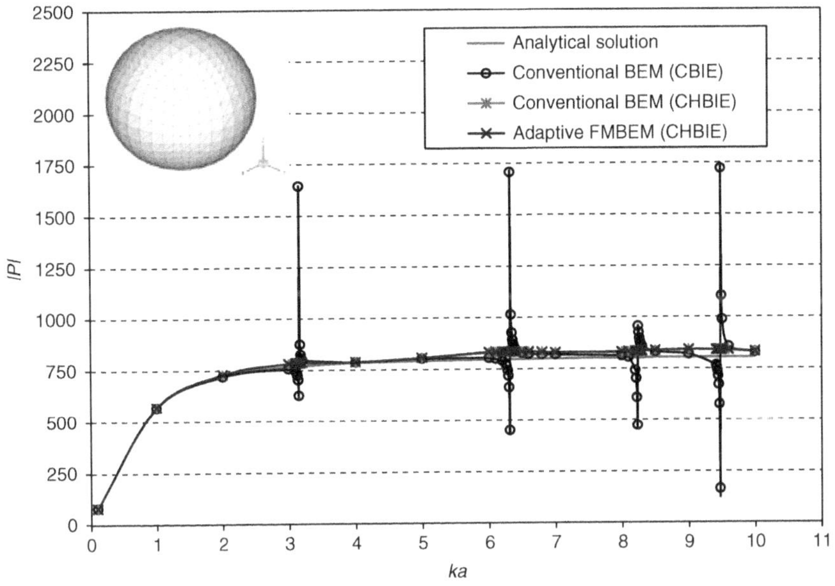

Fig. 4 Frequency sweep plot for the pulsating sphere model

6.3 Scattering from a Rigid Sphere

A rigid sphere with radius $a = 1$ m centered at (0, 0, 0) is used to test the fast multipole BEM code for scattering problems. The sphere is meshed with 1,200 elements and impinged upon by an incident wave of unit amplitude $\phi^I(x, y, z) = e^{-ikz}$, with $ka = \pi$, one of the fictitious eigenfrequencies for the CBIE, and traveling along the negative z axis. Sample field points are evenly distributed on a semicircle of $r = 5a$, centered at (0, 0, 0). The velocity potential curves plotted in Fig. 5 shows that the adaptive FMBEM using Burton-Miller formulation successfully overcomes the non-uniqueness difficulties at this fictitious frequency and yields very accurate results.

6.4 Scattering from Multiple Objects

A multi-scatterer model containing 1,000 randomly distributed capsule-like rigid scatterers in a $2 \times 2 \times 2$ m domain is studied next. Each scatterer is meshed with 200 boundary elements, with a total of 200,000 elements for the entire model. The incident wave is e^{-ikx} with $k = 1$. Sample points are taken at an annular data collection surface with inner and outer radius equal to 5 and 10, respectively. The computed velocity potential distribution contour is shown in Fig. 6 for this discretization. Total CPU time used to solve this large model is 3,352 s using the laptop PC.

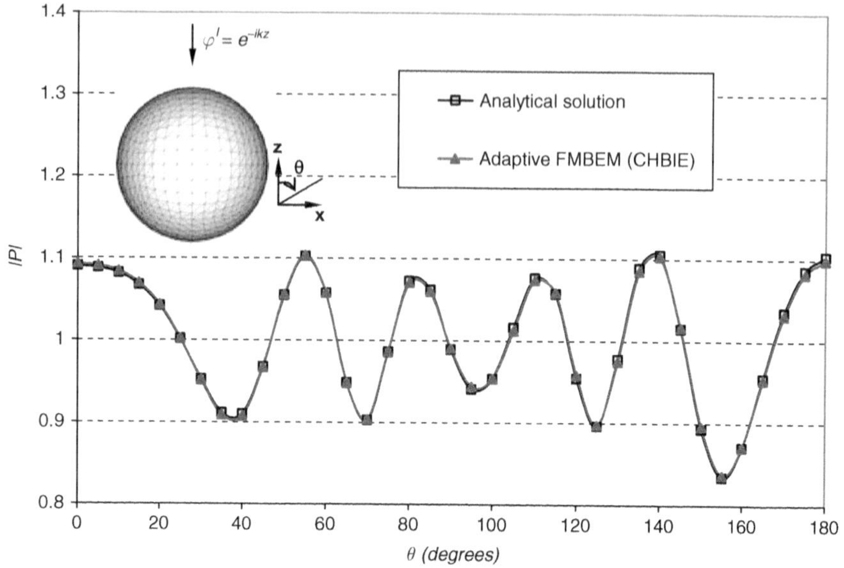

Fig. 5 Scattering from the rigid sphere at the fictitious eigenfrequency $ka = \pi$

Fig. 6 Computed velocity potential for the multiple scatterer model

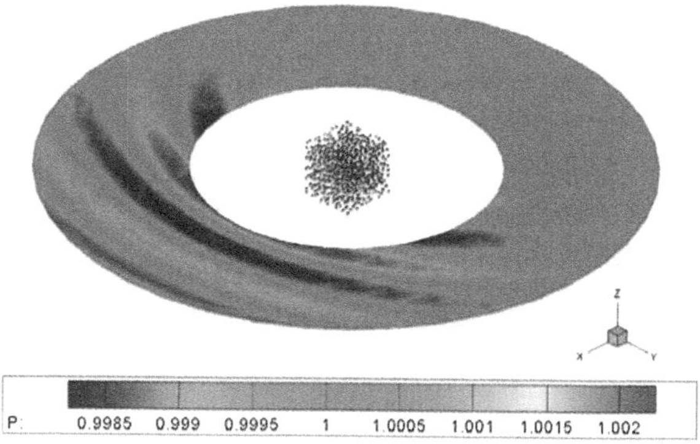

To study the computational efficiency of the fast multipole BEM, the BEM model is rerun with an increasing number of scatterers in the model. The numbers of elements are increased from 1,600 to 200,000, corresponding to 8 to 1,000 scatterers in the model. The total CPU time used to solve these multiple scatterer problems on the laptop PC is shown in Fig. 7, which exhibits a linear behavior and thus suggests the $O(N)$ efficiency of the developed fast multipole BEM code.

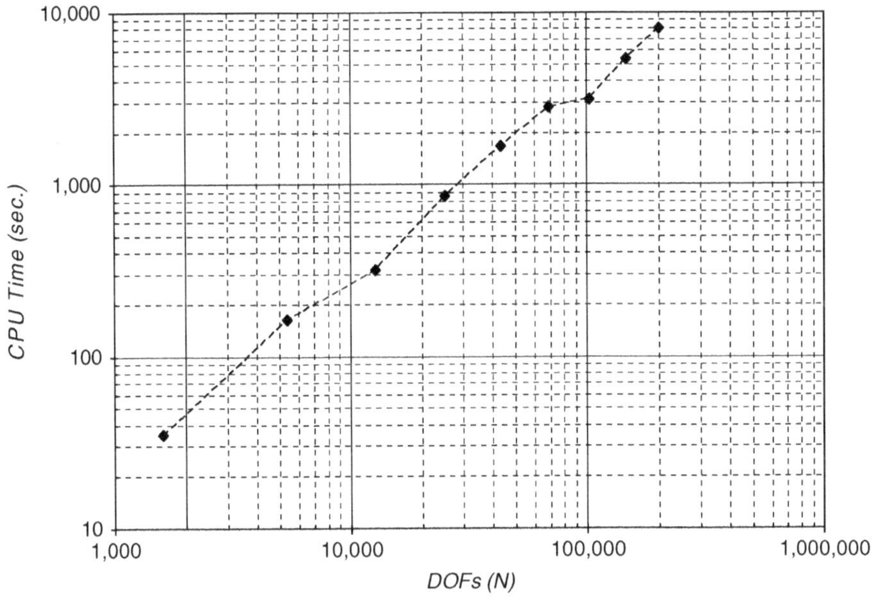

Fig. 7 Total CPU time used to solve the multiple scatterer problem

6.5 Analysis of Sound Barriers – A Half-Space Acoustic Wave Problem

Many of the acoustic problems are present in half spaces, such as noise control problems due to airplane takeoff or landing near an airport, or due to traffic on highway near a residential area. Using the BEM, these half-space acoustic wave problems can also be modeled readily. In these cases, half-space Green's function need to be employed and the same adaptive algorithm for the full-space problems can be employed. Detailed formulations of the adaptive fast multipole BEM for 3-D half-space acoustic wave problems can be found in Shen and Liu (2009).

Figures 8 and 9 show the evaluated sound levels (in dB) for a BEM model of three buildings near a highway without and with a sound barrier, respectively, using the fast multipole BEM for half-space acoustic wave problems (Shen and Liu 2009). The dimensions ($L \times W \times H$) of the three buildings are $30 \times 10 \times 20$, $20 \times 12 \times 15$ and $9.5 \times 9 \times 8$ (in m), respectively. The barrier has a height of 6 m and length of 255.94 m. One source point load with 20 Hz frequency is located 13 m away from the middle point of the barrier and 1 m above the ground. The BEM model contains 56,465 triangular elements. In the case with no sound barrier, the surface of the larger building closest to the source has the maximum sound level of 94 dB, while the smaller building that is furthest away from the source registers the smallest dB, as shown in Fig. 8. After inserting the barrier in the model, the maximum sound level on the surfaces of the buildings is reduced to 90 dB, as shown in Fig. 9. The

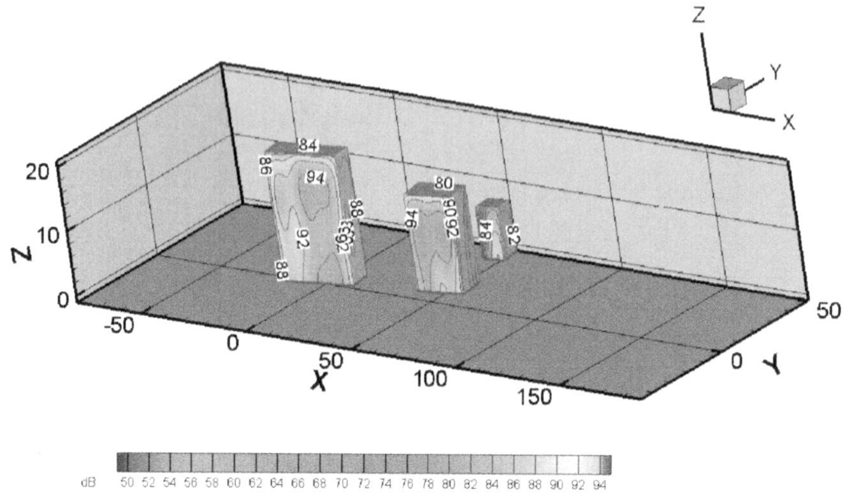

Fig. 8 Noise level (dB) on the buildings without the barrier

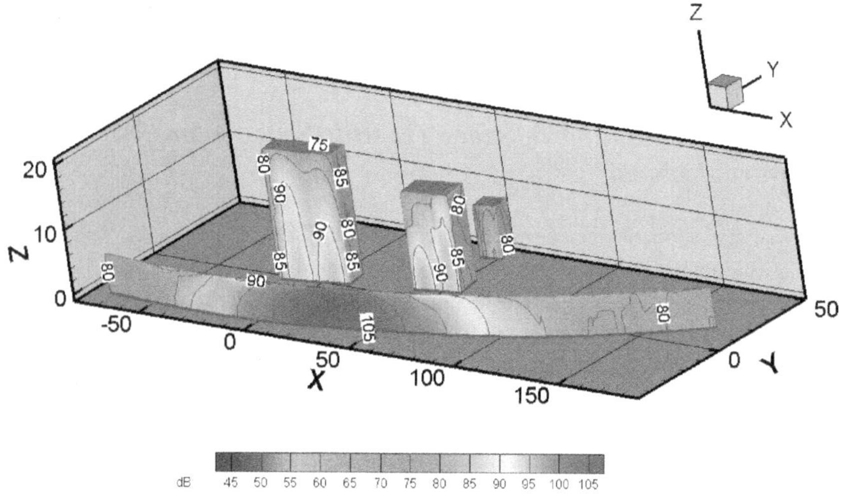

Fig. 9 Noise level (dB) on the buildings with the barrier

effect of the sound barriers in reducing the noise levels near highways is evident from this BEM simulation.

7 Conclusions

Some of the recent development of the fast multipole BEM for both 2-D and 3-D acoustic wave problems are reviewed in this paper. The basic formulations are provided and the numerical examples clearly demonstrate the potentials of the fast multipole BEM for solving large-scale acoustic wave problems. Improvements are still need be made to the fast multipole BEM discussed in this chapter. For example, adaptive tree structures can be implemented which can handle slender structure more effectively. For M2L translations, the recursive relations of the translation operators and use rotation-coaxial translation decomposition of the translation operators given by Gumerov and Duraiswami (2003) can be applied to reduce the computational complexity. The developed fast multipole BEM can also be extended to solve many other coupled acoustic problems, such as acoustic waves interacting with elastic structures (Chen and Liu 1999; Chen et al. 2000), and multi-domain acoustic wave problems as in biological applications.

References

M. Abramowitz and I. A. Stegun, *Handbook of Mathematical Functions with Formulas, Graphs, and Mathematical Tables*, 10th ed (United States Department of Commerce: U.S. Government Printing Office, Washington, D.C., 1972).

S. Amini, "On the choice of the coupling parameter in boundary integral formulations of the exterior acoustic problem," *Appl. Anal.*, **35**, No. 75–92 (1990).

A. J. Burton and G. F. Miller, "The application of integral equation methods to the numerical solution of some exterior boundary-value problems," *Proc. R. Soc. Lond. A*, **323**, No. 201–210 (1971).

J. T. Chen and K. H. Chen, "Applications of the dual integral formulation in conjunction with fast multipole method in large-scale problems for 2D exterior acoustics," *Eng Anal Bound Elem*, **28**, No. 6, 685–709 (2004).

S. H. Chen and Y. J. Liu, "A unified boundary element method for the analysis of sound and shell-like structure interactions. I. Formulation and verification," *J. Acoust. Soc. Am.*, **103**, No. 3, 1247–1254 (1999).

S. H. Chen, Y. J. Liu, and X. Y. Dou, "A unified boundary element method for the analysis of sound and shell-like structure interactions. II. Efficient solution techniques," *J. Acoust. Soc. Am.*, **108**, No. 6, 2738–2745 (2000).

R. D. Ciskowski and C. A. Brebbia, *Boundary Element Methods in Acoustics* (Kluwer Academic Publishers, New York, 1991).

R. Coifman, V. Rokhlin, and S. Wandzura, "The fast multipole method for the wave equation: A pedestrian prescription," *IEEE Antennas Propagat. Mag.*, **35**, No. 3, 7–12 (1993).

K. A. Cunefare and G. Koopmann, "A boundary element method for acoustic radiation valid for all wavenumbers," *J. Acoust. Soc. Am.*, **85**, No. 1, 39–48 (1989).

K. A. Cunefare and G. H. Koopmann, "A boundary element approach to optimization of active noise control sources on three-dimensional structures," *J. Vib. Acoust.*, **113**, July, 387–394 (1991).

E. Darve and P. Havé, "Efficient fast multipole method for low-frequency scattering," *J. Comput. Phys.*, **197**, No. 1, 341–363 (2004).

N. Engheta, W. D. Murphy, V. Rokhlin, and M. S. Vassiliou, "The fast multipole method (FMM) for electromagnetic scattering problems," *IEEE Trans. Ant. Propag.*, **40**, No. 6, 634–641 (1992).

M. Epton and B. Dembart, "Multipole translation theory for the three dimensional Laplace and Helmholtz equations," *SIAM J Sci Comput*, **16**, No. 865–897 (1995).

G. C. Everstine and F. M. Henderson, "Coupled finite element/boundary element approach for fluid structure interaction," *J. Acoust. Soc. Am.*, **87**, No. 5, 1938–1947 (1990).

M. Fischer, U. Gauger, and L. Gaul, "A multipole Galerkin boundary element method for acoustics," *Eng Anal Bound Elem*, **28**, No. 155–162 (2004).

L. Greengard, J. Huang, V. Rokhlin, and S. Wandzura, "Accelerating fast multipole methods for the helmholtz equation at low frequencies," *IEEE Comput. Sci. Eng.*, **5**, No. 3, 32–38 (1998).
L. F. Greengard, The Rapid Evaluation of Potential Fields in Particle Systems (The MIT Press, Cambridge, 1988).
L. F. Greengard and V. Rokhlin, "A fast algorithm for particle simulations," *J. Comput. Phys.*, **73**, No. 2, 325–348 (1987).
N. A. Gumerov and R. Duraiswami, "Recursions for the computation of multipole translation and rotation coefficients for the 3-D Helmholtz equation," *SIAM J. Sci. Comput.*, **25**, No. 4, 1344–1381 (2003).
M. F. Gyure and M. A. Stalzer, "A prescription for the multilevel Helmholtz FMM," *IEEE Comput. Sci. Eng.*, **5**, No. 3, 39–47 (1998).
D. S. Jones, "Integral equations for the exterior acoustic problem," *Q. J. Mech. Appl. Math.*, **27**, No. 129–142 (1974).
R. Kress, "Minimizing the condition number of boundary integral operators in acoustic and electromagnetic scattering," *Quart. J. Mech. Appl. Math.*, **38**, No. 2, 323–341 (1985).
G. Krishnasamy, T. J. Rudolphi, L. W. Schmerr, and F. J. Rizzo, "Hypersingular boundary integral equations: some applications in acoustic and elastic wave scattering," *J. App. Mech.*, **57**, No. 404–414 (1990).
R. E. Kleinman and G. F. Roach, "Boundary integral equations for the three-dimensional Helmholtz equation," *SIAM Rev.*, **16**, No. 214–236 (1974).
S. Koc and W. C. Chew, "Calculation of acoustical scattering from a cluster of scatterers," *J. Acoust. Soc. Am.*, **103**, No. 2, 721–734 (1998).
C. Lu and W. Chew, "Fast algorithm for solving hybrid integral equations," *IEE Proc. H*, **140**, No. 455–460 (1993).
Y. J. Liu, "Development and applications of hypersingular boundary integral equations for 3-D acoustics and elastodynamics", Ph.D., Department of Theoretical and Applied Mechanics, University of Illinois at Urbana-Champaign (1992).
Y. J. Liu and F. J. Rizzo, "A weakly-singular form of the hypersingular boundary integral equation applied to 3-D acoustic wave problems," *Comput. Methods Appl.Mech. Eng.*, **96**, No. 271–287 (1992).
Y. J. Liu and S. H. Chen, "A new form of the hypersingular boundary integral equation for 3-D acoustics and its implementation with C^0 boundary elements," *Comput. Methods Appl.Mech. Eng.*, **173**, No. 3–4, 375–386 (1999).
R. Martinez, "The thin-shape breakdown (TSB) of the Helmholtz integral equation," *J. Acoust. Soc. Am.*, **90**, No. 5, 2728–2738 (1991).
W. L. Meyer, W. A. Bell, B. T. Zinn, and M. P. Stallybrass, "Boundary integral solutions of three dimensional acoustic radiation problems," *J. Sound Vib.*, **59**, No. 245–262 (1978).
N. Nishimura, "Fast multipole accelerated boundary integral equation methods," *Appl. Mech. Rev.*, **55**, No. 4 (July), 299–324 (2002).
V. Rokhlin, "Rapid solution of integral equations of classical potential theory," *J. Comp. Phys.*, **60**, No. 187–207 (1985).
V. Rokhlin, "Rapid solution of integral equations of scattering theory in two dimensions," *J. Comput. Phys.*, **86**, No. 2, 414–439 (1990).
V. Rokhlin, "Diagonal forms of translation operators for the Helmholtz equation in three dimensions," *Appl. Comput. Harmon. Anal.*, **1**, No. 1, 82–93 (1993).
H. A. Schenck, "Improved integral formulation for acoustic radiation problems," *J. Acoust. Soc. Am.*, **44**, No. 41–58 (1968).
A. F. Seybert, B. Soenarko, F. J. Rizzo, and D. J. Shippy, "An advanced computational method for radiation and scattering of acoustic waves in three dimensions," *J. Acoust. Soc. Am.*, **77**, No. 2, 362–368 (1985).
A. F. Seybert and T. K. Rengarajan, "The use of CHIEF to obtain unique solutions for acoustic radiation using boundary integral equations," *J. Acoust. Soc. Am.*, **81**, No. 1299–1306 (1987).
L. Shen and Y. J. Liu, "An adaptive fast multipole boundary element method for three-dimensional acoustic wave problems based on the Burton-Miller formulation," *Comput. Mech.*, **40**, No. 3, 461–472 (2007).

L. Shen and Y. J. Liu, "An adaptive fast multipole boundary element method for 3-D half-space acoustic wave problems," **in review**, (2009).

M. A. Tournour and N. Atalla, "Efficient evaluation of the acoustic radiation using multipole expansion," *Int. J. Numer. Methods Eng.* **46**, No. 6, 825–837 (1999).

F. Ursell, "On the exterior problems of acoustics," *Proc. Cambridge Philos. Soc.*, **74**, No. 117–125 (1973).

R. Wagner and W. Chew, "A ray-propagation fast multipole algorithm," *Microwave Opt. Technol. Lett.*, **7**, No. 435–438 (1994).

T. W. Wu, A. F. Seybert, and G. C. Wan, "On the numerical implementation of a Cauchy principal value integral to insure a unique solution for acoustic radiation and scattering," *J. Acoust. Soc. Am.*, **90**, No. 1, 554–560 (1991).

S.-A. Yang, "Acoustic scattering by a hard and soft body across a wide frequency range by the Helmholtz integral equation method," *J. Acoust. Soc. Am.*, 102, No. 5, Pt. 1, November, 2511–2520 (1997).

K. Yoshida, "Applications of fast multipole method to boundary integral equation method", Ph.D. Dissertation, Department of Global Environment Engineering, Kyoto University (2001).

Some Issues on Formulations for Inhomogeneous Poroelastic Media

George D. Manolis

Abstract In this chapter, we examine formulations for the governing equations of poroelastic media exhibiting a mild type of inhomogeneity in the soil skeleton. The preferred formulation is identified, and subsequent application of algebraic transformations involving both displacement and stress fields yield a system of two coupled equations that have the same mathematical structure as those for an equivalent homogeneous medium. This is obviously an advantage, as analytical solutions for point loads can now be sought along lines laid out in the past for homogeneous poroelastic media. The same holds true for the eigenvalue problem, whereby wave numbers are sought for the types of waves that might propagate in these media and for integral formulations to address boundary-value problems. The aforementioned algebraic transformations lead to a quadratic variation of the material parameters with respect to depth, while a parallel line of work explores the exponential variation case. Finally, the issue of poroelasticity versus thermoelasticity is briefly discussed, as is the range of applicability of these formulations.

1 Introduction

A recent review by Selvadurai (2007) gives a detailed overview of analytical methodologies for problems in poroelasticity, with applications drawn primarily from the field of geomechanics. Close to 130 references are cited, ranging from the early contributions of Lord Kelvin (1878) on coupled multi-phase media to Biot's (1941) classical formulations and moving on to recent developments. Poroelestic formulations assume linear elastic behavior for the soil skeleton, as well as Darcy flow for the fluid moving through the pores, resulting in a coupled system of two partial differential equations. For quasi-static problems, these equations are elliptic-parabolic, and a uniqueness theorem (Altay and Dokmeci 1998) ascertains the well-posedness of the boundary-value problem. Although poroelasticity was

---

G.D. Manolis (✉)
Department of Civil Engineering, Aristotle University, Thessaloniki, GR-54124, Greece
e-mail: gdm@civl.auth.gr

A chapter in honor of Dimitri Beskos' 65th birthday

G.D. Manolis, D. Polyzos (eds.), *Recent Advances in Boundary Element Methods*,
DOI 10.1007/978-1-4020-9710-2_20,

originally developed for fluid-saturated soils, it is now applied to a wide variety of man-made materials. In sum, the literature on poroelasticity is quite extensive, and suffices here to mention the review article by Detournay and Cheng (1993) as well as the books by Coussy (1995) and Selvadurai (1996). Poroelastic problems can be broken down into a number of categories, namely (a) the mechanics of deformable poroelastic bodies (e.g., Rudnicki 1987); (b) consolidation of poroelastic layers (e.g., Booker and Small 1982); (c) point loads in the interior of poroelastic continua that yield fundamental solutions (e.g., Cleary 1977); poroelastic contact problems involving the indentation of porous halfspaces by rigid indentors (e.g., Yue and Selvadurai 1995); (e) subsidence problems encountered in the extraction of water, oil or gas from geological formations (e.g., Saxena 1978) that are currently of much interest to the energy industry and (f) experimental measurements for evaluation of the effective pore size in poroelastic materials (Kubik 2007), as well as other properties of engineering importance. Further complications arise in the presence of anisotropy (Tarn and Lu, 1991) and/or inhomogeneity (Mahmood and Deresiewicz 1980) in the soil skeleton, which are legitimate topics in their own right. Regarding the latter topic, it should be mentioned that the most common model is that of an exponential variation in the shear modulus along the depth and under uniaxial stress conditions (Gibson 1974). Another model is that of the linear shear modulus variation, as prompted by experimental evidence provided by measurements on London clay (Ward et al. 1965). An obvious disadvantage in both cases is that unbounded values result for the shear modulus at large depths, a problem only recently addressed by Vrettos (2008). Recently, there has been an effort to combine inhomogeneity with anisotropy (Wang and Pan 2004), which is a useful analytical effort thwarted by the extreme difficulty in determining, either experimental or through in situ measurements, numerical values for the depth variation of at least five material parameters.

In this work, we introduce a series of algebraic transformations (Manolis and Shaw 1996) in the equations governing transient poroelasticity, comprising the coupled Navier's equations for the inhomogeneous soil skeleton and Darcy's equations for incompressible fluid flow through the pores. All material parameters are assumed to be depth-dependent, with the exception of those appearing in the fluid flow constitutive law, which remain constant. Use of these transformations produces a series of constraint equations on the material parameters that are summarized as follows: a fixed Poisson's ratio, plus quadratic variation of the shear modulus, the density and the filtration. Although these constraints restrict the class of inhomogeneous solids to the above special case, nevertheless the resulting coupled system of partial differential equations in the transformed domain is similar to that obtained for an equivalent homogeneous poroelastic solid in the physical domain. Thus, the governing equations of motion are produced in a form that allows use of techniques developed in the past in the context of homogeneous material structure (Cleary 1977; Manolis and Beskos 1989; Pan 1999) for recovery of fundamental solutions and the subsequent construction of boundary integral equation techniques (Predeleanu 1984) for solving boundary value problems of practical value. Finally, the exponential transformation involving all material parameters in both solid and fluid phases, which is the most common type of inhomogeneity from an analysis point of view, is also explored herein.

2 Basic Poroelastic Formulation

We start with Biot's (1941) formulation for the governing equations of transient poroelasticty, assuming at first that all material parameters are constant:

$$\begin{aligned} \tau_{ij,j} - b(\dot{u}_i - \dot{U}_i) + F_i &= \rho_{11}\ddot{u}_i + \rho_{12}\ddot{U}_i \\ \tau_{,i} + b(\dot{u}_i - \dot{U}_i) + X_i &= \rho_{21}\ddot{u}_i + \rho_{22}\ddot{U}_i \end{aligned} \tag{1}$$

In the above, the former is Navier's equation of motion for the soil skeleton and the latter is Darcy's law of fluid flow through the pores. The constitutive law is expressed as

$$\begin{aligned} \tau_{ij} &= \left(\lambda + Q^2/R\right)\,\delta_{ij}e + 2\mu\varepsilon_{ij} + Q\delta_{ij}\,\mathrm{E} \\ \tau &= Qe + R\,\mathrm{E} \end{aligned} \tag{2}$$

while kinematics defines the strain tensors for the solid and fluid phases as

$$\varepsilon_{ij} = 0.5(u_{i,j} + u_{j,i}),\ E_{ij} = 0.5(U_{i,j} + U_{j,i}) \tag{3}$$

In the above, u_i, τ_{ij} and U_i, τ are the solid and fluid displacement and stress fields, respectively, F_i, X_i are body forces per unit volume acting on the solid and fluid phases, while ρ_{ij} are mass densities. Furthermore, b is the filtration constant, λ, μ are the Lamé constants and Q, R are additional elastic constants pertaining to the fluid phase. Also, $\tau = -fp$, where f is the porosity and p the fluid pressure. In all cases, the summation constant is implied for all indices that range from $1 - 3$ for a 3D continuum, while commas and dots respectively indicate spatial ($\underline{x} = x_i$) and temporal (t) derivatives. In their more general form, the boundary conditions involve the traction vector $t_i = \tau_{ij}n_j + \tau n_i$ and the displacements, broken down as

$$\begin{aligned} &t_i = \upsilon_{ij}n_j - \tau_{ij}n_j + \tau n_i - (1 - f)\ \mathrm{P}_i \mid f\ \mathrm{P}_i, x_i \subset S_1, S_2 \\ &u_i = g_i, \quad U_i = h_i, \quad x_i \in S_3, S_4 \end{aligned} \tag{4}$$

where S is the total surface of the poroelastic body. Finally, the initial conditions can be summarized as follows:

$$u_i\left(\underline{x}, 0\right) = u_i^{\circ}, \quad \dot{u}\left(\underline{x}, 0\right) = v_i^{o}, \quad U_i\left(\underline{x}, 0\right) = U_i^{o}, \quad \dot{U}_i\left(\underline{x}, 0\right) = V_i^{o} \tag{5}$$

In sum, the boundary value problem comprises six coupled differential equations in the three components of both solid and fluid displacement fields. Of course, it is possible to recast this system in different forms, and one such form is explored below. It should finally be mentioned that there are two limiting types of behavior: (i) undrained, whereby the fluid is trapped in the porous solid with the fluid content remaining unchanged, and (ii) drained, corresponding to zero fluid pressure p.

2.1 Alternative Poroelastic Formulation

A reformulation of the above coupled system of equations brings about a mathematical structure that is more convenient, and has the added advantage that it corresponds to transient thermoelasticity, as will be shown in the last section. Specifically, if Eq. (1) are added, then

$$\tau_{ii,j} + \tau_{,i} + Z_i = \rho_1 \ddot{u}_i + \rho_2 \ddot{U}_i \tag{6}$$

where $Z_i = F_i + X_i$ and $\rho_1 = \rho_{11} + \rho_{12}$, $\rho_2 = \rho_{21} + \rho_{22}$. Darcy's equation remains unchanged here, but it is worth noting the following result for the fluid stress may be recovered by processing the constitutive law and the kinematics:

$$(1 + Q/R)\,\tau_{,i} = \left(Q + Q^2/R\right) u_{j,ji} + (Q + R)\, U_{j,ji} \tag{7}$$

As it stands, this formulation, as well as the previous one from which it derives, corresponds to undrained conditions.

3 Governing Equations for Inhomogeneous Poroelastic Media

If the soil skeleton has depth-dependent elastic modulii $\mu = \mu(x_3 = z)$, $\lambda = \lambda(z)$, $\rho = \rho(z)$ as shown in Fig. 1, then it is possible to define an algebraic transformation for the displacement field (Manolis and Shaw 1996) that will render the Navier equations similar to those obtained for an equivalent homogeneous medium. This procedure is augmented here with a transformation for the fluid displacements as

$$u_i\left(\underline{x},t\right) = T\left(\underline{x}\right) u_i^*\left(\underline{x},t\right), \qquad U_i\left(\underline{x},t\right) = T\left(\underline{x}\right) U_i^*\left(\underline{x},t\right) \tag{8}$$

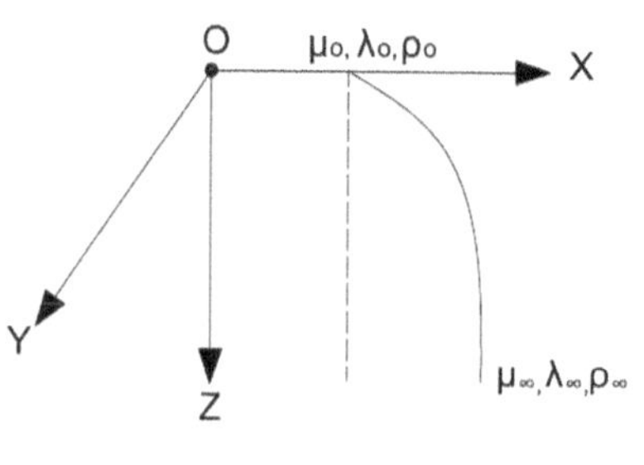

Fig. 1 Inhomogeneous three-dimensional poroelastic medium

We first start with the equilibrium operator in Navier's equation (1) for the inhomogeneous poroelastic medium and recover two components as

$$\tau_{ij,j} = \partial \left\{ \lambda \left(\underline{x}\right) \delta_{ij} e \left(\underline{x}, t\right) + 2\mu \left(\underline{x}\right) \varepsilon_{ij} \left(\underline{x}, t\right) \right\} / \partial x_i + \\ + \left\{ \left(Q^2/R\right) \delta_{ij} e_{,i} \left(\underline{x}, t\right) + Q\delta_{ij} \, \mathrm{E}_{,i} \left(\underline{x}, t\right) \right\} \tag{9}$$

The first component in the above equation pertains to the soil skeleton alone. Following Manolis and Shaw (1996) and substituting the first algebraic transformation of Eq. (8) (see also Appendix 1 for some details) yields

$$\tau_{ij,j} = \left\{ 2\mu\left(z\right) u^*_{j,ij} + \mu\left(z\right) u^*_{i,jj} \right\} \mu^{-1/2}\left(z\right) + \left\{ \left(Q^2 / R\right) u_{j,ij} + QU_{j,ij} \right\} \tag{10}$$

along with a hierarchical system of constraint equations whose consequences can be summarized as $\nu = 0.25$, $\mu(z) = (az+b)^2$, $T(z) = \mu^{-(1/2)} = (az+b)^{-1}$, where constants (a, b) are adjusted so that $\mu(z = 0) = \mu_0$. Next, assuming Q, R both remain constant, the coupling term in the right-hand-side of Eq. (10) may be combined with the fluid stress term to give, along with the aid of Eq. (7), the following result:

$$\left\{ \left(Q / R\right) + 1 \right\} \tau_{,i} \tag{11}$$

We now introduce a third algebraic transformation for the fluid stress gradient, as well as one for all body force terms expressed per unit mass as $Z_i(\boldsymbol{x}, t) = \rho(\boldsymbol{x})\zeta_i(\boldsymbol{x}, t)$:

$$\tau_{,i}\left(\underline{x}, t\right) = \mathrm{T}^{-1}\left(\underline{x}\right) \tau^*_{,i}\left(\underline{x}, t\right), \qquad \zeta_i\left(\underline{x}, t\right) = \mathrm{T}\left(\underline{x}\right) \zeta^*_i\left(\underline{x}, t\right) \tag{12}$$

The transformed fluid stress component now reads as $(1 + Q/R)\mu^{(1/2)}(z)\tau^*_{,i}$. Multiplication of the full Navier equation by $\mu^{1/2}(z)$ yields

$$2\mu\left(z\right) u^*_{j,ij} + \mu\left(z\right) u^*_{i,jj} + \left(Q/R + 1\right) \mu\left(z\right) \tau^*_{,i} + \rho\left(z\right) \zeta^*_i = \\ \rho_1\left(z\right) \ddot{u}^*_i + \rho_2\left(z\right) \ddot{U}^*_i \tag{13}$$

Normalizing both sides of the above equation by the shear modulus $\mu(z)$ and making the final assumption that the material parameters vary proportionally as $\mu(z)/\rho(z) = \mu_0/\rho_0$, then the first governing equation in the transformed domain (superscript *) is obtained in a form similar to that of an equivalent homogeneous medium:

$$2u^*_{j,ij} + u^*_{i,jj} + (Q/R + 1)\tau^*_{,i} + (\rho_0 / \mu_0)\zeta^*_i = (\rho_{10} / \mu_0)\ddot{u}^*_i + (\rho_{20} / \mu_0)\ddot{U}^*_i \tag{14}$$

The second step now is the algebraic transformation of Darcy's equation. No additional transformations are needed and substitution of Eqs. (8) and (12) in Eq. (7) gives

$$\mu^{1/2}(z)\,\tau^*_{,i} + b(z)\,\mu^{-1/2}(z)\left(\dot{u}^*_i - \dot{U}^*_i\right) + \rho_{22}(z)\,\mu^{-1/2}(z)\,\chi^*_i =$$
$$= \rho_{12}(z)\,\mu^{-1/2}(z)\,\ddot{u}_i + \rho_{22}(z)\,\mu^{-1/2}(z)\,\ddot{U}_i \tag{15}$$

Normalizing the above equation by the shear modulus $\mu(z)$ as before and employing the proportional variation of the material parameters for the filtration $b(z)/\rho(z) = b_0/\rho_0$, yields the second governing equation in the transformed domain

$$\tau^*_{,i} + (b_0/\mu_0)\left(\dot{u}^*_i - \dot{U}^*_i\right) + (\rho_{220}/\mu_0)\,\chi^*_i = (\rho_{120}/\mu_0)\,\ddot{u}^*_i + (\rho_{220}/\mu_0)\,\ddot{U}^*_i \tag{16}$$

Thus, Eqs. (14) and (16) now describe the response of the inhomogeneous poroelastic continuum in the algebraically transformed domain (superscript *).

4 The Laplace Transform

In order to recover fundamental solutions for transient poroelastic problems involving the aforementioned class of inhomogeneous solids, the Laplace transform is applied to the coupled system of Eqs. (1). The Laplace domain is also a convenient platform for formulating reciprocal theorems that yield integral equation formulations (Manolis and Beskos 1989), whereby two elastic states (one corresponding to point force solutions and the other being the unknown state sought) are convoluted in time. Both tasks are now much easier because the coupled equations of soil motion and fluid flow attain an elliptic-parabolic character.

More specifically, the direct and inverse Laplace transformations are defined as

$$\bar{f}(s) = \int_0^\infty f(t)e^{-st}dt, \quad f(t) = (1/2\pi i)\int_{c-i\infty}^{c+i\infty} \bar{f}(s)e^{st}ds \tag{17}$$

where the latter formula is a contour integral over the complex plane with c a positive number greater than all singularities of complex function $\bar{f}(s)$. The inverse Laplace transformation is performed numerical for the various solution components using algorithms that perform contour integration (Durbin 1974). Assuming all initial conditions are zero for simplicity, the Laplace transformed system of the governing equations in their alternative formulation given by Eqs. (14) and (16) is shown below as:

$$2\mu_0\bar{u}^*_{j,ij} + \mu_0\bar{u}^*_{i,jj} + (Q/R+1)\bar{\tau}^*_{,i} + \rho_0\bar{\zeta}^*_i = s^2\left(\rho_{10}\bar{u}^*_i + \rho_{20}\bar{U}^*_i\right)$$

$$\mu_0\bar{\tau}^*_{,i} + b_0 s\left(\bar{u}^*_i + \bar{U}^*_i\right) + \rho_{220}\bar{X}^*_i = s^2\left(\rho_{120}\bar{u}^*_i + \rho_{220}\bar{U}^*_i\right) \quad (18)$$

Note that the effect of initials conditions has been omitted in the above system.

5 Exponential-Type Inhomogeneity

The basic model for inhomogeneity involving an exponential variation of the elastic parameters (Gibson 1974) is now applied to the poroelastic case. Specifically, consider the material function $h(\underline{x}) = exp(2\underline{a}\cdot\underline{x})$, where argument $\underline{a}\cdot\underline{x} = a_1x_1+a_2x_2+a_3x_3$ in 3D space, with the $a_k's$ being constants. As before, the material function is identified with the algebraic transformation as $T(\underline{x}) = h^{-(1/2)}(\underline{x})$ and its spatial derivatives are listed below:

$$\left(h^{-1/2}\right)_{,k} = -h^{-1}\left(h^{1/2}\right)_{,k}, \quad h_{,k} = 2a_k h, \quad h^{-1/2}\left(h^{1/2}\right)_{,k} = a_k \quad (19)$$

For any given displacement field $\underline{w}$ in the physical domain, the exponential transformation is $w_j = h^{-(1/2)}(x)\tilde{w}_j$, where the first and second spatial derivatives (strains and curvatures, respectively) are

$$w_{j,k} = h^{-1/2}\left(-a_k\tilde{w}_j + \tilde{w}_{j,k}\right),$$

$$w_{j,kl} = h^{-1/2}\left(a_k a_l\tilde{w}_j - a_l\tilde{w}_{j,k} - a_k\tilde{w}_{j,l} + \tilde{w}_{j,kl}\right) \quad (20)$$

Next, all material parameters of the poroelastic continuum have a proportional variation defined by the same material function h, as given below:

$$\lambda = \lambda_0 h(x), \quad \mu = \mu_0 h(x), \quad Q = Q_0 h(x), \quad R = R_0 h(x) \quad (21)$$

We start with the constitutive laws for the stresses in the solid phase and the fluid pressure in Eqs. (2) written in terms of the displacement fields:

$$\tau_{ij} = \left(\lambda + Q^2/R\right)\delta_{ij}u_{j,j} + \mu(u_{i,j} + u_{j,i}) + Q\delta_{ij}U_{j,j}$$

$$\tau = Qu_{j,j} + RU_{j,j} \quad (22)$$

By applying the exponential transformation for both solid and fluid displacement fields we recover the following forms in the transformed domain (superscript $\sim$):

$$\begin{aligned}
\tau_{ij,j} = {} & \left(\lambda_0 + Q_0^2/R_0\right) h^{1/2} \left[2a_j\delta_{ij}\left(-a_i\tilde{u}_i + \tilde{u}_{i,i}\right)\right. \\
& \left. +\delta_{ij}\left(a_i a_j \tilde{u}_i - a_j\tilde{u}_{i,i} - a_i\tilde{u}_{i,j} + \tilde{u}_{i,ij}\right)\right] \\
& +\mu_0 h^{1/2}\left[2a_j\left(-a_i\tilde{u}_j + \tilde{u}_{i,j} - a_j\tilde{u}_i + \tilde{u}_{j,i}\right)\right. \\
& \left. + \left(a_j^2\tilde{u}_i - a_j\tilde{u}_{i,j} - a_j\tilde{u}_{j,i} + \tilde{u}_{i,jj} + a_i a_j\tilde{u}_j - a_j\tilde{u}_{j,i} - a_i\tilde{u}_{j,j} + \tilde{u}_{j,ij}\right)\right] \\
& +Q_0 h^{1/2}\left[2a_j\,\delta_{ij}\left(-a_i\tilde{U}_i + \tilde{U}_{i,i}\right) + \delta_{ij}\left(a_i a_j\tilde{U}_i - a_j\tilde{U}_{i,i} - a_i\tilde{U}_{i,j} + \tilde{U}_{i,ij}\right)\right] \\
\tau_{,i} = {} & Q_0 h^{1/2}\left[2a_i\left(-a_j\tilde{u}_j + \tilde{u}_{j,j}\right) + \left(a_j a_i\tilde{u}_j - a_i\tilde{u}_{j,j} - a_j\tilde{u}_{j,i} + \tilde{u}_{j,ji}\right)\right] \qquad (23) \\
& +R_0 h^{1/2}\left[2a_i\left(-a_j\tilde{U}_j + \tilde{U}_{j,j}\right) + \left(a_j a_i\tilde{U}_j - a_i\tilde{U}_{j,j} - a_j\tilde{U}_{j,i} + \tilde{U}_{j,ji}\right)\right]
\end{aligned}$$

Next, we will recast the above system in the frequency domain (i.e., under steady-state conditions).

5.1 Frequency Domain Formulation

The exponential transformation for inhomogeneous materials is best developed in the frequency domain. To that end, the terms comprising the diffusion and inertia parts of Eqs. (1) assume the form

$$b(\dot{u}_k - \dot{U}_k) = i\omega b_0 h(\tilde{u}_k - \tilde{U}_k),\;\; \rho_{kl}\ddot{u}_m = -\omega^2\rho_{kl0}\tilde{u}_m,\;\; \rho_{kl}\ddot{U}_m = -\omega^2\rho_{kl0}\tilde{U}_m \qquad (24)$$

with $k, l = 1, 2$. The density terms are now re-defined as

$$\rho_{km}^* = \rho_{km}\omega^2 - i(-1)^{k+m}\omega b \qquad (25)$$

Replacing Eqs. (22), (23) and (24) in Eqs. (1) and canceling the $h^{(1/2)}$ factor yields a matrix system of six differential equations in the six displacement components of the poroelastic continuum in the exponentially transformed domain and under steady-state conditions as

$$L(W) = \Phi, \quad W = (\tilde{u}_i, \tilde{U}_i), \quad \Phi = (-h^{-1/2}F_i, -h^{-1/2}X_i). \qquad (26)$$

The various terms appearing in the above are listed in Appendix 2, where it is observed that the differential operator has the structure $L = A(\partial^2) + N(\partial) + \Gamma$, in which the first partial derivative components (∂) indicate dispersion phenomena. A standard check for the above formulation is to set $h = 1.0$, in which case L reduces to that for the homogeneous poroelastic continuum, namely to $L = A(\partial^2) + \Gamma$.

5.2 The Plane-Strain Case

Here we have that $\underline{u} = \underline{u}(x_1, x_2), \underline{U} = \underline{U}(x_1, x_2)$ and $a_1 = a_3 = 0$. Also, let $\lambda_0 + Q_0^2/R_0 = \mu_0$; then, the differential operator in the exponentially-transformed domain assumes the form given below:

$$A(\partial^2) = \begin{pmatrix} 3\mu_0\partial_1^2 + \mu_0\partial_2^2 & 2\mu_0\partial_{12}^2 & Q_0\partial_1^2 & Q_0\partial_{12}^2 \\ 2\mu_0\partial_{12}^2 & 3\mu_0\partial_2^2 + \mu_0\partial_1^2 & Q_0\partial_{12}^2 & Q_0\partial_2^2 \\ Q_0\partial_1^2 & Q_0\partial_{12}^2 & R_0\partial_1^2 & R_0\partial_{12}^2 \\ Q_0\partial_{12}^2 & Q_0\partial_2^2 & R_0\partial_{12}^2 & R_0\partial_2^2 \end{pmatrix},$$

$$N(\partial) = \begin{pmatrix} 0 & 0 & 0 & -Q_0a_2\partial_1 \\ 0 & 0 & Q_0a_2\partial_1 & 0 \\ 0 & -Q_0a_2\partial_1 & 0 & -R_0a_2\partial_1 \\ Q_0a_2\partial_1 & 0 & R_0a_2\partial_1 & 0 \end{pmatrix}, \text{ and}$$

$$\Gamma = \begin{pmatrix} \rho_{110}^* - \mu_0a_2^2 & 0 & \rho_{120}^* & 0 \\ 0 & \rho_{110}^* - 3\mu_0a_2^2 & 0 & \rho_{120}^* - Q_0a_2^2 \\ \rho_{120}^* & 0 & \rho_{220}^* & 0 \\ 0 & \rho_{120}^* - Q_0a_2^2 & 0 & \rho_{220}^* - R_0a_2^2 \end{pmatrix} \tag{27}$$

In the above 2D case, the inhomogeneous soil skeleton equations are recovered if $Q_0 = R_0 = 0$, a form that involves complex wave numbers (Dineva et al. 2006). Also, note the symmetric structure of A, Γ and the anti-symmetric one of N.

6 Poroelasticity Versus Themoelasticity

A careful reformulation of the equations of poroelasticty, which basically involves writing the constitutive equation for fluid stress (Eq. (2)) in rate form

$$\dot{U}_{i,i} = \dot{\tau}/R - Q\dot{u}_{i,i}/R \tag{28}$$

allows for a parallel solution of the corresponding equations of thermoelasticity (Norris 1994). An intermediate step involves taking the dot product of the gradient of the fluid stress in Darcy's equation as

$$\tau_{,ii} = \rho_{21}\ddot{u}_{i,i} + \rho_{22}\ddot{U}_{i,i} + b(\dot{U}_{i,i} - \dot{u}_{i,i}) - X_{i,i} \tag{29}$$

Combining Eqs. (28) and (29) gives the modified Darcy's equation in form comparable with the coupled heat equation, that is

$$R\tau_{,ii} - b\dot{\tau} + b(Q + R)\dot{u}_{i,i} + RX_{i,i} = \rho_{21}R\ddot{u}_{i,i} + \rho_{22}R\ddot{U}_{i,i} \tag{30}$$

Thus, a comparison of the equations of thermoelasticity and the reformulated ones of poroelasticity in terms of the dependent variables and the material parameters establishes the links shown in Tables 1 and 2 (Manolis and Beskos 1989).

Regarding thermoelasticty, the new parameters introduced are thermal displacements and tractions u_i, t_i, temperature T, normal heat flux $q_n = -kT_{,i}n_i$ with k the thermal conductivity, heat per unit time and volume W, specific heat per unit volume c_ε, material density ρ and finally $m = (3\lambda + 2\mu)\alpha_T$ with α_T the coefficient of thermal expansion. Given that this analogy in no way interferes with the spatial dependence of the material parameters, formulations for inhomogeneous poroelasticty readily extends to inhomogeneous thermoelasticty as well. Finally, the inverse process, i.e., going from thermoelasticty to poroelasticty, is possible only for certain classes of poroelastic materials with $\rho_{12} = \rho_{21} = \rho_{22} = 0$ (Predeleanu 1984). The problem arises from the fact that Eq. (30) involves the fluid accelerations. If those were absent and the aforementioned equations were in terms of (u_i, τ) only, the analogy would be complete. Thus, efforts have been made in establishing relations between u_i, U_i, τ. For instance, Chen and Dargush (1995) used the relation

$$q_i = -\kappa(p_{,i} + \rho_f\ddot{u}_i + (\rho_f/f)\ddot{w}_i), \quad \ddot{w}_i = (\ddot{U}_i - \ddot{u}_i) \tag{31}$$

where p is the pressure, fluid flux q_i is equal to the fluid velocity relative to that of the solid, and κ, ρ_{f}, f are the permeability, fluid density and porosity, respectively.

In closing, the above analogy fits well with the material inhomogeneity of the quadratic type, because in this case the governing equations reduce to a form similar to that of an equivalent homogeneous medium. Finally, it is interesting to speculate on formulations that use the divergence form of Darcy's equation, in lieu of the gradient form given in Eq. (29), in an effort to establish the poroelastic-thermoelastic

Table 1 Analogy between poroelasticity (P) and thermoelasticity (T) variables

Poroelastic	u_i	t_i	$T\alpha$	E	$F_i + X_i$	$RX_{i,i}$
Thermoelastic	u_i	t_i	T	q_n	F_i	W

Table 2 Analogy between poroelasticity (P) and thermoelasticity (T) materials

Poroelastic	λ, μ	ρ_{11}	$Q/R + 1$	R	B	$B(Q + R)$
Thermoelastic	λ, μ	P	$-m$	K	c_ε	$-mT_0$

analogy. This venue may turn out more suitable to use in conjunction with the exponential type of inhomogeneity.

7 Conclusions

This chapter demonstrated the possibility for obtaining convenient formulations in transient poroelasticity for certain classes of inhomogeneous media where the soil skeleton exhibits depth-dependent properties. This is achieved through use of algebraic transformations for the solid and fluid displacement fields, as well as for the fluid stress field. These transformations are valid under a set of constraint equations that limit the type of the inhomogeneous soil skeleton to one with a fixed Poisson's ratio, and quadratic variation in the depth coordinate of the shear modulus, the densities and the filtration parameter. The resulting system of partial differential equations in the algebraically transformed domain (coupled Navier and Darcy equations) now have the same mathematical structure as those for an equivalent homogeneous soil skeleton, thus allowing the recovery of Green's functions in either the Fourier or Laplace transformed domains. One particularly attractive formulation involves four coupled differential equations in terms of the solid phase displacements and the fluid pressure that can be readily correlated with the thermoelastic problem through simple matching of all relevant dependent variables and material parameters. There is, however, a drawback in the quadratic inhomogeneity case because it pre-supposes independent mappings of the displacement and stress fields through the aforementioned algebraic transformations.

The second type of inhomogeneity, namely one involving an exponential variation with depth of all material parameters involving both solid and fluid phases, was also investigated. The most suitable formulation comprises six coupled differential equations in the three displacement components of the solid and fluid phases. Although this is more cumbersome to use because of the presence of diffusion-like terms in the transformed domain, non-the-less it has a symmetric structure that is most useful when solving analytically to recover fundamental solutions by techniques such as the Radon transform.

In closing, a next step would be to recover Green's functions form these formulations that can be used within the context of integral equation formulations, where two material states, one pertaining to the point-force solution and another to the physical problem of interest, are related. Subsequently, a limiting process yields boundary-only integral formulations, whose numerical treatment enables the efficient solution of boundary value problems of engineering importance.

Acknowledgments The author wishes to thank Associate Professor PS Dineva and Professor TV Rangelov of the Bulgarian Academy of Sciences in Sofia for their collaboration on problems involving inhomogeneous media over the past years.

Appendix 1. The Algebraic Transformation

In here, we list some results relevant to the algebraic transformations in conjunction with the dependent variable fields of the inhomogeneous poroelastic medium. Specifically, the spatial derivatives of the soil skeleton displacement field are

$$\begin{aligned} u_{i,j} &= \mathrm{T}u^{*}_{i,j} + \mathrm{T},_j\, u^{*}_{i} \\ u_{j,ij} &= \mathrm{T}u^{*}_{j,ij} + \mathrm{T},_j\, u^{*}_{j,i} + \mathrm{T},_i\, u^{*}_{j,j} + \mathrm{T},_{ij}\, u^{*}_{j} \\ u_{i,jj} &= \mathrm{T}u^{*}_{i,jj} + 2\mathrm{T},_j\, u^{*}_{i,j} + \mathrm{T},_{jj}\, u,^{*}_{i} \end{aligned} \tag{32}$$

Similar results can be recovered for the fluid displacements and stress. Finally, the spatial derivatives of transformation $T(\boldsymbol{x})$ itself are as follows:

$$\begin{aligned} \mathrm{T}\left(\underline{x}\right) &= \mu^{-1/2}\left(\underline{x}\right) \\ \mathrm{T},_i &= -0.5\mu^{-3/2}\mu,_i \\ \mathrm{T},_{ij} &= 0.75\mu^{-5/2}\mu,_i\,\mu,_j - 0.5\mu^{-3/2}\mu,_{ij} \end{aligned} \tag{33}$$

Appendix 2. Matrix Differential Equation for Exponential Case

The general structure of the matrix differential equation for the exponentially inhomogeneous case is $L = A(\partial^2) + N(\partial) + \Gamma$, and its coefficients l_{ij} are symmetric. They are listed below as follows:

1st *row:*

$$\begin{aligned} l_{11} &= \left(\lambda_0 + Q_0^2/R_0 + 2\mu_0\right)\partial_1^2 + \mu_0\partial_2^2 + \mu_0\partial_3^2 - \left(\lambda_0 + Q_0^2/R_0 + 2\mu_0\right)a_1^2 \\ &\quad - \mu_0 a_2^2 - \mu_0 a_3^2 + \rho^{*}_{110} \\ l_{12} &= \left(\lambda_0 + Q_0^2/R_0 + \mu_0\right)\partial_{12}^2 + a_1\left(\lambda_0 + Q_0^2/R_0 - \mu_0\right)\partial_2 \\ &\quad - a_2\left(\lambda_0 + Q_0^2/R_0 - \mu_0\right)\partial_1 - a_1 a_2\left(\lambda_0 + Q_0^2/R_0 + \mu_0\right) \\ l_{13} &= \left(\lambda_0 + Q_0^2/R_0 + \mu_0\right)\partial_{13}^2 + a_1\left(\lambda_0 + Q_0^2/R_0 - \mu_0\right)\partial_3 \\ &\quad - a_3\left(\lambda_0 + Q_0^2/R_0 - \mu_0\right)\partial_1 - a_1 a_3\left(\lambda_0 + Q_0^2/R_0 + \mu_0\right) \\ l_{14} &= Q_0\left(\partial_1^2 - a_1^2 + \rho^{*}_{120}\right) \\ l_{15} &= Q_0\left(\partial_{12}^2 + a_1\partial_2 - a_2\partial_1 + a_1 a_2\right) \\ l_{16} &= Q_0\left(\partial_{13}^2 + a_1\partial_3 - a_3\partial_1 + a_1 a_3\right) \end{aligned}$$

2nd *row:*

$$
\begin{aligned}
l_{22} &= \left(\lambda_0 + Q_0^2/R_0 + 2\mu_0\right)\partial_2^2 + \mu_0\partial_1^2 + \mu_0\partial_3^2 - \left(\lambda_0 + Q_0^2/R_0 + 2\mu_0\right)a_2^2 \\
&\quad - \mu_0 a_1^2 - \mu_0 a_3^2 + \rho_{110}^* \\
l_{23} &= \left(\lambda_0 + Q_0^2/R_0 + \mu_0\right)\partial_{23}^2 + a_2\left(\lambda_0 + Q_0^2/R_0 - \mu_0\right)\partial_3 \\
&\quad - a_3\left(\lambda_0 + Q_0^2/R_0 - \mu_0\right)\partial_2 - a_2 a_3\left(\lambda_0 + Q_0^2/R_0 + \mu_0\right) \\
l_{24} &= Q_0\left(\partial_{12}^2 - a_1\partial_2 + a_2\partial_1 + a_1 a_2\right) \\
l_{25} &= Q_0\left(\partial_2^2 - a_2^2\right) + \rho_{120}^* \\
l_{26} &= Q_0\left(\partial_{23}^2 + a_2\partial_3 - a_3\partial_2 + a_2 a_3\right)
\end{aligned}
$$

3rd *row:*

$$
\begin{aligned}
l_{33} &= \left(\lambda_0 + Q_0^2/R_0 + 2\mu_0\right)\partial_3^2 + \mu_0\partial_1^2 + \mu_0\partial_2^2 \\
&\quad - \left(\lambda_0 + Q_0^2/R_0 + 2\mu_0\right)a_3^2 - \mu_0 a_1^2 - \mu_0 a_2^2 + \rho_{110}^* \\
l_{34} &= Q_0\left(\partial_{13}^2 - a_1\partial_3 + a_3\partial_1 + a_1 a_3\right) \\
l_{35} &= Q_0\left(\partial_{23}^2 - a_2\partial_3 + a_3\partial_2 + a_2 a_3\right) \\
l_{36} &= Q_0\left(\partial_3^2 - a_3^2\right) + \rho_{120}^*
\end{aligned}
$$

4th *row:*

$$
\begin{aligned}
l_{44} &= R_0\left(\partial_1^2 - a_1^2\right) + \rho_{220}^* \\
l_{45} &= R_0\left(\partial_{12}^2 + a_1\partial_2 - a_2\partial_1 + a_1 a_2\right) \\
l_{46} &= R_0\left(\partial_{13}^2 + a_1\partial_3 - a_3\partial_1 + a_1 a_3\right)
\end{aligned}
$$

5th *row:*

$$
\begin{aligned}
l_{55} &= R_0\left(\partial_2^2 - a_2^2\right) + \rho_{220}^* \\
l_{56} &= R_0\left(\partial_{23}^2 + a_2\partial_3 - a_3\partial_2 + a_2 a_3\right)
\end{aligned}
$$

6th *row:*

$$
l_{66} = R_0\left(\partial_3^2 - a_3^2\right) + \rho_{220}^* \tag{34}
$$

References

Altay GA, Dokmeci, MC (1998) A uniqueness theorem in Biot's poroelasticity theory. ZAMP 49:838–847

Biot MA (1941) General theory of three-dimensional consolidation. Appl. Phys. 12:155–164

Booker JR, Small JC (1982) Finite layer analysis of soil consolidation I. Int. J. Num. Anal. Meth. Geom. 6:151–171

Chen J, Dargush GF (1995) Boundary element method for dynamic poroelastic and thermoelestaic analyses. Int. J. Sol. Struct. 32: 2257–2278

Cleary MP (1977) Fundamental solutions for a fluid-saturated porous solid. Int. J. Solid. Struct. 13: 785–806

Coussy O (1995) Mechanics of porous media. Wiley, New York

Detournay E, Cheng AHD (1993) Fundamentals of poroelasticity. In: Hudson J (ed) Comprehensive rock engineering, Pergamon, Oxford

Dineva PS, Datcheva M, Shanz T (2006) BIEM for seismic wave propagation in fluid saturated multilayered media. In: HF Schweiger (ed), Proceedings of the Sixth European Conference on Numerical Methods in Geotechnical Engineering, Taylor and Francis/Balkema, Rotterdam, pp. 257–265

Durbin F (1974) Numerical inversion of Laplace transforms: an efficient improvement to Dubner and Abate's methods. Comp. J. 17:371–376

Gibson RE (1974) The analytical method in soil mechanics. Geotech. 24:115–140

Kubik J (2007) Ultrasonic waves in saturated porous media: Pore/grain size identification. In: J Katsikadelis (ed) Proceedings of the Sixth GGP Conference on Progress in Mechanics, NTUA Publication, Athens, pp. 71–72

Mahmood MS, Deresiewicz H (1980) Settlement of inhomogeneous consolidating soils I. The single drained layer under confined compression. Int. J. Num. Anal. Meth. Geom. 4:57–72

Manolis GD, Beskos DE (1989) Integral formulations and fundamental solutions of dynamic poreoelasticity and thermoelasticity. Acta Mech. 76:89–104; (1990) Errata. Acta Mech 83: 223–226

Manolis GD, Shaw RP (1996) Green's functions for the vector wave equation in a mildly heterogeneous continuum. Wave Motion 24:59–83

Norris AN (1994) Dynamic Green's functions in anisotropic, piezoelectric, thermoelastic and poroelastic solids. Proc. Roy. Soc. Lond. A 447: 175–188

Pan E (1999) Green's functions in layered poroelastic half-spaces. Int. J. Num. Anal. Meth. Geom. 23:1631–1653

Predeleanu M (1984) Development of boundary element method to dynamic problems for porous media. Appl. Math. Mod. 8:378–382

Rudnicki JW (1987) Plane-strain dislocations in linear elastic diffusive solids. ASME J. Appl. Mech. 54: 545–552

Saxena SK, ed (1978) Evaluation and prediction of subsidence. ASCE, New York

Selvadurai APS, ed (1996) Mechanics of porous media. Kluwer, Dordrect

Selvadurai APS (2007) The analytical methods in geomechanics. Appl. Mech. Rev. 60:87–106

Tarn JQ, Lu CC (1991) Analysis of subsidence due to a point sink in an anisotropic porous elastic halfspace. Int. J. Num. Anal. Meth. Geom. 15:572–592

Thompson W, Lord Kelvin (1878) On the thermoelastic, thermomagnetic and pyroelctric properties of matter. Q. J. Math. 1:4–27

Vrettos C (2008) Green's functions for harmonic surface loads on elastic solids with depth-degrading stiffness. EABE J. Spec. Iss. on BEM/MRM for Inh. Sol., 32:1037–1045

Wang CD, Pan E (2004) Stresses due to vertical subsurface loading for an inhomogeneous cross-anisotropic halfspace. Int. J. Num. Anal. Meth. Geom. 28:1233–1255

Ward WH, Marsland A, Samuels SG (1965) Properties of the London clay at the Ashford common shaft. Geotech. 15:324–344

Yue ZQ, Selvadurai APS (1995) Contact problems for saturated poroelastic solid. J. Eng. Mech. 121: 502–512

Axisymmetric Acoustic Modelling by Time-Domain Boundary Element Techniques

Webe João Mansur, Arnaldo Warszawski and Delfim Soares Jr.

Abstract In this work, a numerical time-domain approach to model acoustic wave propagation in axisymmetric bodies is developed. The acoustic medium is modeled by the Boundary Element Method (BEM), whose time convolution integrals are evaluated analytically, employing the concept of finite part integrals. All singularities for space integration, present at the expressions generated by time integration, are treated adequately. Some applications are presented in order to demonstrate the validity of the analytical expressions generated for the BEM, and the results obtained with the present approach are compared with those generated by applying numerical time integration.

1 Introduction

The Boundary Element Method (BEM) has been widely used in the last decades to solve transient wave propagation problems in acoustic media, governed by the scalar wave equation. The computational advantage of only discretizing the boundary of the medium as well as the possibility of representing infinite media without mesh truncation are two characteristics that make BEM fairly suitable to acoustic problems modelling.

Initially, boundary element procedures were developed to consider axisymmetric acoustic wave propagation considering only frequency-domain formulations. In this context, Grannell et al. (1994) deduced a high precision method to solve the Neumann problem for the Helmholtz equation and Tsinopoulos et al. (1999) and Soenarko (1993) dealt with acoustic wave propagation problems with non-axisymmetric boundary conditions. Regarding time-domain analysis, numerical integrations of 3D fundamental solutions around an axisymmetric axis are among the first approaches developed (Israil et al. 1992; Beskos 1997). In a recent work, Czygan and Von Estorff (2003) presented a procedure where integration of the 3D

W.J. Mansur (✉)
Department of Civil Engineering, COPPE – Federal University of Rio de Janeiro, CP 68506, CEP 21945-970, Rio de Janeiro, RJ, Brazil
e-mail: webe@coc.ufrj.br

G.D. Manolis, D. Polyzos (eds.), *Recent Advances in Boundary Element Methods*,
DOI 10.1007/978-1-4020-9710-2_21,

fundamental solution around the axis of axisymmetry is carried out analytically, thus obtaining the axisymmetric fundamental solution. The time integration of the resulting expressions, however, was evaluated numerically. In the present work, these integrals are evaluated analytically, generating expressions that are integrated numerically in space, along the discretizated boundary. In some cases, the concept of finite part integrals, introduced by Hadamard (1952) and used by Mansur and Carrer (1993) to perform analytical integration of two-dimensional kernels, is required.

2 Basic Concepts

The scalar wave equation (Morse and Feshbach 1953), which governs wave propagation in acoustic media, is given by:

$$\nabla^2 p - \frac{1}{c^2}\frac{\partial^2 p}{\partial t^2} = 0 \tag{1}$$

where c is the medium wave propagation velocity and $p(X,t)$ is the fluid pressure (X represents a generic point in the domain and$t$stands for time). The problem boundary conditions are given by (null initial conditions are considered here):

$$\begin{aligned} p\,(X,t) &= \bar{p},\ \ X \in \Gamma_p \\ \frac{\partial p}{\partial n}\,(X,t) &= \bar{q},\ \ X \in \Gamma_q \end{aligned} \tag{2}$$

where Γ_p and Γ_q are regions of the boundary where essential and natural boundary conditions are prescribed, respectively (n stands for the coordinate in the direction of the outward unit vector normal to the boundary at X).

Using a weighted residual sentence, as presented in Mansur (1983), equation (1) can be manipulated, providing the following integral equation:

$$\begin{aligned} c\,(\xi)\,p\,(\xi,t) = &\int_0^{t^+}\int_\Gamma p^*\,(X,t,\xi,\tau)\,q\,(X,\tau)\,d\Gamma d\tau \\ &- \int_0^{t^+}\int_\Gamma q^*\,(X,t,\xi,\tau)\,p\,(X,\tau)\,d\Gamma\,d\tau \end{aligned} \tag{3}$$

where the coefficient $c(\xi)$ is a geometrical parameter (as it is in a time-independent case). The three-dimensional fundamental solution and its spatial derivative are given by:

$$p^* = (c/(4\pi R))\,\delta\left[c\,(t-\tau)-R\right] \tag{4}$$

$$q^* = \partial p^*/\partial n = (\partial r/\partial n)(1/4\pi R)\left(-(c/R)\delta\left[c\,(t-\tau)-R\right]+\dot{\delta}\left[c\,(t-\tau)-R\right]\right) \tag{5}$$

where δ is the Dirac Delta function and R is the distance between the source and the field point.

The equations above refer to the three-dimensional case. For the particular case of axisymmetry, Czygan and Von Estorff (2003) separated the integrals along the three-dimensional boundary into integrals along the axisymmetric 2D boundary generatrix followed by integrals around the axis of axisymmetry. The infinitesimal boundary element was represented by:

$$d\Gamma = x_1\,d\theta\,d\Gamma_{axi} \tag{6}$$

where, as depicted in Fig. 1, x_1 is an "axis 1" coordinate and the axisymmetric boundary is contained in "plane 1–2".

Due to axisymmetry, the problem and, therefore, its solution, does not depend on the θ angle around the axis of axisymmetry. Thus, the integral along θ can be eliminated through integration of the three-dimensional fundamental solution, obtaining the axisymmetric fundamental solution (Czygan and Von Estorff 2003):

$$p^*_{axi} = \frac{1}{2\pi}\int_0^{2\pi} p^*_{3d}\,d\theta = \frac{c}{4\pi^2}\frac{H\left[c\,(t-\tau)-r\right]\;H\left[\sqrt{4x_1\xi_1+r^2}-c\,(t-\tau)\right]}{\left\{\left[c^2\,(t-\tau)^2-r^2\right]\left[x_1\xi_1-\frac{1}{4}\left(c^2\,(t-\tau)^2-r^2\right)\right]\right\}^{1/2}} \tag{7}$$

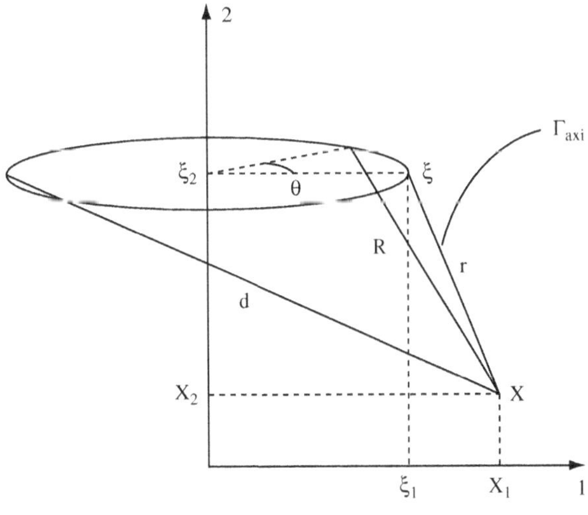

Fig. 1 Geometrical description of the axisymmetric problem

where r relates to R by the following formula:

$$R(r,\theta) = \sqrt{r^2 + 2x_1\xi_1(1-\cos\theta)} \tag{8}$$

r being the shortest distance between the field point and the source ring generated by the rotation of the source point located on the axisymmetric 2D boundary. The greatest distance d is given by:

$$d = \sqrt{r^2 + 4x_1\xi_1} \tag{9}$$

The expression on the right-hand-side of equation (7) is non-null only for $r < c(t-\tau) < d$; i.e., only for a point already reached by the wave emitted by its nearest point on the source ring and not reached by the wave emitted by its farthest point on this source ring (thus, the axisymmetric case is an intermediate case between the two-dimensional and three-dimensional cases).

The derivative of the axisymmetric fundamental solution is given by:

$$q^*_{axi} = \left\{ [n_1 x_1 + n_2 (x_2 - \xi_2)] \left[x_1 \xi_1 - (1/2)\left(c^2 (t-\tau)^2 - r^2\right)\right] - n_1 x_1 \xi_1^2 \right\} \mathrm{x}$$

$$\mathrm{x} \frac{c}{4\pi^2} \frac{H[c(t-\tau)-r]\, H\left[\sqrt{4x_1\xi_1 + r^2} - c(t-\tau)\right]}{\left\{ \left[c^2 (t-\tau)^2 - r^2\right] \left[x_1\xi_1 - \frac{1}{4}\left(c^2 (t-\tau)^2 - r^2\right)\right] \right\}^{3/2}} \tag{10}$$

where n_1 and n_2 are the components of the outward unit vector normal to the boundary, with respect to axes 1 and 2 shown in Fig. 1.

When the source point ξ or the field point X is located along the axis of axisymmetry, the axisymmetric fundamental solution coincides with the three-dimensional fundamental solution.

3 Time Integration

In order to perform numerical analyses using equation (3), it is necessary to discretize the model into boundary elements and time steps. In order to do so, pressure and flux values, at any boundary point, at any instant, are approximated by means of discrete values at a number of boundary nodes, at certain discrete time instants, as indicated below:

$$p(X,t) = \sum_{j=1}^{NN} \sum_{m=1}^{n+1} \phi^m(t) \eta^j(X) p_j^m \tag{11}$$

$$q(X,t) = \sum_{j=1}^{NN} \sum_{m=1}^{n+1} \theta^m(t) \nu^j(X) q_j^m \tag{12}$$

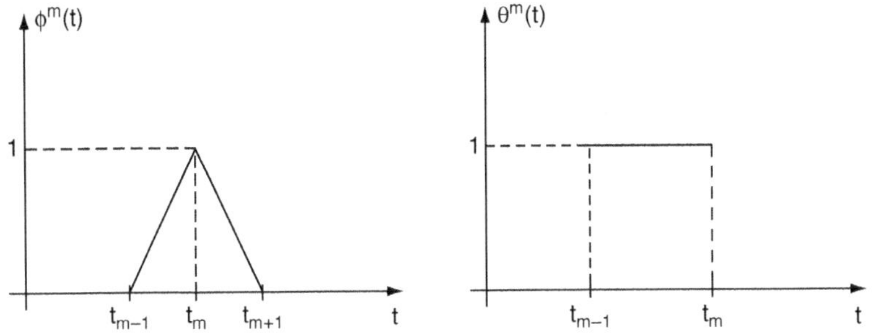

Fig. 2 Time interpolation functions $\phi^m(t)$ and $\theta^m(t)$

where $\phi^m(t)$ and $\theta^m(t)$ are time interpolation functions and $\eta^j(X)$ and $\nu^j(X)$ are space interpolation functions, for p and q, respectively. The integer n represents the number of time steps up to the current time of the analysis and NN stands for the number of boundary nodes. p_j^m and q_j^m are, respectively, pressure and flux values at node j and at discrete time m. In this work, linear and piecewise constant interpolation functions are adopted for pressure and flux time discretizations, respectively (see Fig. 2); for spatial discretization, linear shape functions are considered (see Fig. 3).

By introducing equations (11) and (12) into equation (3) (adapted for the axisymmetric case), the following algebraic system of equations can be obtained:

$$C_{ij} p_j^{n+1} = \sum_{m=1}^{n+1} \sum_{j=1}^{NN} G_{ij}^{(n+1)m} q_j^m - \sum_{m=1}^{n+1} \sum_{j=1}^{NN} H_{ij}^{(n+1)m} p_j^m \tag{13}$$

where:

$$H_{ij}^{(n+1)m} = \int_\Gamma \eta^j(X) x_1(X) \int_0^{t_{n+1}} q^*(X, t_{n+1}; \xi_i, \tau) \phi^m(\tau) d\tau d\Gamma_{axi}(X) \tag{14}$$

$$G_{ij}^{(n+1)m} = \int_\Gamma \nu^j(X) x_1(X) \int_0^{t_{n+1}} p^*(X, t_{n+1}; \xi_i, \tau) \theta^m(\tau) d\tau d\Gamma_{axi}(X) \tag{15}$$

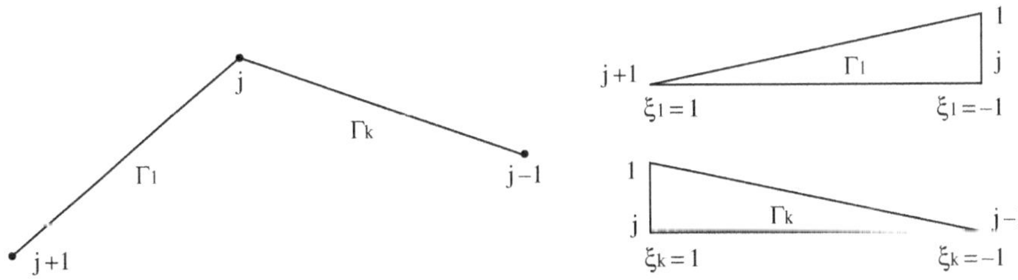

Fig. 3 Space interpolation functions $\eta^j(X)$ and $\nu^j(X)$ along elements adjacent to node j

The main objective here is to find analytical expressions for the time integrals described in equations (14) and (15). Since $\phi^m(\tau)$ is non-null only for $t_{m-1} \leq \tau \leq t_{m+1}$, the integrals are evaluated only within this interval for each matrix $\mathbf{H}^{(n+1)m}$. Moreover, for each time step, integrals are calculated only along the initial time interval ($t_1 \leq \tau \leq t_2$), because the other terms are already known through calculations performed in previous steps: the terms $\mathbf{H}^{(n+1)m}$ are equal to $\mathbf{H}^{n(m-1)}$, due to the translation property. This integration methodology produces parcels which are added to matrices $\mathbf{H}^{(n+1)1}$ and $\mathbf{H}^{(n+1)2}$, according to the branches of the interpolation functions in focus. Only one time integral must be evaluated to compute the $\mathbf{G}^{(n+1)m}$ entries, as piecewise constant time interpolation functions $\theta^m(\tau)$ are employed.

3.1 The Axisymmetric Case

Aiming to simplify the analytical integration procedure, the integration variable is changed from τ to u, according to the following expression:

$$u_\tau = c\,(t_{n+1} - \tau) \tag{16}$$

Taking into account the above-defined variable (and, as a consequence, the definitions: $u_i = c(t_{n+1} - t_i)$; $u_f = c(t_{n+1} - t_f)$ etc.), six possible cases may happen:

- **Case 1**: $r > u_i$ (integration results null);
- **Case 2**: $r > u_f$ and $d > u_i \geq r$ (integration from r to u_i);
- **Case 3**: $r > u_f$ and $u_i \geq d$ (integration from $r$to d);
- **Case 4**: $u_f > r$ and $d > u_i$ (integration from u_f to u_i);
- **Case 5**: $d > u_f \geq r$ and $u_i \geq d$ (integration from u_f to d);
- **Case 6**: $u_f > d$ (integration results null).

The final expressions to be integrated are then:

$$g\,(X, \xi, t_n) = 1/(2\pi^2) \int_{u_f}^{u_i} \left[\left(u^2 - r^2\right)\,\left(d^2 - u^2\right)\right]^{-1/2} du \tag{17}$$

$$h_I\,(X, \xi, t_n) = 1/(c\Delta t \pi^2) \int_{u_f}^{u_i} \left[2C_1 - C_2\left(u^2 - r^2\right)\right]\left[u - u_f\right] \left[\left(u^2 - r^2\right)\,\left(d^2 - u^2\right)\right]^{-3/2} du \tag{18}$$

$$h_F\,(X, \xi, t_n) = 1/(c\Delta t \pi^2) \int_{u_f}^{u_i} \left[2C_1 - C_2\left(u^2 - r^2\right)\right]\left[u_i - u\right] \left[\left(u^2 - r^2\right)\,\left(d^2 - u^2\right)\right]^{-3/2} du \tag{19}$$

where:

$$C_1 = [n_1(x_1 - \xi_1) + n_2(x_2 - \xi_2)]\; x_1\xi_1 = (\partial r/\partial n) r x_1 \xi_1 \tag{20}$$

$$C_2 = [n_1 x_1 + n_2(x_2 - \xi_2)] \tag{21}$$

The integrals described by equation (17) are easily evaluated, employing the concept of elliptical integrals (Abramowitz and Stegun 1968) and the corresponding results are presented in the Appendix.

A more complex evaluation process is required, however, concerning the integrals described by equations (18) and (19): depending on the integration limits, non-integrable singular integrands may occur. In this case, such integrals can be understood only through the concept of finite part integral (Hadamard 1952).

The first step treating equation (18) is to separate its integrand into four distinct parcels (analogous procedure is adopted for equation (19)). The four parcels, as well as their integration procedures considering the previous discussed limit cases, are presented in the sequence.

3.1.1 Parcel 1

The first parcel is represented by:

$$2C_1 u_f \left[\left(u^2 - r^2\right)\left(d^2 - u^2\right)\right]^{-3/2} \tag{22}$$

and its treatment considering each case of analysis is described as follows:

Case 2

The integrand presents a non-integrable singularity at the lower limit of integration (r). The following procedure is adopted:

$$\int_r^{u_i} \left[\left(u^2 - r^2\right)\left(d^2 - u^2\right)\right]^{-3/2} du = \int_r^{u_i} \frac{A(u) - A(r)}{(u - r)^{3/2}}\, du + \int_r^{u_i} \frac{A(r)}{(u - r)^{3/2}} du \tag{23}$$

where:

$$A(u) = \left[\left(d^2 - u^2\right)(u + r)\right]^{-3/2} \tag{24}$$

The first term on the right-hand-side of equation (23) can be integrated analytically, while for the second term, the following expression is used, which corresponds to the finite part of the integral:

$$\int_r^{u_i} A(r)\,(u - r)^{-3/2} du = -2A(r)(u_i - r)^{-1/2} \tag{25}$$

Case 3

In this case, the integrand presents non-integrable singularities corresponding to both integration limits. However, it is possible to take advantage from other case results (results presented previously or later on) by using the following property:

$$\int_{r}^{d} f(u)du = \int_{r}^{u_f} f(u)du + \int_{u_f}^{d} f(u)du \tag{26}$$

Taking equation (26) into account, it is just necessary to sum two parcels: the one equivalent to case 2, changing u_i by u_f, and the one corresponding to case 5, presented later.

Case 4

This integral can be evaluated in the usual Riemann sense, since the integrand does not present singularities along the integration interval. However, in order to take advantage of the results obtained for case 2, the following property is considered:

$$\int_{u_f}^{u_i} f(u)du = \int_{r}^{u_i} f(u)du - \int_{r}^{u_f} f(u)du \tag{27}$$

Therefore, the integral indicated on the left-hand-side of equation (27) is equal to the result obtained for case 2 minus an equivalent expression, in which u_i is replaced by u_f.

Case 5

Again, the integrand shown in equation (22) presents a non-integrable singularity, now at the upper integration limit (d). The following procedure is adopted:

$$\int_{u_f}^{d} \left[\left(u^2 - r^2\right)\left(d^2 - u^2\right)\right]^{-3/2} du = \int_{u_f}^{d} \frac{B(u) - B(r)}{(d-u)^{3/2}} du + \int_{u_f}^{d} \frac{B(d)}{(d-u)^{3/2}} du \tag{28}$$

where:

$$B(u) = \left[\left(u^2 - r^2\right)(d+u)\right]^{-3/2} \tag{29}$$

As in case 2, the first parcel on the right-hand-side of equation (28) can be integrated analytically, while, for the second parcel, equation (30) is employed, which corresponds to the finite part of the integral:

$$\int_{u_f}^{d} B(d)\,(d-u)^{-3/2}\,du = -2B(d)\,(d-u_f)^{-1/2} \tag{30}$$

3.1.2 Parcel 2

The second parcel is represented by:

$$2C_1 u\,\left[\left(u^2-r^2\right)\,\left(d^2-u^2\right)\right]^{-3/2} \tag{31}$$

and the procedures for all the cases of analysis, regarding the present parcel, are analogous to those of Parcel 1, considering the expressions for $A(u)$ and $B(u)$ multiplied by u.

3.1.3 Parcel 3

The third parcel is represented by:

$$-C_2\,u_f\,\left(u^2-r^2\right)\,\left[\left(u^2-r^2\right)\,\left(d^2-u^2\right)\right]^{-3/2} \tag{32}$$

and its treatment considering Case 2 and Case 5 is described below. For Case 3 and Case 4, the procedures adopted for the previous parcels are repeated.

Case 2

For this case, the singularity present in expression (32) at the lower limit r is integrable, because the factor that contains this singularity is raised to the 1/2 power. Thus, the integration is evaluated analytically in the usual Riemann sense.

Case 5

The procedures adopted for this case, in the previous parcels, are repeated, taking into account the following modification in the expression of $B(u)$:

$$B(u) = \left(u^2-r^2\right)^{-1/2}(d+u)^{-3/2} \tag{33}$$

3.1.4 Parcel 4

The forth parcel is represented by:

$$-C_2\,u\,\left(u^2-r^2\right)\,\left[\left(u^2-r^2\right)\,\left(d^2-u^2\right)\right]^{-3/2} \tag{34}$$

and the procedures for all the cases, regarding the present parcel, are analogous to those of Parcel 3, considering the expression for $B(u)$ multiplied by u.

3.2 The Three-Dimensional Case

When the source point is located along the axis of axisymmetry, the final expressions that arise are those of the three-dimensional case, namely:

$$g(X, \xi, t_n) = 1/(4\pi r) \tag{35}$$

$$h_I(X, \xi, t_n) = 1/(4\pi r c \Delta t)(\partial r/\partial n)(u_f/r) \tag{36}$$

$$h_F(X, \xi, t_n) = 1/(4\pi r c \Delta t)(\partial r/\partial n)(-u_i/r) \tag{37}$$

which are employed for the case $u_i \geq r \geq u_f$.

4 Space Integration

It is important to notice that spatial numerical integration along boundary elements produces more accurate results when the elements are subdivided into segments, each segment containing only points belonging to the corresponding case of analysis. In the present work, the expressions resulting from time integration (previous section) are numerically integrated in space by the Gauss-Legendre method. In some cases, however, the integrand presents singularities that, although integrable, can deteriorate the accuracy of the numerical integration process. These singularities are discussed next.

4.1 Singularities at $r = 0$

Elements that contain the source point require special integration procedures, as the use of standard numerical integration algorithms (e.g., Gauss-Legendre, Simpson etc.) lead to highly inaccurate results. Analytical integration is then recommended; however, when the kernel expression is too complex, as it is the present case, the use of semi-analytical procedures (as described here) may be advisable.

The first addressed case is that of elliptic integrals of the first kind present in the expressions of the Appendix. When $r = 0$, it can be immediately inferred that both elliptic integral arguments equal to 1. In this case, elliptic integrals tend to infinity. Considering that the difference between the incomplete (F) and complete (K) elliptic integrals of the first kind is finite at $r = 0$ and that, in this situation:

$$K\left(\frac{\sqrt{d^2 - r^2}}{d}\right) = -\frac{d}{2\sqrt{d^2 - r^2}} \ln\left(\frac{r^2}{16\left(d^2 - r^2\right)}\right) \tag{38}$$

Equation (39) can be obtained:

$$\int_{\Gamma_i} \frac{1}{d} F(a,b) \frac{l-r}{l} x_1 dr = \int_{\Gamma_i} \left[\frac{1}{d} F(a,b) \frac{l-r}{l} x_1 + \frac{\ln r}{2} \right] dr - \int_{\Gamma_i} \frac{\ln r}{2} dr \quad (39)$$

where a and b represent the arguments of the incomplete elliptic integral. The first parcel on the right-hand-side of equation (39) can be integrated numerically, because it is not singular when $r = 0$, and the second parcel is integrated analytically.

4.2 Singularities at Fundamental Solution Wave Fronts

These singularities appear in the expressions of the Appendix when r or d coincides with some wave front (u_i or u_f). They are always caused by the presence, in the denominator, of the square root of $(d^2 - u^2_{f/i})$ or $(u^2_{f/i} - r^2)$. In the present work, it is discussed the complete treatment procedure considering two illustrative situations, being the remainder analogous.

4.2.1 First Situation

This first situation is related to expression (A–2a) when $r = u_i$. In this expression, only the following parcel of the integral along the boundary, which origins from equation (A–2a) first parcel, is singular at $r = u_i$:

$$\int_0^{l_S} -\frac{\sqrt{d^2 - u_i^2}}{\left(d^2 - r^2\right)^2 u_i \sqrt{u_i^2 - r^2}} x_1 \eta^j(x) C_1 dx \quad (40)$$

where the integration limits refer to the segment in which the boundary element was subdivided, being l_s this segment length. Again, it is used here a semi-analytical procedure, through which the following equation is obtained:

$$\int_0^{l_S} -\frac{\sqrt{d^2 - u_i^2}}{u_i \left(d^2 - r^2\right)^2 \sqrt{u_i^2 - r^2}} x_1 \eta^j(x) C_1(x) dx = \int_0^{l_S} \frac{A(x) - A(u_i)}{\sqrt{u_i^2 - r^2}} dx + \int_0^{l_S} \frac{A(u_i)}{\sqrt{u_i^2 - r^2}} dx \quad (41)$$

where:

$$A(x) = -\frac{\sqrt{d^2 - u_i^2}}{\left(d^2 - r^2\right)^2 u_i} x_1 \eta^j(x) C_1(x)$$
$$A(u_i) = -\frac{(4\xi_1)^{-3/2}}{u_i} \sqrt{x_1(u_i)} \eta^j(u_i) C_1'(u_i) \tag{42}$$

and $C_1'(x) = C_1(x)/x_1$.

The first parcel on the right-hand-side of equation (41) can be integrated numerically, once the integrand has a finite limit at $r = u_i$. To integrate analytically the second parcel, an integration variable change is advisable, where x is substituted by r. However, the Jacobian of the transformation (derivative *dx/dr*, given by equation (43)) can change its signal within the segment (the changing point is that where the radius r is minimum). Thus, it must be identified, for each segment, if the radius only increases (positive derivative), if it only decreases (negative derivative), or if the derivative signal changes. Considering the Jacobian signal, the integration way is kept, changing only the integration variable.

$$dx/dr = \pm r/\sqrt{r^2 - h_r^2} \tag{43}$$

In equation (43), h_r is the smallest distance between the source point and the straight line that contains the segment over which the integration is being evaluated. Considering the above-discussed variable change, analytical integration becomes elementary:

$$\int_0^{l_S} \frac{A(u_i)}{\sqrt{u_i^2 - r^2}} dx = \int_{r_i}^{r_f} \frac{A(u_i)}{\sqrt{u_i^2 - r^2}} \frac{\pm r\, dr}{\sqrt{r^2 - h_r^2}} = A(u_i) \left[\pm \mathrm{sen}^{-1} \sqrt{\frac{r^2 - h^2}{u_i^2 - h^2}} \right]_{r_i}^{r_f} \tag{44}$$

If a signal change occurs along the element:

$$\int_{r_i}^{r_f} \frac{A(u_i)}{\sqrt{u_i^2 - r^2}} \frac{\pm r dr}{\sqrt{r^2 - h_r^2}} = -\int_{r_i}^{h_r} \frac{A(u_i)}{\sqrt{u_i^2 - r^2}} \frac{r dr}{\sqrt{r^2 - h_r^2}} + \int_{h_r}^{r_f} \frac{A(u_i)}{\sqrt{u_i^2 - r^2}} \frac{r dr}{\sqrt{r^2 - h_r^2}} =$$
$$= A(u_i) \left[\mathrm{sen}^{-1} \sqrt{\frac{r^2 - h_r^2}{u_i^2 - h_r^2}} \right]_{r_i}^{r_f} \tag{45}$$

Expressions for $A(x)$ and $A(u_i)$ or $A(u_f)$ corresponding to other singularities at wave fronts at r may be deduced analogously (Warszawski 2005).

4.2.2 Second Situation

This second situation is related to expression (A–2a) when $d = u_i$. The singularity is present only in the following integral, which origins from the first parcel of equation (A–2a):

$$\int_0^{l_S} \frac{\sqrt{u_i^2 - r^2}}{u_i\left(d^2 - r^2\right)^2 \sqrt{d^2 - u_i^2}} x_1 \eta^j(x) C_1 dx \tag{46}$$

Employing once again the procedure applied to equation (40):

$$\int_0^{l_S} \frac{\sqrt{u_i^2 - r^2}}{u_i\left(d^2 - r^2\right)^2 \sqrt{d^2 - u_i^2}} x_1 \eta^j(x) C_1 dx = \int_0^{l_S} \frac{B(x) - B(u_i)}{\sqrt{d^2 - u_i^2}} dx + \int_0^{l_S} \frac{B(u_i)}{\sqrt{d^2 - u_i^2}} dx \tag{47}$$

where:

$$B(x) = \frac{\sqrt{u_i^2 - r^2}}{\left(d^2 - r^2\right)^2 u_i} x_1 \eta^j(x) C_1(x)$$
$$B(u_i) = \frac{(4\xi_1)^{-3/2}}{u_i} \sqrt{x_1(u_i)} \eta^j(u_i) C_1'(u_i) \tag{48}$$

In order to perform the analytical integration of the second term on the right-hand-side of equation (47), a variable change from x to d is considered and its Jacobian is given by:

$$dd/dx = \pm d/\sqrt{d^2 - h_d^2} \tag{49}$$

As in the previous situation, the derivative *dd/dx* may take positive or negative value; however, in the present situation, signal changes along the segment cannot occur. This is due to the fact that along the segment extension, d cannot equal h_d (smallest distance between the farthest point on the ring source and the straight line that contains the segment), since, for cases 2 and 4, d must be greater than u_i, and

u_i must be greater than h_d; the same happens for case 5, regarding u_f. Thus, the integral is calculated according to equation (50):

$$\int_0^{l_S} \frac{B(u_i)}{\sqrt{d-u_i^2}} dx = \int_{r_i}^{r_f} \frac{B(u_i)}{\sqrt{d-u_i^2}} \frac{\pm d}{\sqrt{d^2-h_d^2}} dd$$
$$= B(u_i)\left[\pm \ln\left(\sqrt{d^2-h_d^2}+\sqrt{d-u_i^2}\right)\right]_{r_r}^{r_f} \quad (50)$$

Expressions for $B(x)$ and $B(u_i)$ or $B(u_f)$ corresponding to other singularities at wave fronts at d may be deduced analogously (Warszawski 2005).

5 Time-Marching Scheme

In order to avoid typical Time-Domain Boundary Element Method instabilities, the linear θ method (Yu et al. 1998) is also considered here, alternatively to the time-marching scheme represented by equation (13).

In the linear θ method, the analysis is carried out at time instant $t_{n+\theta} = t_n + \theta\Delta t(\theta \geq 1)$, instead of at time t_{n+1}. Matrices **G** and **H** are then evaluated considering time $t_{n+\theta}$ instead of time t_{n+1}, and equation (13) can be rewritten as:

$$(C_{ij} + H_{ij}^{(n+\theta)(n+\theta)})p_j^{n+\theta} - G_{ij}^{(n+\theta)(n+\theta)}q_j^{n+\theta} = \sum_{m=1}^{n}\sum_{j=1}^{NN} G_{ij}^{(n+\theta)m} q_j^m$$
$$-\sum_{m=1}^{n}\sum_{j=1}^{NN} H_{ij}^{(n+\theta)m} p_j^m \quad (51)$$

Once results are computed at time $t_{n+\theta}$, they are interpolated to time t_{n+1}, taking into account the time interpolation functions adopted. Since linear and piecewise constant time interpolation functions are being considered (see Fig. 2), the variables are evaluated at time t_{n+1} as follows:

$$p_i^{n+1} = \frac{1}{\theta} p_i^{n+\theta} + \frac{\theta - 1}{\theta} p_i^n \quad (52)$$

$$q_i^{n+1} = q_i^{n+\theta} \quad (53)$$

6 Numerical Application

In this section, the wave propagation through a prismatic circular rod is analysed. The rod is subjected to non-null natural boundary condition at one of its extremities $(q(t) = H(t-0))$ and to null essential boundary condition at the opposite end. The rod dimensions are: $a = 6$ m, $b = 12$ m and $c = 1$ m, as shown in Fig. 4. Boundary

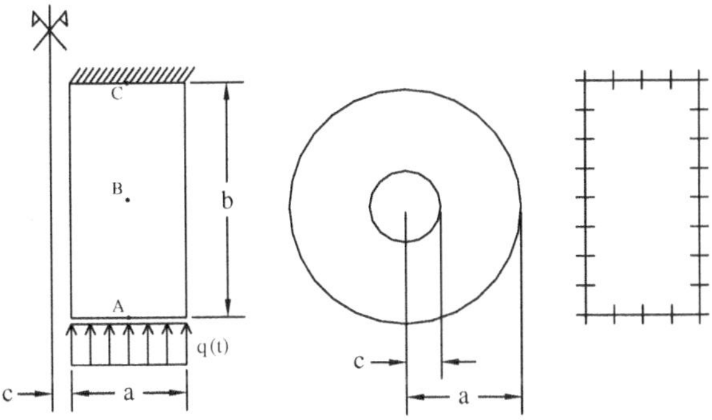

Fig. 4 Sketch of the model; transversal section and boundary element mesh

elements measuring 1.5 m are employed to discretize the model, composing a 24-element mesh. The wave propagation velocity of the medium is $c = 1.0\,\text{m/s}$. The time discretization is selected according to a β parameter: $\beta = c\Delta t/l$, where Δt is the time-step of the analysis and l is the boundary element length.

In Fig. 5, numerical results are compared with the corresponding analytical solution (Nowacki 1963). For $\beta = 0.2$, numerical results become unstable. This occurs because, the smaller the β value, the shorter the distance travelled by the wave front during one time-step interval; thus, the element segment to be integrated concentrates along its extension integrands with high gradients, reducing the numerical integration accuracy. Therefore, to obtain more accurate results, the time-step cannot be arbitrarily reduced, since unsuitable values of β may be reached: time-step and element length must be simultaneously refined.

In the sequence, the period of analysis is extended and the solution behaviour is monitored according to the adopted β parameter. When the period of analysis is increased up to 576 s (12 response cycles), numerical stability is verified only when $\beta = 0.8$, as depicted in Fig. 6(b). Numerical damping is also verified for $\beta = 0.8$; this damping is inherent to the present time-domain BEM approach, and, in fact, when it occurs, the method stability increases (see Fig. 6).

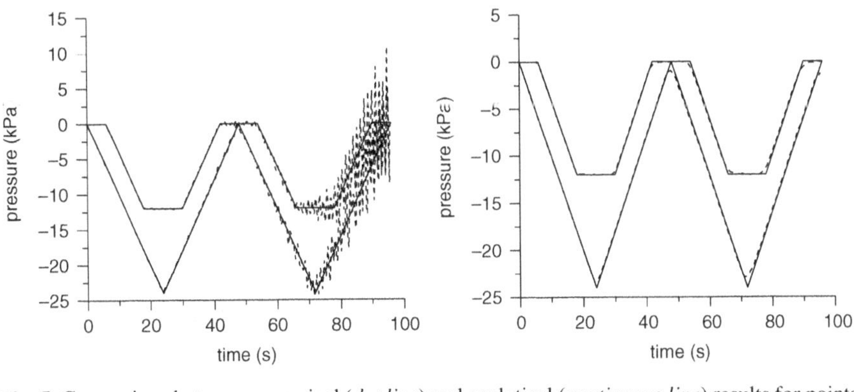

Fig. 5 Comparison between numerical (*dot line*) and analytical (*continuous line*) results for points A and B – $\beta = 0.2$ and $\beta = 0.6$

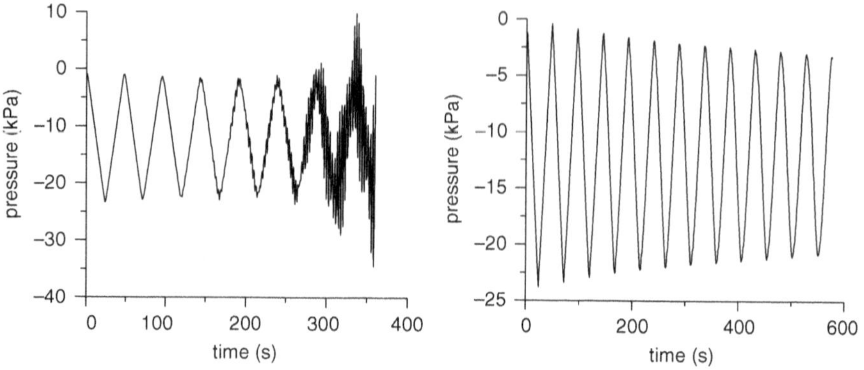

Fig. 6 Pressure at point A– $\beta = 0.6$ and $\beta = 0.8$

To eliminate the present instabilities, the linear θ method is adopted as an alternative time-marching scheme. Time history results for pressure at point A, considering β equal to 0.6 and 0.8 and θ equal to 1.05 and 1.10 are depicted in Fig. 7. When θ is increased, numerical damping is introduced and stability is reached.

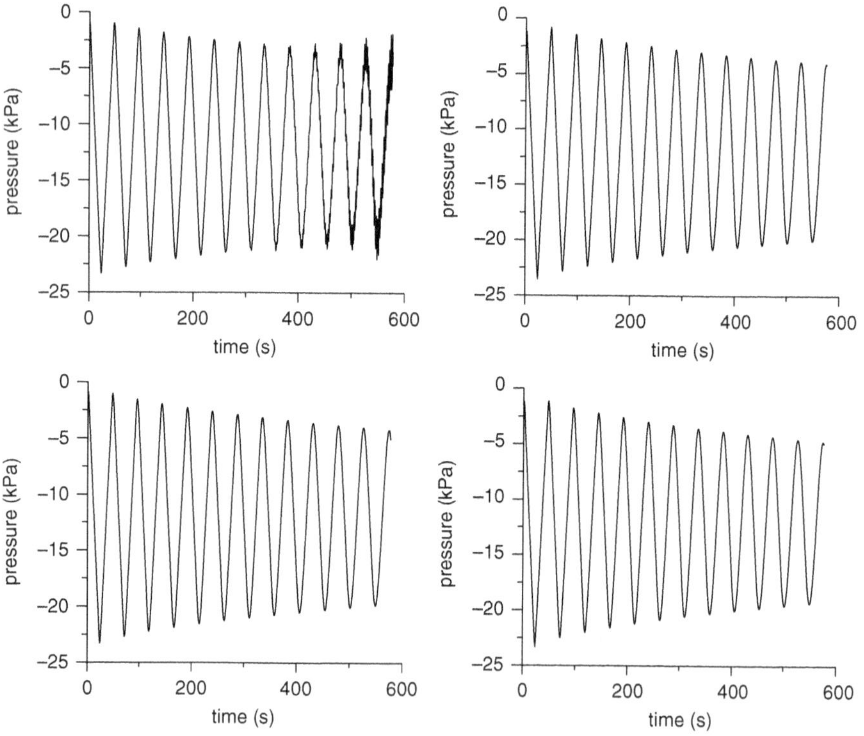

Fig. 7 Pressure at point A considering different θ and β values – $\beta = 0.6$ and $\theta = 1.05$; $\beta = 0.8$ and $\theta = 1.05$; $\beta = 0.6$ and $\theta = 1.10$; $\beta = 0.8$ and $\theta = 1.10$

7 Conclusions

The present work is concerned with the time-domain formulation of the Boundary Element Method applied to model axisymmetric problems governed by the scalar wave equation. The time integrations required by the time-domain BEM approach were carried out analytically, obtaining expressions to be integrated, subsequently, along the boundary. Singularities of boundary integrals were removed analytically and the remaining kernels were settled quite suitable for numerical integration by usual quadrature rules.

The analytical time integrations produced numerical kernels, which lead to results more accurate than those obtained when time integrations are carried out numerically. The proposed numerical application provided good results, and validated the expressions obtained from the analytical time integration, as well as other described features concerning the computational algorithm presented.

Appendix: Analytical Expressions for Time Integrals

Expressions for g, h_I andh_F are described next, considering the possible cases previously discussed.

$$g = E0/(2d\pi^2) \tag{54}$$

$$h_i = +\left[2C_1(u_i E_1 - E_2) - C_2(u_i E_3 - E_4)\right] /(c\Delta t\pi^2) \tag{55}$$

$$h_f = -\left[2C_1(u_f E_1 - E_2) - C_2(u_f E_3 - E_4)\right] /(c\Delta t\pi^2) \tag{56}$$

Case 2

$$E0 = F_A^i \tag{57}$$

$$E1 = V_B^i + v_B F_A^i - v_A E_A^i \tag{58}$$

$$E2 = V_B^i \tag{59}$$

$$E3 = V_C^i/u_i + v_C(F_A^i - E_A^i) \tag{60}$$

$$E4 = V_C^i \tag{61}$$

Case 3

$$E0 = K_D \tag{62}$$

$$E1 = v_B K_D - v_A E_C^i \tag{63}$$

$$E2 = 0 \tag{64}$$

$$E3 = v_C(K_D - E_D) \tag{65}$$

$$E4 = 0 \tag{66}$$

Case 4

$$E0 = F_A^i - F_A^f \quad (67)$$

$$E1 = V_B^i - V_B^f + \upsilon_B(F_A^i - F_A^f) + \upsilon_A(E_A^f - E_A^i) \quad (68)$$

$$E2 = V_B^i - V_B^f \quad (69)$$

$$E3 = V_C^i/u_i - V_C^f/u_f + \upsilon_C(F_A^i - F_A^f) + \upsilon_C(E_A^f - E_A^i) \quad (70)$$

$$E4 = V_C^i - V_C^f \quad (71)$$

Case 5

$$E0 = F_B^i \quad (72)$$

$$E1 = -u_f V_A^f + \upsilon_B F_B^i - \upsilon_A E_B^i \quad (73)$$

$$E2 = -V_B^f \quad (74)$$

$$E3 = -u_f V_C^f/d^2 + \upsilon_C(F_B^f - E_B^f) \quad (75)$$

$$E4 = -V_C^f \quad (76)$$

where the variables adopted in the above expressions are given by:

$$\Lambda_A^m = \Lambda\left((d/u_m)\sqrt{(u_m^2 - r^2)/(d^2 - r^2)}, \sqrt{(d^2 - r^2)}/d\right) \quad (77)$$

$$\Lambda_B^m = \Lambda\left((d/u_m)\sqrt{(d^2 - u_m^2)/(d^2 - r^2)}, \sqrt{(d^2 - r^2)}/d\right) \quad (78)$$

$$\Lambda_C^m = \Lambda\left((d/u_m)\sqrt{(u_m^2 - r^2)/(d^2 - r^2)}\right) \quad (79)$$

$$V_A^m = \left[r^2(u_m^2 - r^2) - d^2(d^2 - u_m^2)\right]\left[(d^2 - r^2)d^2r^2\sqrt{(d^2 - u_m^2)(u_m^2 - r^2)}\right]^{-1} \quad (80)$$

$$V_B^m = \left[(u_m^2 - r^2) - (d^2 - u_m^2)\right]\left[(d^2 - r^2)\sqrt{(d^2 - u_m^2)(u_m^2 - r^2)}\right]^{-1} \quad (81)$$

$$V_C^m = \sqrt{(u_m^2 - r^2)}\left[(d^2 - r^2)\sqrt{(d^2 - u_m^2)}\right]^{-1} \quad (82)$$

$$\upsilon_A = (d^2 + r^2)\left[(d^2 - r^2)^2 d\, r^2\right]^{-1} \quad (83)$$

$$\upsilon_B = 2\left[(d^2 - r^2)^2 d\right]^{-1} \quad (84)$$

$$\upsilon_C = \left[(d^2 - r^2)\, d\right]^{-1} \quad (85)$$

and Λ (equations (6)) may represent F (meaning incomplete elliptic integral of the first kind), K (meaning complete elliptic integral of the first kind) or E (meaning complete or incomplete elliptic integral of the second kind).

References

Abramowitz M, Stegun IA (1968) Handbook of mathematical functions with formulas, graphs and mathematical tables. Dover Publications, New York.

Beskos DE (1997) Boundary element methods in dynamic analysis. Applied Mechanics Reviews 50:149–197.

Czygan O, Von Estorff O (2003) An analytical fundamental solution of the transient scalar wave equation for axisymmetric systems. Computer Methods in Applied Mechanics and Engineering 192:3657–3671.

Grannell JJ, Shirron JJ, Couchman LS (1994) A hierarchic p-version boundary–element method for axisymmetric acoustic scattering. Journal of the Acoustical Society of America 5:2320–2329.

Hadamard J (1952) Lecture on Cauchy's problem in linear partial differential equations. Dover Publications, New York.

Israil ASM, Banerjee PK, Wang HC (1992) Time-domain formulations of BEM for two-dimensional, axisymmetric and three-dimensional scalar wave propagation. In: Advanced Dynamic Analysis by Boundary Element Method – v. 7, Chapter 3, Elsevier Applied Science: London, 75–113.

Mansur WJ (1983) A time-stepping technique to solve wave propagation problems using the boundary element method. Ph.D. Thesis, University of Southampton, England.

Mansur WJ, Carrer JAM (1993) Two-dimensional transient BEM analysis for the scalar wave equation: kernels. Engineering Analysis with Boundary Elements 12:283–288.

Morse PM, Feshbach H (1953) Methods of theoretical physics. McGraw-Hill, New York.

Nowacki W (1963) Dynamic of elastic systems. John Wiley & Sons, New York.

Soenarko B (1993) A boundary element formulation for radiation of acoustic waves from axisymmetric bodies with arbitrary boundary conditions. Journal of the Acoustic Society of America 93:631–639.

Tsinopoulos SV, Agnantiaris JP, Polyzos D (1999) An advanced boundary element/fast Fourier transform axisymmetric formulation for acoustic radiation and wave scattering problems. Journal of the Acoustical Society of America 105:1517–1526.

Warszawski A (2005) BEM-FEM coupling for axisymmetric acoustic elastodynamic problems in time domain. M.Sc. Dissertation (in Portuguese), COPPE/UFRJ, Brazil.

Yu G, Mansur WJ, Carrer JAM, Gong L (1998) A linear θ method applied to 2D time-domain BEM. Communications in Numerical Methods in Engineering 14:1171–1179.

Fluid-Structure Interaction by a Duhamel-BEM / FEM Coupling

Andre Pereira and Gernot Beer

Abstract A pure interface coupling formulation is developed for time domain analysis of coupled fluid-structure systems. Finite elements are applied to model the structure as an elastic continuum, while the fluid region is modeled as an acoustic media by the Boundary Element Method. The coefficient matrices for the fluid-structure interface are determined by applying unit impulses at the boundary of the fluid regions using the concept of Duhamel integrals, which are numerically approximated by means of the Convolution Quadrature Method. The proposed approach, greatly simplifies the assembly of sub-regions and the coupling to finite elements. The stability and accuracy of the proposed method are verified on some selected numerical examples.

1 Introduction

In many engineering applications, such as fluid-structure interaction analysis, the influences of both media on each other must be considered. Moreover, the correct representation of infinite or semi-infinite extension of the fluid domain is of great importance. The Boundary Element Method (BEM) is well suited for this task since it implicitly fulfils the radiation condition (Dominguez 1993). Therefore, to model the fluid a special boundary element formulation is proposed, while finite elements are used to represent the structure (with the possibility to consider non-linear formulation).

In order to perform the coupling, the interface variables are evaluated using a direct and strong procedure. The boundary element region is modelled extending the idea of a substructure technique for elastodynamics, called Duhamel-BEM (Moser 2005), to acoustic fluids. This technique is based on the generalization of Duhamel integrals (Clough and Penzien 1993), their numerical approximation by means of the Convolution Quadrature Method (CQM) and the extension for scalar

A. Pereira (✉)
Institute for Structural Analysis, Graz University of Technology
Lessingstrasse 25/II, 8010 Graz, Austria
e-mail: andre.pereira@tugraz.at

G.D. Manolis, D. Polyzos (eds.), *Recent Advances in Boundary Element Methods*,
DOI 10.1007/978-1-4020-9710-2_22, © Springer Science+Business Media B.V. 2009

wave propagation problems. The CQM (Lubich 1988) is a technique which approximates convolution integrals, in this case the Duhamel integrals, by a quadrature rule whose weights are determined with Laplace transformed Green's functions and a multi-step method. Therefore, the Laplace domain BEM formulation based on a direct collocation approach is used for the spatial representation of the fluid.

Most of the published work concerning the BEM applied to transient fluid-structure interaction problems is based on pure time domain formulations, while very few use frequency or transformed domain formulations. Detailed review on dynamic analysis with BEM can be found, e.g., in two comprehensive reviews reported by Beskos (1987, 1997) including the main improvements and the major developments in that area until 1997. Yu et al. (2002) show that instability problems represent one of the major drawbacks of the pure time domain BEM formulations. To overcome these problems, a more stable time domain formulation was introduced by Lie and Yu (2001) based on the linear θ method. However, a new idea based on a transformed domain formulation (which combines the advantages of both Laplace and time domain formulations) is proposed here, yielding a very stable and accurate approach. The formulation results in an interface mass matrix for the BEM which allows an easy coupling to standard finite elements.

Finally, to investigate the accuracy and the stability of the proposed coupling scheme, two selected problems are analyzed and the results obtained are compared with the results from the literature.

2 Fluid Modeling by the Duhamel-BEM Approach

Small amplitude acoustic waves propagate through an ideal homogeneous fluid of density ρ_f and speed of sound c, according to the linear wave equation

$$\nabla^2 p = \frac{1}{c^2}\frac{\partial^2 p}{\partial t^2} \tag{1}$$

where ∇^2 is the Laplacian operator and p is the acoustic pressure in the fluid at time t. Equation (1) must be accompanied by appropriate boundary and initial conditions. It describes phenomena in which energy is propagated by waves and has applications in problems of sound propagation, the sloshing of liquids in a container, and fluid-structure interaction. In the discretized form of Eq. (1) only nodal pressure appear as degree of freedom.

In order to formulate the coupled problem, consider a linear isotropic body (domain Ω), consisting of two homogeneous regions (sub-domains), shown in Fig. 1, with a fluid sub-domain Ω^f coupled to a solid sub-domain Ω^s.

In the notation adopted here, the first of the two subscripts denote the location of a quantity, whereas the second one denotes its cause. The letters D, N and I refer, respectively, to the quantities at the Dirichlet boundary Γ_{D} (pressure known), Neumann boundary Γ_{N} (flux known) and coupled interface Γ_{I} (both unknown) ($\Gamma = \Gamma_{\mathrm{D}} + \Gamma_{\mathrm{N}} + \Gamma_{\mathrm{I}}$). When necessary, superscripts f and s denote quantities

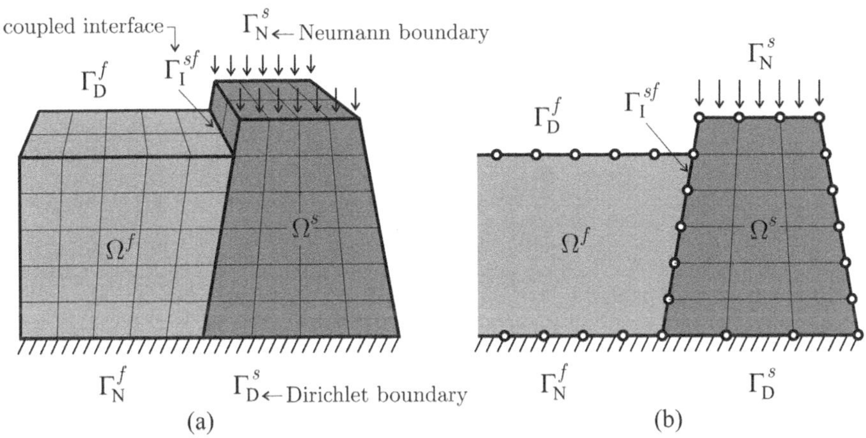

Fig. 1 Coupled fluid-structure system: (**a**) three- and (**b**) two-dimensional models

belonging, respectively, to the fluid and structure sub-domains. Also, overbars are used to denote prescribed quantities.

Assuming linear behaviour, allowing for the superposition of effects, the time-dependent pressure component at the interface of the fluid region Γ_I^f can be split up into three parts, i.e.

$$\mathbf{p}_I(t) = \mathbf{p}_{II}(t) + \mathbf{p}_{IN}(t) + \mathbf{p}_{ID}(t) \tag{2}$$

where $\mathbf{p}_{II}(t)$ is the pressure vector at the interface caused by the time-dependent interface fluxes $\mathbf{q}_I(t)$; $\mathbf{p}_{IN}(t)$ is the pressure vector at the interface due to the transient fluxes $\bar{\mathbf{q}}_N(t)$ on ${\Gamma_N}^f$; and $\mathbf{p}_{ID}(t)$ is the pressure vector at the interface due to the transient pressures $\bar{\mathbf{p}}_D(t)$ on ${\Gamma_D}^f$.

Since the superposition law holds, $\mathbf{p}_{II}(t)$ can be represented by a summation of convolution integrals, using the concept of the Duhamel integral (Clough and Penzien 1993). Consequently, each term of Eq. (2) may be expressed in the respective form

$$\mathbf{p}_I(t) = \int_0^t \mathbf{A}_{II}(t,\tau)\,\mathbf{q}_I(\tau)d\tau + \int_0^t \mathbf{A}_{IN}(t,\tau)\,\bar{\mathbf{q}}_N(\tau)d\tau + \int_0^t \mathbf{A}_{ID}(t,\tau)\,\bar{\mathbf{p}}_D(\tau)d\tau \tag{3}$$

In Eq. (3), the matrix $\mathbf{A}_{II}(t,\tau)$ is a unit flux impulse response matrix, where its coefficients $A_{II}(t,\tau)_{ij}$ may be interpreted as the time dependent pressure at the interface node i due to an impulsive unit flux $\delta(t-\tau)$ applied at time τ at the interface node j, with zero flux at all other interface nodes and zero boundary conditions at ${\Gamma_N}^f$ and ${\Gamma_D}^f$ in the same boundary element region Ω^f, as illustrated in Fig. 2. Here, $\delta(t-\tau)$ denotes the Dirac delta function and $i,j = 1,\ldots,N_{\mathrm{Do}\Gamma}$.

In analogous manner, the matrix $\mathbf{A}_{IN}(t,\tau)$ of Eq. (3) represents the time dependent pressures at the interface due to the impulsive unit flux $\delta(t-\tau)$ applied at

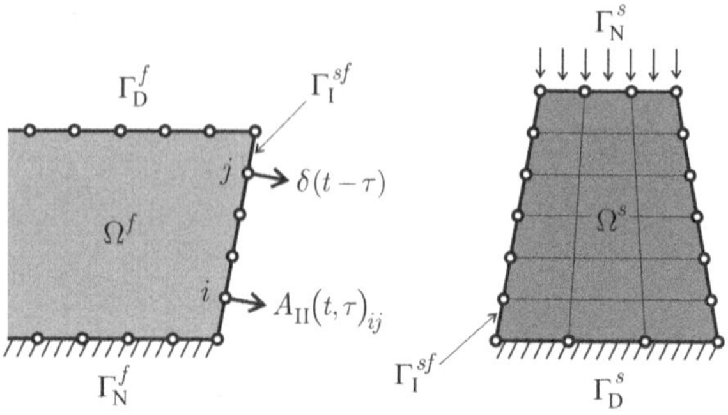

Fig. 2 Application of an impulsive unit pressure at time $t = \tau$ at the coupled interface

time τ along the element degree of freedoms at $\Gamma_N{}^f$ of region Ω^f, with zero flux along the interface; the matrix $\mathbf{A}_{ID}(t, \tau)$ represents the time dependent pressures at the interface due to the impulsive unit pressure $\delta(t - \tau)$ applied at time τ along the nodes at $\Gamma_D{}^f$ of region Ω^f, with zero flux along the interface. Therefore, the integrals in Eq. (3) can be understood as a generalization of the concept of Duhamel integrals.

There is no closed form solution for $\mathbf{A}_{II}(t, \tau)$, $\mathbf{A}_{IN}(t, \tau)$ and $\mathbf{A}_{ID}(t, \tau)$ available in the time domain. However, the matrices can be numerically computed in the Laplace domain without any numerical difficulties. An additional advantage is that, one can benefit from all advantages of the Laplace domain boundary element formulation, such as the availability of a wide range of fundamental solutions. For these reasons, the fluid is modelled in Laplace domain.

The Laplace transform of a Dirac impulse is given by $L\{\delta(t - \tau)\} = e^{-s\tau}$. However, for wave propagation problems, the translation principle holds which states that the response of an acoustic fluid due to an impulse does not depend on absolute time, but it is rather a function of the time difference between the instant in which the impulse is applied and the instant in which the response is observed, see Eq. (4). Without loss of generality one can therefore apply the impulses at absolute time $\tau = 0$, yielding $L\{\delta(t - \tau)\} = 1$. Thus, the Laplace transform of $\mathbf{A}_{II}(t, \tau)$, $\mathbf{A}_{IN}(t, \tau)$ and $\mathbf{A}_{ID}(t, \tau)$, i.e.,

$$\begin{aligned}
\mathbf{A}_{II}(t,\tau) &= \mathbf{A}_{II}(t+\Delta t, \tau+\Delta t) = \mathbf{A}_{II}(t-\tau) \Rightarrow L\{\mathbf{A}_{II}(t-\tau)\} = \hat{\mathbf{A}}_{II}(s) \\
\mathbf{A}_{IN}(t,\tau) &= \mathbf{A}_{IN}(t+\Delta t, \tau+\Delta t) = \mathbf{A}_{IN}(t-\tau) \Rightarrow L\{\mathbf{A}_{IN}(t-\tau)\} = \hat{\mathbf{A}}_{IN}(s) \quad (4) \\
\mathbf{A}_{ID}(t,\tau) &= \mathbf{A}_{ID}(t+\Delta t, \tau+\Delta t) = \mathbf{A}_{ID}(t-\tau) \Rightarrow L\{\mathbf{A}_{ID}(t-\tau)\} = \hat{\mathbf{A}}_{ID}(s)
\end{aligned}$$

may be obtained by simply applying unit pressure and unit flux boundary conditions in the Laplace domain. From a more mathematical point of view, the new technique can be seen as a special transformed domain formulation or indirect time-domain boundary integral method (Antes 1988).

2.1 Time Discretization by the CQM

In order to allow for a numerical evaluation of these Duhamel or convolution integrals, occurring in Eq. (3), the Convolution Quadrature Method (CQM) developed by Lubich (1988) is adopted. The convolution integrals are approximated by a quadrature rule (similar to Gauss quadrature), whose weights can be determined exclusively by the Laplace transformed function and a linear multi-step method.

The time t is divided into n equal time steps Δt, so that $t = n\Delta t$. The CQM approximates convolution integrals, as those integrals presented in Eq. (3), as follows

$$v(n\Delta t) = \int_{\tau=0}^{t} h(t-\tau)\, g(\tau)\, d\tau \approx \sum_{k=0} \omega_{n-k}\left(\hat{h}(s), \Delta t\right) g(k\Delta t) \tag{5}$$

where $\hat{h}(s)$ are the Laplace transforms of the function $h(t)$, and ω_n are the convolution weights (or integration weights) which are the coefficient of the power series

$$\hat{h}\left(\frac{\gamma(z)}{\Delta t}\right) = \sum_{m=0}^{\infty} \omega_m(\hat{h}, \Delta t)\, z^m \tag{6}$$

with the complex variable z and $s = \gamma(z)/\Delta t$.

The coefficients of a power series are usually calculated with Cauchy's integral formula. After a polar coordinate transformation, this integral is approximated by a trapezoidal rule with L equal steps $2\pi/L$, which leads to

$$\omega_n(\hat{h}, \Delta t) = \frac{1}{2\pi i} \int_{|z|=\mathcal{R}} \hat{h}\left(\frac{\gamma(z)}{\Delta t}\right) z^{-n-1} dz \approx \frac{\mathcal{R}^{-n}}{L} \sum_{l=0}^{L-1} \hat{h}\left(\frac{\gamma(\mathcal{R}e^{il(2\pi/L)})}{\Delta t}\right) e^{-inl\frac{2\pi}{L}} \tag{7}$$

where $\mathcal{R}$ is the radius of a circle in the domain of analyticity of $\hat{h}(z)$, and $\gamma(z)$ is the quotient of characteristic polynomials of the underlying multi-step method, which needs to satisfy certain stability conditions. The evaluation of Eq. (7) can be performed very efficiently by using the Fast Fourier Transform (FFT). Therefore, to compute the convolution weights, it is sufficient to know $\hat{h}(s)$ at least approximately for a set of discrete Laplace parameters.

Lubich (1988) suggests as appropriate choice for $\gamma(z)$ the P-order backward differentiation formula

$$\gamma(z) = \sum_{m=1}^{P} \frac{1}{m}(1-z)^m \tag{8}$$

with $P = 2$ showing to be the best choice as shown by Schanz (2001), when dealing with problems as those which will be studied here. In addition, if an error ε is assumed in the computation of $\hat{h}(z)$ and choosing $L = n$ and $\mathcal{R}^n = \sqrt{\varepsilon}$, it is suggested the use of $\varepsilon = 10^{-10}$ (Schanz 2001). More details on the CQM can be found in Lubich (1988) and Schanz (2001).

In the case of the proposed methodology, the CQM requires only the knowledge of $\mathbf{A}_{\mathrm{II}}(s)$, $\mathbf{A}_{\mathrm{IN}}(s)$ and $\mathbf{A}_{\mathrm{ID}}(s)$, Eq. (4), for a discrete set of Laplace parameters. Thus, the convolution integrals of Eq. (3) take the following discretized form ($t = n\Delta t$ and $\tau = k\Delta t$), with their respective integration weight matrices,

$$\begin{aligned} \int_0^t \mathbf{A}_{\mathrm{II}}(t-\tau)\,\mathbf{q}_{\mathrm{I}}(\tau)\,d\tau &\approx \sum_{k=0}^{n}\left\{\omega_{n-k}^{\mathrm{A_{II}}}\,\mathbf{q}_{\mathrm{I}}(k\Delta t)\right\},\\ \int_0^t \mathbf{A}_{\mathrm{IN}}(t-\tau)\,\bar{\mathbf{q}}_{\mathrm{N}}(\tau)\,d\tau &\approx \sum_{k=0}^{n}\left\{\omega_{n-k}^{\mathrm{A_{IN}}}\,\bar{\mathbf{q}}_{\mathrm{N}}(k\Delta t)\right\},\\ \int_0^t \mathbf{A}_{\mathrm{ID}}(t-\tau)\,\bar{\mathbf{p}}_{\mathrm{D}}(\tau)\,d\tau &\approx \sum_{k=0}^{n}\left\{\omega_{n-k}^{\mathrm{A_{ID}}}\,\bar{\mathbf{p}}_{\mathrm{D}}(k\Delta t)\right\}, \end{aligned} \tag{9}$$

where

$$\begin{aligned} \omega_n^{\mathrm{A_{II}}} &= \frac{\mathcal{R}^{-n}}{L}\sum_{l=0}^{L-1}\hat{\mathbf{A}}_{\mathrm{II}}\left(\frac{\gamma}{\Delta t}\right)e^{-inl\frac{2\pi}{L}}\\ \omega_n^{\mathrm{A_{IN}}} &= \frac{\mathcal{R}^{-n}}{L}\sum_{l=0}^{L-1}\hat{\mathbf{A}}_{\mathrm{IN}}\left(\frac{\gamma}{\Delta t}\right)e^{-inl\frac{2\pi}{L}}\\ \boldsymbol{\omega}_n^{\mathrm{A_{ID}}} &= \frac{\mathcal{R}^{-n}}{L}\sum_{l=0}^{L-1}\hat{\mathbf{A}}_{\mathrm{ID}}\left(\frac{\gamma}{\Delta t}\right)e^{-inl\frac{2\pi}{L}} \end{aligned} \tag{10}$$

which are the convolution weights.

Substitution of Eq. (9) into Eq. (3) and extracting $\mathbf{p}_{\mathrm{I}}(n\Delta t)$ yields for each fluid region

$$\mathbf{p}_{\mathrm{I}}(n\Delta t) \approx \omega_0^{\mathrm{A_{II}}}\mathbf{q}_{\mathrm{I}}(n\Delta t) + \bar{\mathbf{p}}_{\mathrm{I}}(n\Delta t) \tag{11}$$

with

$$\bar{\mathbf{p}}_{\mathrm{I}}(n\Delta t) = \sum_{k=0}^{n-1}\omega_{n-k}^{\mathrm{A_{II}}}\mathbf{q}_{\mathrm{I}}(n\Delta t) + \sum_{k=0}^{n}\omega_{n-k}^{\mathrm{A_{IN}}}\bar{\mathbf{q}}_{\mathrm{N}}(n\Delta t) + \sum_{k=0}^{n}\omega_{n-k}^{\mathrm{A_{ID}}}\bar{\mathbf{p}}_{\mathrm{D}}(n\Delta t) \tag{12}$$

which contains only known values.

For the specific case of systems consisted of only fluid regions, the necessary coupling conditions may be obtained by assuming bonded contact between coupled regions and invoking equilibrium and compatibility. In the case of two coupled fluid regions, one can write

$$\begin{aligned}\mathbf{p}_{\mathrm{I}}^{1} + \mathbf{p}_{\mathrm{I}}^{2} &= \mathbf{0} \\ \mathbf{q}_{\mathrm{I}}^{1} = \mathbf{q}_{\mathrm{I}}^{2} &= \mathbf{q}_{\mathrm{I}}\end{aligned} \tag{13}$$

Using Eqs. (13), the convolution weight matrices $\omega_0^{\mathrm{A_{II}}}$ and the interface pressure vectors $\bar{\mathbf{p}}_{\mathrm{I}}(t)$ of all regions may be also assembled into a global system of equations, i.e.

$$\{\omega_0^{\mathrm{A_{II}}}\}\,\{\mathbf{q}_{\mathrm{I}}(n\Delta t)\} + \{\mathbf{p}_{\mathrm{I}}(n\Delta t)\} = \mathbf{0} \tag{14}$$

with $\{\omega_0^{\mathrm{A_{II}}}\}$ being the assembled convolution weight matrix and $\{\mathbf{p}_{\mathrm{I}}(n\Delta t)\}$ being the assembled interface pressure vector. Equation (14) can be solved for each time step to obtain the interface fluxes $\{\mathbf{q}_{\mathrm{I}}(n\Delta t)\}$. Note, however, that $\{\omega_0^{\mathrm{A_{II}}}\}$ has to be inverted only once.

2.2 *Spatial Discretization by Laplace Domain BEM*

The computation of the convolution weights for the fluid regions by using Eq. (10) can be carried out with a Laplace domain boundary element formulation (a direct collocation approach is used here).

The time-dependent differential equation (1) can be transformed to a time-independent one by means of the Laplace transform. The Laplace-transform $\hat{g}(s)$ of a function $g = g(t)$ is defined by

$$\hat{g}(s) = \mathcal{L}\{g(t)\} = \int_0^\infty g(t)e^{-st}dt \tag{15}$$

where $s \in \mathrm{C}$ is the complex Laplace transform parameter, and hats ($\hat{}$) denote Laplace transformed quantities.

Application of the Laplace transform to Eq. (1) under the assumption of zero initial conditions, i.e., $p(t = 0) = 0$ and $\dot{p}(t = 0) = 0$, yields

$$\nabla^2 \hat{p} = \left(\frac{s}{c}\right)^2 \hat{p} \tag{16}$$

which is the so-called Helmholtz equation.

Applying the Betti's reciprocal theorem to the scalar wave equation, the Somigliana's identity for a boundary element region in Laplace domain (Banerjee 1993), relating the pressure at a point ξ in a homogeneous domain Ω

with the pressures $\hat{p}(s,x)$ and fluxes $\hat{q}(s, \mathbf{x}) = \partial \hat{p}(s, \mathbf{x})/\partial n$ at the boundary $\Gamma = \partial\Omega$, is given by

$$\hat{p}(s, \xi) = \int_\Gamma \hat{P}(s, \mathbf{x}, \xi)\hat{q}(s, \mathbf{x})d\Gamma(\mathbf{x}) - \int_\Gamma \hat{Q}(s, \mathbf{x}, \xi)\hat{p}(s, \mathbf{x})d\Gamma(\mathbf{x}) \tag{17}$$

with $\mathbf{x} \in \Gamma$. $\hat{P}(s, \mathbf{x}, \xi)$ and $\hat{Q}(s, \mathbf{x}, \xi)$ are the pressure and flux fundamental solutions of the Laplace domain wave equation. They are given, e.g., in Dominguez (1993) and Banerjee (1993).

In order to solve Eq. (17), the boundary of Ω has to be discretized into elements. Introducing isoparametric boundary elements into Eq. (17), and by collocating this equation at each boundary node, a system of equation

$$\{\hat{\mathbf{Q}}\}\{\hat{\mathbf{p}}\} = \{\hat{\mathbf{P}}\}\{\hat{\mathbf{q}}\} \tag{18}$$

is obtained. $\{\hat{\mathbf{Q}}\}$ and $\{\hat{\mathbf{P}}\}$ are the coefficient matrices assembled from the element contributions, as explained in detail, e.g., in Beer (2001) and $\{\hat{\mathbf{p}}\}$ and $\{\hat{\mathbf{q}}\}$ are the pressures and fluxes at all nodes.

Therefore, the fluid regions are represented by a Laplace domain boundary element formulation for wave propagation in acoustic media, by applying unit impulses on the boundary and obtaining the responses for the interface pressures due to each impulse (Fig. 2). Then, the fluid quantities are transformed back to the time domain by means of a special transformation technique (i.e., the CQM), as explained subsequently.

3 Structure Modeling by the Newmark-FEM Approach

The FEM is used here to represent the solid regions (structure). A key feature of the proposed approach is that the coupling to finite elements becomes relatively easy.

The structure behaves as an elastic media, which is modelled by the dynamic equilibrium equation, arising from the finite element discretization (damping effects are neglected here), i.e.

$$\mathbf{M}\ddot{\mathbf{u}} + \mathbf{K}\mathbf{u} = \mathbf{f} \tag{19}$$

where $\mathbf{M}$ and $\mathbf{K}$ are, respectively, mass and stiffness matrices, $\ddot{\mathbf{u}}$ and $\mathbf{u}$ are, respectively, the acceleration vector and the displacement vector, and $\mathbf{f}$ is the nodal force vector.

Using the Newmark time stepping algorithm (Newmark 1959), one obtains the following approximations

$$\dot{\mathbf{u}}(n\Delta t) = \dot{\mathbf{u}}((n-1)\Delta t) + [(1-\gamma)\ddot{\mathbf{u}}((n-1)\Delta t) + \gamma\ddot{\mathbf{u}}(n\Delta t)]\,\Delta t$$

$$\mathbf{u}(n\Delta t) = \mathbf{u}((n-1)\Delta t) + \dot{\mathbf{u}}((n-1)\Delta t)\,\Delta t + \left[\left(\frac{1}{2}-\beta\right)\ddot{\mathbf{u}}((n-1)\Delta t) + \beta\ddot{\mathbf{u}}(n\Delta t)\right]\Delta t^2 \quad (20)$$

with time discretized into N time steps of equal duration Δt.

Substitution of Eq. (20), with $\beta = 1/4$ and $\gamma = 1/2$ (constant average acceleration), into Eq. (19) results in the following equation

$$\tilde{\mathbf{M}}\,\ddot{\mathbf{u}}(n\Delta t) = \tilde{\mathbf{f}}(n\Delta t) \quad (21)$$

with

$$\tilde{\mathbf{M}} = \mathbf{M} + \frac{\Delta t^2}{4}\mathbf{K} \quad (22)$$

and

$$\tilde{\mathbf{f}}(n\Delta t) = \mathbf{f}(n\Delta t) - \mathbf{K}\left\{\mathbf{u}((n-1)\Delta t) + \Delta t\dot{\mathbf{u}}((n-1)\Delta t) + \frac{\Delta t^2}{4}\ddot{\mathbf{u}}((n-1)\Delta t)\right\} \quad (23)$$

Therefore, a standard finite element formulation for elastodynamics may be used to represent the structure. The relationship between the structure and the fluid quantities (such as the force $\mathbf{f}$ and the fluid pressure p, respectively) are stated in the following.

4 Fluid-Structure Coupling Procedure

Now, splitting the nodal accelerations of the structure sub-domain Ω^s into coupled accelerations $\ddot{u}_{\mathrm{I}}$ and uncoupled ones $\ddot{u}_{\mathrm{F}}$, the Eq. (21) is rewritten as

$$\begin{bmatrix} \tilde{\mathbf{M}}_{\mathrm{II}} & \tilde{\mathbf{M}}_{\mathrm{IF}} \\ \tilde{\mathbf{M}}_{\mathrm{FI}} & \tilde{\mathbf{M}}_{\mathrm{FF}} \end{bmatrix} \begin{Bmatrix} \ddot{\mathbf{u}}_{\mathrm{I}}^s(n\Delta t) \\ \ddot{\mathbf{u}}_{\mathrm{F}}(n\Delta t) \end{Bmatrix} - \begin{Bmatrix} \tilde{\mathbf{f}}_{\mathrm{I}}^s(n\Delta t) \\ \tilde{\mathbf{f}}_{\mathrm{F}}(n\Delta t) \end{Bmatrix} = \begin{Bmatrix} \mathbf{f}_{\mathrm{I}}^s(n\Delta t) \\ \mathbf{0} \end{Bmatrix} \quad (24)$$

where subscript F denotes quantities at the finite element uncoupled boundary ($\Gamma_{\mathrm{F}} = \Gamma_{\mathrm{N}}{}^s + \Gamma_{\mathrm{D}}{}^s$); $\mathbf{f}_{\mathrm{I}}{}^s(n\Delta t)$ is the coupling nodal force vector, containing the interaction forces on $\Gamma_I{}^s$. As in general the tangential forces exerted by the fluid are negligible the vector $\mathbf{f}_I^s$ can be given in terms of pressure p along Γ_I as

$$\mathbf{f}_I^s = \int_{\Gamma_I} \bar{\mathbf{N}}^T \mathbf{n}\, p\, d\Gamma \tag{25}$$

where $\mathbf{n}$ is a unit normal vector to the interface, and $\bar{\mathbf{N}}$ represents the same shape functions which interpolate the nodal displacements.

Therefore, Eq. (25) can be written as

$$\mathbf{f}_I{}^S = \int_{\Gamma_I} \bar{\mathbf{N}}^T \mathbf{n}\, \mathbf{N}\, \mathbf{p}_I^s d\Gamma = \int_{\Gamma_I} \bar{\mathbf{N}}^T \mathbf{n}\, \mathbf{N}\, \mathbf{p}_I^f d\Gamma = \mathbf{T}\mathbf{p}_I \tag{26}$$

In analogous manner, one can write for the coupling nodal flux vector $\mathbf{q}_I$ the following expression (Zienkiewicz and Bettess, 1978)

$$\mathbf{q}_I^f = \int_{\Gamma_I} \mathbf{N}^T \rho\, \mathbf{n}^T \bar{\mathbf{N}}\, \ddot{\mathbf{u}}_I^f d\Gamma = \int_{\Gamma_I} \mathbf{N}^T \rho\, \mathbf{n}^T \bar{\mathbf{N}}\, \ddot{\mathbf{u}}_I^s d\Gamma = \rho\, \mathbf{T}^T \ddot{\mathbf{u}}_I \tag{27}$$

From now on it is possible to couple the fluid (boundary elements) to the structure (finite elements), just transforming Eq. (11) by means of Eqs. (26) and (27). In this way, the interaction force vector $\mathbf{f}_I{}^f$, containing all interaction forces generated by the boundary element region coupled to a finite element region along the interface can be expressed in terms of the interface accelerations, i.e.

$$\tilde{\mathbf{M}}_{II}^f\, \ddot{\mathbf{u}}_I^f(n\Delta t) = \mathbf{f}_I^f(n\Delta t) + \tilde{\mathbf{f}}_I^f(n\Delta t) \tag{28}$$

where

$$\tilde{\mathbf{M}}_{II}^f = \mathbf{T}\, {}^{sys}\omega_0^{A_{II}}\, \rho\, \mathbf{T}^T \qquad \text{and} \qquad \tilde{\mathbf{f}}_I^f(n\Delta t) = \mathbf{T}\, \mathbf{p}_I^f(n\Delta t) \tag{29}$$

Assuming bonded contact between coupled regions and invoking equilibrium and compatibility, the necessary coupling conditions at the interface should be considered

$$\begin{aligned} \mathbf{f}_I^s(n\Delta t) + \mathbf{f}_I^f(n\Delta t) &= \mathbf{0} \quad \text{(equilibrium)} \\ \ddot{\mathbf{u}}_I^s(n\Delta t) = \ddot{\mathbf{u}}_I^f(n\Delta t) &= \ddot{\mathbf{u}}_I(n\Delta t) \quad \text{(compatibility)} \end{aligned} \tag{30}$$

Finally, introducing Eqs. (24) and (28) into Eqs. (30), one obtains the following system of equations

$$\begin{bmatrix} \tilde{\mathbf{M}}_{II}^s + \tilde{\mathbf{M}}_{II}^f & \tilde{\mathbf{M}}_{IF} \\ \tilde{\mathbf{M}}_{FI} & \tilde{\mathbf{M}}_{FF} \end{bmatrix} \begin{Bmatrix} \ddot{\mathbf{u}}_I(n\Delta t) \\ \ddot{\mathbf{u}}_F(n\Delta t) \end{Bmatrix} = \begin{Bmatrix} \tilde{\mathbf{f}}_I^s(n\Delta t) + \tilde{\mathbf{f}}_I^f(n\Delta t) \\ \tilde{\mathbf{f}}_F(n\Delta t) \end{Bmatrix}. \tag{31}$$

Equation (31) must be solved for each time step but this involves a re-solution with different right hand side only. It is important to note, that the system becomes also non-symmetric due to the characteristics of the employed BEM formulation. The coupling strategy presented here may be easily implemented by means of very simple adaptations of an existing finite element code and this is one of the main advantages of the approach proposed in this work. Efficiency may be greatly improved here by implementing a simple a-priori cut off strategy, as proposed by Hackbusch et al. (2005).

A global strategy for implementation of the proposed fluid-structure coupling approach is schematically described by Fig. 3. First, the problem geometry is approximated as well as the variation of field quantities (e.g., displacement, traction, pressure, flux etc) using a suitable set of computational convenient simple functions. Next, the set of matrices which represent the system are computed depending on the region specification (e.g., fluid or structure). At the end of this process, the time

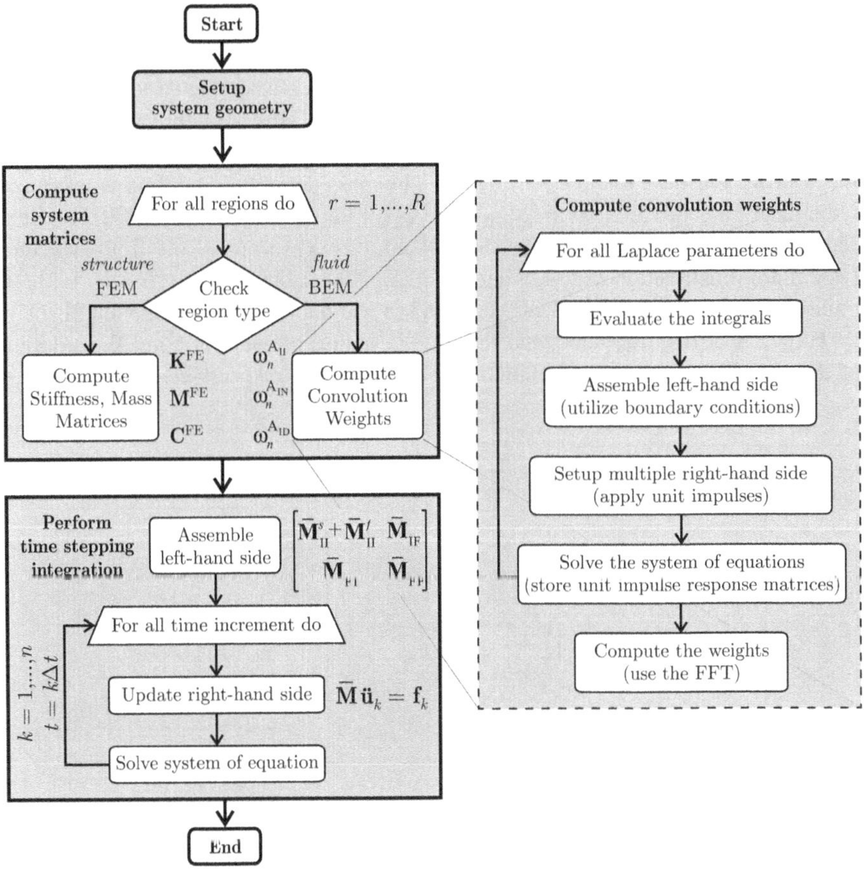

Fig. 3 Flow chart for the proposed coupling approach considering fluid regions

integration can be performed, by using Newmark for finite elements and CQM for boundary elements, resulting in a scheme which requires just the update of the right hand side for a new increment of time.

5 Numerical Results

In order to investigate the accuracy and the stability of the proposed technique, two problems are analyzed and the obtained results are compared to the ones from the literature.

5.1 One Dimensional Fluid-Structure Rod

A benchmark example frequently used to validate fluid-structure interaction formulations is the wave propagation in a rod with fluid container, as shown in Fig. 4. The purpose of this example is to check the results obtained by the proposed formulation against the analytical solution of the problem.

The rod is fixed at the bottom and the fluid surface is subjected to a Heaviside compression load. The material properties of the elastic structure are given by: Young's modulus E; Poisson's ratio $\nu = 0$; mass density ρ^s; and compression wave speed of the elastic domain $c_p{}^2 = E/\rho^s$. The mass density of the fluid is taken the same that of the structure $\rho^f = \rho^s$, as well as for the wave speed of the fluid $c = c_p$. The boundary conditions and the discretization used to model the fluid-structure system are illustrated in Fig. 4, where $a = 2b$. The structure is discretized with 16 finite elements, and the fluid with 16 boundary elements with the same length L.

The vertical displacements and the fluid-structure pressure at point **B**, midpoint of the coupled interface, are plotted in Figs. 5 and 6, respectively. The time step

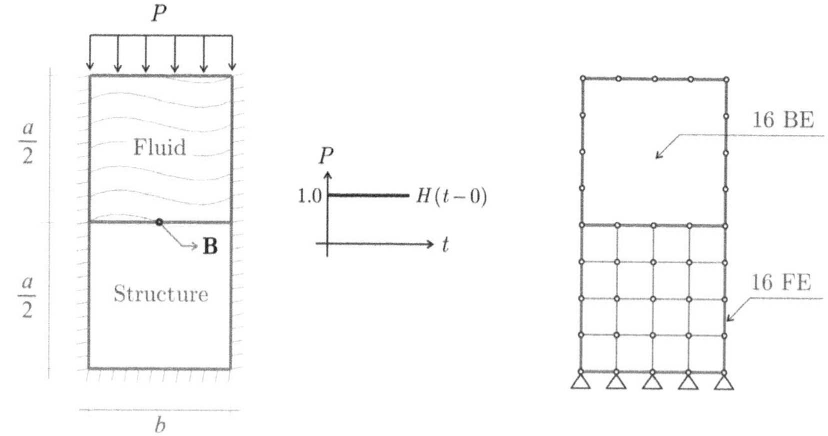

Fig. 4 Step function excitation of a fluid-structure rod with a BEM/FEM discretization

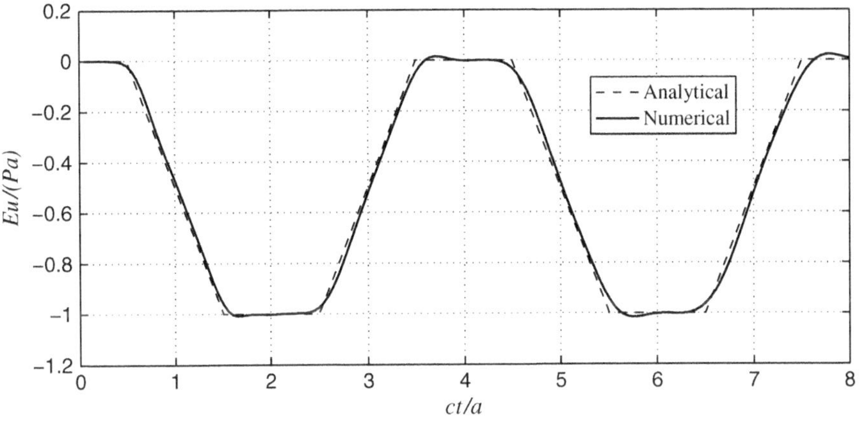

Fig. 5 Vertical displacement at point **B** in the fluid-structure rod

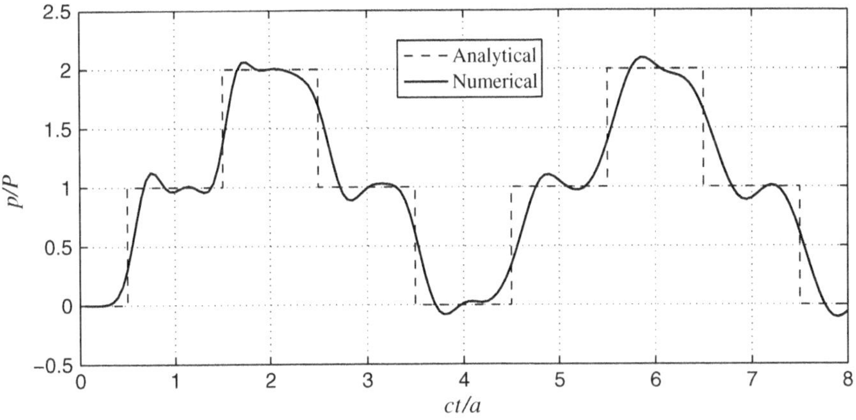

Fig. 6 Fluid-structure pressure at point **B** in the fluid-structure rod

size can be estimated by $\beta = c\Delta t/L$, with β restricted to a very small range where stable and satisfactory results are achieved ($0.2 < \beta < 0.6$). The results for the rod are obtained with $\beta = 0.4$.

It can be observed, that the results are stable and in good agreement with the correspondent analytical solution. The same example is analyzed by Yu et al. (2002), which use a stable pure time domain formulation for fluid-structure interaction problems. In the same work, the authors show that for the standard BEM/FEM coupling procedure an immediate instability appears for the fluid-structure pressure at point **B**. Note that in the proposed formulation, this instability problem does not occur.

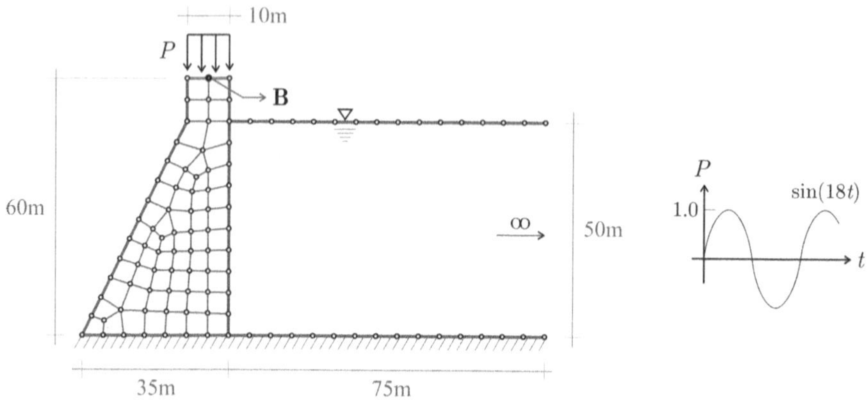

Fig. 7 Dam-reservoir system: geometry, discretization and loading

5.2 Dam-Water Interaction

In order to demonstrate the relevance and applicability of the proposed formulation to realistic problems, a dam-reservoir system is simulated dynamically, as shown in Fig. 7. An identical problem is analyzed by Estorff and Antes (1991). Therefore, the results obtained by Estorff and Antes (1991) may be used here as reference solutions to the proposed formulation.

The dam is considered to be linear elastic and is discretized with 54 iso-parametric linear finite elements while the adjacent semi-infinite water reservoir is modelled with 40 linear boundary elements. The material properties of the dam are: Young's modulus $E = 3.437 \times 10^6$ kN/m^2; Poisson's ratio $\nu = 0.25$; and mass density $\rho^s = 2.0$ t/m^3. The adjacent water is characterized by its pressure wave velocity $c = 1,436$ m/s and density of $\rho^f = 1.0$ t/m^3. At the free surface of the fluid region the hydrodynamic pressure are prescribed to be zero, as are the accelerations along the bottom of the reservoir.

Figure 8 illustrates the transient response for the vertical displacements of point **B** at the crest of the dam due to a vertical sinusoidal load of $P = \sin(18t)$. In Estorff and Antes (1991), the authors perform a serie of numerical experiments varying the water level, to study the influence of this level on the dam. However, just the highest level (50 m) was considered here. For this case, the results obtained by the proposed formulation are stable and in good agreement with the results found in the reference solution (Estorff and Antes 1991).

6 Conclusions

A novel technique for the analysis of dynamic fluid-structure interaction problems with BEM/FEM coupling has been presented. It is based on simple engineering concepts, i.e., the concept of stiffness (mass) matrix that is widely used in structural

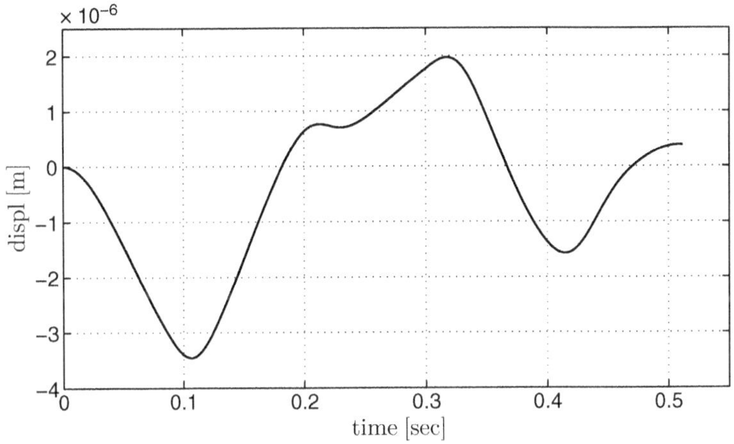

Fig. 8 Vertical displacement at point **B** in the dam-reservoir system

engineering and the idea of Duhamel integrals. In addition, the combination of the Laplace domain BEM with a powerful mathematical tool, namely the CQM, allows the numerical modelling of the fluid regions in time domain, whereas the structure is modelled by standard finite element procedures.

The main advantage of this approach is that it requires only the discretization of the fluid boundary, taking into account its semi-infinite extent. Moreover, there is an additional benefit from calculating in the Laplace domain without worrying about the time discretization, and still obtaining stable and accurate results.

It has been shown that satisfactory results are obtained and that the algorithm is very stable with respect to time step size. Therefore, the proposed formulation may be used as a powerful simulation tool for fluid-structure systems, e.g. dams subjected to earthquake.

Acknowledgments The authors gratefully acknowledge the support of the Austrian national science fund FWF Proj. Number P17527.

References

Antes H (1988) Anwendungen der Methode der Randelemente in der Elasto- und Flu- iddynamik. Mathematische Methoden in der Ingenieurtechnik, Teubner Verlag, Germany.

Banerjee PK (1993) The Boundary Element Methods in Engineering. McGraw-Hill, London.

Beer G (2001) Programming the Boundary Element Method, Wiley, New York.

Beskos DE (1987) Boundary element methods in dynamic analysis. Appl Mech Rev 40:1–23.

Beskos DE (1997) Boundary element methods in dynamic analysis. Part II (1986–1996). Appl Mech Rev 50:149–197.

Clough RW, Penzien J (1993) Dynamics of Structures. McGraw-Hill, London.

Dominguez J (1993) Boundary elements in dynamics. Computational Mechanics Publications, Southampton, UK.

Estorff O von, Antes H (1991) On FEM-BEM coupling for fluid-structure interaction analyses in the time domain. Int. J. for Num. Meth. in Engg. 31:1151–1168.

Hackbusch W, Kress W, Sauter SA (2005) Sparse Convolution Quadrature for Time Domain Boundary Integral Formulations of the Wave Equation. Preprint 116/2005, Max Planck Institute for Mathematics in the Sciences, Leipzig, December 2005.

Lie ST, Yu G (2001) Multi-domain fluid-structure interaction analysis with a stable time domain BEM/FEM coupling procedure. Engineering Computations 19:6–21.

Lubich Ch (1988) Convolution quadrature and discretized operational calculus I. Numerische Mathematik 52:129–145.

Moser W (2005) Transient coupled BEM-BEM and BEM-FEM analyses. PhD thesis TUGraz, Austria.

Newmark NM (1959) A method of computation for structural dynamics. ASCE J Eng Mech Div 85:67–94.

Schanz M (2001) Wave propagation in viscoelastic and poroelastic continua. Springer-Verlag, Berlin, Germany.

Yu GY, Lie ST, Fan SC (2002) Stable boundary element method/finite element method procedure for dynamic fluid-structure interaction. ASCE J Eng Mech 128:909–915.

Zienkiewicz OC, Bettess P (1978) Fluid-structure dynamic interaction and wave forces. An introduction to numerical treatment. Int. J. for Num. Meth. in Engg. 13:1–16.

BEM Solution of Creep Fracture Problems Using Strain Energy Density Rate Concept

C.P. Providakis

Abstract This paper presents a procedure for analyzing crack-tip regions of creeping, cracked structural components using the strain energy density rate concept and boundary element methodology. The investigated structural components are considered to undergo time-dependent, two-dimensional creep deformation and to be subjected to remote loading conditions while theirs deformation is assumed to be described by the elastic power law creep model. Some special boundary elements were used to simulate the time-dependent singular behavior of stress and strain fields at the crack tip of the investigated creeping materials. The strain energy density rate theory is applied to determine the direction of the crack initiation for a center cracked plate in tension which is subjected to Mode I loading conditions.

1 Introduction

In fracture mechanics problems, a main task is the analysis of the crack tip singular stress and strain fields and evaluation of some important fracture parameters which affect crack propagation. The problem become much more complicated if time-dependent inelastic deformations in creeping cracked structural components is considered. This is a particular concern in the design of aircraft engines and steam turbines where the high temperature prevails. In these high level of temperatures the time-dependent creep fracture phenomenon can be considered as of multi-scale nature, particularly when physical size is scaled down to the dimensions of the material microstructure. For a dominant crack in metallic components that undergo creep deformation, the creation of macrocrack surface along the main crack (Mode I) path should be distinguished from the creation of microcrack surfaces off to the side of main crack where the creep enclaves are located. In this sense, creep fracture could be also considered as a multiscale process.

More than two decades ago, the strain energy density criterion was proposed by Sih (1973) as a fracture criterion in contrast to the conventional theory of G and K of the Griffith's energy release rate assumptions in elastic fracture mechanics. This

C.P. Providakis (✉)
Department of Applied Sciences, Technical University of Crete, GR-73100 Chania, Greece
e-mail: cpprov@mred.tuc.gr

G.D. Manolis, D. Polyzos (eds.), *Recent Advances in Boundary Element Methods*,

DOI 10.1007/978-1-4020-9710-2_23,

provided an alternative approach to failure prediction for the same stress solution. The distinctions were emphasized in the works of Sih (1991). The strain energy density criterion gained momentum and credibility in engineering. A review on the use of this criterion can be found in Gdoutos (1984) and Carpinteri (1986). Recently Providakis (2006), presented a methodology based on the rate of the strain energy density to predict the crack initiation in creeping material under elevated temperature. The main idea was based on the fact that since the strain energy density criterion applies equally well to material with nonlinear behavior that are dissipative or non-dissipative, there are no difficulties to extend it to include creeping behavior of metallic components. In general, the strain energy density rate function fluctuates near the locations where abrupt changes of creeping level take place. The peaks and valleys of the fluctuation can be identified with failure by material creeping and fracture. The concept was applied to the prediction of many crack initiation problems in two-dimensional metal structural components without any problem or limitation imposed on the physical behavior of the material.

It is well known that for cases of realistic and practical problems in time-dependent fracture analysis of creeping cracked components the use of numerical solutions such as finite element method (FEM) and boundary element method (BEM) become imperative. For a review on the subject on can consult Beskos (1987).

In the search for an accurate, yet generalized, computational method for evaluating singular crack tip stress and strain fields, the singular element approach in conjunction with boundary element method (BEM) has been properly used in various fracture mechanics applications. Several researchers have contributed to this field: Blandford et al. (1981) were the first who introduced the traction singular quarter-point boundary element approach in combination with a multi-domain formulation to the solution of both symmetrical and non-symmetrical crack problems. Thereafter, this approach has been extensively used in the application of the boundary element method to two- and three- dimensional crack problems. An extension of the quarter-point element technique was used by Hantschel et al. (1990) who made an attempt to model crack tip fields arising in two-dimensional elastoplastic cracked panels by introducing some special singular boundary elements which took into account the HRR singularity field as presented in Hutchinson (1968) and Rice and Rosengren (1968) for locations near the crack tip.

In connection with the boundary element determination of near crack tip stress and strain fields in cracked structural components undergoing two-dimensional inelastic deformation one should mention the works of Bassani and McClintock (1981) and those of Professor Mukherjee and his co-workers for Mode I and II in Mukherjee and Morjaria (1981) and Morjaria and Mukherjee (1981) and Mode III in Morjaria and Mukherjee (1982). A more comprehensive review in BEM solutions of inelastic could be found in the review article of Aliabadi (1977).

Fracture analyses which take into account time-dependent inelastic deformation problems arising in creeping cracked structural components are, in the BEM literature, almost non-existent. To the authors knowledge only the work of Mukherjee's research team, as it is described above, addresses the problem with the use of the Green's function approach. This Green's function approach, although accurate, is however limited to problems with a single crack modelled as a very thin ellipse. The main difficulty for the solution of this kind of fracture problems for both FEM and FEM methodologies is that, while at zero time the form of the stress singularity near crack tip is the well known elastic singularity $r^{-1/2}$ (where r is the distance from the crack tip) for any other time step the order of singularity changes to $r^{-1/(m+1)}$ in relation to the creep exponent m. Thus, it is clear that for this kind of problem the singularity order is variable and for a proper numerical solution of the problem one should be able to change the singularity order in a consistent time-dependent manner. Providakis and Kourtakis (2002) adopted a new special singular boundary element to take into account the variability of the singularity order generated in this kind of problems.

In the present paper, the strain energy density rate concept is applied as a fracture criterion in association with the use of a previously developed, by the present authors, creep strain-traction singular element (CR-STSE) to determine the crack initiation involved in creeping cracked two-dimensional plates. Numerical examples are presented for a central cracked plate (CCP) and a square plate with central hole. The creep constitutive model used in the numerical calculations is the Norton power law creep model (Nortran 1929) but any other creep constitutive model having similar mathematical structure can be easily implemented in the proposed algorithm.

2 Asymptotic Crack-Tip Fields in a Creeping Material

The material behavior in this paper is described by the elastic-nonlinear viscous constitutive relation according to the Norton power law relation (Nortran 1929)

$$\dot{\varepsilon} = \frac{\dot{\sigma}}{E} + \dot{\varepsilon}_0(\frac{\sigma}{\sigma_0})^m \tag{1}$$

where E is the elasticity modulus, σ_0 is a reference stress, $\dot{\varepsilon}_0$ is a reference creep strain rate and m is the creep exponent. Under the assumption of multiaxial stress states, the extension of equation (1) can be read as

$$\begin{aligned} \dot{\varepsilon}_{ij} &= \dot{\varepsilon}^e_{ij} + \dot{\varepsilon}^n_{ij} \\ \dot{\varepsilon}^e_{ij} &= \frac{1+\nu}{E}\dot{S}_{ij} + \frac{1-2\nu}{3E}\dot{\sigma}_{kk}\delta_{ij} \\ \dot{\varepsilon}^n_{ij} &= \frac{3}{2}\dot{\varepsilon}_0\left(\frac{\sigma_e}{\sigma_0}\right)^{m-1}\frac{S_{ij}}{\sigma_0} \end{aligned} \tag{2}$$

where S_{ij} are the components of the deviatoric stress tensor and $S_{ij} = \sigma_{ij} - \sigma_{kk}\delta_{ij}/3$ and σ_e is the Misses effective stress defined by $\sigma_e = \left((3/2)S_{ij}S_{ij}\right)^{1/2}$. From the inspection of (1) and (2) it could be noted that if there is a singular crack tip field at time t = 0 the elastic singularity fields prevail at the crack tip. In subsequent time step and at distances sufficient close to the crack tip the creep strain part of the total strain rate is much larger than the elastic strain rates and it seems to control the crack tip fields (m > 1). Thus, the constitutive equations (1) and (2) become power law creep relationships. Using the Hoff analogy (Hoff 1954) to contrast the power-law creep relation with the power-law hardening relation, Riedel and Rice (1988) and Ohji et al. (1979) presented the HRR-type singularity fields for power-law creep material described by the equations

$$\begin{aligned}\sigma_{ij} &= \sigma_0 \left(\frac{C(t)}{\dot{\varepsilon}_0\sigma_0 I_n r}\right)^{\frac{1}{n+1}} \tilde{\sigma}_{ij}(\theta)\\ \dot{\varepsilon}_{ij} &= \dot{\varepsilon}_0 \left(\frac{C(t)}{\dot{\varepsilon}_0\sigma_0 I_n r}\right)^{\frac{n}{n+1}} \tilde{\varepsilon}_{ij}(\theta)\\ \dot{u}_i &= \dot{\varepsilon}_0 r \left(\frac{C(t)}{\dot{\varepsilon}_0\sigma_0 I_n r}\right)^{\frac{n}{n+1}} \tilde{u}_{ij}(\theta)\end{aligned} \tag{3}$$

where the radial distance r from the crack tip and the angle θ in relation to the x axis are shown in Fig. 1. The dimensionless constants I_n and the θ-variation functions of the suitably normalized functions $\tilde{\sigma}_{ij}$, $\tilde{\varepsilon}_{ij}$ and $\tilde{u}_{ij}$ depend on the creep exponent m and have been tabulated in Shih (1983).

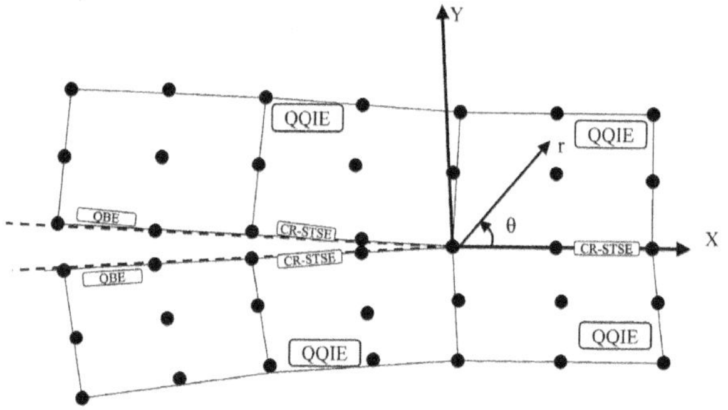

Fig. 1 Geometry of the crack-tip and creep strain-traction singular element configuration

3 Boundary Integral Equations

The Navier equation for the displacement rates of a structural component undergoing plane strain deformation and under the presence of non-elastic strains can be written as

$$\dot{u}_{i,jj} + \frac{1+\nu}{1-\nu}\dot{u}_{k,ki} = -\frac{\dot{F}_i}{G} + 2\dot{\varepsilon}^n_{ij,j} + \frac{2(1+\nu)}{1-2\nu}(\alpha\dot{T})_{,i} \quad (i, j, k = 1, 2) \tag{4}$$

where F_i is the prescribed body force per unit volume, G, ν and α are the shear modulus, Poisson's ratio and coefficient of linear thermal expansion, respectively, u_i is the displacement vector. Suitable traction and displacement rate boundary conditions must be prescribed. The integral representation of the solution of a point P on the boundary of the body (with $\dot{F}_i = 0$) has the following initial strain form

$$\begin{aligned}(\delta_{ij} - C)\,\dot{u}_i\,(P) = &\int_\Gamma \left[U_{ij}(P,Q)\dot{t}_j(Q) - T_{ij}(P,Q)\dot{u}_j(Q)\right] d\Gamma_q \\ &+ \int_\Omega \left[\Sigma_{jki}(P,q)\dot{\varepsilon}^n_{jk}(q) + +\tilde{\Sigma}_{jki}(P,q)\delta_{jk}\alpha\dot{T}(q)\right] d\Omega_q\end{aligned} \tag{5}$$

where δ_{ij} is Kronecker delta, P,Q are boundary points, q is an interior point, Γ and Ω are the boundary and the surface of the body, respectively. The kernels U_{ij},T_{ij},Σ_{jki} and $\widetilde{\Sigma}_{jki}$ are known singular solutions due to a point load in an infinite elastic solid in plane strain (e.g., Mukherjee 1977). The traction and displacement rates are denoted by $\dot{t}$ and $\dot{t}$, respectively. The coefficients C_{ij} are known functions of the included angle at the boundary corner at P, the angle between the bisector of the corner angle and the x-axis. Equation (5) is a system of integral equations for the unknown traction and displacements rates in terms of their prescribed values on the boundary, and the non-elastic strain rates. The unknown quantities only appear on the boundary of the body and the surface integrals are known at any time through the constitutive equations.

The stress rates can be obtained by direct differentiation of equation (5) resulting in

$$\begin{aligned}\dot{\sigma}_{ij}\,(p) = &\int_\Gamma \left[\bar{U}_{ijk}\,(p,Q)\,\dot{t}_k\,(Q) - \bar{T}_{ijk}\,(p,Q)\,\dot{u}_k\,(Q)\right] d\Gamma_q - 2G\dot{\varepsilon}^n_{ij}(p) \\ &- 3K\alpha\dot{T}(p)\delta_{ij} + \int_\Omega \left[\bar{\Sigma}_{ijkl}\,(p,q)\,\dot{\varepsilon}^n_{kl}\,(q) + \bar{\tilde{\Sigma}}_{ijkl}(p,q)\delta_{kl}\alpha\dot{T}(q)\right] d\Omega_q \\ &(i,j,k,l = 1,2)\end{aligned} \tag{6}$$

where G and K are the shear and bulk modulus, respectively; $\bar{\Sigma}_{ijkl}$ and $\bar{\tilde{\Sigma}}_{ijkl}$ are inelastic and temperature effect kernel functions, respectively, which are also defined in the work of Mukherjee (1977).

4 Special Boundary Element Implementation and Solution Procedure

The integral equations (5) and (6) are expressed in this paper by discretizing the boundary and the interior into a number of standard three-noded quadratic boundary elements and nine-noded quadratic quadrilateral interior surface elements, respectively, provided that they are not adjacent to the crack tip.

By following the procedure developed in Providakis and Kourtakis (2002) to produce a special element which presents the HRR-type singularity of equations (3) at the crack tip (Fig. 1), one can obtain the following new set of shape functions N_a^u which depend upon the creep exponent m

$$N_1^u = 2^{\frac{1}{1+m}}\left[\left(\frac{r}{l}\right)^{1+\frac{1}{1+m}} - \left(\frac{r}{l}\right)^{\frac{1}{1+m}}\right] + \left(\frac{r}{l}\right)$$

$$N_2^u = 2^{1+\frac{1}{1+m}}\left[\left(\frac{r}{l}\right)^{\frac{1}{1+m}} - \left(\frac{r}{l}\right)^{1+\frac{1}{1+m}}\right] \tag{7}$$

$$N_3^u = 2^{\frac{1}{1+m}}\left[\left(\frac{r}{l}\right)^{1+\frac{1}{1+m}} - \left(\frac{r}{l}\right)^{\frac{1}{1+m}}\right] - \left(\frac{r}{l}\right) + 1$$

where l is the length of the new special quadratic element, the distance r $= l -$ x and the ratio can be defined in terms of the intrinsic coordinate ζ as (r/l)=(1-ζ)/2. By taking the derivatives of the new shape functions (7) one can observe that these derivatives display a $r^{-m/(m+1)}$ singularity near the crack tip which is the actual situation for the strain rate singularities according to (3).

Since in boundary element methodology displacement and tractions are independently represented the above derived singular element for the simulation of crack tip behavior of displacement rates, fails to model the expected from equations (3) crack tip behavior of tractions which displays an order of $-1/(m+1)$ singularity. Thus, for the proper simulation of the traction rate singularity different shape functions are derived by the use of the derivatives of the shape functions (7) and finally modified to the following separate forms N_a^t in terms of creep exponent m

$$N_1^t = 2^{\frac{m}{1+m}}\left[\left(\frac{l}{r}\right)^{\frac{1}{1+m}} - \left(\frac{r}{l}\right)^{\frac{m}{1+m}}\right] - 2 + 2\left(\frac{r}{l}\right)$$

$$N_2^t = 2^{\frac{m}{1+m}}\left[\left(\frac{l}{r}\right)^{\frac{1}{1+m}} - \left(\frac{r}{l}\right)^{\frac{m}{1+m}}\right]$$

$$N_3^t = 2^{\frac{m}{1+m}}\left[-\left(\frac{l}{r}\right)^{\frac{1}{1+m}} + \left(\frac{r}{l}\right)^{\frac{m}{1+m}}\right] + 1 \tag{8}$$

where now r $=$ x and the ratio $(r/l) = (1 + \zeta)/2$. A simultaneous simulation of displacement and traction rate fields, by the use of the shape functions (7) and (8), respectively, yields to the proposed, in the present BEM approach, creep strain-traction singular element (CR-STSE) (Fig. 1).

Then, by applying a boundary nodal point collocation procedure to the discretized versions of equations (5) and (6) one can obtain the following system of equations in matrix form

$$[A]\{\dot{u}\} = [B]\{\dot{\tau}\} + [E]\{\dot{\varepsilon}^n\} + [T]\{\dot{b}^T\} \tag{9}$$

$$\{\dot{\sigma}\} = [\tilde{B}]\{\dot{\tau}\} + [\tilde{E}]\{\dot{\varepsilon}_n\} + [\tilde{T}]\{\dot{b}^T\} \tag{10}$$

However, the vector $\{\dot{\varepsilon}^n\}$ is known at any time through the constitutive equations and the stress rates of equation (6) while the vector $\{\dot{b}^T\}$ could be easily computed through the known values of temperature profile for the whole structural component. Half of the total number of components of $\{\dot{u}\}$ and $\{\dot{\tau}\}$ are prescribed through the boundary condition while the other half are unknowns.

Then, the initial distribution of the nonelastic strain has to be prescribed. Thus, the only existed strains at time step t = 0 are elastic and then, the thermal and initial stresses and displacements can be obtained from the solution of the corresponding elastic problem. By the use of equations (9) and (10) the displacement and stress rates can be obtained at time step t = 0 while the rates of change of the nonelastic strains can be computed from constitutive equations. Thus, the initial rates of all the relevant variables are now known and their values at a new time Δt can be obtained by integrating forward in time. The rates are then obtained at time Δt and so on, and finally the time histories of all the variables can be computed. Another important task in this approach is the choice of a suitable time integration scheme. For the purposes of the present paper, an Euler type algorithm with automatic time-step control is employed.

5 The Strain Energy Rate Definition

For power law creep materials the strain energy density rate (SEDR) can be analytically determined as

$$\dot{W} = \int_0^{\dot{\varepsilon}^n} \sigma d\dot{\varepsilon}^n = \frac{m}{m+1}\sigma_e \dot{\varepsilon}^n \tag{11}$$

The SEDR was estimated according to the boundary element procedure developed previously by solving in time the system of equations (9) and (10) and then by using the analytic equation (11) for each time step.

6 Numerical Examples

First consider a central cracked plate (CCP) specimen with height (h) and width (w) = 40.6 × 20.3 cm, as shown in Fig. 2, made by a power law creeping material (superalloy Inconel 800H at 650°C) with properties E = 153.7 GPa, σ_0 = 417.04 MPa, ν = 0.33, creep exponent m = 5 and the parameter

$$B = \frac{\dot{\varepsilon}_0}{\sigma_0^m} = 2.1x10^{-32}x6894.73 \quad (Pa)^{-5}/h \tag{12}$$

The specimen contains a central crack of depth a = 0.125 w. The specimen is subjected to a remote uniform load of 129.2 MPa which is suddenly applied. The symmetry of the specimen was used and thus a quarter of the plate was analyzed.

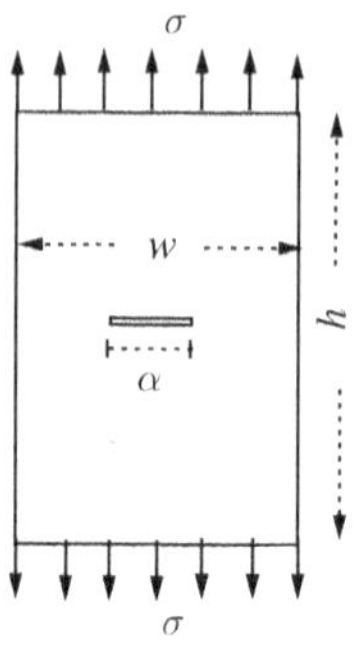

Fig. 2 Central cracked plate geometry

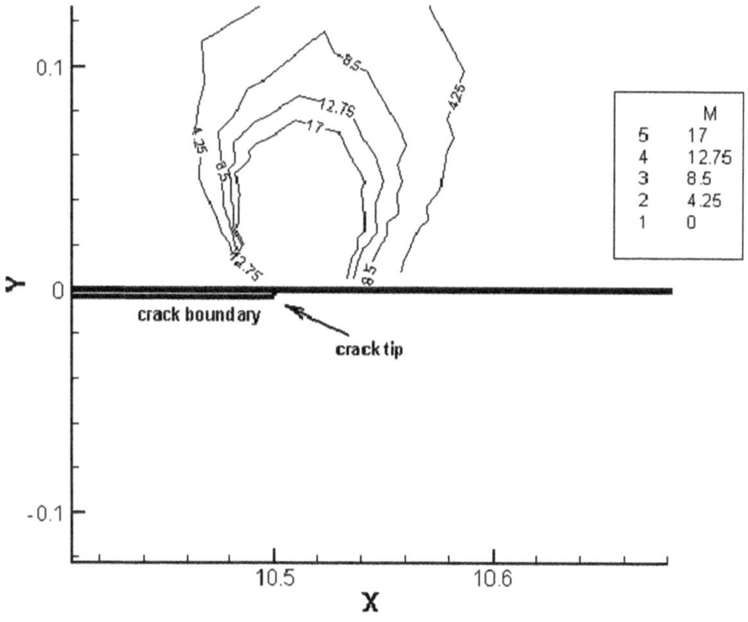

Fig. 3 Strain energy density rate contours (t = 0.0 h)

The computer software established according to the BEM methodology presented in this paper can provide sufficient data related to stress and strain distribution history. Based on these data and after using equation (11) the strain energy density rate distribution could be predicted at each time step. Typical diagrams of contours of strain energy density rate distributions for a quadrant of the specimen and for two different time steps (t = 0.0 and 5.01 h) are shown in Figs. 3 and 4, respectively.

Next, consider a square plate with a circular hole at its center. The plate dimensions are: side length 0.254 m and hole radius 0.0254 m as shown in Fig. 5. This plate

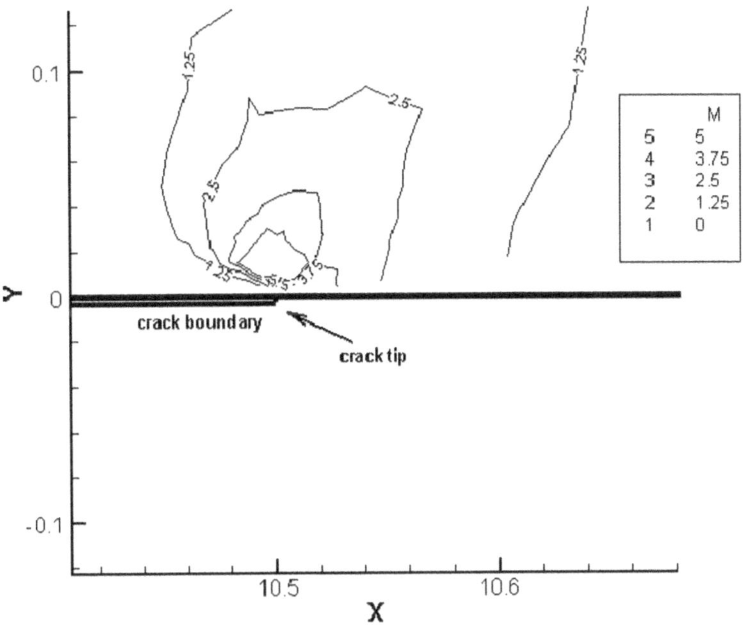

Fig. 4 Strain energy density rate contours (t = 5.01 h)

Fig. 5 Geometry of the square plate with a central hole

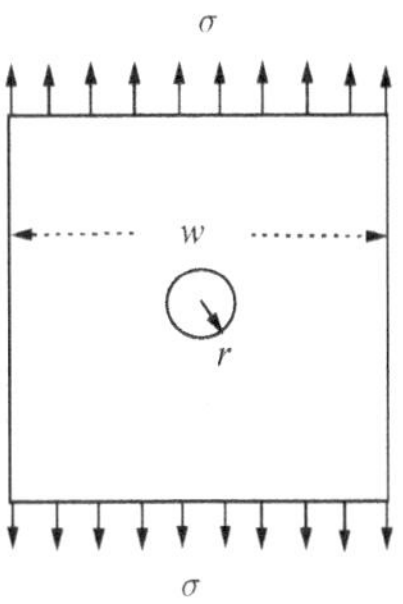

specimen is made by a power law creeping material with properties: E − 206 GPa, ν = 0.3, creep exponent n = 4 and creep parameter

$$B = \frac{\dot{\varepsilon}_0}{\sigma_0^m} = 4.42x10^{-16} \quad (MPa)^{-4}/h \tag{13}$$

The plate is loaded by a suddenly applied remote loading of magnitude 100 MPa. Typical diagrams of the surface distribution of strain energy density rate values around the hole perimeter are shown in Figs. 6 and 7.

It could be noted from the inspection of the Figs. 6 and 7 that the strain energy density rate decreases with increasing time. This trend was expected since the

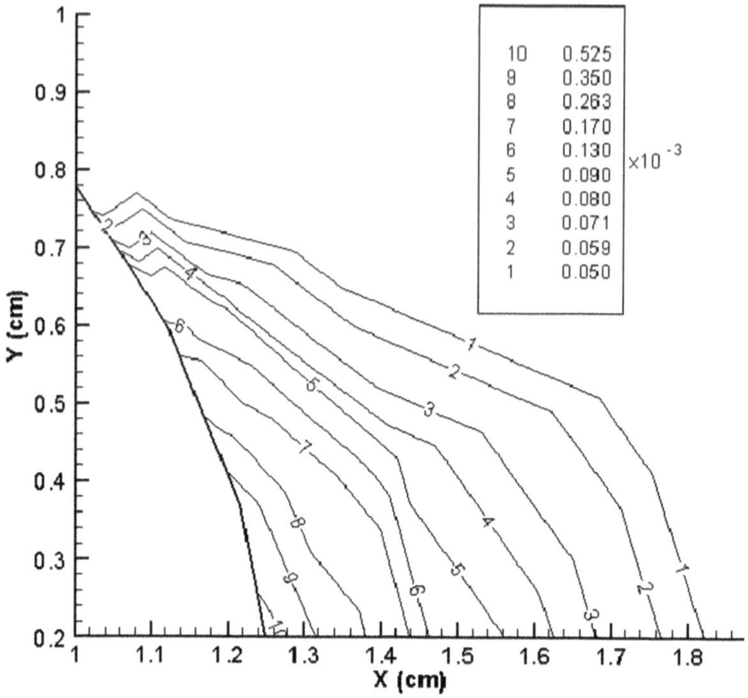

Fig. 6 Distribution of strain energy density rate values along the hole perimeter (t = 0 h)

applied load remains fixed in time and the crack is assumed to be stationary. It could be also noted that all curves posses a minimum at angle $\theta_o = 0$.

Taking into account the strain energy density criterion this indicates that crack would initiate at $\theta_o = 0$ and along the axis of symmetry of load symmetry under present crack mode I.

7 Conclusions

Application of the strain energy density rate concept is performed using the boundary element methodology to investigate the time-dependent crack analysis problem of creeping structural components. The present boundary element procedure is based on the implementation of a special singular boundary element for the estimation of the strain energy density rate distribution close to crack tip fields arising in creeping structural components undergoing Mode I deformation under the effect of remote loading condition.

Encouraged by the results of the present work, it could be concluded that the strain energy density concept could be further used to effectively investigate the behaviour of creeping materials.

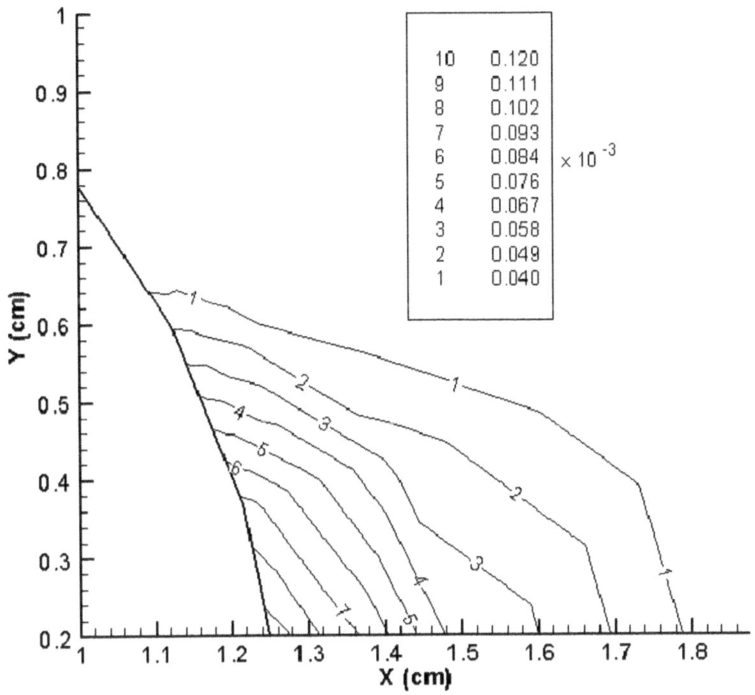

Fig. 7 Distribution of strain energy density rate values along the hole perimeter (t = 400 h)

References

Aliabadi MH (1977) Boundary element formulations in fracture mechanics. Appl Mech Rev 50(2):83–96

Bassani JL, McClintock FA (1981) Creep relaxation of stress around a crack tip. Int J Solids Struct 17:479–492

Beskos DE (1987) Numerical methods in dynamic fracture mechanics. EUR-11300 EN. Joint Research Center of EC Ispra Establishment. Ispra. Italy

Blandford GE, Ingraffea AR, Ligget JA (1981) Two dimensional stress intensity factor computations using the boundary element method. Int J Num Meth Engng 17:387–404

Carpinteri A (1986) Crack growth and material damage in concrete: plastic collapse and brittle fracture. In: Engineering Applications of Fracture Mechanics V, Martinus Nijhof, Netherlands

Gdoutos EE (1984) Problems of mixed mode crack propagation. Engineering Applications of Fracture Mechanics II. Martinus Nijhof, Netherlands

Hantschel T, Busch M, Kuna M, Maschke HG (1990) Solution of elastic-plastic crack problems by an advanced boundary element method. Numerical Methods. In: Luxmoore AR, D.R.J. Owen DRJ (eds.) Fracture Mechanics. Pine Ridge Press, Swansea

Hoff NJ (1954) Approximate analysis of structures in presence of moderately large creep deformations. Q Appl Mech 12:49–55

Hutchinson JW (1968) Singular behavior at the end of a tensile crack in a hardening material. J Mech Phys Solids 16:13–31

Morjaria M, Mukherjee S (1981) Numerical analysis of planar, time-dependent inelastic deformation of plates with cracks by the boundary element method. Int J Solids Struct. 17:127–143

Morjaria M, Mukherjee S (1982) Numerical solution of stresses near crack tips in time-dependent inelastic fracture mechanics. Int J Fract 8(4):293–310

Mukherjee S (1977) Corrected boundary integral equations in planar thermoelastoplasticity. Int J Solids Struct. 13:331–335

Mukherjee S, Morjaria M (1981) A boundary element formulation for planar time-dependent inelastic deformation of plates with cutouts. Int J Solids Struct 17:115–126

Nortran FH (1929) The Creep of Steel at High Temperatures. McGraw-Hill, New York.

Ohji K, Ogura K, Kubo S (1979) Stress-strain field and modified J-integral in the vicinity of crack tip under transient creep conditions. Jp Soc Mech Eng 790(13):18–20.

Providakis CP, Kourtakis S (2002) Time-dependent creep fracture using singular boundary elements. Comp Mech 29:298–306

Providakis CP (2006) A strain energy density rate approach to the BEM analysis of creep fracture problems. Struct Durability Health Mon 2(4):249–254

Rice JR, Rosengren GF (1968) Plane strain deformation near a crack tip in a power-law hardening material. J Mech Phys Solids 16:1–12

Riedel H, Rice JR (1988) Tensile cracks in creeping solids. In: Proc. of Twelve Conference on Fracture Mechanics. ASTM STP 700, 112–130.

Shih CF (1983) Tables of Hutchinson-Rice-Rosengren singular field quantities. MRL E-147, Division of Engineering, Brown University.

Sih GC (1973) Some basic problems in fracture mechanics and new concepts. J Eng Fract Mech 5:365–377.

Sih GC (1991) Mechanics of Fracture Initiation and Propagation. Kluwer Academic Publishers, Dordrecht

MFS with RBF for Thin Plate Bending Problems on Elastic Foundation

Qing-Hua Qin, Hui Wang and V. Kompis

Abstract In this chapter a meshless method, based on the method of fundamental solutions (MFS) and radial basis functions (RBF), is developed to solve thin plate bending on an elastic foundation. In the presented algorithm, the analog equation method (AEM) is firstly used to convert the original governing equation to an equivalent thin plate bending equation without elastic foundations, which can be solved by the MFS and RBF interpolation, and then the satisfaction of the original governing equation and boundary conditions can determine all unknown coefficients. In order to fully reflect the practical boundary conditions of plate problems, the fundamental solution of biharmonic operator with augmented fundamental solution of Laplace operator are employed in the computation. Finally, several numerical examples are considered to investigate the accuracy and convergence of the proposed method.

1 Introduction

Thin plate structures are widely used in engineering practice for the design of aircraft, ship, and ground structures. Numerical study of their behaviour under various loadings conditions is, therefore, essential. Apart from a few thin plate bending problems with simple transverse loads or simple boundary conditions, a general solution is difficult to obtain analytically. Some numerical methods such as finite element method (FEM) (Martin and Carey 1989), boundary element method (BEM) (Bittnar and Sejnoha 1996), hybrid-Trefftz finite element method (HT-FEM) (Qin 2000), and method of fundamental solution (MFS) (Kupradze and Aleksidze 1964), are, thus, developed to analyze bending deformation of thin plate structures under various transverse loads and boundary conditions.

As one of the numerical methods above, MFS, developed in 1964 (Kupradze and Aleksidze 1964), is a boundary-type meshless method, which is based on the

Q-H. Qin (✉)
Department of Engineering, Australian National University, Canberra, ACT, Australia, 0200
e-mail: Qinghua.qin@anu.edu.au

G.D. Manolis, D. Polyzos (eds.), *Recent Advances in Boundary Element Methods*,
DOI 10.1007/978-1-4020-9710-2_24, © Springer Science+Business Media B.V. 2009

combination of set of fundamental solutions with different sources. The typical feature of MFS is that the approximated field satisfies, a prior, the governing partial different equations (PDE) in the domain and the satisfaction of boundary conditions is used to determine the unknown coefficients. This feature makes MFS to be suitable for analysing homogeneous boundary value problems (BVP) (Fairweather and Karageorghis 1998). The methods similar to the MFS are virtual boundary element/collocation method (Sun et al. 1999; Yao and Wang 2005), the F-Ttrefftz method (Karthik and Palghat 1999), the charge simulation method (Katsurada 1994; Rajamohan and Raamachandran 1999), and the singularity method (Nitsche and Brenner (1990). Additionally, Wang et al. (Wang et al. 2005; Wang and Qin 2006; Wang et al. 2006; Wang and Qin 2007) combined the MFS and RBF and used for analyzing steady and transient heat conduction, linear and nonlinear potential problems.

However, for thin plate bending problems, the existence of transverse load and elastic foundation terms makes it difficult to employ the MFS directly. Besides, the fundamental solution is difficult to obtain or very complex for some plate bending problems such as dynamic thin plate on elastic foundations and anisotropic plate bending problems. The standard MFS is, thus, not suitable for analysing this category of problems. As a result, new technologies are proposed to treat such problems (Misra et al. 2007; Ferreira 2003; Leitao 2001; Liu et al. 2006). For example, Kansa's method and symmetric Hermite method based on RBF were used to analyze some special thin plate bending problems (Misra et al. 2007; Ferreira 2003; Leitao 2001; Liu et al. 2006). It is noted, however, that special treatments of collocation are needed in both the standard MFS (Rajamohan and Raamachandran 1999) (when using the traditional fundamental solution of biharmonic operator) and Kansa's method in order to satisfy the specified boundary conditions, because there are two known quantities at each point on the boundary for thin plate bending problems. Additionally, the analog equation method (AEM) (Nerantzaki and Katsikadelis 1996) is used in the process of BEM to solve thin plate bending with variable thickness.

In this chapter, the mixture of AEM, RBF and MFS are employed to solve the thin plate bending on elastic foundations. Noting the feature of the governing equation of thin plate, that is the fourth-order equation, the AEM is first used to convert the original governing equation into an equivalent biharmonic equation with fictitious transverse load. Its particular and homogeneous solutions are, then, constructed by means of RBF and MFS, which uses the improved fundamental solution, respectively. Finally, satisfaction of boundary conditions and the original governing equation can be used to determine all unknown coefficients. In contrast to the standard MFS and Kansa's method, the presented method is more effective to treat various transverse load and boundary conditions.

The outline of the chapter is arranged as follows. Section 2 gives a description of basic equations of thin plate bending on elastic foundations. Coupled MFS with AEM, and RBF for plate bending problems are presented in Section 3. Finally, several numerical examples are considered in Section 4 and some conclusions are made in Section 5.

2 Basic Equations of Thin Plate Bending

Consider a thin plate under an arbitrary transverse loads as shown in Fig. 1. It is assumed that the thickness h of the thin plate is in the range of $1/20 \sim 1/100$ of its span approximately.

Under the assumption above the Kirchhoff thin plate bending theory can be employed. The governing equation of thin plate on an elastic foundation under arbitrary transverse load $p(\mathbf{x})$ is, thus, written as (Zhang 1984)

$$D\nabla^4 w(\mathbf{x}) + k_w w(\mathbf{x}) = p(\mathbf{x}) \tag{1}$$

where $w(\mathbf{x})$ denotes the lateral deflection of interest at the point $\mathbf{x} = (x_1, x_2) \in \Omega \subset \mathbb{R}^2$, D is the flexural rigidity defined by

$$D = \frac{Eh^3}{12\left(1 - \nu^2\right)} \tag{2}$$

where E is Young's modulus, ν Poisson's ratio, h plate thickness, k_w the parameter of Winkler foundation, and ∇^4 is the biharmonic differential operator defined by

$$\nabla^4 = \frac{\partial^4}{\partial x_1^4} + 2\frac{\partial^4}{\partial x_1^2 \partial x_2^2} + \frac{\partial^4}{\partial x_2^4} \tag{3}$$

What follows is to establish a linear equation system of thin plate bending for determining the unknown deflection $w(\mathbf{x})$ which satisfies Eq. (1) and boundary conditions listed in Table 1. The boundary conditions in Table 1 are described by two displacement components (w, θ_n) and two internal forces (M_n, V_n).

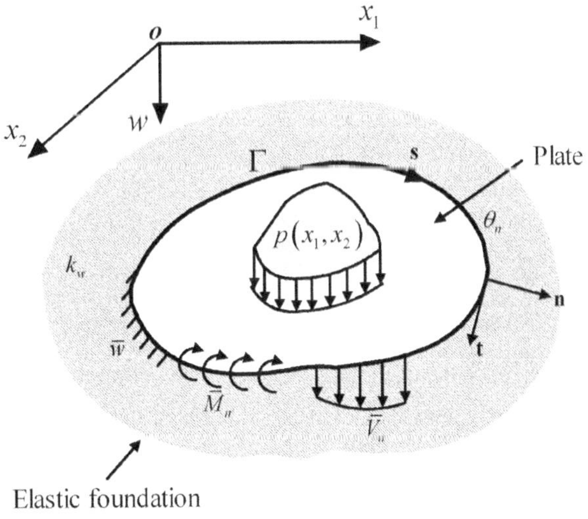

Fig. 1 Configuration of thin plate bending on elastic foundation under transverse distributed loads

Table 1 Common boundary conditions in thin plate bending

Types of support	Mathematical expressions
Simple support	$w = 0, \quad M_n = 0$
Fixed edge	$w = 0, \quad \theta_n = 0$
Free edge	$V_n = 0, \quad M_n = 0$

The variables θ_n, M_n, and V_n are, respectively, outward normal derivative of deflection, bending moment, and Kirchhoff's equivalent shear force. They can be expressed in terms of deflection $w(\mathbf{x})$ as

$$\begin{aligned} \theta_n &= w_{,i} n_i, \\ M_n &= -D\left[\nu w_{,ii} + (1-\nu) w_{,ij} n_i n_j\right] \\ V_n &= Q_n + \frac{\partial M_{nt}}{\partial s} = -D\left[w_{,ijj} n_i + (1-\nu) w_{,ijk} n_i t_j t_k\right] \end{aligned} \tag{4}$$

where $\mathbf{n} = [n_1, n_2]$ and $\mathbf{t} = [-n_2, n_1]$ are the outward unit normal vector and tangential vector on the boundary, respectively, s is the arc length along the boundary measured from a certain boundary point, and

$$M_{nt} = -D(1-\nu) w_{,ij} n_i t_j, \quad Q_n = -D w_{,ijj} n_i \tag{5}$$

3 Formulation

In this section, a meshless formulation for thin plate bending with an elastic foundation is presented by means of the combination use of analog equation method (AEM), method of fundamental solutions (MFS) and radial basis functions (RBF). With the proposed meshless method it is easy and simple for solving plate bending problems with various transverse loads and boundary conditions.

3.1 Analog Equation Method (AEM)

Following the way in Nerantzaki and Katsikadelis (1996), the fourth-order plate bending equation can be written in terms of biharmonic operator as (Nerantzaki and Katsikadelis 1996)

$$D\nabla^4 w(\mathbf{x}) = \tilde{p}(\mathbf{x}) \tag{6}$$

where $\tilde{p}(\mathbf{x})$ is fictitious transverse load including the term with the unknown deflection. The equation above is a plate bending equation without elastic foundation and its fundamental solution is available in the literature. The fictitious transverse load $\tilde{p}(\mathbf{x})$ can be expressed in terms of RBFs.

The solution to Eq. (6) is firstly divided into two parts: homogeneous solution and particular solution, which satisfy the following equations, respectively,

$$\begin{cases} D\nabla^4 w_h(\mathbf{x}) = 0 \\ D\nabla^4 w_p(\mathbf{x}) = \tilde{p}(\mathbf{x}) \end{cases} \tag{7}$$

Specially, for the case of thin plate bending problems without elastic foundation ($k_w = 0$) we have $\tilde{p}(\mathbf{x}) = p(\mathbf{x})$. The procedure of AEM is unnecessary in this case.

3.2 Method of Fundamental Solutions (MFS)

For a well-posed thin plate bending problem, there are two known and two unknown quantities at each point on the boundary. Therefore, we need two equations to determine the two unknowns at each point. Considering this feature the corresponding MFS is constructed based the following two fundamental solutions.

It's well known that the general solution of a biharmonic equation can be expressed in the following form

$$w_h(\mathbf{x}) = A + r^2 B \tag{8}$$

where A and B are two independent functions satisfying the Laplace equation, respectively,

$$\nabla^2 A = 0, \quad \nabla^2 B = 0 \tag{9}$$

So, we can combine the fundamental solutions of biharmonic operator and Laplace operator to fulfill the character of boundary conditions mentioned above, that is

$$w_h(\mathbf{x}) = \sum_{i=1}^{N_S} \left[\phi_{1i} w_1^*(\mathbf{x}, \mathbf{y}_i) + \phi_{2i} w_2^*(\mathbf{x}, \mathbf{y}_i)\right] \quad \mathbf{x} \in \Omega, \quad \mathbf{y}_i \notin \Omega \tag{10}$$

where N_S are source points outside the domain, $w_1^*(\mathbf{x}, \mathbf{y})$ and $w_2^*(\mathbf{x}, \mathbf{y})$ are fundamental solutions of biharmonic operator and Laplace operator, respectively, which can be written as

$$w_1^*(\mathbf{x}, \mathbf{y}) = -\frac{1}{8\pi D} r^2 \ln r$$
$$w_2^*(\mathbf{x}, \mathbf{y}) = -\frac{1}{2\pi D} \ln r$$

with $r = \|\mathbf{x} - \mathbf{y}\|$.

Unlike the approaches in Long and Zhang (2002) and Sun and Yao (1997) constructing fundamental solutions to adapt the requirement of boundary conditions

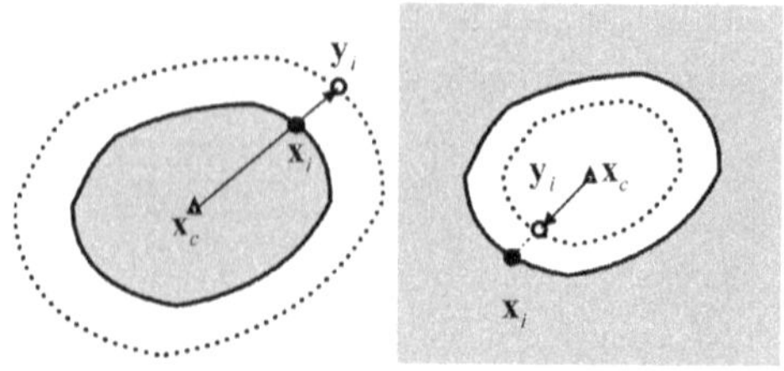

(a) interior domain (b) exterior domain

Fig. 2 Configuration of source points

in thin plate bending, the proposed approach is simpler and more convenient in practical applications.

It is easy to verify that Eq. (10) satisfies the first equation in Eq. (7).

The proper location of the source points is an important issue in the MFS with respect to the accuracy of numerical solutions. Here the position of the source points can be evaluated by means of the following equation (Young et al. 2006):

$$\mathbf{y} = \mathbf{x}_b + \gamma (\mathbf{x}_b - \mathbf{x}_c) \tag{11}$$

where $\mathbf{y}$ are the spatial coordinates of a particular source point, $\mathbf{x}_b$ the spatial coordinates of related boundary points, and $\mathbf{x}_c$ the central coordinates of the solution domain. γ is a dimensionless real parameter, which is positive for the case of external boundary and negative for the case of internal boundary (see Fig. 2).

3.3 Radial Basis Function (RBF)

In order to obtain the particular solution corresponding to the fictitious transverse load $\tilde{p}(\mathbf{x})$, the radial basis function approximation of $\tilde{p}(\mathbf{x})$ is written in the form (Golberg et al. 1999).

$$\tilde{p}(\mathbf{x}) = \sum_{j=1}^{N_I} \alpha_j \phi_j(\mathbf{x}) \tag{12}$$

where the set of radial basis functions $\phi_j(\mathbf{x})$ is taken as $\phi(r_j)$ where $r_j = \left\| \mathbf{x} - \mathbf{x}_j \right\|$. $\phi(r_j)$ is defined in Table 2.

Table 2 Particular solutions for the biharmonic equation

	Power spline (PS) RBF	Thin plate spline (TPS) RBF
ϕ	r^{2n-1}	$r^{2n}\ln r$
$D\Phi$	$\dfrac{r^{2n+3}}{(2n+1)^2(2n+3)^2}$	$\dfrac{r^{2n+4}}{16(n+1)^2(n+2)^2}\left[\ln r - \dfrac{2n+3}{(n+1)(n+2)}\right]$

Similarly, the particular solution $w_p(\mathbf{x})$ is expressed by the linear combination of approximated particular solutions $\Phi_j(\mathbf{x}) = \Phi(r_j)$, that is

$$w_p(\mathbf{x}) = \sum_{j=1}^{N_I} \alpha_j \Phi_j(\mathbf{x}) \tag{13}$$

The satisfaction of the relation of $w_p(\mathbf{x})$ and $\tilde{p}(\mathbf{x})$ in Eq. (7) requires

$$D\nabla^4 \Phi(r_j) = \phi(r_j) \tag{14}$$

Therefore, once the expression of radial basis function $\phi(r_j)$ is given, the approximated particular solutions $\Phi(r_j)$ can be determined from Eq. (14).

3.4 Solution of Deflection

The solution $w(\mathbf{x})$ can be obtained by putting the obtained homogeneous and particular parts together and written as

$$w(\mathbf{x}) = \sum_{j=1}^{N_I} \alpha_j \Phi_j(\mathbf{x}) + \sum_{i=1}^{N_S} \left[\phi_{1i} w_1^*(\mathbf{x}, \mathbf{y}_i) + \phi_{2i} w_2^*(\mathbf{x}, \mathbf{y}_i)\right] \tag{15}$$

The unknowns α_j, ϕ_{1i}, and ϕ_{2i} can be determined by substituting Eq. (15) into the original governing equation (1) at N_I interpolation points and boundary conditions (4) at N_S boundary points. For example, the substitution of Eq. (15) into Eqs. (1) and (4) yields following system of linear equations

$$(D\nabla^4 + k_w)\mathbf{w}(\mathbf{x})|_{\mathbf{x}=\mathbf{x}_i} \{\mathbf{A}\} = p(\mathbf{x}_i) \quad (i = 1, 2, \cdots N_I)$$

$$\left[\begin{matrix} \mathbf{w}(\mathbf{x}) \\ -D(\nu\mathbf{w}(\mathbf{x})_{,kk} + (1-\nu)\mathbf{w}(\mathbf{x})_{,kl} n_k n_l) \end{matrix}\right]\Bigg|_{\mathbf{x}=\mathbf{x}_i} \{\mathbf{A}\} = \left\{\begin{matrix} \bar{w}(\mathbf{x}_i) \\ \bar{\theta}_n(\mathbf{x}_i) \end{matrix}\right\} (i = 1, 2, \cdots, N_S) \tag{16}$$

for a simply-supported plate, where

$$\mathbf{w}(\mathbf{x})|_{\mathbf{x}=\mathbf{x}_i} = \{\Phi_1(\mathbf{x}_i)\ \Phi_2(\mathbf{x}_i) \cdots \Phi_{N_I}(\mathbf{x}_i)\ w_1^*(\mathbf{x}_i, \mathbf{y}_1)$$
$$w_2^*(\mathbf{x}_i, \mathbf{y}_1)\ w_1^*(\mathbf{x}_i, \mathbf{y}_{N_S}) \cdots w_2^*(\mathbf{x}_i, \mathbf{y}_{N_S})\}$$

$$\{\mathbf{A}\} = \{\alpha_1\ \alpha_2 \cdots \alpha_{N_I}\ \phi_{11}\ \phi_{21} \cdots \phi_{1N_S}\ \phi_{2N_S}\}^{\mathrm{T}}$$

Once all unknown coefficients are determined, the deflection w, rotation θ_n, moment M_n and reaction force V_n can be calculated by using Eqs. (4) and (15).

4 Numerical Examples

In this section, two numerical examples are considered to investigate the performance of the proposed algorithm. In order to provide a more quantitative understanding of results, the average relative error (Arerr) is introduced as

$$Arerr\,(f) = \sqrt{\frac{\sum\limits_{i=1}^{N} (f_{numerical} - f_{exact})_i^2}{\sum\limits_{i=1}^{N} (f_{exact})_i^2}} \tag{17}$$

where N is the number of test points and $(f)_i$ is an arbitrary field function such as deflection at point i.

The first example is a thin plate bending problem without elastic foundations, and is designed to demonstrate the convergence, stability, and feasibility of the proposed formulation. The second one is a typical thin plate bending resting on a Winkler elastic foundation.

Example 1 (Simply-supported square plate).

Consider a square plate subjected to uniformly distributed load q_0 (see Fig. 3). All four edges of the plate are simply-supported, i.e. $w = 0$ and $M_n = 0$ along all edges.

The analytical solution for this problem is

$$w\,(x_1, x_2) = \sum_{m=1}^{\infty}\sum_{n=1}^{\infty} A_{mn} \sin\frac{m\pi x_1}{a} \sin\frac{n\pi x_2}{a} \tag{18}$$

with

$$A_{mn} = \frac{q_{mn}}{D\pi^4 \left(\frac{m^2+n^2}{a^2}\right)^2}$$

$$q_{mn} = \frac{4q_0\,[-1+\cos(m\pi)]\,[-1+\cos(n\pi)]}{mn\pi^2}$$

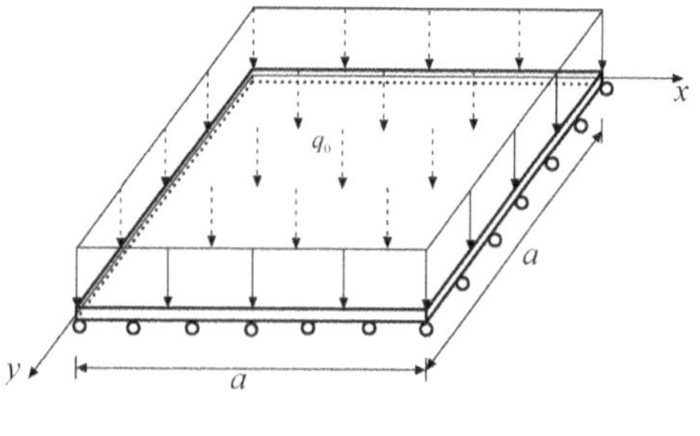

Fig. 3 Simply-supported square plate subjected to uniformly transverse load

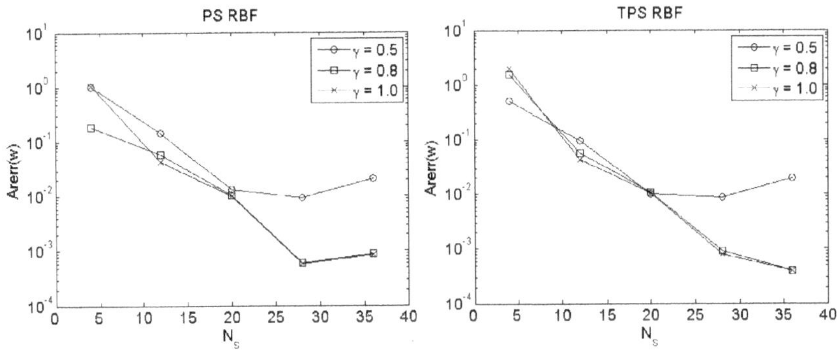

Fig. 4 Demonstration of convergence with the increase of N_S and $N_I = 64$

In the computation, the related parameters are given as $a = 1$ m, $h = 0.01$ m, $q_0 = 1$ kN/m^2, $E = 2.1 \times 10^4$ MPa, $\nu = 0.3$, $k_w = 0$. Results of deflection in Fig. 4 show that Thin Plate Spline (TPS) RBF is more stable than Power Spline (PS) RBF, especially for larger N_S. Meanwhile, it is found that the location of source points can affect convergent performance. The larger value of γ, the better accuracy and convergent performance. Based on the convergent performance shown in Fig. 4, TPS RBF and $\gamma = 0.8$ are chosen in late computation. It is also found from Fig. 5 that a good convergence is achieved with the increase of N_I. All these results show a good convergence and accuracy of the proposed algorithm.

The distribution of moment along $y = 0.5$ is plotted in Fig. 6, in which $N_S = 36$ and $N_I = 64$ are used. It is found a good agreement between numerical results and analytical solutions as displayed.

Example 2 (Square plate on a Winkler elastic foundation). Consider the same square plate as in Example 1. The same boundary and transverse uniformly load are

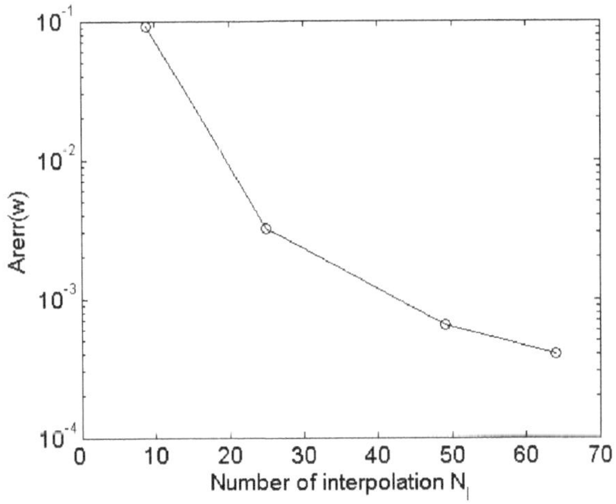

Fig. 5 Demonstration of convergence with the increase of N_I and $N_S = 36$

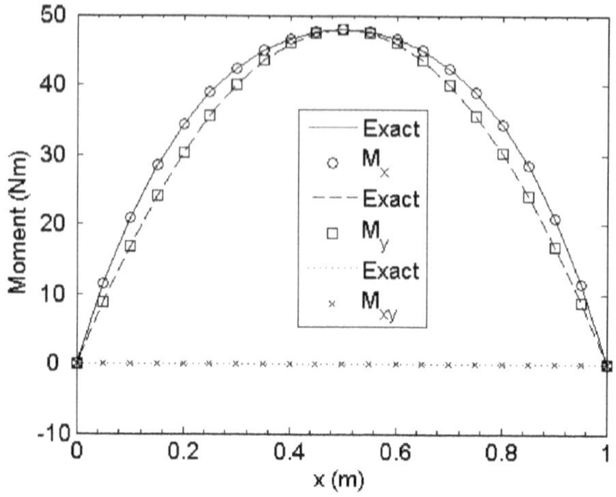

Fig. 6 Distribution of moment along with $y = 0.5$

again used. The parameter of Winkler foundation k_w is taken to be 4.9×10^7 N/m^3. In this case, the analytical solution of deflection is the same as that of Eq. (18) except for

$$A_{mn} = \frac{q_{mn}}{D\pi^4 \left(\frac{m^2+n^2}{a^2}\right)^2 + k_w}$$

Due to the symmetry of the problem, one quarter of the solution domain is considered. The distribution of deflection and moment along $y = 0.5$ is evaluated with $N_S = 36$ and $N_I = 121$ and the corresponding results are shown in Fig. 7. It can be seen from Fig. 7 that the proposed MFS-based method provides a very good accuracy approximation for the corresponding analytical solution.

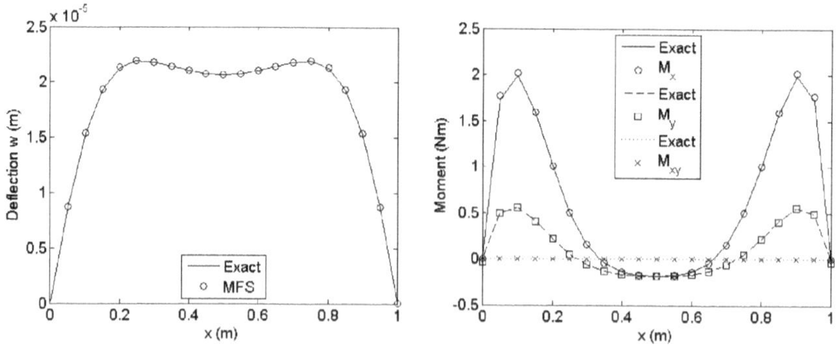

Fig. 7 Distribution of deflection and moment along $y = 0.5$

5 Remarks and Conclusions

In this chapter, a meshless method based on the combination use of AEM, RBF and MFS are developed to solve the thin plate bending on elastic foundations. According to the feature of the governing equation of thin plate, the AEM is first used to convert the original governing equation into an equivalent biharmonic equation with unknown fictitious transverse load, and then the corresponding particular and homogeneous solutions are constructed by means of RBF and MFS, both of which employ the improved fundamental solution described in Section 3.2. Finally, the satisfaction of boundary conditions and the original governing equation can be used to determine all unknown coefficients. Numerical results show a good accuracy and convergence of the proposed method.

From the solution procedure in Section 3, the presented method is more convenient for treating various transverse load and boundary conditions, and can easily be applied to other thin plate problems, such as variable thickness thin plate problems.

References

Bittnar Z, Sejnoha J (1996) Numerical Methods in Structural Mechanics, ASCE Press, New York

Fairweather G, Karageorghis A (1998) The method of fundamental solutions for elliptic boundary value problems. Adv Comput Math 9:69–95

Ferreira AJM (2003) A formulation of the multiquadric radial basis function method for the analysis of laminated composite plates. Comput Struct 59:385–392

Golberg MA, Chen CS, Bowman H (1999) Some recent results and proposals for the use of radial basis functions in the BEM. Eng Anal Bound Elem 23:285–296

Karthik B, Palghat AR (1999) A particular solution Trefftz method for non-linear Poisson problems in heat and mass transfer. J Comput Phys 150:239–267

Katsurada M (1994) Charge simulation method using exterior mapping functions. Japan J Indust Appl Math 11: 47–61

Kupradze VD, Aleksidze MA (1964) The method of functional equations for the approximate solution of certain boundary value problems. USSR Comput Math Phys 4:82–126

Leitao VMA (2001) A meshless method for Kirchhoff plate bending problems. Int J Numer Meth Engng 52:1107–1130

Liu Y, Hon YC, Liew KM (2006) A meshfree Hermite-type radial point interpolation method for Kirchhoff plate problems. Int J Numer Meth Engng 66:1153–1178

Long SY, Zhang Q (2002) Analysis of thin plates by the local boundary integral equation (LBIE) method. Eng Anal Bound Elem 26:707–718

Martin HC, Carey GF (1989) The Finite Element method (4th Edition). McGraw-Hill, London

Misra RK, Sandeepb K, Misra A (2007) Analysis of anisotropic plate using multiquadric radial basis function. Eng Anal Bound Elem 31:28–34

Nerantzaki MS, Katsikadelis JT (1996) An analog equation solution to dynamic analysis of plates with variable thickness. Eng Anal Bound Elem 17:145–152

Nitsche LC, Brenner H (1990) Hydrodynamics of particulate motion in sinusoidal pores via a singularity method. AIChE J 36:1403–1419

Qin QH (2000) The Trefftz Finite and Boundary Element Method. WIT Press, Southampton

Rajamohan C, Raamachandran J (1999) Bending of anisotropic plates by charge simulation method. Adv Eng Softw 30:369–373

Sun HC, Yao WA (1997) Virtual boundary element-linear complementary equations for solving the elastic obstacle problems of thin plate. Finite Elem Anal Des. 27:153–161
Sun HC, Zhang LZ, Xu Q et al. (1999) Nonsingularity Boundary Element Methods. Dalian University of Technology Press, Dalian (in Chinese)
Wang H, Qin QH (2006) A meshless method for generalized linear or nonlinear Poisson-type problems. Eng Anal Bound Elem 30:515–521
Wang H, Qin QH (2007) Some problems with the method of fundamental solution using radial basis functions. Acta Mechanica Solida Sin 20:21–29
Wang H, Qin QH, Kang YL (2005) A new meshless method for steady-state heat conduction problems in anisotropic and inhomogeneous media. Arch Appl Mech 74:563-579
Wang H, Qin QH, Kang YL (2006) A meshless model for transient heat conduction in functionally graded materials. Comput Mech 38:51–60
Yao WA, Wang H (2005) Virtual boundary element integral method for 2-D piezoelectric media. Finite Elem Anal Des 41:9–10
Young DL, Jane SJ, Fan CM, et al. (2006) The method of fundamental solutions for 2D and 3D Stokes problems. J Comput Phys 211:1–8
Zhang FF (1984) The elastic thin plates. Science Press of China, Beijing (in Chinese)

Time Domain B-Spline BEM Methods for Wave Propagation in 3-D Solids and Fluids Including Dynamic Interaction Effects of Coupled Media

Dimitris C. Rizos

Abstract This chapter summarizes the theoretical developments of the author's research team during the last 15 years in the area of a new family of direct time domain Boundary Element Methods that are based on the B-Spline fundamental solutions for wave propagation in solid and fluid media, and the resulting methodologies for solving problems involving dynamic interaction of coupled media. Special element types have been developed to accommodate rigid assemblages and fluid-solid interfaces and are summarized herein. Coupling procedures of the proposed BEM methods with the FEM are also outlined. A number of examples reported in the literature by the author and his coworkers are included at the end of the chapter.

1 Introduction

Investigations of problems involving wave propagation in solids and fluids and the Dynamic Interaction of Coupled Media (DICM) has been the focus of numerous research activities. Solution methodologies can be classified into the *direct* methods and the *multi-step* methods, Rizos and Karabalis (1999). Extensive literature reviews within the framework of analytic and numerical solutions for wave propagation in solids and fluids including DICM based on past and current research of the author and his co-workers (e.g. Karabalis et al. 1996, 2003; Rizos and Karabalis 1999, 1998, 1994; Karabalis and Rizos 1993) Stehmeyer and Rizos (2006), and Rizos and Zhou (2006a) have established the following:

1. Closed form analytic and semi-analytic solutions are used in multi-step methods. They are restricted to problems of very simple geometry and use oversimplified

D.C. Rizos (✉)
Department of Civil and Environmental Engineering, University of South Carolina, Columbia, SC 29208, USA
e-mail: rizos@engr.sc.edu

A chapter in honor of Dimitri Beskos' 65th birthday

G.D. Manolis, D. Polyzos (eds.), *Recent Advances in Boundary Element Methods*,
DOI 10.1007/978-1-4020-9710-2_25, © Springer Science+Business Media B.V. 2009

assumptions. Consequently, they cannot accommodate sophisticated nonlinear material models. Numerical solutions are better suited for DICM analysis and are commonly used in direct methods. The Finite Element Method (FEM), the Boundary Element Method (BEM), and hybrid FEM-BEM are among the most popular techniques. General literature reviews on BEM, FEM and coupled BE-FE methods for problems in transient analysis and SSI have been presented by Beskos (1997), Mackerle (1996, 1998, 1999) and Rizos and Karabalis (1999), among others.

2. The FEM methods and their advantages are well-established. The nature of the FEM implies that the solution domain must be finite. However, cases involving infinite media, such as the soil region, impose the radiation condition at infinity that cannot be represented according to FEM standard procedures that truncate the infinite extends by introducing fictitious boundaries. For such cases, special types of elements have been developed that model the infinite extents at the fictitious boundaries. These elements can be classified in four types, i.e., the infinite elements (e.g. Chow and Smith 1982), the absorbing elements (e.g. Lysmer et al. 1975), the hyperelements, (e.g. Kausel and Tassoulas 1981), and elements based on hybrid models (e.g. Tzong 1981).
3. Boundary Element Methods for wave propagation and dynamic interaction analysis of coupled media are newer techniques compared to the FEM and also well established. Time domain BEM formulations for 3-D DICM and general wave propagation problems in the direct time domain have been developed in the last three decades. BEM formulations based on the full space fundamental solutions can be classified in three major types: (i) Dirac-δ fundamental solutions (e.g., Banerjee et al. 1986, 1987), (ii) Step Impulse fundamental solutions (Karabalis and Beskos 1984,1986,1987), and (iii) High order B-Spline fundamental solutions (Rizos 1993; Rizos and Karabalis 1994,1998; Rizos and Zhou 2006a; Zhou 2005). The BEM formulation based on the B-Spline fundamental solutions have been developed by the author and have shown superior accuracy and efficiency (Rizos and Loya 2002; Rizos and Karabalis 1995, 1997; Karabalis and Rizos 1995; Rizos and Zhou 2006b).
4. Coupled FEM-BEM formulations combine the advantages of the FEM and BEM methods and eliminate the shortcomings, providing an attractive numerical tool for DICM problems. The idea of coupling the FEM and the BEM finds its origins in the work of McDonal and Wexley in the early 1970s on the microwave theory. Zienkiewicz et al. (1977) presented the first organized formulation for analysis of solids. The collocation formulations, related to the present work, can be classified in two types; the Boundary Element (BE) approach(Brebbia and Georgiou 1979) and the Finite Element (FE) approach (Zienkiewicz et al. 1977). Most of the coupled FEMBEM applications reported in the literature for DICM are of the FE or BE type mentioned above and are expressed in the frequency domain, e.g. Bielak et al. (1984), and Chuhan et al. (1993), among others. Only a few publications have dealt with the time domain BEM-FEM techniques for DICM and wave propagation problems. Karabalis and Beskos (1985) and

Spyrakos and Beskos (1986) have reported on 2-D and 3-D flexible foundations modeled with finite elements and coupled with the boundary elements following the BE approach. The more generally applicable coupling formulation presented by Fukui (1987) is developed for 2-D scattering of anti-plane waves. Estorff (1990) and Estorff and Kausel (1989) reported on a FEM-BEM coupling with application to 2-D plane-strain. The work of Estorff and Firuziaan (2000), Czygan and Estorff (2002), Estorff and Hagen (2005) and Soares et al. (2007) appears as the most general one, and addresses non-linear behavior of the FE domain. Such an approach yields non-symmetric and non-positive definite system matrix that requires the development of special solution strategies involving iterations.

This chapter summarizes the theoretical developments of the author's research team during the last fifteen years in the area of a new family of direct time domain Boundary Element Methods for wave propagation in solid and fluid media, and the resulting methodologies for solving problems involving DICM problems. B-Spline polynomials have been adopted to derive the B-Spline fundamental solutions to the associated boundary integral equations. The concept of the B-Spline Impulse Response Functions (BIRF) of the dynamic domains is presented. Special element types have been developed to accommodate rigid assemblages and fluid-solid interfaces and are summarized herein. Coupling procedures of the proposed BEM methods with other methods, and predominantly with the FEM, are also outlined. A number of examples reported in the literature are included at the end of the chapter that demonstrate the different applications of the proposed methodology.

2 Fundamentals

2.1 Governing Equations of Motion

For wave propagation in solid media, the kinematic relations, equilibrium equations, and the material constitutive law provide a system of equations that describes the dynamic response of a linear elastic homogeneous and isotropic solid at any time t. These equations may be combined to obtain the Navier displacement equations of motion expressed as,

$$(\lambda + 2\mu)\,\nabla\nabla\cdot\mathbf{u}(\mathbf{x},t) - \mu\,\nabla\times\nabla\times\mathbf{u}(\mathbf{x},t) + \rho(\mathbf{x})\,\mathbf{b}(\mathbf{x},t) = \rho\,\ddot{\mathbf{u}}(\mathbf{x},t) \qquad (1)$$

where $\mathbf{u}(\mathbf{x},t)$ is the displacement field, $\mathbf{b}(\mathbf{x},t)$ represents the body forces per unit mass, λ and μ are the Lame material constants and $\rho(\mathbf{x})$ is the mass density of the medium. Equivalently, considering an inviscid and compressible fluid in which the amplitude of the motion is small, the wave equation is derived by combining Navier-Stokes equation, and the continuity equation as,

$$c^2\nabla^2 p(\mathbf{x},t) - \ddot{p}(\mathbf{x},t) = 0, \quad c^2 = \frac{B}{\rho} \tag{2}$$

where $p(\mathbf{x}, t)$ represents the pressure field in excess of the hydrostatic value, ρ is the mass density of the fluid, B denotes the bulk modulus, c is the velocity of the traveling pressure wave, and dots indicate differentiation with respect to time.

2.2 Integral Representation

Following well-established procedures, e.g. Erringen and Suhubi (1975), Rizos and Karabalis (1994, 1998), and Bonnet (1995), the differential governing equations for wave propagation in solids, Equation (1), or fluids, Equation (2) can be expressed in a Boundary Integral Equation (BIE) form, which, for zero initial conditions and in the absence of body forces, take the form,

$$c(\xi)\, p(\xi,t) = \int_S \left[G(\mathbf{x},t,\xi\,|q(\mathbf{x},t)) - T(\mathbf{x},t,\xi\,|p(\mathbf{x},t))\right] dS \tag{3}$$

where $\mathbf{x}$ and ξ are points in the domain or its boundary, the integration is over the boundary, S, of the domain, and G and T are the associated fundamental solutions. The term $c(\xi)$ is known as the "jump term" and depends on the location of point ξ and the smoothness of the boundary in its neighborhood of point ξ. The field variables $p(\mathbf{x}, t)$ and $q(\mathbf{x}, t)$ correspond to either the displacement and traction, respectively, in a solid medium or the potential and its normal derivative in a fluid medium. A fundamental solution is defined as any dynamic state in the domain of interest that satisfies the associated governing Equations (1) or (2) and is developed based on a point source

$$f(\mathbf{x},t) = \delta(\xi - \mathbf{x})\, g(t) \tag{4}$$

acting at a point ξ (source point) in a fixed direction. The time modulation $g(t)$ has to be at least twice continuously differentiable and vanishing for $t < 0$. This work employs the family of the cardinal B^k-spline polynomials of any order k to represent the time variation of the forces and has been reported by Rizos (1993), Rizos and Karabalis (1994, 1998), and Rizos and Zhou (2006a,b) i.e. $g(t) = B_i^k(t-\tau)$. Other approaches reported in the literature adopt the Dirac delta function, $g(t) = \delta(t-\tau)$ that yields the Green's functions or a step impulse variation, i.e., $g(t) = (H(t-\tau) - H(t))$, as first reported in Karabalis and Beskos (1984).

The B-Spline polynomials are introduced next followed by the B-Spline fundamental solutions for wave propagation in solids and fluids.

3 B-Spline BEM Formulations

3.1 B-Spline Polynomials

The B-Spline functions are piece-wise smooth polynomials of order k derived on the basis of a knot sequence $t_n,\ n = 1 \cdots N$. B-Spline polynomials of any order k are generated by the recursive formula,

$$B_n^k(t) = \begin{cases} 1 & t_n < t < t_{n+1} \\ 0 & elsewhere \end{cases} \quad k = 1 \tag{5}$$

$$B_n^k(t) = \left(\frac{t - t_n}{t_{n+k-1} - t_n}\right) B_n^{k-1}(t) + \left(\frac{t_{n+k} - t}{t_{n+k} - t_{n+1}}\right) B_{n+1}^{k-1}(t)\ k > 1 \tag{6}$$

For an equally spaced knot sequence $t_n = ndt$, the B-Spline polynomials have two properties that are of prime interest to the proposed methodology: (i) any B_n polynomial can be expressed by a representative (base) one appropriately shifted in time, i.e., $B_n^k(t) = B_0^k(t - ndt)$, and (ii) the base polynomial B_0^k is non zero only within small interval, i.e., $B_0^k(t) > 0\ \forall\ t \in (0, \Delta t)$. The interval $\Delta t = kdt$ is known as the B-Spline support and depends on the order k. In this work the 4th order B-Spline polynomial is used since it is the lowest order polynomial that satisfies the continuity conditions. B-Spline polynomials are commonly used in function interpolation. Hence, any function $f(t)$ can be represented as a linear combination of B-Spline polynomials as,

$$f(t) \approx \sum_{i=1}^{j} f(\tau_i) B_0^k(t - t_i),\ t \in \left[t_1, t_j\right], \\ \tau_i = \frac{t_i + t_{i+1} + \cdots + t_{i+k-1}}{k - 1},\ k > 1 \tag{7}$$

Rizos (1993) and Rizos and Zhou (2005) employed this function to generate the higher order B-Spline fundamental solutions for 3D wave propagation in solids and fluids discussed next.

3.2 B-Spline Fundamental Solutions

In this section the B-Spline fundamental solutions for wave propagation in solids and fluids are presented under the assumptions of linearized conditions and homogeneity. In view of Equation (6), the B_o^k-Spline fundamental solutions, $G_{ij}(\underline{x}, t; \underline{\xi}$

$|B_o^k(\tau).)$, and $T_{ij}(\underline{x}, t; \underline{\xi}|B_o^k(\tau))$ are derived for a source and a receiver point of the elastodynamic domain as,

$$G_{ij}\left(\underline{x}, t; \xi \left|B_o^k(\tau)\right.\right) = G_{ij}^B = \frac{1}{4\pi\rho}\left\{a_{ij}(r)\, F\left(\tau, r, c_1, c_2\right) + b_{ij}(r)\, B_o^k\left(\tau - \frac{r}{c_1\Delta t}\right) + c_{ij}(r)\, B_o^k\left(\tau - \frac{r}{c_2\Delta t}\right)\right\} \tag{8}$$

$$\begin{aligned} T_{ij}\left(\underline{x}, t; \xi \left|B_o^k(\tau)\right.\right) &= T_{ij}^B \\ &= \frac{1}{4\pi}\left\{d_{ij}(r)\, F(\tau, r, c_1, c_2) + e_{ij}(r)\, B_o^k\left(\tau - \frac{r}{c_1\Delta t}\right)\right. \\ &\quad + f_{ij}(r)\, B_o^k\left(\tau - \frac{r}{c_2\Delta t}\right) + g_{ij}(r)\, \dot{B}_o^k\left(\tau - \frac{r}{c_1\Delta t}\right) \\ &\quad \left. + h_{ij}(r)\, \dot{B}_o^k\left(\tau - \frac{r}{c_2\Delta t}\right)\right\} \end{aligned} \tag{9}$$

where c_1 and c_2 are the pressure and shear wave velocities, respectively, $r = |\mathbf{x} - \xi|$ is the distance between the points under consideration, and dots indicate derivatives with respect to time. The term $F(\tau, r, c_1, c_2)$ represents the combined effects of the pressure and shear waves and can be evaluated analytically (Rizos and Karabalis 1994). The coefficients a_{ij} through h_{ij} depend on spatial parameters only and are reported in Rizos (1993). In a similar way, the B_o^k-Spline fundamental solutions $G_{ij}\left(\underline{x}, t; \underline{\xi} \left|B_o^k(\tau)\right.\right)$, and $T_{ij}\left(\underline{x}, t; \underline{\xi} \left|B_o^k(\tau)\right.\right)$ are derived for a source and a receiver point of the hydrodynamic domain as, Rizos and Zhou (2006a):

$$G^B = G\left(\mathbf{x}, t, \xi \left|\delta(\mathbf{x} - \xi)\, B_0^4(t)\right.\right) = \frac{1}{4\pi r} B_0^k\left(\frac{t - \frac{r}{c}}{\Delta t}\right) \tag{10}$$

$$\begin{aligned} T^B = T\left(\mathbf{x}, t, \xi \left|\delta(\mathbf{x} - \xi)\, B_0^4(t)\right.\right) &= \frac{\partial G\left(\mathbf{x}, t, \xi \left|\delta(\mathbf{x} - \xi)\, B_0^k(t)\right.\right)}{\partial \mathbf{n}} \\ &= \frac{-1}{4\pi r^2}\left\{B_0^k\left(\frac{t - \frac{r}{c}}{\Delta t}\right) + \frac{r}{c}\dot{B}_0^k\left(\frac{t - \frac{r}{c}}{\Delta t}\right)\right\}\frac{\partial r}{\partial \mathbf{n}} \end{aligned} \tag{11}$$

It is evident that the fundamental solutions are singular when the source point coincides with the receiver point at time t for $0 < t < \Delta t$, Rizos (1993).

3.3 Space and Time Discretizations

Following the procedures reported in Rizos (1993), Rizos and Karabalis (1998) and Rizos and Zhou (2006a) the bounding surface of the solution domain is discretized into NE number of surface elements for a total of M nodes and NN number of degrees of freedom. Consequently, the geometry and field variables can be interpolated at any time, t, based on a set of shape functions N_i and corresponding nodal values as,

$$\mathbf{x} = N_i \mathbf{x}_i, \quad p(\mathbf{x}, t) = N_i p_i(t), \quad q(\mathbf{x}, t) = N_i q_i(t) \tag{12}$$

Introducing the spatial discretization indicated in Equations (12), the B-Spline fundamental solutions, Equations (8), (9), (10), and (11), and the time knot sequence that defines the fourth order B_0^4-Spline polynomial of the fundamental solution into the BIE Equation (3), the latter is written in a discrete form in space and time as:

$$c(\xi)\, p\left(\xi, t^N\right) = \sum_{n=1}^{N+1} \sum_{el=1}^{NE} \left\{ \int_{S_{el}} G^B\left((N_i \mathbf{x}_i), t_n, \xi \,\middle|\, B_0^4\right) (N_i q_i(t_{N-n+2}))\, dS_{el} \right.$$
$$\left. - \int_{S_{el}} T^B\left((N_i \mathbf{x}_i), t_n, \xi \,\middle|\, B_0^4\right) (N_i p_i(t_{N-n+2}))\, dS_{el} \right\} \tag{13}$$

where the spatial integration is defined over the area, S_{el}, of every element, el, and the summation with respect to $el = 1, 2, \ldots, NE$ indicates superposition over all elements. The summation with respect to $n = 1, 2, \ldots N + 1$ pertains to superposition over all time steps up to step $N + 1$ and represents a convolution time integral in a discrete form. It should be noted that the proposed method does not require the explicit use of any time integration algorithms. By writing Equation (13) for every boundary point, ξ, and appropriately collecting the nodal values, the desired algebraic system of equations is derived which can be solved in a time marching scheme. The system of equations is cast in a matrix form at step N as,

$$\frac{1}{2}\mathbf{p}^N = \sum_{n=1}^{N+1} \mathbf{G}^N \mathbf{q}^{N-n+2} - \mathbf{T}^N \mathbf{p}^{N-n+2} \tag{14}$$

where $\mathbf{p}$ and $\mathbf{q}$ are column matrices of size NN containing degree of freedom values of the field variables and their normal derivative, respectively, and superscripts indicate the time step at which quantities are evaluated. The coefficient matrices $\mathbf{G}^N$ and $\mathbf{T}^N$ are of size $NN \times NN$ and represent the influence of a boundary degree of freedom on another boundary degree of freedom. The integrations involved in the evaluation of these matrices are performed numerically using the techniques discussed in Rizos (1993), and Rizos and Zhou (2006a). In problems where the formulations need to be expressed in terms of forces an appropriate transformation is reported in Rizos (2000).

4 Solution Procedures

4.1 B-Spline Impulse Response Function Concept

The B-Spline Impulse Response Function (BIRF) of the boundary of the dynamic domain is obtained from the solutions of Equations (14) by applying a unit B-Spline impulse excitation, $\mathbf{q}_j = B_o^k(t)$ of duration Δt at a single degree of freedom j only. Criteria for the selection of the B-Spline support are discussed in Rizos and Karabalis (1998), Rizos and Zhou (2006a) and O'Brien and Rizos (2005). In view of the tension free boundary condition on the surface of the domain, the incident excitation field will generate only a scattered field. Noting that the B-Spline time modulation of the excitation is embedded in the fundamental solutions, Equation (14) yields the BIRF, $\mathbf{b}_j^N = \mathbf{p}\left|_{\mathbf{q}_j}\right.$, as,

$$\frac{1}{2}\mathbf{b}_j^N = \mathbf{G}^N\mathbf{q}_j - \sum_{n=1}^{N}\mathbf{T}^n\mathbf{b}_j^{N-n+2} = \mathbf{g}_j^N - \mathbf{T}^1\mathbf{b}_j^{N+1} - \mathbf{T}^2\mathbf{b}_j^N - \sum_{n=3}^{N}\mathbf{T}^n\mathbf{b}_j^{N-n+2} \quad (15)$$

where $\mathbf{g}_j^N$ is the column of matrix $\mathbf{G}$ that corresponds to the "loaded" *jth* degree of freedom. The time stepping scheme expressed in Equation (15) is an implicit one. The corresponding explicit scheme is obtained by interpolating the BIRF at time t^N in a consistent manner using cubic B-Spline polynomials, Equation (7), i.e.:

$$\mathbf{b}^N = \frac{1}{6}\mathbf{b}^{N-1} + \frac{2}{3}\mathbf{b}^N + \frac{1}{6}\mathbf{b}^{N+1} \Rightarrow \mathbf{b}^{N+1} = 2\mathbf{b}^{N+1} - \mathbf{b}^{N-1} \quad (16)$$

Substitution of Equation (16) into Equation (15) yields the explicit scheme as:

$$\mathbf{b}_j^N = \left(\frac{1}{2}\mathbf{I} + 2\mathbf{T}^1 + \mathbf{T}^2\right)^{-1}\left\{\mathbf{g}_j^N - \sum_{n=3}^{N+1}\mathbf{T}^n\mathbf{b}_j^{N-n+2} + \mathbf{T}^1\mathbf{b}_j^{N-1}\right\} \quad (17)$$

where $\mathbf{I}$ is the identity matrix. Equation (17) represents the BIRF of the free surface of the dynamic domain due to the application of a single impulse at degree of freedom j. Thus, the BIRF vectors, $\mathbf{b}_j^N$, can be computed for all degrees of freedom $j = 1, 2 \ldots NN$ and collected in a matrix form as:

$$\mathbf{B}^N = \left\lfloor \mathbf{b}_1^N, \mathbf{b}_2^N, \ldots, \mathbf{b}_j^N, \ldots \mathbf{b}_{NN}^N \right\rfloor \quad (18)$$

Equation (18) is the BIRF of the system expressed in a discrete form in space and time. Each element B_{ij} in the matrix represents the impulse response of degree of freedom i at time step N due to a "unit" excitation applied at degree of freedom j. This matrix is a characteristic of the system and needs to be computed only once for the specific geometry of the free surface of the solution domain. In the general case, the B-Spline impulse response is computed for all degrees of freedom of the problem and the impulse response matrix, $\mathbf{B}^N$, is square. In most practical cases, however, the BIRF of the system needs to be computed only for those degree of freedom that are expected to carry a forced excitation during the solution, yielding a rectangular matrix or a square matrix of significantly reduced size. Because of the small, finite duration of the B-Spline impulse excitation, and the wave attenuation, the number of time steps N that the BIRF matrices are obtained for, is small compared to the response of the system to general excitations and limited to only a few time steps, as discussed in O'Brien and Rizos (2005) and Rizos and Zhou (2006a,b). Once this characteristic response of the system is known, the system can be analyzed for any arbitrary transient loading, $\mathbf{q}(t = t_n) = \mathbf{q}^n$, and the solution is obtained as a mere superposition of the BIRF functions as,

$$\mathbf{p}^N = \sum_{n=1}^{N+1} \mathbf{B}^n \mathbf{q}^{N-n+2} \tag{19}$$

where it is implied that the BIRF are calculated for a 4th order B-Spline impulse excitation. The proposed approach is very efficient especially when multiple load cases are considered, since the BIRF functions are independent of the external excitation and are typically of shorter duration than the external excitation.

4.2 Treatment of Rigid Bodies

In order to accommodate rigid bodies in contact with solids or fluids, a *rigid surface boundary element* is derived within the framework of the B-Spline BEM and is summarized in this section, while details of the formulation can be found in Rizos (2000), Rizos and Zhou (2006a,b) and Zhou (2005). The rigid body is assumed massless and, therefore, only kinematic interaction effects are considered. However, the mass of the rigid body and the inertia interaction effects can be routinely handled through coupling of the BEM with appropriate FEM procedures, as discussed in a subsequent Section 5. The formulation is based on the constrained equations of the rigid body and compatibility of displacements and force equilibrium between the rigid surface and the solid or fluid medium at their interface. To this end, the motion of any point Q on a rigid surface, shown for example in Fig. 1.

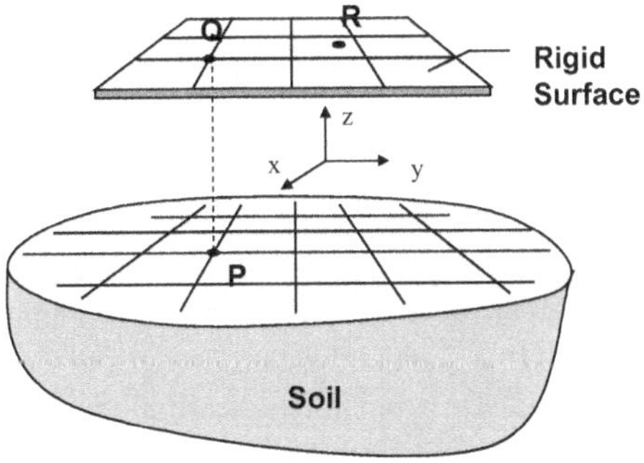

Fig. 1 Rigid boundary element – definitions (Rizos 2000)

can be described by the translation Δ_R and rotation $\mathbf{\Theta}_R$ of a reference point R as,

$$\mathbf{u}_Q = \mathbf{I}\Delta_R + \mathbf{H}\mathbf{\Theta}_R \tag{20}$$

where I is the identity matrix, **I** is the identity matrix and **H** is defined in Rizos (2000). If the motion is described by the acceleration, then Equation (20) is simply differentiated twice with respect to time. Similarly, the forces, $\mathbf{F}_Q$, on any rigid surface point Q are related to the forces, $\mathbf{F}_R$, and moments $\mathbf{M}_R$ applied on the reference point R as,

$$\mathbf{F}_R = -\mathbf{I}\,\mathbf{F}_Q, \quad \mathbf{M}_R = \mathbf{H}\,\mathbf{F}_Q \tag{21}$$

For every point Q on the rigid surface there is a corresponding point P on the soil, or fluid, surface whose motion is described by the displacements, $\mathbf{u}_P$, or normal particle acceleration, respectively. Under the assumption that the rigid surface remains always in contact with the dynamic medium, compatibility of displacements and equilibrium of forces must be satisfied,

$$\mathbf{u}_Q = \mathbf{u}_P \text{ (a)}, \quad \mathbf{F}_P + \mathbf{F}_Q = \mathbf{0} \text{ (b)} \tag{22}$$

It should be noted that not all of the degrees of freedom x, y and z need be coupled, or only the normal, or tagential components may be coupled, in which case appropriate projections of the displacement, or acceleration vectors on the specified direction needs to be performed. An expression similar to Equation (22a) can be written in terms of normal particle acceleration if the later is chosen to represent the motion, as in the case of interaction of the rigid body with a fluid medium. It should be emphasized that when a rigid surface is in contact with a dynamic medium, the forces developed at the interface are unknown. In view of the compatibility and equilibrium Equations (22) and BEM Equations (14), the rigid surface boundary element is defined by three nodes, i.e., node P on the dynamic medium, node Q on the rigid surface and the reference node R of the rigid surface and the corresponding *rigid surface boundary element* equations are cast in the following matrix form as

$$\begin{Bmatrix} [\quad] \\ \mathbf{0} \\ \mathbf{F}_R \\ \mathbf{M}_R \end{Bmatrix} = \begin{bmatrix} [\quad] & [\quad] & \mathbf{0} & \mathbf{0} \\ -\mathbf{I} & \mathbf{0} & \mathbf{I} & \mathbf{H} \\ \mathbf{0} & \mathbf{I} & \mathbf{0} & \mathbf{0} \\ \mathbf{0} & \mathbf{H}^T & \mathbf{0} & \mathbf{0} \end{bmatrix} \begin{Bmatrix} \mathbf{u}_P^N \\ \mathbf{F}_P^N \\ \Delta_R^N \\ \mathbf{\Theta}_R^N \end{Bmatrix} \tag{23}$$

where the quantities in brackets indicate information provided by the BEM Equations (14) for node P on the solid domain in contact with the rigid surface (Rizos 2000). It is evident that a "*rigid surface boundary element*" can be formulated based on equations (23) for all points in contact with the rigid surface and directly superimposed to the system of equations (14).

4.3 Fluid-Solid Interfaces

For problems involving interaction of fluid with solid media a *fluid-solid interface* element is derived in order to couple the field variables in the two media, Zhou (2005), Rizos and Zhou (2006a). This interface is typically defined between the *Wet Surface* (*WS*) on the solid and the corresponding area in the fluid medium. For every point *WS* on the wet surface of the solid, there is a corresponding point F on the fluid surface whose motion is described by the acceleration in the direction of the outward normal to the fluid boundary $\ddot{u}_{F_\mathbf{n}}$. Under the assumption that the rigid body remains always in contact with the fluid in the normal direction and there is no friction between the solid and the fluid, compatibility of the normal component of the acceleration requires

$$\ddot{u}_{F\mathbf{n}} = -\ddot{u}_{WS\mathbf{n}} = -\ddot{\mathbf{u}}_{WS} \cdot \mathbf{n} \tag{24}$$

where $\mathbf{n}$, represents the outward normal vector at the point on the wet surface and the dot indicates the projection of the acceleration vector on the outward normal. At the interface between the solid and the fluid media, flux field at fluid nodes on the wet surface is related to the acceleration as (Zhou 2005):

$$\left\{\frac{\partial p}{\partial \mathbf{n}}\right\}_F = \mathbf{q}_F = -\rho \mathbf{a}_F \tag{25}$$

where $\mathbf{a}_F = \left[\ddot{u}_{F_1\mathbf{n}} \quad \ddot{u}_{F_2\mathbf{n}} \quad \cdots \quad \ddot{u}_{F_{MM}\mathbf{n}}\right]^T$, $\mathbf{q}_F = \left(\frac{\partial p}{\partial \mathbf{n}}\right)_F$ is the vector of the nodal flux normal to the wet surface, and ρ, is the mass density of the fluid medium. In view of Equations (19) and (25), the pressure distribution on the wet surface nodes can be computed by superposition of appropriate BIRF functions as,

$$\mathbf{p}_F^N = \sum_{n=1}^{N+1} \mathbf{B}^{N-n+2} \mathbf{q}_F^N = -\rho \sum_{n=1}^{N+1} \mathbf{B}^{N-n+2} \mathbf{a}_F^N \tag{26}$$

where superscripts indicate time step, $\mathbf{p}_F$ is the pressure at each node on the fluid surface caused by the applied acceleration, and $\mathbf{B}$ is the B-Spline impulse response matrix of the fluid surface. Due to the assumption that the solid remains always in contact with the fluid in the normal direction the pressure distribution along the wet surface of the solid medium can be computed as

$$\mathbf{p}_{WS}^N = -\mathbf{p}_F^N = \rho \sum_{n-1}^{N+1} \mathbf{B}^{N-n+2} \mathbf{a}_F^N \tag{27}$$

and, therefore, can be directly accounted for in the solid domain.

4.4 Coupling with the FEM and Other Methods

The developed BEM methodology can be coupled with the FEM and other time marching methodologies in a straightforward manner. This section briefly discusses the coupling procedures with the FEM method for wave propagation in solids while details of the approach can be found in Rizos and Wang (2002), Rizos and Karabalis (1999), O'Brien and Rizos (2005), Zhou (2005), Rizos and Zhou (2006a,b) and Stehmeyer and Rizos (2007).

4.4.1 Modified BEM

The starting point for coupling the BEM with other techniques is the computation of the response of dynamic system to arbitrary excitations, as indicated in Equation (19), repeated here in a force (**P**) – displacement (**u**) form for demonstration purposes,

$$\mathbf{u}^N = \sum_{n=1}^{N+1} \mathbf{B}^n \mathbf{P}^{N-n+2} \tag{28}$$

It is implied, therefore, that the BIRF function of the common interface is already computed. Vector **P** is the vector of driving forces applied at interface nodes, and it consists of the contact forces transmitted from the FEM solution domain. These forces are typically not known at forward steps of the solution and need to be interpolated in a consistent manner using a cubic B-Spline polynomial following the work of Rizos and Loya (2002) as,

$$\mathbf{P}^{N+1} = 2\mathbf{P}^N - \mathbf{P}^{N-1} \tag{29}$$

Substituting Equation (29) into Equation (28) results in

$$\mathbf{u}^N = \left(2\mathbf{B}^1 + \mathbf{B}^2\right)\mathbf{P}^N + \sum_{n=3}^{N+1} \mathbf{B}^n \mathbf{P}^{N-n+2} - \mathbf{B}^1\mathbf{P}^{N-1} = \mathbf{F}\mathbf{P}^N + \mathbf{H}^N \tag{30}$$

where $\mathbf{F} = 2\mathbf{B}^1 + \mathbf{B}^2$ is an equivalent flexibility matrix, and $\mathbf{H}^N = \sum_{n=3}^{N+1} \mathbf{B}^n \mathbf{P}^{N-n+2} - \mathbf{B}^1\mathbf{P}^{N-1}$ is a load history vector. The flexibility matrix is independent of time, and only needs to be computed once. The history matrix contains the effects of the time history of the response of the system on the current time step of the solution; this vector need be updated at each time step. Equation (30) can be solved for the displacement of the interface nodes. Once the displacements are computed, velocities and accelerations are determined through a finite difference approximation, as

$$\dot{\mathbf{u}}^N = \frac{\mathbf{u}^{N+1} - \mathbf{u}^N}{\Delta t}, \quad \ddot{\mathbf{u}}^N = \frac{\dot{\mathbf{u}}^{N+1} - \dot{\mathbf{u}}^N}{\Delta t} \tag{31}$$

Equations (30) and (31) represent the BEM solver that is adopted in the proposed coupled BEM-FEM methodology.

4.4.2 Finite Element Method

Standard FEM procedures are used for the coupling with the BEM method. The dynamic equations are solved directly in time using Newmark's method. Following well established procedures the governing equations of motion can be rearranged in a system of algebraic equations as,

$$\mathbf{D}\mathbf{u}^N = \mathbf{P}^N + \mathbf{H}^N \tag{32}$$

where $\mathbf{D}$ is the dynamic matrix, $\mathbf{u}^N$ is the displacement vector and $\mathbf{H}^N$ is a modification to the nodal force vector, $\mathbf{P}^N$, at step N. In order to develop the FEM solver suitable for the proposed coupled FEM-BEM scheme, Equation (32) is rearranged so that a solution can be reached in a single step. To this end, the system of Equation (32) is partitioned with respect to the free, f, and interface, s, degrees of freedom and after separation of known from unknown quantities is written as:

$$\begin{bmatrix} \mathbf{D}_{ff} & 0 \\ \mathbf{D}_{sf} & -\mathbf{I} \end{bmatrix} \begin{Bmatrix} \mathbf{u}_f \\ \mathbf{R}_s \end{Bmatrix} = \begin{Bmatrix} \mathbf{P}_f \\ 0 \end{Bmatrix} + \begin{Bmatrix} \mathbf{P}_f^{eq} \\ \mathbf{P}_s^{eq} \end{Bmatrix} + \begin{Bmatrix} \mathbf{H}_f \\ \mathbf{H}_s \end{Bmatrix} - \begin{bmatrix} \mathbf{D}_{fs} \\ \mathbf{D}_{ss} \end{bmatrix} \{\mathbf{u}_s\} \tag{33}$$

where the vector of nodal forces applied on all degrees of freedom consists of the vector of concentrated forces and moments applied directly on the nodes, $\mathbf{P}^{nod}$, and the nodal equivalent forces and moments, $\mathbf{P}^{eq}$, due to loading applied on elements. At the interface degrees of freedom the unknown force reactions, $\mathbf{R}_s$, are also accounted for in the equilibrium of the system. The displacements at the interface degrees of freedom, $\mathbf{u}_s$, are assumed known at all times. Equation (33) can now be solved for the unknown displacements and support (interface) reactions at step N, while the displacement at the interface degrees of freedom, $\mathbf{u}_s$, are assumed known as provided by the BEM solver. Equation (33) represents the FEM solver that is adopted in the proposed coupled BEM-FEM methodology.

4.4.3 Coupled BEM-FEM

The BEM and FEM discussed previously are coupled in a staggered solution approach. The two domains are connected to each other at their interface degrees of freedom and it is assumed that they remain always in contact. As a result, forces and displacements of the FEM domain at the interface must be introduced into the BEM domain. As the BEM domain is loaded, waves are generated and the resulting displacements due to wave propagation must be applied to the FEM model through the

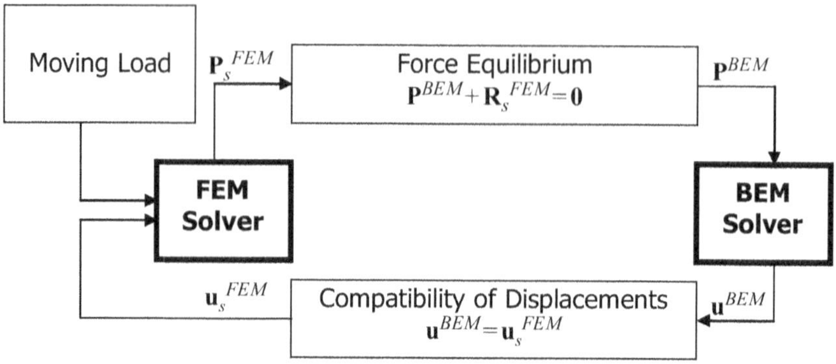

Fig. 2 BEM-FEM coupling scheme, O'Brien and Rizos (2005)

common interface degrees of freedom. Considering compatibility of displacements and force equilibrium, the following relationships must be satisfied at every time step of the solution.

$$\mathbf{u}_s^{BEM} = \mathbf{u}_s^{FEM} \text{ (a)} \quad \mathbf{P}_s^{BEM} + \mathbf{R}_s^{FEM} = 0 \text{ (b)} \tag{34}$$

where subscript s pertains to the interface degrees of freedom, Rizos and Loya (2002). A graphical representation of the staggered solution scheme is shown in Fig. 2. Initially, at time $t = 0$, known external loads are applied to the FEM domain. Initial velocity and acceleration can also be specified in addition to forces and displacements. The FEM solver computes the support reactions at the interface nodes from the loading conditions and structural configuration according to Equation (33). From force equilibrium, the support reactions at the contact degrees of freedom are now used as input to the BEM solver. The BEM solver in turn computes the displacements, velocities and accelerations at the contact nodes. This information is then passed back to the FEM solver as initial conditions for the next time step.

This coupling scheme implies that the two solvers need to assume a common time step of analysis in order to enforce the compatibility and equilibrium conditions at the interface. Although each solver when considered independently is stable, the coupled procedure may become unstable because now the support conditions in the FEM solver become time dependant. Consequently, the two solvers may require different time steps in order to improve convergence of the method. Therefore, the initial conditions at each step of the first solver need to be updated at time nodes where a solution from the other solver is not available. To resolve this issue a modification of the coupled BEM-FEM scheme is introduced that accommodates the selection of different time steps in each method. This modification is based on a linear interpolation scheme and is discussed in O'Brien and Rizos (2005).

5 Demonstrative Examples

5.1 Hydrodynamic Forces Due to Ship Maneuvering

A demonstration example of a marine vessel maneuvering in real time referenced in Zhou (2005) and Rizos and Zhou (2008) is summarized in this section. The wet surface of a marine vessel is idealized for the purposes of this example as a rigid rectangular box of length L=500 ft, width (beam) B=50 ft and depth (draft) D=50 ft. The mass of the vessel is m=10 $\times$ 10^6 lb and it is assumed lumped at a reference point, RP, located at the free surface level at the intersection of the two vertical planes of symmetry of the vessel, as shown in Fig. 3. The wave velocity in the water is taken as 4,862 ft/s. The water free surface is modeled by the BEM and the rigid vessel along with the inertia interaction effects by the FEM.

The BIRF of the reference point, $\mathbf{B}_{RP}$, pertaining to hydrodynamic forces and moments that develop on the rigid vessel due to a B-Spline impulse acceleration applied at the reference point in the longitudinal direction is computed first. Subsequently, the effects of propulsion forces on the acceleration of the vessel are computed. A step impulse propulsion force of duration $\Delta t_p = 0.4$ s, and amplitude $F_{po} = 100 \times 10^6$ lb is applied at the RP. The force is applied at time t=0.0225 s in order to demonstrate that the condition of quiescent past is satisfied. The time history of the acceleration of the vessel due to the application of this load is shown in Fig. 4. where the transient effects are evident.

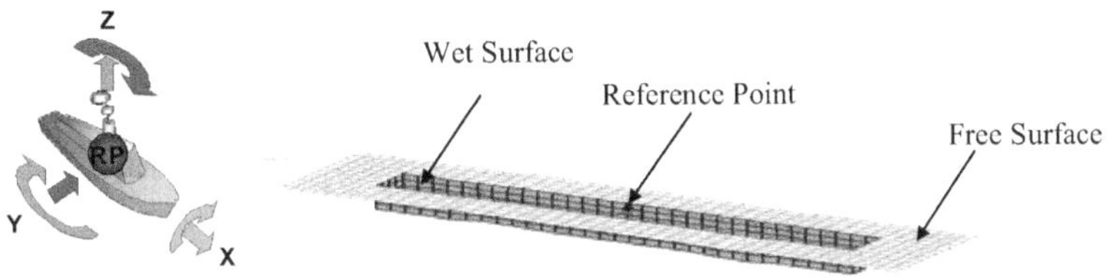

Fig. 3 Geometry and discretization of wet surface and free field (Zhou 2005)

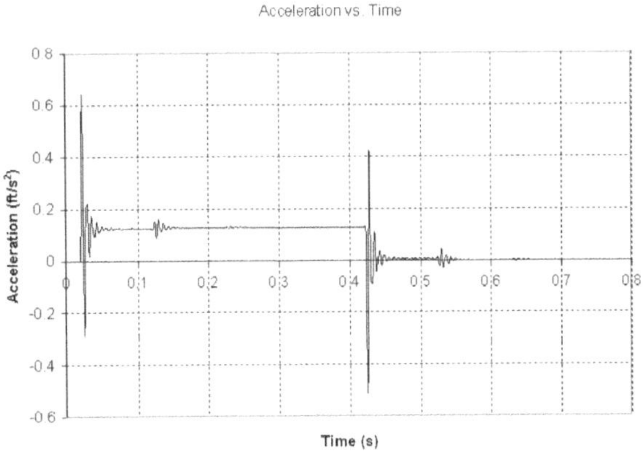

Fig. 4 Time history of vessel acceleration due to step impulse propulsion force, Rizos and Zhou (2008)

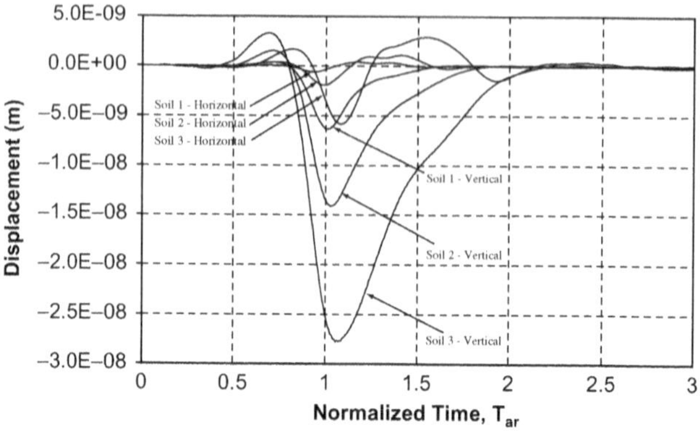

Fig. 5 Vertical and horizontal response of sleeper due to unit moving load at typical HST speeds for three soil conditions, O'Brien (2004)

5.2 *Vibrations Induced by the Passage of Trains*

This section summarizes the example presented in O'Brien (2004) that pertains to the investigation of the response of a railroad track system due to a single vertical unit load moving with a constant velocity along the rail. The track system consists of 20 sleepers, making it 18.145 m long. The sleepers are assumed rigid and three soil types are considered with shear wave velocities C_{s1}=131.6, C_{s2}=76.0 and C_{s3}=46.0 m/s, respectively. The soil and its kinematic interaction with the sleepers is modeled by the BEM, while the FEM is used to model the rails and the mass of the sleepers. A moving load velocity of 76.4 m/s (275 km/h) is chosen to demonstrate the system response due to a load traveling at normal service speeds of high speed trains that may exceed the shear wave velocity in the soil (supercritical speeds). This velocity is subcritical for Soil 1, close to critical for Soil 2, and supercritical for Soil 3. The horizontal and vertical response of the 10th sleeper, (observation sleeper) is monitored and plotted in Fig. 5. The time axis is normalized with respect to the arrival time of the load at the observation sleeper, i.e., $T_{ar} = 1.0$ indicates the time at which the load is directly above the observation sleeper, while values $T_{ar} > 1.0$ indicate times after the load has passed over the observation sleeper. The figure shows a clear shifting of the peak vertical response away from $\mathrm{T}_{ar} = 1.0$ as the load progresses from sub to supercritical. The same holds true for the horizontal displacements. The horizontal tie motion also indicates the severity of the displacements in the soft soil. The predicted horizontal motion in Soil 3 is on the same order of magnitude as the peak vertical displacements in the stiff soil.

Acknowledgments The author is thankful to Professor Dimitri Beskos for his lifelong mentorship and introducing him in the world of Boundary Elements. The author is equally grateful to Professor Dimitri Karabalis for his endless support, guidance and encouragement through the years. This work could not have been completed without the contributions of the author's graduate students James Wang, Kushal Loya, Saiying Zhou, John O'Brien and Edward Sthemeyer.

References

Banerjee PK, Ahmad S, and Manolis GD (1986) Transient elastodynamic analysis of 3-D problems by Boundary Element Method.. Earthquake Engineering and Structural Dynamics, 14:933–949.

Banerjee PK, Ahmad S, and Manolis GD (1987) Advanced elastodynamic analysis. In: DE Beskos (ed): Boundary Element Methods in Mechanics pp. 258–284, Elsevier, Amsterdam.

Beskos DE, (1997) Boundary element methods in dynamic analysis: Part II 1986–1996. ASME Applied Mechanics Reviews, 50(3):149–197.

Bielak J, MacCamy RC, and McGhee DS (1984) On the coupling of finite element and boundary integral methods. In: SK Datta (ed.): Earthquake Source Modeling, Ground Motion, and Structural Response, AMD-60, 115-132, ASME, New York.

Bonnet M, (1995) Boundary Integral Equation Methods for Solids and Fluids, John Wiley & Sons, New York.

Brebbia CA and Georgiou P (1979) Combination of boundary and finite elements for elastostatics. Applied Mathematical Modeling, 3:212–220.

Chow YK and Smith IM (1982) Infinite elements for dynamic foundation analysis. Numerical Methods in Geomechanics., 1:15–22.

Chuhan Z, Feng J, and Guanglun W (1993) A method of FE-BE-IBE coupling for seismic interaction of arch dam-canyons. In M Tanaka et al. (Eds.): Boundary Element Methods. Elsevier, Amsterdam.

Czygan O and Estorff O von (2002) Fluid-structure interaction by coupling BEM and nonlinear FEM. Engineering Analysis with Boundary Elements, 26(9):773–779

Erringen AC and Suhubi ES (1975) Elastodynamics – Vol. II, Linear Theory, Academic Press, New York.

Estorff O (1990) Soil-structure interaction analysis by a combination of boundary and finite elments. In: SA Savidis (Ed.): Earthquake Resistant Constructions and Design. Balkem, Rotterdam.

Estorff O and Firuziaan (2000) Coupled BEM/FEM approach for nonlinear soil/structure interaction, Engineering Analysis with Boundary Elements, 24:715–725.

Estorff von O and Hagen C (2005) Iterative coupling of FEM and BEM in 3D transient elastodynamics. Engineering Analysis with Boundary Elements, 29(8):775–787

Estorff O and Kausel E (1989) Coupling of boundary and finite elements for soil-structure interaction problems. Earthquake Engineering and Structural Dynamics, 18:1065–1075.

Fukui T (1987) Time marching BE-FE method in 2-D elastodynamic problem. International Conference BEM IX Stuttgart.

Karabalis DL and Beskos DE (1984) Dynamic response of rigid surface foundations by time domain boundary element method. Earthquake Engineering and Structural Dynamics, 12:73–93.

Karabalis DL and Beskos DE (1985) Dynamic response of 3-D flexible surface foundations by time domain finite and boundary element methods. International Journal of Soil Dynamics and Earthquake Engineering, 4:91–101.

Karabalis DL and Beskos DE (1986) Dynamic analysis of 3-D embedded foundations by time domain boundary element method. Computer Methods in Applied Mechanics and Engineering, 56:91–119.

Karabalis DL and Beskos DE, (1987) Dynamic soil – structure interaction, Chapter 11. In: DE Beskos (Ed.): Boundary Element Methods in Mechanics, North Holland, Amsterdam.

Karabalis DL, Cokkinides GJ, Rizos DC, and Tassoulas JL (1996) Finite Element Transient Analysis (FETA) V1.00, Theoretical and User's Manual, (Prepared for Westinghouse Savannah River Company), Department of Civil Engineering, University of South Carolina.

Karabalis DL, Karavasilis TL, Dimas AA and Rizos DC (2003) Dynamic fluid-structure-soil interaction: application to spherical tanks. INDEPTH Document IDP-TR-UP-2-01, Version 1.0, European Commission.

Karabalis DL and Rizos DC (1993), Dynamic analysis of 3-D foundations. Chapter 6, pp. 177–208. In: GD Manolis and TG Davis (Eds): Boundary Element Techniques in Geomechanics, Computational Mechanics Publications, Southampton.

Karabalis DL and Rizos DC (1995) On the efficiency and accuracy of a new advanced time domain BEM for 3-D elastodynamics. In: D Hui and S Michaelidis (Eds) Society of Engineering Science 32nd Annual Technical Meeting, (Univ. of New Orleans and Tulane Univ., New Orleans).

Kausel E and Tassoulas JL, (1981): Transmitting boundaries: a close form comparison. Bulletin Seismic. Society of America, 71:143–159.

Lysmer J, Seed HB, Udaka T, Huang RN, and Tsai CF (1975) Efficient finite element analysis of seismic soil structure interaction.Earthquake Engineering Research Center, Report No. EERC 75-34, University of California, Berkeley.

Mackerle J (1996) FEM and BEM in geomechanics: foundations and soil-structure interaction – a b(1992–1994). Finite Elements in Analysis and Design, 22(3):249–263.

Mackerle J (1998) Finite element and boundary element analysis of bridges, roads and pavements – A bibliography (1994–1997). Finite Elements in Analysis and Design, 29(1):65–83.

Mackerle J (1999) Earthquake analysis of structures: FEM and BEM approaches A bibliography (19951998). Finite Elements in Analysis and Design, 32:113–124.

O'Brien J (2004) Train induced vibrations in a ground-track system. MSc Thesis, Dept. of Civil and Environmental Engineering, University of South Carolina, Columbia, SC, USA.

O'Brien J and Rizos DC (2005) A Staggered BEM-FEM methodology for simulation of high speed train induced ibrations. Soil Dynamics and Earthquake Engineering, 25:289–301.

Rizos DC (1993) Advanced time domain boundary element method for general 3d elastodynamic problems. PhD thesis, Dep. of Civil and Environmental Engineering, University of South Carolina, Columbia, South Carolina, USA.

Rizos DC (2000) A rigid surface boundary element for 3-D soil structure interaction analysis in the direct time domain. Computational Mechanics, 26(6):582–591.

Rizos DC and Karabalis DL (1994) An advanced direct time domain BEM formulation for 3-D elastodynamic problems. Computational Mechanics, 15(3):249–269.

Rizos DC and Karabalis DL (1995) Applications of an advanced time domain BEM to 3-D Problems in geomechanics. In: S.N. Atluri, G. Yagawa and T.A. Cruse (eEds) Computational Mechanics '95 Theory and Applications, pp. 3080—3085, Springer-Verlag, Berlin.

Rizos DC and Karabalis DL (1997) A time domain BEM for 3-D elastodynamic analysis using the B-Spline fundamental solutions,. Iin: F.G. Benitez (eEd.), Fundamental Solutions in Boundary Elements: Formulation and Integration, SAND (Camas).

Rizos DC and Karabalis DL (1998) A time domain BEM for 3-D elastodynamic analysis using the B-Spline fundamental solutions. Computational Mechanics, 22:108–115.

Rizos DC and Karabalis DL (1999) Soil-fluid-structure interaction. Invited chapter in: E Kausel and GD Manolis (Eds): Wave Motion Problems in Earthquake Engineering, Computational Mechanics Publications, Southampton.

Rizos DC and Loya K. (2002) Dynamic and seismic analysis of foundations based on free field B-Spline characteristic response histories. Journal of Engineering Mechanics ASCE, 128(4):438–448.

Rizos DC and Wang J (2002) Coupled BEM-FEM solutions for direct time domain soilstructure interaction analysis. Engineering Analysis with Boundary Elements, 26(10):877–888.

Rizos DC and Wang ZY (1999) Applications of Staggered BEM-FEM solutions to soil structure interaction. In: N Jones and R Ghanem (Eds): Proceedings of the 13th Engineering Mechanics Division Specialty Conference, Johns Hopkins University, Baltimore (CD-ROM).

Rizos DC and Zhou S (2005) An Advanced BEM for 3D scalar wave propagation and fluid-structure interaction analysis. BEM/MRM 27, Orlando Florida.

Rizos DC and Zhou S (2006a) An advanced direct time domain BEM for 3-D wave propagation in acoustic media", Journal of Sound and Vibration, 293(1–2):196–212.

Rizos DC and Zhou S (2006b) Efficiency and stability studies of a time domain boundary element methodology for fluid-structure interaction analysis", 15th U.S. National Congress on Theoretical and Applied Mechanics (USNCTAM06) Boulder Colorado.

Rizos DC and Zhou S (2008) A B-spline BEM for fluid-structure-interaction analysis of rigid structures. Engineering Analysis with Boundary Elements (under review).

Soares Jr. D, Mansur WJ, and Estorff O. von (2007) An efficient time-domain FEM/BEM coupling approach based on FEM implicit Green's functions and truncation of BEM time convolution process. Computer Methods in Applied Mechanics and Engineering, 196(9–12):1816–1826.

Spyrakos CC and Beskos DE (1986) Dynamic response of flexible strip foundations by boundary and finite elements. Soil Dynamics and Earthquake Engineering, 5:84–96.

Stehmeyer E and Rizos DC (2006) A simplified B-Spline impulse response function (BIRF) model for the transient SSI analysis of rigid foundations. Soil Dynamics and Earthquake Engineering, 26(5):421–434.

Stehmeyer E and Rizos DC (2007) Considering dynamic soil structure interaction (SSI) effects on seismic isolation retrofit efficiency and the importance of natural frequency ratio. Soil Dynamics and Earthquake Engineering, In Press, Corrected Proof Available online 12 September 2007.

Tzong TJ, Gupta S, and Penzien J, (1981) Two-dimensional hybrid modeling of soil-structure interaction, Report No. EERC 81/11, University of California, Berkeley.

Zhou S (2005) An advanced boundary element method for ship/water dynamic interaction analysis. PhD Dissertation, Dept. of Civil and Environmental Engineering University of South Carolina, Columbia, SC, USA.

Zienkiewicz OC, Kelly DW, and Bettess P (1977) The coupling of the finite element method and boundary solution procedures. International Journal for Numerical methods in Engineering, 11:355–375.

A BEM Solution to the Nonlinear Inelastic Uniform Torsion Problem of Composite Bars

Evangelos J. Sapountzakis and Vasileios J. Tsipiras

Abstract In this chapter the elastic-plastic uniform torsion analysis of composite cylindrical bars of arbitrary doubly symmetric cross section consisting of materials in contact, each of which can surround a finite number of inclusions, taking into account the effect of geometric nonlinearity is presented employing the boundary element method. The bar is axially elastically supported at the centroids of its end cross sections, treating the cases of free axial boundary conditions (vanishing axial force), restrained axial shortening or given axial force as special ones.

1 Introduction

Designs based on elastic analysis are likely to be extremely conservative not only due to the significant difference between first yield in a cross section and full plasticity but also due to the unaccounted for yet significant reserves of strength that are not mobilized in redundant members until after inelastic redistribution takes place. Besides, since thin-walled open sections have low torsional stiffness, the torsional deformations can be of such magnitudes that it is not adequate to treat the angles of cross-section rotation as small. Thus, both material inelasticity and geometric nonlinearity are important for investigating the ultimate strengths of beams that fail by torsion. Moreover, in recent years composite structural elements consisting of a relatively weak matrix reinforced by stronger inclusions or of different materials in contact are of increasing technological importance in engineering. Composite structures can produce very elegant solutions to complex structural engineering challenges, while composite beams or columns offer many significant advantages, such as high load capacity with small cross-section and economic material use, simple connection to other members as for steel construction, good fire resistance etc. Steel beams or columns totally encased in concrete are most common examples.

E.J. Sapountzakis (✉)
School of Civil Engineering, National Technical University of Athens, Zografou Campus, GR-157 80, Greece
e-mail: cvsapoun@central.ntua.gr

A chapter in honor of Dimitri Beskos' 65th birthday

G.D. Manolis, D. Polyzos (eds.), *Recent Advances in Boundary Element Methods*,
DOI 10.1007/978-1-4020-9710-2_26, © Springer Science+Business Media B.V. 2009

Several researchers have dealt with the elastic-plastic uniform torsional behaviour of homogeneous beams, while the corresponding problem of composite bars has not yet been examined. Also, to the authors' knowledge the boundary element method has not yet been used for the numerical analysis of the aforementioned problem.

In this chapter, the elastic-plastic uniform torsion analysis of composite cylindrical bars of arbitrary doubly symmetric cross section consisting of materials in contact, each of which can surround a finite number of inclusions, taking into account the effect of geometric nonlinearity is presented employing the boundary element method. The bar is axially elastically supported at the centroids of its end cross sections, treating the cases of free axial boundary conditions (vanishing axial force), restrained axial shortening or given axial force as special ones. The stress-strain relationships for the materials are assumed to be elastic-plastic-strain hardening. The incremental torque-rotation relationship is computed based on the finite displacement (finite rotation) theory, that is the transverse displacement components are expressed so as to be valid for large rotations and the longitudinal normal strain includes the second-order geometric nonlinear term often described as the "Wagner strain". The torsional rigidity of the cross section is evaluated directly employing the primary warping function of the cross section (Sapountzakis and Mokos 2003) depending on both its shape and the progress of the plastic region. A boundary value problem with respect to the aforementioned function is formulated and solved employing a BEM approach. The proposed formulation procedure is based on the assumption of no local or lateral torsional buckling or distortion and includes the following essential features and novel aspects compared with previous ones:

i. Large deflections and rotations are taken into account, that is the strain-displacement relationships contain higher order displacement terms.
ii. For the first time in the literature, the influence of the second Piola – Kirchhoff normal stress component to the plastic/elastic moment ratio in uniform inelastic torsion is demonstrated.
iii. The influence of the axial restraint to the torsional stiffness of the bar is demonstrated, while the values of the arising axial force of the bar are also presented for each load step.
iv. For each one of the materials of the cross section, material inelasticity is taken into account, that is the elastic-plastic incremental stress-strain relationship is derived from the von Mises yield criterion, a strain flow rule and a strain hardening rule. Integrations of stress resultants for every iterative step and restoration of equilibrium for every converged incremental step are performed numerically using a set of monitoring stations distributed over the area of the cross section.
v. The present formulation is applicable to bars of arbitrary composite cross section, while the case of a homogeneous cross section can be treated as a special one.

vi. The presented formulation does not stand on the assumption of a thin-walled structure and therefore the cross section's torsional rigidity is evaluated exactly without using the so-called Saint – Venant's torsional constant.
vii. The boundary conditions at the interfaces between different material regions have been taken into account.
viii. The proposed method can be efficiently applied to composite beams of thin or thick walled cross section and to laminated composite beams, without the restrictions of the "refined models".
ix. The developed procedure retains most of the advantages of a BEM solution over a pure domain discretization method, although it requires domain discretization, which is used only to evaluate integrals.

Numerical results are presented to illustrate the method and demonstrate its efficiency and accuracy. The contribution of the normal stresses and the effect of axial restraint is investigated by numerical examples with great practical interest, while the case of a homogeneous cross section is treated as a special one.

2 Statement of the Problem

Consider a bar of length l with an arbitrarily shaped doubly symmetric composite cross section, consisting of materials in contact, each of which can surround a finite number of inclusions, occupying the regions $\Omega_j (j = 1, 2, \ldots, K)$ of the $\bar{y}, \bar{z}$ plane (Fig. 1). Let also the boundaries of the nonintersecting regions Ω_j be denoted by $\Gamma_j (j = 1, 2, \ldots, K.)$. These boundary curves are piecewise smooth, i.e. they may have a finite number of corners. The normal stress-strain relationships for the materials are assumed to be elastic-plastic-strain hardening with initial modulus of elasticity and shear modulus E_j, G_j, respectively and initial yield stress $(\sigma_{Y0})_j$, $(j = 1, 2, \ldots, K)$ (Fig. 2).

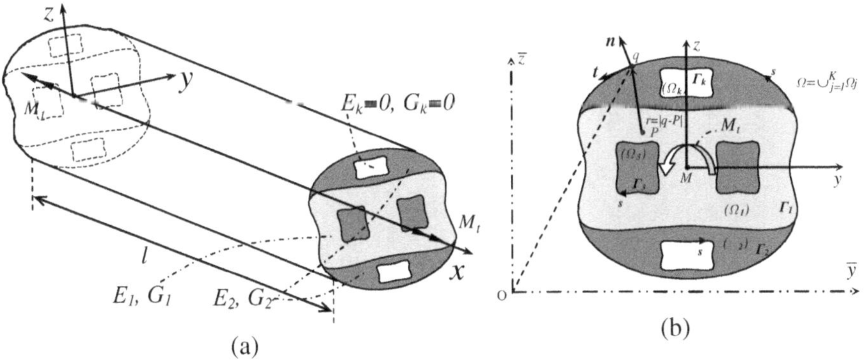

Fig. 1 Prismatic bar subjected to a twisting moment (**a**) with a composite cross section of arbitrary shape occupying the two dimensional region Ω (**b**)

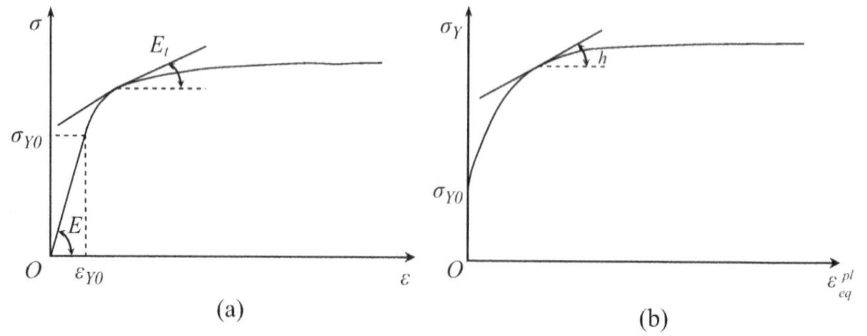

Fig. 2 Normal stress – strain (**a**) and yield stress – equivalent plastic strain (**b**) relationships

When the bar is subjected to uniform torque arising from two concentrated torsional moments M_t at its ends while the warping is elastically restrained only at their corresponding end cross sections' centroids, the angle of twist per unit length $\theta'(x)$ remains constant along the bar. Under these conditions, the bar is leaded to uniform torsion and the displacement components in the x, y and z directions are given in terms of the angle of twist $\theta(x)$ and its derivative with respect to x, as

$$u = u_s(x) + \theta' \cdot \phi_M^P(y, z) \tag{1a}$$

$$v = -z \cdot sin\theta - y \cdot (1 - cos\theta) \tag{1b}$$

$$w = y \cdot sin\theta - z \cdot (1 - cos\theta) \tag{1c}$$

where the transverse displacement components v, w are valid for large rotations (Chen and Trahair 1992), while $\phi_M^P(y, z)$ is the primary warping function (Sapountzakis and Mokos 2003) and $u_s(x)$ is the axial shortening of the bar.

Substituting Eqs. (1) in the non-linear (Green) strain-displacement relations of the non-vanishing strains

$$\varepsilon_{xx} = \frac{\partial u}{\partial x} + \frac{1}{2} \cdot \left[\left(\frac{\partial u}{\partial x} \right)^2 + \left(\frac{\partial v}{\partial x} \right)^2 + \left(\frac{\partial w}{\partial x} \right)^2 \right] \tag{2a}$$

$$\gamma_{xy} = \frac{\partial u}{\partial y} + \frac{\partial v}{\partial x} + \frac{\partial u}{\partial y} \cdot \frac{\partial u}{\partial x} + \frac{\partial v}{\partial y} \cdot \frac{\partial v}{\partial x} + \frac{\partial w}{\partial y} \cdot \frac{\partial w}{\partial x} \tag{2b}$$

$$\gamma_{xz} = \frac{\partial w}{\partial x} + \frac{\partial u}{\partial z} + \frac{\partial u}{\partial z} \cdot \frac{\partial u}{\partial x} + \frac{\partial v}{\partial z} \cdot \frac{\partial v}{\partial x} + \frac{\partial w}{\partial z} \cdot \frac{\partial w}{\partial x} \tag{2c}$$

assuming moderate large deflections ($(\partial u/\partial x)^2 \ll \partial u/\partial x, (\partial u/\partial x)(\partial u/\partial y) \ll (\partial v/\partial x) + (\partial u/\partial y), (\partial u/\partial x)(\partial u/\partial z) \ll (\partial w/\partial x) + (\partial u/\partial z)$) and having in mind that $\theta'(x)$, $u_s'(x)$ remain constant along the bar ($\theta''(x) = 0$, $u_s''(x) = 0$), the non vanishing total strain resultants of the j-th material ($j = 1, 2, \ldots, K$) are given as

$$(\varepsilon_{xx})_j = u_s' + \frac{1}{2} \cdot (y^2 + z^2) \cdot (\theta')^2 \tag{3a}$$

$$(\gamma_{xy})_j = \theta' \cdot \left(\left(\frac{\partial \phi_M^P}{\partial y} \right)_j - z \right) \tag{3b}$$

$$(\gamma_{zx})_j = \theta' \cdot \left(\left(\frac{\partial \phi_M^P}{\partial z} \right)_j + y \right) \tag{3c}$$

where the second-order geometrically nonlinear term in the right hand side of Eq. (3a) is often described as the "Wagner strain". Assuming that throughout an infinitesimal strain increment the primary warping function remains constant and employing Eqs. (3) the strain rates of the j-th material ($j = 1, 2, \ldots, K$) are defined as

$$(d\varepsilon_{xx})_j = du_s' + (y^2 + z^2) \cdot \theta' \cdot d\theta' \tag{4a}$$

$$(d\gamma_{xy})_j = d\theta' \cdot \left(\left(\frac{\partial \phi_M^P}{\partial y} \right)_j - z \right) \quad (d\gamma_{xz})_j \tag{4b}$$

$$(d\gamma_{xy})_j = d\theta' \cdot \left(\left(\frac{\partial \phi_M^P}{\partial z} \right)_j + y \right) \tag{4c}$$

where $d(\cdot)$ denotes infinitesimal increments over time (rates).

Considering strains to be small, employing the second Piola – Kirchhoff stress tensor and assuming isotropic and homogeneous materials for zero Poisson ratio $\nu = 0$, the stress rates of the j-th material ($j = 1, 2, \ldots, K$) are defined in terms of the strain rates as

$$\begin{Bmatrix} (dS_{xx})_j \\ (dS_{xy})_j \\ (dS_{xz})_j \end{Bmatrix} = \begin{bmatrix} E_j & 0 & 0 \\ 0 & G_j & 0 \\ 0 & 0 & G_j \end{bmatrix} \begin{Bmatrix} (d\varepsilon_{xx}^{el})_j \\ (d\gamma_{xy}^{el})_j \\ (d\gamma_{xz}^{el})_j \end{Bmatrix} \tag{5}$$

or in vector form

$$(d\mathbf{S})_j = \mathbf{C}_j d\varepsilon_j^{el} \tag{6}$$

as long as the j-th material remains elastic, that is

$$d\varepsilon_j = d\varepsilon_j^{el} \tag{7}$$

If plastic flow occurs

$$d\varepsilon_j = d\varepsilon_j^{el} + d\varepsilon_j^{pl}, \qquad \left(d\varepsilon_j^{pl}\right)^T = \left\{ \left(d\varepsilon_{xx}^{pl}\right)_j \left(d\gamma_{xy}^{pl}\right)_j \left(d\gamma_{xz}^{pl}\right)_j \right\} \tag{8}$$

a Von Mises yield criterion and an associated flow rule for the j-th material are considered. According to the yield condition, this is given as

$$f_j = \left[(S_{xx})_j^2 + 3\left((S_{xy})_j^2 + (S_{xz})_j^2 \right) \right]^{1/2} - \sigma_{Yj}\left(\left(\varepsilon_{eq}^{pl}\right)_j \right) = 0 \tag{9}$$

where σ_{Yj} is the yield stress of the j-th material and $\left(\varepsilon_{eq}^{pl}\right)_j$ is its equivalent plastic strain, the rate of which is defined as

$$\left(d\varepsilon_{eq}^{pl}\right)_j = \sqrt{\frac{2}{3} \cdot \left(d\varepsilon_j^{pl}\right)^T \cdot d\varepsilon_j^{pl}} \tag{10}$$

It can be shown that $\left(d\varepsilon_{eq}^{pl}\right)_j = d\lambda_j$, where $d\lambda_j$ is the proportionality factor. Moreover, the plastic modulus h_j is defined as

$$h_j = \frac{d\sigma_{Yj}}{d\left(\varepsilon_{eq}^{pl}\right)_j} \tag{11}$$

and derived from a tension test as (Fig. 2)

$$h_j = \frac{E_{tj}}{1 - \frac{E_{tj}}{E_j}} \tag{12}$$

Thus, for linear hardening it is $\Delta\sigma_{Yj} = h_j \cdot \Delta\lambda_j$, where $\Delta(\cdot)$ denotes a finite increment over time. According to the associated flow rule the plastic strain rates are given as

$$\left\{ \left(d\varepsilon_{xx}^{pl}\right)_j \left(d\gamma_{xy}^{pl}\right)_j \left(d\gamma_{xz}^{pl}\right)_j \right\}^T = d\lambda_j \cdot \left\{ \frac{\partial f_j}{\partial (S_{xx})_j} \; \frac{\partial f_j}{\partial (S_{xy})_j} \; \frac{\partial f_j}{\partial (S_{xz})_j} \right\}^T \tag{13}$$

Knowing from Eq. (8) that total strain resultants are equal with the sum of their elastic and their plastic part and substituting Eqs. (4) in Eq. (5) we obtain the following expressions for the stress rates in the inelastic case

$$(dS_{xx})_j = E_j\left((d\varepsilon_{xx})_j - \left(d\varepsilon_{xx}^{pl}\right)_j\right) = E_j du_s' + E_j\left(y^2+z^2\right)\theta' d\theta' - E_j\left(d\varepsilon_{xx}^{pl}\right)_j$$

$$\left(dS_{xy}\right)_j = G_j \cdot \left(\left(d\gamma_{xy}\right)_j - \left(d\gamma_{xy}^{pl}\right)_j\right) = G_j \cdot d\theta' \cdot \left(\left(\frac{\partial \phi_M^P}{\partial y}\right)_j - z\right) - G_j \cdot \left(d\gamma_{xy}^{pl}\right)_j$$

$$(dS_{xz})_j = G_j \cdot \left((d\gamma_{xz})_j - \left(d\gamma_{xz}^{pl}\right)_j\right) = G_j \cdot d\theta' \cdot \left(\left(\frac{\partial \phi_M^P}{\partial z}\right)_j + y\right) - G_j \cdot \left(d\gamma_{xz}^{pl}\right)_j$$

(14a,b,c)

noting that for the elastic case the plastic terms in the above relations are ignored.

Applying the stress components (14a,b,c) in the rate form of the first equation of equilibrium neglecting the body forces (Washizu 1975)

$$\frac{\partial (dS_{xx})_j}{\partial x} + \frac{\partial \left(dS_{xy}\right)_j}{\partial y} + \frac{\partial (dS_{xz})_j}{\partial z} = 0 \tag{15}$$

the following governing equation for the primary warping function $\left(\phi_M^P\right)_j$ in the $\Omega_j(j = 1, 2, \ldots, K)$ region is obtained as

$$\left(\nabla^2 \phi_M^P\right)_j = 0 \quad \text{at the elastic part of the } \Omega_j(j = 1, 2, \ldots, K) \text{ region} \tag{16a}$$

$$\left(\nabla^2 \phi_M^P\right)_j = \frac{1}{d\theta'} \cdot \left(\frac{\partial \left(d\gamma_{xy}^{pl}\right)_j}{\partial y} + \frac{\partial \left(d\gamma_{xz}^{pl}\right)_j}{\partial z}\right) \quad \text{at the plastic part of the } \Omega_j \text{ region} \tag{16b}$$

where $(\nabla^2)_j \equiv (\partial^2/\partial y^2.)_j + (\partial^2/\partial z^2.)_j$ is the Laplace operator and $\Omega = \cup_{j=1}^{K}\Omega_j$ denotes the whole region of the composite cross section.

The boundary conditions of the primary warping function will be derived from the following physical considerations:

- The traction vector in the direction of the normal vector $\boldsymbol{n}$ vanishes on the free surface of the beam, that is

$$\left(dS_{xy}\right)_j \cdot n_y + (dS_{xz})_j \cdot n_z = 0 \tag{17a}$$

- The traction vectors in the direction of the normal vector $\boldsymbol{n}$ on the interfaces separating the j and i different materials are equal in magnitude and opposite in direction, that is

$$(dt_x)_j = (dt_x)_i \text{ or } \left(dS_{xy}\right)_j \cdot n_y + (dS_{xz})_j \cdot n_z = \left(dS_{xy}\right)_i \cdot n_y + (dS_{xz})_i \cdot n_z \tag{17b}$$

- The displacement components remain continuous across the interfaces, since it is assumed that the materials are firmly bonded together

where $n_y = \cos\beta$, $n_z = \sin\beta$ are the direction cosines of the normal vector $\boldsymbol{n}$ to the boundaries Γ_j ($j = 1, 2, \ldots, K$), with $\beta = \widehat{y, \boldsymbol{n}}$ (see Fig. 1). It is worth noting that on both sides of the equality of (17b) the normal vector **n** points in one and the same direction, while the third physical consideration ensures the continuity of the primary warping function $\left(\phi_M^P\right)_j$ inside the region Ω_j (j =1,2,. . . ,K) as well as across the boundaries separating different materials $\left(\left(\phi_M^P\right)_j = \left(\phi_M^P\right)_i\right)$. Substituting Eqs. (14a,b,cb, c) in Eqs. (17a, b) the boundary conditions for the primary warping function $\left(\phi_M^P\right)_j$ are obtained as

$$G_j \left(\frac{\partial \phi_M^P}{\partial n}\right)_j - G_i \left(\frac{\partial \phi_M^P}{\partial n}\right)_i = \left(G_j - G_i\right)\left(z \cdot n_y - y \cdot n_z\right)$$

at the elastic part of the boundary (18a)

$$G_j \cdot \left(\frac{\partial \phi_M^P}{\partial n}\right)_j - G_i \cdot \left(\frac{\partial \phi_M^P}{\partial n}\right)_i = \left(G_j - G_i\right) \cdot \left(z \cdot n_y - y \cdot n_z\right)$$
$$+ G_j \cdot \frac{\left(d\gamma_{xn}^{pl}\right)_j}{d\theta'} - G_i \cdot \frac{\left(d\gamma_{xn}^{pl}\right)_i}{d\theta'}$$

at the plastic part of the boundary (18b)

where $\left(d\gamma_{xn}^{pl}\right)_k = \left(d\gamma_{xy}^{pl}\right)_k \cdot n_y + \left(d\gamma_{xz}^{pl}\right)_k \cdot n_z (k = i, j)$, G_i is the shear modulus of the Ω_i region at the common part of the boundaries of Ω_j and Ω_i regions, or $G_i = 0$ at the free part of the boundary of Ω_j region, while $(\partial/\partial n)_j \equiv n_y \left(\partial / \partial y\right)_j + n_z \left(\partial / \partial z\right)_j$ denotes the directional derivative normal to the boundary Γ_j. The vector **n** normal to the boundary Γ_j is positive if it points to the exterior of the Ω_j region. It is worth here noting that the normal derivatives across the interior boundaries vary discontinuously.

Thus, the primary warping function $\left(\phi_M^P\right)_j$ in each incremental loading step will be evaluated from the solution of the Neumann problem described by the governing equations (16) inside the two dimensional region Ω_j, subjected to the boundary conditions (18) on its boundary Γ_j, respectively.

2.1 Equations of Global Equilibrium

In the elastic case neglecting body forces the principle of virtual work for virtual strains yields

$$\int_V \left(S_{xx} \cdot \delta\varepsilon_{xx} + S_{xy} \cdot \delta\gamma_{xy} + S_{xz} \cdot \delta\gamma_{xz}\right) dV = \int_A \left(t_x \cdot \delta u + t_y \cdot \delta v + t_z \cdot \delta w\right) dA \tag{19}$$

where $\delta(\cdot)$ denotes virtual quantities, t_x, t_y, t_z are the components of the traction vector with respect to the undeformed surface of the bar including the cross section ends, A is the surface area and V is the volume of the bar at its initial undeformed state. Considering that the proposed formulation is valid for large rotations the external torque M_t and the external axial force N are defined as

$$M_t = \sum_{j=1}^{K} \int_{\Omega_j} \left[\left(P_{yx}\right)_j \cdot (-z \cdot \cos\theta - y \cdot \sin\theta) + (P_{zx})_j \cdot (y \cdot \cos\theta - z \cdot \sin\theta) \right] d\Omega_j \tag{20a}$$

$$N = \sum_{j=1}^{K} \int_{\Omega_j} (P_{xx})_j \, d\Omega_j \tag{20b}$$

where $(P_{xx})_j$, $\left(P_{yx}\right)_j$, $(P_{zx})_j$ are components of the first Piola – Kirchhoff stress tensor. Employing Eq. (15) having in mind that the lateral surface of the bar in uniform torsion is free of tractions, the stress – strain relationship (Eq. (5)), the virtual form of Eqs. (1) and knowing that both the angle of twist per unit length θ' and the axial force N are constant throughout the bar, Eqs. (19), (20) after some manipulation yield

$$S_N = E_1 A \cdot u_s' + \frac{1}{2} E_1 I_P \cdot \left(\theta'\right)^2 \tag{21a}$$

$$M_t = E_1 I_P \cdot \theta' \cdot u_s' + \frac{1}{2} E_1 I_{PP} \cdot \left(\theta'\right)^3 + G_1 I_t \cdot \theta' \tag{21b}$$

where S_N is the axial force stress resultant, while the most general relation between S_N and the axial shortening of the bar u_s can be written as

$$\alpha_1 S_N + \alpha_2 u_s = \alpha_3 \quad \text{at the bar ends} \tag{22}$$

where $\alpha_i (i = 1, 2, 3)$ are constants specified at the bar ends. The boundary conditions (22) are the most general axial boundary conditions including also the elastic support. It is apparent that all types of the conventional boundary conditions (restrained, free or elastic support) can be derived form this relation by specifying appropriately the constants $\alpha_i (i = 1, 2, 3)$. Moreover, in Eqs. (21)

$$A = \sum_{j=1}^{K} \frac{E_j}{E_1} \int_{\Omega_j} d\Omega_j$$

$$I_t = \sum_{j=1}^{K} \frac{G_j}{G_1} \int_{\Omega_j} \left[\left(\left(\frac{\partial \phi_M^P}{\partial y} \right)_j - z \right)^2 + \left(\left(\frac{\partial \phi_M^P}{\partial z} \right)_j + y \right)^2 \right] d\Omega_j \tag{23a,b,c,d}$$

$$I_P = \sum_{j=1}^{K} \frac{E_j}{E_1} \int_{\Omega_j} (y^2 + z^2) \, d\Omega_j \quad I_{PP} = \sum_{j=1}^{K} \frac{E_j}{E_1} \int_{\Omega_j} (y^2 + z^2)^2 \, d\Omega_j$$

where the first material is considered as reference material. It is worth here noting that the reduction of Eqs. (23) using the modulus of elasticity E_1 or the shear modulus G_1 of the first material, could be achieved using any other material, considering it as reference material. Equation (21b) after the elimination of the axial force N and the axial shortening of the bar u_s employing Eq. (21a) and the boundary conditions (22) gives the governing differential equation of the non-linear elastic uniform torsion problem.

For the inelastic case the principle of virtual work for virtual incremental strains yields

$$\int_V \left[(S_{xx}+\Delta S_{xx})\,\delta\Delta\varepsilon_{xx}+\left(S_{xy}+\Delta S_{xy}\right)\delta\Delta\gamma_{xy}+(S_{xz}+\Delta S_{xz})\,\delta\Delta\gamma_{xz}\right]dV$$
$$= M_t\,\delta\Delta\theta + N\,\delta\Delta u_s \tag{24}$$

Moreover, employing Eqs. (3) the incremental normal and shearing strains, valid for finite increments are given as

$$(\Delta\varepsilon_{xx})_j = \Delta u_s{}' + \frac{1}{2}\cdot\left(y^2+z^2\right)\cdot\left(2\cdot\theta'+\Delta\theta'\right)\cdot\Delta\theta' \tag{25a}$$

$$\left(\Delta\gamma_{xy}\right)_j = \Delta\theta'\cdot\left(\left(\frac{\partial\phi_M^P}{\partial y}\right)_j - z\right) \tag{25b}$$

$$(\Delta\gamma_{xz})_j = \Delta\theta'\cdot\left(\left(\frac{\partial\phi_M^P}{\partial z}\right)_j + y\right) \tag{25c}$$

Using these equations and having in mind that plastic flow is taken into account in the computation of $\left(\phi_M^P\right)_j$ and its components, after some manipulation the following equilibrium equations are obtained

$$E_1A\cdot\Delta u_s' + E_1I_P\cdot\theta'\Delta\theta' = \Delta S_N \tag{26a}$$

$$E_1I_P\cdot\theta'\Delta u_s' + \left[E_1I_{PP}\cdot\left(\theta'\right)^2 + G_1I_t + W\right]\cdot\Delta\theta' = M_t - S_{M_t} \tag{26b}$$

where

$$W = \sum_{j=1}^{K}\int_{\Omega_j}(S_{xx})_j\cdot\left(y^2+z^2\right)d\Omega_j \tag{27}$$

is the geometrically non-linear component of the tangent stiffness matrix ΔS_N is the incremental axial force stress resultant and S_N, S_{M_t} are the axial force and torsional moment stress resultants, respectively given as

$$S_N = \sum_{j=1}^{K} \int_{\Omega_j} (S_{xx})_j \, d\Omega_j \tag{28a}$$

$$\begin{aligned} S_{M_t} = & \left[\sum_{j=1}^{K} \int_{\Omega_j} (S_{xx})_j \cdot \left(y^2 + z^2\right) d\Omega_j \right] \cdot \theta' + \\ & + \sum_{j=1}^{K} \int_{\Omega_j} \left[(S_{xy})_j \cdot \left(\left(\frac{\partial \phi_M^P}{\partial y} \right)_j - z \right) + (S_{xz})_j \cdot \left(\left(\frac{\partial \phi_M^P}{\partial z} \right)_j + y \right) \right] d\Omega_j \end{aligned} \tag{28b}$$

Considering the axial boundary conditions (22) in their incremental form, equilibrium Eqs. (26) can be solved for the unknown displacement components $\Delta u_s'$, $\Delta\theta'$. These new components are then used to compute the stress resultants S_N, S_{M_t} and the procedure goes on until the out-of-balance forces i.e. $N - S_N$, $M_t - S_{M_t}$ are approximately equal to zero.

2.2 *Integration of the Inelastic Rate Equations*

Let us consider a load station m where equilibrium has been restored. Thus, all the constitutive quantities such as stresses $\{S_j\}_m^T = \{(S_{xx})_j \; (S_{xy})_j (S_{xz})_j\}_m$, total strains $\{\varepsilon_j\}_m^T = \{(\varepsilon_{xx})_j \; (\gamma_{xy})_j \; (\gamma_{xz})_j\}_m$, plastic strains $\{\varepsilon_j^{pl}\}_m^T = \{(\varepsilon_{xx}^{pl})_j \; (\gamma_{xy}^{pl})_j \; (\gamma_{xz}^{pl})_j\}_m$ and internal variables $\{q_j\}_m^T = \{(\varepsilon_{eq}^{pl})_j\}_m$ of the j-th material ($j = 1, 2, \ldots, K$) are known. In order to update the aforementioned constitutive quantities into their corresponding values on the next load station $m+1$ (taking into account that the incremental displacement vector $\{\Delta r\}_m^T = \{\Delta u_s' \; \Delta\theta'\}_m$ is known) the cutting-plane algorithm proposed by Simo and Ortiz [27] is employed due to its simplicity and convergence benefits. This algorithm is a two-step one described in the following.

First step: Elastic prediction Having in mind that the incremental displacement vector $\{\Delta r\}_m$ is known and using Eqs. (25) the incremental total strain vector $\{\Delta\varepsilon_j\}_m^T = \{(\Delta\varepsilon_{xx})_j \; (\Delta\gamma_{xy})_j \; (\Delta\gamma_{xz})_j\}_m$ is easily computed. The trial stresses of the next load station are then computed employing the relation

$$\left\{S_j^{Tr}\right\}_{m+1} = \left\{S_j\right\}_m + \left[D_j^e\right] \cdot \left\{\Delta\varepsilon_j\right\}_m \tag{29}$$

where

$$\left\{S_j^{Tr}\right\}_{m+1}^T = \left\{ \left(S_{xx}^{Tr}\right)_j \; \left(S_{xy}^{Tr}\right)_j \; \left(S_{xz}^{Tr}\right)_j \right\}_{m+1} \tag{30}$$

and $[D_j^e]$ is the elastic constitutive matrix of the j-th material ($j = 1,2,\ldots,K$) given in Eq. (5). It is worth here noting that the assumption that plasticity does not cause any damage effects on the material has been employed, thus $[D_j^e]$ is constant throughout the whole loading history. Using Eq. (9) the trial yield function $f_j^{Tr} = f_j(\{S_j^{Tr}\}_{m+1}, \{(\varepsilon_{eq}^{pl})_j\}_m)$ is then evaluated. If $f_j^{Tr} \leq 0$ then no plastic flow occurs and the constitutive quantities are updated as

$$\{S_j\}_{m+1} = \{S_j^{Tr}\}_{m+1} \qquad \{\varepsilon_j\}_{m+1} = \{\varepsilon_j\}_m + \{\Delta\varepsilon_j\}_m \tag{31a,b}$$

$$\left\{\varepsilon_j^{pl}\right\}_{m+1} = \left\{\varepsilon_j^{pl}\right\}_m \quad \left\{\left(\varepsilon_{eq}^{pl}\right)_j\right\}_{m+1} = \left\{\left(\varepsilon_{eq}^{pl}\right)_j\right\}_m \tag{31c,d}$$

If $f_j^{Tr} > 0$, the yield criterion is violated and the algorithm proceeds to step two.

Second step: Plastic correction. A Newton – Raphson approach is used so as to return the stresses to the yield surface. The k-th iteration of this approach is described as follows.

The yield function is linearized around its current value as

$$f_j^{(k+1)} = f_j^{(k)} + \left\{\left(\frac{\partial f_j}{\partial\{S_j\}}\right)^{(k)}\right\}^T \{\delta S_j\} + \left(\frac{\partial f_j}{\partial\left(\varepsilon_{eq}^{pl}\right)_j}\right)^{(k)} \left(\delta\varepsilon_{eq}^{pl}\right)_j \tag{32}$$

Employing the plastic flow rule (Eq. (13)), $\{\delta\varepsilon_j^{pl}\}$, $(\delta\varepsilon_{eq}^{pl})_j$, $\{\delta S_j\}$ are computed as

$$\begin{aligned} &\left\{\delta\varepsilon_j^{pl}\right\} = \delta\lambda_j \cdot \left(\frac{\partial f_j}{\partial\{S_j\}}\right)^{(k)} \qquad \left(\delta\varepsilon_{eq}^{pl}\right)_j = \delta\lambda_j \\ &\{\delta S_j\} = -\left[D_j^e\right] \cdot \left\{\delta\varepsilon_j^{pl}\right\} \end{aligned} \tag{33a,b,c}$$

Substituting Eqs. (33b, c) into Eq. (32) we obtain

$$f_j^{(k+1)} = f_j^{(k)} - \delta\lambda_j \cdot \left(\frac{\partial f_j}{\partial\{S_j\}}^{(k)}\right)^T \left[D_j^e\right] \cdot \frac{\partial f_j}{\partial\{S_j\}}^{(k)} + \frac{\partial f_j}{\partial\left(\varepsilon_{eq}^{pl}\right)_j}^{(k)} \cdot \delta\lambda_j \tag{34}$$

from which after using the relations

$$\left(\frac{\partial f_j}{\partial \{S_j\}}^{(k)}\right)^T \left[D_j^e\right] \cdot \frac{\partial f_j}{\partial \{S_j\}}^{(k)} = 3 \cdot G_j \tag{35}$$

$$\left(\frac{\partial f_j}{\partial \left(\varepsilon_{eq}^{pl}\right)_j}\right)^{(k)} = \left(\frac{\partial f_j}{\partial \sigma_{Yj}}\right)^{(k)} \left(\frac{\partial \sigma_{Yj}}{\partial \left(\varepsilon_{eq}^{pl}\right)_j}\right)^{(k)} = -h_j\left(\left(\varepsilon_{eq}^{pl}\right)_j^{(k)}\right) \tag{36}$$

arising from the Von Mises yield criterion and the adopted strain hardening rule, the proportionality factor $\delta\lambda_j$ is obtained as

$$\delta\lambda_j = \frac{f_j^{(k)}}{3 \cdot G_j + h_j^{(k)}} \tag{37}$$

where $h_j = h_j((\varepsilon_{eq}^{pl})_j^{(k)})$ is the plastic modulus of the k-th iteration.

3 Incremental – Iterative Solution Algorithm

The modified Newton-Raphson method is usually used for the incremental-iterative solution of the nonlinear analysis, that is the tangent stiffness matrix is kept constant throughout the whole increment. In the present study, load control over the incremental steps is used and load stations are chosen according to the load history and convergence requirements. Before the restoration of equilibrium, the internal stress resultants S_N, S_{M_t} and the external loadings $N = 0$, M_t respectively, are not in equilibrium and hence the out-of-balance forces do not vanish. The stress resultants are computed with the incremental and not the iterative strains because experience has shown that this strategy is more accurate for the inelastic case. The generalized cutting-plane algorithm is adopted to integrate the rate equations of elastoplasticity because of its simplicity and fast convergence. Due to geometrical non-linearity, load steps must be considered from the beginning of the loading history, even though the material is initially elastic, that is the elastic – plastic transition cannot be determined in a single load step as in the geometrically linear case. Finally, in order to obtain the correct tangential stiffness matrix (due to the presence of the torsion constant I_t) the plastic strains of the previous load increment are introduced in Eqs. (16b), (18b) instead of using the elastic ϕ_M^P, I_t (initial stiffness method), which leads to serious errors since the shear stress distribution over the cross-section is not modified.

4 Integral Representations for the Primary Warping Function

The evaluation of the primary warping function is accomplished using BEM (Katsikadelis 2002), as this is presented in (Sapountzakis and Mokos 2003).

Fig. 3 Composite cross section of example 1

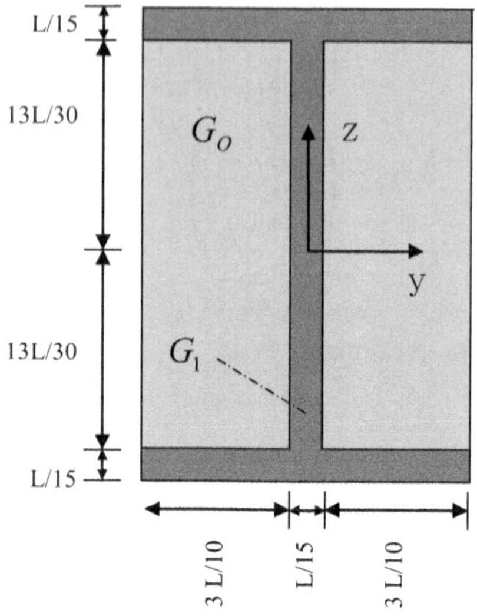

5 Numerical Example

The composite cross section of Fig. 3 (L =30 cm) of elastic-perfectly plastic materials ($h = 0$), consisting of a steel-I beam ($E = 21 \times 10^7$kN/m^2, G=87.5 $\times$ 10^6 kN/m^2, $\sigma_Y = 28.5 \times 10^4$ kN/m^2) partially encased in concrete (E=32.3184 $\times$ 10^6 kN/m^2, $G = 13.466 \times 10^6$ kN/m^2, $\sigma_Y = 2.0 \times 10^4$ kN/m^2) has been studied. In order to examine the effect of combined material and geometrical nonlinearities, the

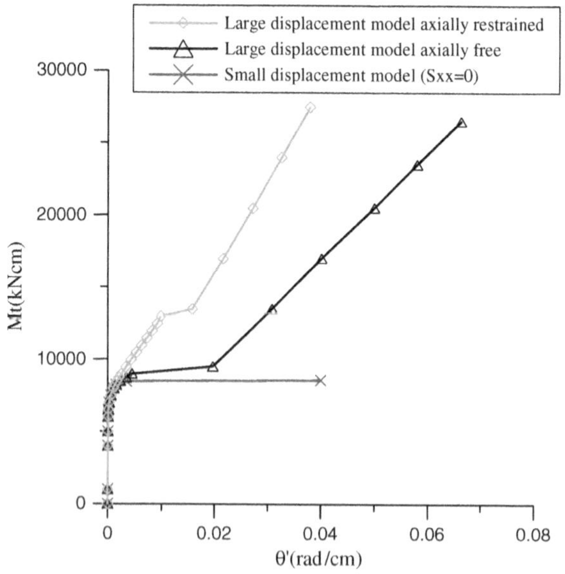

Fig. 4 Torque – twist curves for the large and the small displacement models

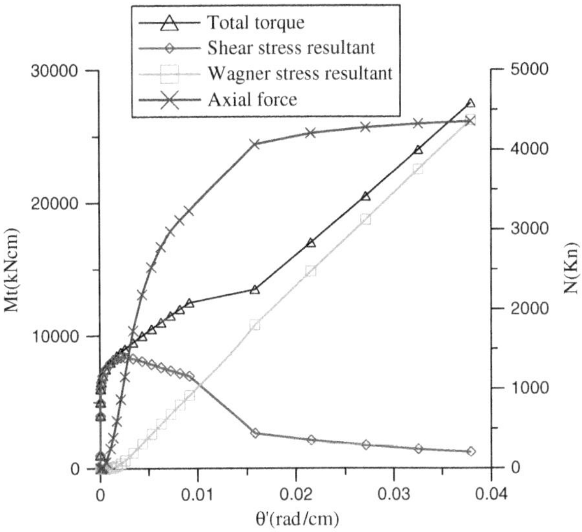

Fig. 5 Shear and Wagner stress resultants forming the total torque – twist curve and variation of the axial force

variation of the angle of twist per unit length $\theta'(x)$ of the composite cross section with respect to the applied external torque M_t is presented in Fig. 4 for the large displacement model for the cases of axially free or restrained boundary conditions. The significant influence of the axial boundary condition to the torsional rigidity is observed. In the same figure the corresponding curve of the small displacement model ($S_{xx} = 0$) is presented to verify both the slight increase of the torsional rigidity in low values of the angle of twist per unit length due to the geometrical non-linearity, which in higher values is proven to be significant due to the predominant action of the "Wagner term". Moreover, in Fig. 5 for the case of axially restrained beam the contribution of both the shear and the normal stress components of the second Piola – Kirchhoff stress tensor to the total undertaken torque together with the axial force for various values of the angle of twist per unit length $\theta'(x)$ are presented demonstrating the negligible values of the Wagner stress resultant in the elastic range which increase without reaching a maximum value after plastic flow occurs. This resultant becomes very large (larger than that of the shear stresses) and the tension stresses induced may lead to tension rapture. In Table 1 the elastic torque M_t^{el}, the fully plastic torque M_t^{pl} (that is the torque incremental step for which plastic flow occurs for the entire cross section) and the shape factor κ of

Table 1 Elastic torque (kNcm), fully plastic torque (kNcm) and shape factor

		Large displacements		
	Small displacements	Free	Restrained	Variation (%)
M_t^{el}	4120	4120	4120	0.00
M_t^{pl}	8585	10000	13500	35.00
$\kappa = M_t^{pl}/M_t^{el}$	2.084	2.427	3.277	35.00

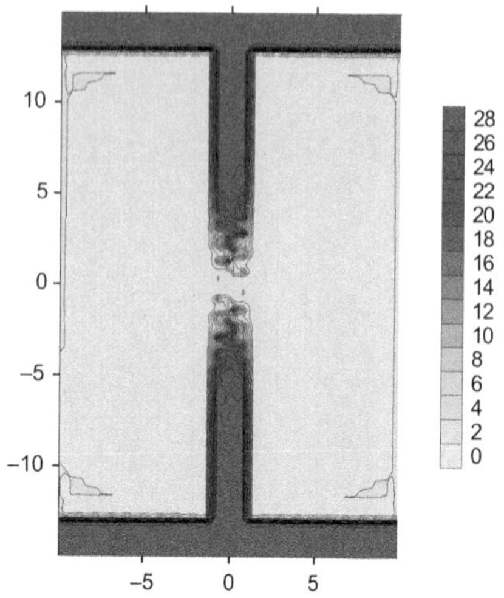

Fig. 6 Normal component of the stress tensor for the composite cross section

the composite cross section are presented for both the large (for the cases of free and restrained beam) and the small displacement models. The conclusions already drawn from Fig. 4 are easily verified. Finally, in Fig. 6 the contour lines of the normal component of the second Piola – Kirchhoff stress tensor S_{xx} are presented for the applied external torque $M_t = 24,000\,\mathrm{kNcm}$.

6 Conclusions

This chapter presents the elastic-plastic uniform torsion analysis of composite cylindrical bars of arbitrary doubly symmetric cross section, taking into account the effect of geometric nonlinearity and employing the boundary element method. The bar is axially elastically supported at the centroids of its end cross sections, treating the cases of free axial boundary conditions (vanishing axial force), restrained axial shortening or given axial force as special ones. The developed procedure retains most of the advantages of a BEM solution over a FEM approach, since the employed area discretization is used only for numerical integration of domain integrals and does not lead to a greater number of equilibrium equations.

References

Chen G, Trahair NS (1992) Inelastic nonuniform torsion of steel I-beams. J. Constr. Steel. Res. 23: 189–207

Katsikadelis JT (2002) Boundary Elements: Theory and Applications. Elsevier, Amsterdam-London

Sapountzakis EJ, Mokos VG (2003) Warping shear stresses in nonuniform torsion of composite bars by BEM. Comp. Meth. Appl. Mech. Engrg. 192: 4337–4353

Washizu K (1975) Variational Methods in Elasticity and Plasticity. Pergamon Press, Oxford

Time Domain BEM: Numerical Aspects of Collocation and Galerkin Formulations

Martin Schanz, Thomas Rüberg and Lars Kielhorn

Abstract Time domain Boundary Element formulations are very well suited to treat wave propagation phenomena in semi-infinte domains, e.g., to simulate phenomena in earthquake engineering. Beside an analytical integration within each time step there is the formulation based on the Convolution Quadrature Method which utilizes the Laplace domain fundamental solutions. Within this technique not only the extension to inelastic material behavior is easy also the formulation of a symmetric Galerkin procedure can be established because the regularisation has to be performed only for the Laplace domain kernels.

Here, a collocation method is presented resulting in a saddle point formulation as well as a symmetric Galerkin procedure. Both techniques are numerically compared also with the usual collocation formulation. Both new formulations show a better numerical behavior, i.e., the condition numbers of the respective matrices are improved. This yields also to a better stability in time.

1 Introduction

The Boundary Element Method (BEM) in time domain is especially important to treat wave propagation problems in semi-infinite domains. In this application the main advantage of this method becomes obvious, i.e., its ability to model the Sommerfeld radiation condition correctly. Certainly this is not the only advantage of a time domain BEM but very often the main motivation as, e.g., in earthquake engineering.

The first boundary integral formulation for elastodynamics was published by Cruse and Rizzo (1968). This formulation performs in Laplace domain with a subsequent inverse transformation to the time domain to achieve results for the transient behavior. The corresponding formulation in Fourier domain, i.e., frequency domain, was presented by Domínguez (1978). The first boundary element formulation directly in the time domain was developed by Mansur for the scalar wave equation

M. Schanz (✉)
Institute of Applied Mechanics, Graz University of Technology, Austria
e-mail: m.schanz@tugraz.at

G.D. Manolis, D. Polyzos (eds.), *Recent Advances in Boundary Element Methods*,
DOI 10.1007/978-1-4020-9710-2_27,

and for elastodynamics with zero initial conditions (Mansur, 1983). The extension of this formulation to non-zero initial conditions was presented by Antes (1985). Detailed information about this procedure may be found in the book of Domínguez (1993). A comparative study of these possibilities to treat elastodynamic problems with BEM is given by Manolis (1983). A completely different approach to handle dynamic problems utilizing static fundamental solutions is the so-called dual reciprocity BEM. This method was introduced by Nardini and Brebbia (1982) and details may be found in the monograph of Partridge et al. (1992). A very detailed review on elastodynamic boundary element formulations and a list of applications can be found in two articles of Beskos (1987, 1997).

The above listed methodologies to treat elastodynamic problems with the BEM show mainly the two ways: direct in time domain or via an inverse transformation in Laplace domain. Mostly, the latter is used, e.g., (Ahmad and Manolis, 1987). Since all numerical inversion formulas depend on a proper choice of their parameters (Narayanan and Beskos, 1982), a direct evaluation in time domain seems to be preferable. Also, it is more natural to work in the real time domain and observe the phenomenon as it evolves. But, as all time-stepping procedures, such a formulation requires an adequate choice of the time step size. An improper chosen time step size leads to instabilities or numerical damping. Four procedures to improve the stability of the classical dynamic time-stepping BE formulation can be quoted: the first employs modified numerical time marching procedures, e.g., Antes and Jäger (1995) for acoustics, Peirce and Siebrits (1997) for elastodynamics; the second employs a modified fundamental solution, e.g., Rizos and Karabalis (1994) for elastodynamics; the third employs an additional integral equation for velocities (Mansur et al., 1998); and the last uses weighting methods, e.g., Yu et al. (1998) for elastodynamics and Yu et al. (2000) for acoustics.

Beside these improved approaches there exist the possibility to solve the convolution integral in the boundary integral equation with the so-called Convolution Quadrature Method (CQM) proposed by Lubich (1988). It utilizes the Laplace domain fundamental solution and results not only in a more stable time stepping procedure but also damping effects in case of visco- or poroelasticity can be taken into account (see Schanz and Antes (1997) or Schanz (2001b)). This methodology is used in the following to establish a collocation based BEM. Different to the usual collocation methods, here, a saddle point formulation is proposed.

Additionally, a symmetric Galerkin formulation in time domain is presented. For Galerkin type BE formulations see the overview given by Bonnet et al. (1998). Mostly, those formulations are established for elastostatics. In elastodynamics in Laplace domain Frangi and Novati (1998) have published a symmetric Galerkin formulation in 2D. Here, the time domain formulation based on the CQM is presented for 3D acoustics and elastodynamics utilizing the advantage that the Laplace domain fundamental solutions can be used and, therefore, the regularisation is much more simple compared to a pure time domain approach. Finally, only a weakly singular formulation is achieved. A numerical comparison of both proposed time domain formulations with the classical collocation approach closes the paper.

Throughout this chapter vectors and tensors are denoted by bold symbols and matrices by sans serif and upright symbols. An empty space in a matrix is used to avoid writing tedious zeros.

2 Time Domain Boundary Integral Equations

The hyperbolic partial differential equations to be considered in this work are the *scalar wave equation* and the *elastodynamic system*. The former is given by

$$\frac{\partial^2 p}{\partial t^2}(\mathbf{x}, t) - c^2(\Delta p)(\mathbf{x}, t) = 0 \tag{1}$$

and describes, for instance, the changes in the acoustic pressure $p(\mathbf{x}, t)$ in an idealized fluid. $\mathbf{x}$ and t are the position in the three-dimensional Euclidean space $\mathbb{R}^3$ and the time point from the interval $(0, \infty)$. In equation (1), c denotes the speed of wave propagation, i.e., $c^2 = \kappa/\rho$ with the compressibility κ and the mass density ρ of the fluid.

The dynamic variation of the displacement field $\mathbf{u}(\mathbf{x}, t)$ of an elastic solid under the assumptions of linear elasticity is governed by the system of equations

$$\frac{\partial^2 \mathbf{u}}{\partial t^2}(\mathbf{x}, t) - c_1^2 \nabla(\nabla \cdot \mathbf{u}(\mathbf{x}, t)) + c_2^2 \nabla \times (\nabla \times \mathbf{u}(\mathbf{x}, t)) = \mathbf{0}\,. \tag{2}$$

The material properties of the solid are represented by c_1 and c_2, which are the speeds of propagation of the pressure and the shear waves, respectively. These speeds are given by $c_1^2 = (\lambda + \mu)/\rho$ and $c_2^2 = \mu/\rho$ with the Lamé parameters λ and μ.

2.1 Hyperbolic Initial Boundary Value Problem

In order to unify the notation, equations (1) and (2) are represented by

$$\rho \frac{\partial^2 u}{\partial t^2}(\mathbf{x}, t) + (\mathcal{L}u)(\mathbf{x}, t) = 0, \tag{3}$$

where $\mathcal{L}$ is an elliptic partial differential operator. u is now a general unknown function replacing either the acoustic pressure p or the displacement field $\mathbf{u}$. Note that equation (3) has been obtained by multiplying either (1) or (2) by the mass density ρ.

Using this notation, the considered hyperbolic initial boundary value problems are of the form

$$\begin{aligned}
\left[\left(\rho\frac{\partial^2}{\partial t^2}+\mathcal{L}\right)u\right](\mathbf{x},t) &= 0 && (\mathbf{x},t)\in\Omega\times(0,\infty)\\
u(\mathbf{y},t) &= g_D(\mathbf{y},t) && (\mathbf{y},t)\in\Gamma_D\times(0,\infty)\\
q(\mathbf{y},t) &= g_N(\mathbf{y},t) && (\mathbf{y},t)\in\Gamma_N\times(0,\infty)\\
u(\mathbf{x},t) &= 0 && (\mathbf{x},t)\in\Omega\times(-\infty,0]\,.
\end{aligned} \tag{4}$$

The first statement in (4) requires the fulfillment of the partial differential equation in the spatial domain Ω for all times $0 < t < \infty$. This spatial domain Ω has the boundary Γ which is subdivided into two disjoint sets Γ_D and Γ_N at which boundary conditions are prescribed. The Dirichlet boundary condition is the second statement of (4) and assigns a given datum g_D to the unknown u on the part Γ_D of the boundary. Similarly, the Neumann boundary condition is the third statement in which the datum g_N is assigned to the function q on the boundary part Γ_N. This new function q represents either the acoustic flux q_n or the surface traction $\mathbf{t}$ depending on the case of the scalar wave equation (1) or the elastodynamic system (2) as the underlying model. These quantities can be expressed by

$$q_n(\mathbf{y},t) = \lim_{\Omega\ni\mathbf{x}\to\mathbf{y}\in\Gamma}\left[\nabla p(\mathbf{x},t)\cdot\mathbf{n}(\mathbf{y})\right] \tag{5a}$$

$$\mathbf{t}(\mathbf{y},t) = \lim_{\Omega\ni\mathbf{x}\to\mathbf{y}\in\Gamma}\left[\boldsymbol{\sigma}(\mathbf{x},t)\cdot\mathbf{n}(\mathbf{y})\right], \tag{5b}$$

where $\boldsymbol{\sigma}$ is the stress tensor depending on the displacement field $\mathbf{u}$ according to the strain-displacement relationship and Hooke's law. Both boundary conditions have to hold for all times. Finally, in the last statement of (4) the condition of a quiescent past is given which implies the homogeneous initial conditions

$$u(\mathbf{x},0) = 0 \qquad \text{and} \qquad \frac{\partial u}{\partial t}(\mathbf{x},0) = 0\,. \tag{6}$$

For later purposes it is convenient to express the relations (5) by means of the traction operator

$$q(\mathbf{y},t) = (\mathcal{T}u)(\mathbf{y},t)\,. \tag{7}$$

Hence, depending on the underlying model, the operator $\mathcal{T}$ maps either the acoustic pressure p to the surface flux q_n or the displacement field $\mathbf{u}$ to the surface traction $\mathbf{t}$. See Kupradze et al. (1979) for an explicit expression of the operator $\mathcal{T}$ for the case of elasticity.

2.2 Dynamic Integral Equations

The starting point of the boundary integral formulations of this work is the representation formula. This can be derived from the dynamic reciprocal identity (Wheeler and Sternberg, 1968)

$$\int_\Omega \left[\left(\rho \frac{\partial^2}{\partial t^2} + \mathcal{L} \right) u \right] * v \,\mathrm{d}\mathbf{x} + \int_\Gamma q * v \,\mathrm{d}s = \int_\Omega \left[\left(\rho \frac{\partial^2}{\partial t^2} + \mathcal{L} \right) v \right] * u \mathrm{d}\mathbf{x} + \int_\Gamma (\mathcal{T} v) * u \,\mathrm{d}s. \tag{8}$$

In this expression, v is another function fulfilling the corresponding initial boundary value problem (4) with different prescribed boundary data. Moreover, the Riemann convolution has been used which is defined as

$$(g * h)(\mathbf{x}, t) = \int_0^t g(\mathbf{x}, t - \tau) h(\tau) \mathrm{d}\tau. \tag{9}$$

Inserting into identity (8) the fundamental solution u^* of equation (3) for the test function v, yields the representation formula

$$\begin{aligned} u(\mathbf{x}, t) = & \int_0^t \int_\Gamma u^*(\mathbf{x} - \mathbf{y}, t - \tau) q(\mathbf{y}, \tau) \mathrm{d}s_\mathbf{y} \mathrm{d}\tau \\ & - \int_0^t \int_\Gamma (\mathcal{T}_\mathbf{y} u^*)(\mathbf{x} - \mathbf{y}, t - \tau) u(\mathbf{y}, \tau) \mathrm{d}s_\mathbf{y} \mathrm{d}\tau. \end{aligned} \tag{10}$$

Here, the surface measure $\mathrm{d}s_\mathbf{y}$ carries its subscript in order to emphasize that the integration variable is $\mathbf{y}$. Similarly, $\mathcal{T}_\mathbf{y}$ indicates that the derivatives involved in the computation of the surface flux or traction due to equations (5) are taken with respect to the variable $\mathbf{y}$. Explicit expressions for the used fundamental solutions can be found, for instance, in Kausel (2006). By means of equation (10), the unknown u is given at any point $\mathbf{x}$ inside the domain Ω and at any time $0 < t < \infty$ if the boundary data $u(\mathbf{y}, \tau)$ and $q(\mathbf{y}, \tau)$ are known for all points $\mathbf{y}$ of the boundary Γ and times $0 < \tau < t$.

The first boundary integral equation is obtained by taking expression (10) to the boundary. Using operator notation, this boundary integral equation reads

$$(\mathcal{V} * q)(\mathbf{x}, t) = \mathcal{C}(\mathbf{x}) u(\mathbf{x}, t) + (\mathcal{K} * u)(\mathbf{x}, t) \tag{11}$$

for points $\mathbf{x}$ on the boundary Γ. The introduced operators are the single layer operator $\mathcal{V}$, the integral-free term $\mathcal{C}$, and the double layer operator $\mathcal{K}$ which are defined as

$$(\mathcal{V} * q)(\mathbf{x}, t) = \int_0^t \int_\Gamma u^*(\mathbf{x} - \mathbf{y}, t - \tau) q(\mathbf{y}, \tau) \mathrm{d}s_\mathbf{y} \mathrm{d}\tau \tag{12a}$$

$$\mathcal{C}(\mathbf{x}) u(\mathbf{x}, t) = \mathcal{I} + \lim_{\varepsilon \to 0} \int_{\partial B_\varepsilon(\mathbf{x}) \cap \Omega} (\mathcal{T}_\mathbf{y} u^*)^\top (\mathbf{x} - \mathbf{y}, 0) u(\mathbf{x}, t) \mathrm{d}s_\mathbf{y} \tag{12b}$$

$$(\mathcal{K} * u)(\mathbf{x}, t) = \lim_{\varepsilon \to 0} \int_0^t \int_{\Gamma \setminus B_\varepsilon(\mathbf{x})} (\mathcal{T}_\mathbf{y} u^*)^\top (\mathbf{x} - \mathbf{y}, t - \tau) u(\mathbf{y}, \tau) \mathrm{d}s_\mathbf{y} \mathrm{d}\tau. \tag{12c}$$

In these expressions, $B_\varepsilon(\mathbf{x})$ denotes a ball of radius ε centered at $\mathbf{x}$ and $\partial B_\varepsilon(\mathbf{x})$ is its surface. Note that the single layer operator (12a) involves a weakly singular integral and, in case of the elastodynamic system, the double layer operator (12c) has to be understood in the sense of a principal value.

Application of the traction operator $\mathcal{T}_\mathbf{x}$ to the dynamic representation formula (10) yields the second boundary integral equation

$$(\mathcal{D} * u)(\mathbf{x}, t) = \mathcal{C}'(\mathbf{x}) q(\mathbf{x}, t) - (\mathcal{K}' * q)(\mathbf{x}, t), \qquad \mathcal{C}'(\mathbf{x}) := \mathcal{I} - \mathcal{C}(\mathbf{x}). \tag{13}$$

The newly introduced operators are the hypersingular operator $\mathcal{D}$ and the adjoint double layer operator $\mathcal{K}'$. They are defined as

$$(\mathcal{D} * u)(\mathbf{x}, t) = -\lim_{\varepsilon \to 0} \int_0^t \mathcal{T}_\mathbf{x} \int_{\Gamma \setminus B_\varepsilon(\mathbf{x})} (\mathcal{T}_\mathbf{y} u^*)^\top (\mathbf{x} - \mathbf{y}, t - \tau) u(\mathbf{y}, \tau) \mathrm{d}s_\mathbf{y} \mathrm{d}\tau \tag{14a}$$

$$(\mathcal{K}' * q)(\mathbf{x}, t) = \lim_{\varepsilon \to 0} \int_0^t \int_{\Gamma \setminus B_\varepsilon(\mathbf{x})} (\mathcal{T}_\mathbf{x} u^*)(\mathbf{x} - \mathbf{y}, t - \tau) q(\mathbf{y}, \tau) \mathrm{d}s_\mathbf{y} \mathrm{d}\tau. \tag{14b}$$

The hypersingular operator has to be understood in the sense of a finite part.

2.3 Mixed Initial Boundary Value Problems

For the solution of mixed initial boundary value problems of the form (4), either a non-symmetric formulation by means of the first boundary integral equation (11) or a symmetric formulation using both first and second boundary integral equations, (11) and (13), is considered.

Non-symmetric formulation. At first, a continuous extension $\tilde{g}_D$ of the prescribed Dirichlet datum g_D to the whole boundary Γ is introduced such that

$$\tilde{g}_D(\mathbf{x}, t) = g_D(\mathbf{x}, t), \qquad (\mathbf{x}, t) \in \Gamma_D \times (0, \infty). \tag{15}$$

A new unknown $\tilde{u}$ is thus defined by

$$\tilde{u} = u - \tilde{g}_D. \tag{16}$$

Inserting $u = \tilde{u} + \tilde{g}_D$ into the first boundary integral equation (11) yields

$$\mathcal{V} * q = \mathcal{C}\tilde{u} + \mathcal{K} * \tilde{u} + \mathcal{C}\tilde{g}_D + \mathcal{K} * \tilde{g}_D, . \tag{17}$$

For the solution of the mixed initial boundary value problem (4) the non-symmetric formulation has the form

$$\begin{aligned} \mathcal{V} * q - \mathcal{C}\tilde{u} - \mathcal{K} * \tilde{u} &= f_D \quad (\mathbf{x}, t) \in \Gamma \times (0, \infty) \\ q &= g_N \quad (\mathbf{x}, t) \in \Gamma_N \times (0, \infty), \end{aligned} \tag{18}$$

with the abbreviation $f_D = \mathcal{C}\tilde{g}_D + \mathcal{K} * \tilde{g}_D$. Effectively, in this formulation the Neumann boundary condition is employed as a side condition, whereas the given Dirichlet datum is directly fulfilled by the solution u.

Symmetric formulation. In order to establish a symmetric formulation, the first boundary integral equation (11) is used only on the Dirichlet boundary Γ_D whereas the second one (13) is used only on the Neumann part Γ_N. Taking the prescribed boundary conditions (4) into account, yields

$$\begin{aligned} \mathcal{V} * q - \mathcal{K} * u &= \mathcal{C} g_D, \qquad (\mathbf{x}, t) \in \Gamma_D \times (0, \infty) \\ \mathcal{K}' * q + \mathcal{D} * u &= \mathcal{C}' g_N, \qquad (\mathbf{x}, t) \in \Gamma_N \times (0, \infty). \end{aligned} \tag{19}$$

Further, the Dirichlet datum u and the Neumann datum q are decomposed into

$$u = \tilde{u} + \tilde{g}_D \qquad \text{and} \qquad q = \tilde{q} + \tilde{g}_N. \tag{20}$$

In this decomposition, arbitrary but fixed extensions, $\tilde{g}_D$ and $\tilde{g}_N$, of the given Dirichlet and Neumann data, g_D and g_N, are introduced such that

$$\begin{aligned} \tilde{g}_D(\mathbf{x}, t) &= g_D(\mathbf{x}, t), \qquad (\mathbf{x}, t) \in \Gamma_D \times (0, \infty) \\ \tilde{g}_N(\mathbf{x}, t) &= g_N(\mathbf{x}, t), \qquad (\mathbf{x}, t) \in \Gamma_N \times (0, \infty) \end{aligned} \tag{21}$$

holds. The extension $\tilde{g}_D$ of the given Dirichlet datum has to be continuous due to regularity requirements (Steinbach, 2008).

Inserting the decompositions (20) into (19) leads to the symmetric formulation for the unknowns $\tilde{u}$ and $\tilde{q}$

$$\begin{aligned} \mathcal{V} * \tilde{q} - \mathcal{K} * \tilde{u} &= f_D, \quad (\mathbf{x}, t) \in \Gamma_D \times (0, \infty) \\ \mathcal{D} * \tilde{u} + \mathcal{K}' * \tilde{q} &= f_N, \quad (\mathbf{x}, t) \in \Gamma_N \times (0, \infty) \end{aligned} \tag{22}$$

with the abbreviations

$$
\begin{aligned}
f_D &= \mathcal{C}\tilde{g}_D + \mathcal{K} * \tilde{g}_D - \mathcal{V} * \tilde{g}_N \\
f_N &= \mathcal{C}'\tilde{g}_N - \mathcal{K}' * \tilde{g}_N - \mathcal{D} * \tilde{g}_D.
\end{aligned}
\tag{23}
$$

3 Boundary Element Formulations

3.1 Spatial and Temporal Discretizations

Let the boundary Γ of the considered domain be represented in the computation by an approximation Γ_h which is the union of geometrical elements

$$
\Gamma_h = \bigcup_{e=1}^{N_e} \tau_e . \tag{24}
$$

τ_e denote finite elements, e.g., surface triangles as in this work, and their total number is N_e. Now, the boundary functions $\tilde{u}$ and q are approximated with shape functions ψ_j, which are defined with respect to the geometry partitioning (24), and time dependent coefficients u_i and q_j

$$
u_h(\mathbf{y},t) = \sum_{i=1}^{N} u_i(t)\phi_i(\mathbf{y}) \qquad \text{and} \qquad q_h(\mathbf{y},t) = \sum_{j=1}^{M} q_j(t)\psi_j(\mathbf{y}). \tag{25}
$$

For simplicity, consider the application of the single layer operator as given in (12a). Inserting the approximation of q due to (25) yields

$$
\begin{aligned}
&\int_0^t \int_\Gamma u^*(\mathbf{x}-\mathbf{y},t-\tau)\left(\sum_{j=1}^{M} q_j(\tau)\psi_j(\mathbf{y})\right) \mathrm{d}s_{\mathbf{y}}\mathrm{d}\tau \\
&\quad = \sum_{j=1}^{M} \int_0^t \left(\int_\Gamma u^*(\mathbf{x}-\mathbf{y},t-\tau)\psi_j(\mathbf{y})\mathrm{d}s_{\mathbf{y}}\right) q_j(\tau)\mathrm{d}\tau = \sum_{j=1}^{M} V_j * q_j .
\end{aligned}
\tag{26}
$$

Note that the introduced abbreviation V_j is is still a function of $\mathbf{x}$ and t.

As pointed out in the introduction, the preferred method of temporal discretization is here the Convolution Quadrature Method. This method has been introduced by Lubich (1988) and is used for the temporal discretization of boundary integral equations, e.g., by Schanz (2001b). Its basic idea is to approximate the convolution (26) by a quadrature formula on an equidistant time grid of step size Δt, i.e., $0 = t_0 < \Delta t = t_1 < \cdots < n\Delta t = t_n$,

$$(V_j * q_j)(\mathbf{x}, t_n) \approx \sum_{\nu=0}^{n} \omega_{n-\nu}(\Delta t, \gamma, \hat{u}^*, \psi_j) q_j(\nu \Delta t). \tag{27}$$

In this expression, the quadrature weights $\omega_{n-\nu}$ depend on the step size Δt, the characteristic polynomial γ of an underlying multistep method, the Laplace transformed fundamental solution $\hat{u}^*$, and the shape function ψ_j. Confer Schanz (2001b) for the technical details on the computation of these quadrature weights $\omega_{n-\nu}$. For simplicity, let $\omega_{n-\nu}(\mathcal{V}, \psi_j)$ denote the quadrature weight of equation (27) due to the application of the single layer operator and, similarly, $\omega_{n-\nu}(\mathcal{K}, \phi_i)$, $\omega_{n-\nu}(\mathcal{K}', \psi_j)$, and $\omega_{n-\nu}(\mathcal{D}, \phi_i)$ are the corresponding quadrature weights resulting from the approximation of the application of the double layer, the adjoint double layer, and the hypersingular operator, respectively.

Inserting the approximations (26) and (27) into the first integral equation (17) yields the residual

$$\begin{aligned} R_1(\mathbf{x}, t_n) = & \sum_{\nu=0}^{n} \left(\sum_{j=1}^{M} \omega_{n-\nu}(\mathcal{V}, \psi_j) q_j - \sum_{i=1}^{N} \omega_{n-\nu}(\mathcal{K}, \phi_i) u_i \right) \\ & - \sum_{i=1}^{N} \mathcal{C} \phi_i u_i - f_{D,n}. \end{aligned} \tag{28}$$

In this expression, $f_{D,n}$ denotes the approximation f_D in equation (18) at time point t_n. Similarly, inserting the corresponding approximations into the second boundary integral equation (13) gives the residual

$$\begin{aligned} R_2(\mathbf{x}, t_n) = & \sum_{\nu=0}^{n} \left(\sum_{i=1}^{N} \omega_{n-\nu}(\mathcal{D}, \phi_i) u_i + \sum_{j=1}^{M} \omega_{n-\nu}(\mathcal{K}', \psi_j) q_j \right) \\ & - \sum_{j=1}^{M} \mathcal{C}' \psi_j q_j - f_{N,n} \end{aligned} \tag{29}$$

3.2 Collocation Method

The considered collocation method is based on the first integral equation only. The main idea is to require that the residual R_1 vanishes on a certain set of points $\{\mathbf{x}_k^*\}_{k=1}^{K}$, which are the K collocation points. In this work, the number of collocation points is chosen as $K = M$ such that there are as many collocation points as coefficients q_j and, therefore, the discretization and projection of the single layer operator yields a square matrix. In fact, the discretized and collocated first boundary integral equation now reads

$$\mathsf{V}_0\mathsf{q}_n - (\mathsf{C}+\mathsf{K}_0)\mathsf{u}_n = \mathsf{f}_{D,n} - \sum_{\nu=1}^{n}\left(\mathsf{V}_\nu\mathsf{q}_{n-\nu} - \mathsf{K}_\nu\mathsf{u}_{n-\nu}\right) \tag{30}$$

with the following matrix entries

$$\begin{aligned} \mathsf{V}_\nu[k,j] &= (\omega_\nu(\mathcal{V},\psi_j))(\mathbf{x}_k^*) & \mathsf{K}_\nu[k,i] &= (\omega_\nu(\mathcal{K},\phi_i))(\mathbf{x}_k^*) \\ \mathsf{C}[k,i] &= (\mathcal{C}\phi_i)(\mathbf{x}_k^*) & \mathsf{f}_{D,n}[k] &= f_{D,n}(\mathbf{x}_k^*) \\ \mathsf{q}_n[j] &= q_j(n\Delta t) & \mathsf{u}_n[i] &= u_i(n\Delta t). \end{aligned} \tag{31}$$

Equation (30) represents a fully discretized version of the first equation of the continuous system (18). Now, it remains to incorporate the prescribed Neumann datum as in the second equation of system (18). This is carried out by requiring this condition to be fulfilled at every time point t_n if weighted by the shape functions ϕ_i. Therefore, one obtains the system

$$\mathsf{B}\mathsf{q}_n = \mathsf{f}_{N,n}. \tag{32}$$

In this expression, the matrix entries are

$$\mathsf{B}[i,j] = \int_\Gamma \phi_i\psi_j \mathrm{d}s \quad \text{and} \quad \mathsf{f}_{N,n}[i] = \int_\Gamma \phi_i g_N(\cdot, n\Delta t)\mathrm{d}s. \tag{33}$$

Combining systems (30) and (32) yields the non-symmetric block system of equations for the vectors of coefficients q_n and u_n for the time point $t_n = n\Delta t$

$$\begin{pmatrix} \mathsf{V}_0 & -(\mathsf{C}+\mathsf{K}_0) \\ \mathsf{B} & \end{pmatrix}\begin{pmatrix} \mathsf{q}_n \\ \mathsf{u}_n \end{pmatrix} = \begin{pmatrix} \mathsf{f}_{D,n} \\ \mathsf{f}_{N,n} \end{pmatrix} - \sum_{\nu=1}^{n}\begin{pmatrix} \mathsf{V}_\nu & -\mathsf{K}_\nu \\ & \end{pmatrix}\begin{pmatrix} \mathsf{q}_{n-\nu} \\ \mathsf{u}_{n-\nu} \end{pmatrix}. \tag{34}$$

3.3 Galerkin Method

Contrary to the collocation method where the residual is demanded to vanish at specific points, in Galerkin based approaches the residual is required to be orthogonal on some set of appropriate test functions. Moreover, to obtain symmetric system matrices the test functions are chosen to be equivalent to the approximations for the unknown primary and secondary variable field, i.e., the sets of test functions $\{\phi_i\}_{i=1}^N$ and $\{\psi_j\}_{j=1}^M$ fulfill

$$\begin{aligned}\langle R_1, \psi_j\rangle_{\Gamma_D} &:= \int_{\Gamma_D} R_1(\mathbf{x}, t_n)\psi_j(\mathbf{x})\mathrm{d}s_{\mathbf{x}} = 0\\ \langle R_2, \phi_i\rangle_{\Gamma_N} &:= \int_{\Gamma_N} R_2(\mathbf{x}, t_n)\phi_i(\mathbf{x})\mathrm{d}s_{\mathbf{x}} = 0.\end{aligned} \tag{35}$$

Thus, the fully discretized system of boundary integral equations for a time $t_n = n\Delta t$ reads as

$$\begin{pmatrix} \mathsf{V}_0 & -\mathsf{K}_0 \\ \mathsf{K}_0' & \mathsf{D}_0 \end{pmatrix}\begin{pmatrix} \mathsf{q}_n \\ \mathsf{u}_n \end{pmatrix} = \begin{pmatrix} \mathsf{f}_{D,n} \\ \mathsf{f}_{N,n} \end{pmatrix} - \sum_{\nu=1}^{n}\begin{pmatrix} \mathsf{V}_\nu & -\mathsf{K}_\nu \\ \mathsf{K}_\nu' & \mathsf{D}_\nu \end{pmatrix}\begin{pmatrix} \mathsf{q}_{n-\nu} \\ \mathsf{u}_{n-\nu} \end{pmatrix} \tag{36}$$

with the following matrix entries

$$\begin{aligned}\mathsf{V}_\nu[k,j] &= \langle \omega_\nu(\mathcal{V},\psi_j),\psi_k\rangle_{\Gamma_D} & \mathsf{K}_\nu[k,i] &= \langle \omega_\nu(\mathcal{K},\phi_i),\psi_k\rangle_{\Gamma_D}\\ \mathsf{K}_\nu'[\ell,j] &= \langle \omega_\nu(\mathcal{K}',\psi_j),\phi_\ell\rangle_{\Gamma_N} & \mathsf{D}_\nu[\ell,i] &= \langle \omega_\nu(\mathcal{D},\phi_i),\phi_\ell\rangle_{\Gamma_N}\\ \mathsf{f}_{D,n}[k] &= \langle f_{D,n},\psi_k\rangle_{\Gamma_D} & \mathsf{f}_{N,n}[\ell] &= \langle f_{N,n},\phi_\ell\rangle_{\Gamma_N}\\ \mathsf{q}_n[j] &= q_j(n\Delta t) & \mathsf{u}_n[i] &= u_i(n\Delta t).\end{aligned} \tag{37}$$

Note, the left-hand side of (36) constitutes a skew-symmetric system of block matrices since $\mathsf{V}_\nu = \mathsf{V}_\nu^\top$, $\mathsf{D}_\nu = \mathsf{D}_\nu^\top$, and $\mathsf{K}_\nu' = -\mathsf{K}_\nu^\top$ hold.

Finally, for the numerical treatment of the hypersingular operator the finite part has to be either evaluated by an analytical integration or by an appropriate regularization. Here, a regularization based on the application of Stokes theorem is used (Kielhorn and Schanz, 2007). The basic ideas concerning this regularization approach may also be found in Kupradze et al. (1979) or Nedelec (1982).

3.4 Direct Solution Algorithm

The systems (34) and (36) are solved by a direct solution method. Therefore, the first block equation is solved for the coefficient vector q_n and inserted in the second block equation. This yields the Schur complement system

$$\mathsf{S}\mathsf{u}_n = \mathsf{g}_n. \tag{38}$$

In case of the non-symmetric collocation method, the system matrix in this equation is of the form

$$\mathsf{S} = \mathsf{B}\mathsf{V}_0^{-1}(\mathsf{C} + \mathsf{K}_0) \tag{39}$$

and itself non-symmetric. On the other hand, the symmetric Galerkin method gives the Schur complement

$$\mathsf{S} = \mathsf{D}_0 + \mathsf{K}_0'\mathsf{V}_0^{-1}\mathsf{K}_0 \tag{40}$$

which is obviously a symmetric matrix. The intermediate system (38) is obtained by using either a LU-decomposition in the non-symmetric case or a Cholesky decomposition in the symmetric case of the discretized single layer operator V_0. Finally, system (38) is solved by similar decompositions in order to compute the unknown coefficients u_n. Then the coefficients q_n are found using the first block equation.

4 Numerical Results

Next, numerical studies are given for the both boundary element formulations presented within this work. Moreover, to emphasize the improvements and drawbacks, results are also given for the boundary element formulation stated in Schanz (2001b). The boundary element method given there is based on the standard collocation scheme.

In the numerical examples, the approximation of u due to (25) is piecewise linear continuous on the surface triangles. In case of the standard collocation, the collocation points are located on the triangles vertices and the approximation of q is also piecewise linear continuous. Contrary, in the collocation formulation here, q can be approximated independently of u. Unfortunately, piecewise constant approximations are ruled out due to stability requirements of the matrix B. Therefore, the approximation is chosen piecewise linear discontinuous and the collocation points are located *inside* the triangles. The proposed Galerkin method uses piecewise constant shape functions for the approximation of q.

For the numerical studies two material models corresponding to the partial differential equations (1) and (2) are chosen. While the acoustic model represents air ($\kappa = 1.42 \cdot 10^5\,\mathrm{kN/m^2}$, $\rho = 1.2\,\mathrm{kg/m^3}$) the elastodynamic model reflects steel ($\lambda = 0$, $\mu = 1.06 \cdot 10^{11}\,\mathrm{kN/m^2}$, $\rho = 7850\,\mathrm{kg/m^3}$). Thereby an artificial value $\lambda = 0$ is used being equivalent to a Poisson's ratio of $\nu = 0$ just in order to compare the results against a 1-d analytical solution of longitudinal waves in an elastodynamic column (Graff, 1975).

The used geometry model represents a 3-d column of size $3\,\mathrm{m} \times 1\,\mathrm{m} \times 1\,\mathrm{m}$. For both material models a homogeneous Dirichlet datum $g_D = 0$ is prescribed at the left end of the column. The right end is excited by pressure jump of $g_N = -H(t)$ according to a unit step function $H(t)$. The remaining surfaces are of homogeneous Neumann type, i.e., $g_N = 0$ holds.

In the elastodynamic case the above given boundary conditions are of vector type. In this case, the inhomogeneous Neumann condition is meant to be a traction acting in longitudinal direction, i.e., $\mathbf{g}_N$ has to be identified with $\mathbf{g}_N = -H(t)\mathbf{e}_1$. Moreover, the remaining boundary conditions $\mathbf{g}_D$ and $\mathbf{g}_N$ are assumed to be homogeneous in all three directions.

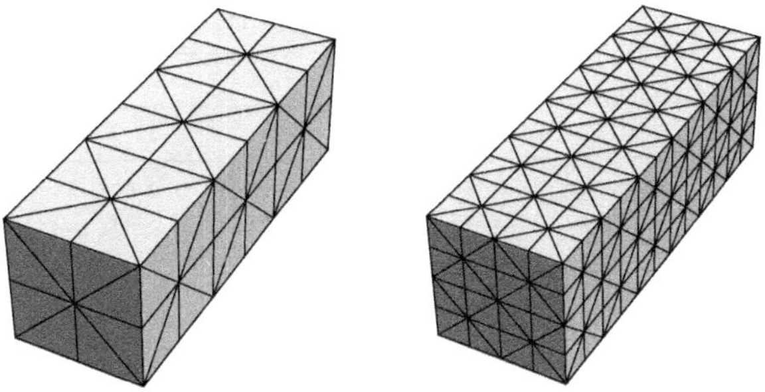

Fig. 1 Spatial discretizations of the considered column

Figure 1 shows two regular meshes for the described problems. The first one will be referred to as mesh 1 and is made up of 112 elements and 58 nodes. The second one, mesh 2, consists of 448 elements and 226 nodes. In order to compare results for different time and spatial discretizations the dimensionless value $\beta = c\Delta t/h$ is introduced. This value depends on the time step size Δt, the mesh size h, and the the wave velocity c. In elastodynamics, this velocity has to be identified with the velocity c_1 of the compression wave. The mesh size h is chosen to be the cathetus of the triangles, i.e., $h = 0.5$ m for mesh 1 and $h = 0.25$ m for mesh 2.

Table 1 shows the sizes of the system matrices for the two considered spatial discretizations. A denotes the system matrix occurring in the standard collocation approach (Schanz, 2001a) which results finally in a mixture of single- and double-layer entries. Moreover, V represents the discretized single layer potential of the first time step, and S marks the Schur complement system. Finally, the subscripts C and G indicates whether the collocation or the Galerkin scheme is under consideration. While the dimensions of the single layer potential vary between the non-symmetric and the symmetric formulation, the Schur complement system is always of the same size and, therefore, $\dim(\mathsf{S}_C) = \dim(\mathsf{S}_G) = \dim(\mathsf{S})$ holds.

The Tables 2a and 2b depict the condition numbers of the system matrices in the first time step according to different spatial and temporal discretizations as well as to the different numerical methods mentioned here. The condition numbers are given as estimates in the 1-norm and are computed using standard LAPACK routines. Further, to improve the numerical behavior of the standard collocation method dimensionless variables are introduced. This variable transformation is based on an

Table 1 Dimensions of the system matrices for different meshes

	Mesh	dim(A)	dim(V_C)	dim(V_G)	dim(S)
Elastodynamic	1	174	1008	24	147
	2	678	4032	96	603
Acoustic	1	58	336	8	49
	2	226	1344	32	201

Table 2 Condition numbers of the system matrices for the two meshes and different time steps

(a) Elastodynamic

Mesh	1	1	1	2	2	2
β	0.2	0.4	0.8	0.2	0.4	0.8
cond(V_C)	9.32	15.2	30.9	9.19	16.1	28.4
cond(S_C)	13.1	14.8	17.4	12.0	14.3	15.9
cond(V_G)	1.91	2.49	3.60	2.13	3.09	4.66
cond(S_G)	11.5	10.0	9.48	10.7	9.91	8.53
cond(A)	267	187	181	540	353	311

(b) Acoustic

Mesh	1	1	1	2	2	2
β	0.2	0.4	0.8	0.2	0.4	0.8
cond(V_C)	7.49	14.7	28.3	7.71	13.1	25.9
cond(S_C)	8.43	7.06	5.89	8.74	7.23	5.46
cond(V_G)	1.53	2.15	3.25	1.79	2.52	4.40
cond(S_G)	7.58	5.56	3.64	8.59	6.86	4.67
cond(A)	75.1	51.1	46.4	153	101	86.0

appropriate scaling of the spatial, temporal, and material parameters (Schanz and Kielhorn, 2005). All other computations are done without any modifications to the input data.

At first, it can be stated that the condition numbers which are based on the numerical approaches presented here are better than those obtained by the standard collocation scheme. A circumstance, which is attributed to the mixture of single- and double-layer entries in those collocation methods. Moreover, those condition numbers are almost doubled when the finer grid is considered. On the opposite, the conditions numbers for both the single layer potential and the Schur complement resulting from the collocation formulation presented within this work seem to be independent of the underlying spatial discretization. Nearly the same holds for the symmetric Galerkin method. Although the condition numbers are slightly increasing for the finer grid, the spatial discretization influences the condition numbers only weakly.

The condition numbers decrease with an increasing time step size in case of the standard collocation. Contrary, the condition numbers concerning the single layer potentials V_C and V_G decrease with smaller time step sizes in case of the formulations presented here. Concerning the Schur complement systems S_C and S_G the situation seems to change. Here, the condition numbers increase with smaller time step sizes in both the numerical methods and the different material models. Only the elastodynamic case using the new collocation scheme marks an exception. The condition numbers seem to decrease with a finer time discretization in this case.

Next, only elastodynamic results are reported because the acoustic results are comparable. Figure 2 depicts the displacement and traction solution with respect to all numerical methods mentioned here. Moreover, those solutions are based on mesh 1 and vary in the temporal discretizations.

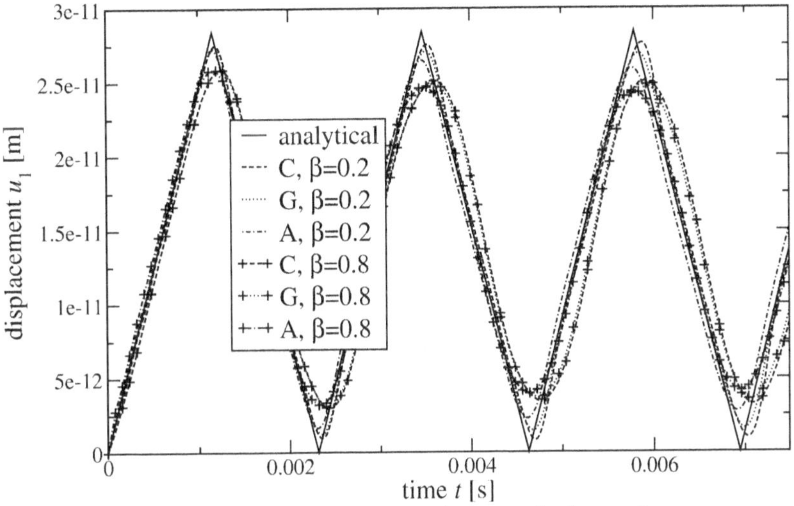

(a) Displacement solution at the free end

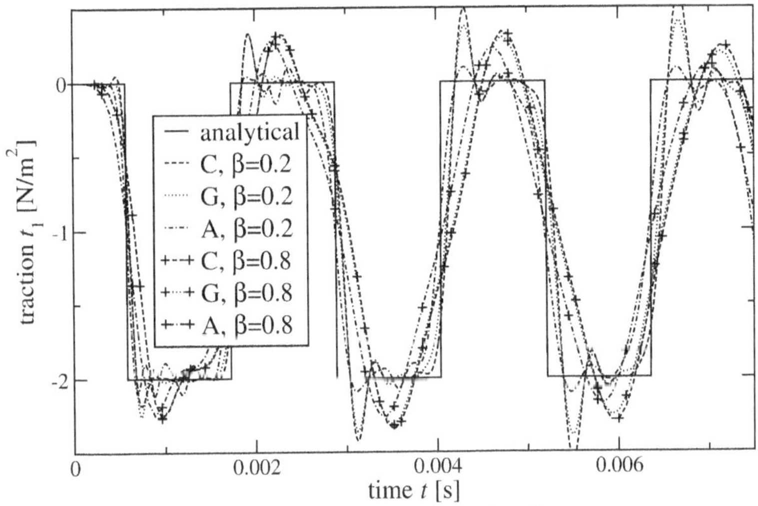

(b) Traction solution at the fixed end

Fig. 2 Elastodynamic solutions for different time step sizes (Mesh 1)

First of all, it can be stated that, in general, a finer time discretization yields better results for all considered numerical methods. Nevertheless, there are some differences between them. For instance, considering the displacement solution in

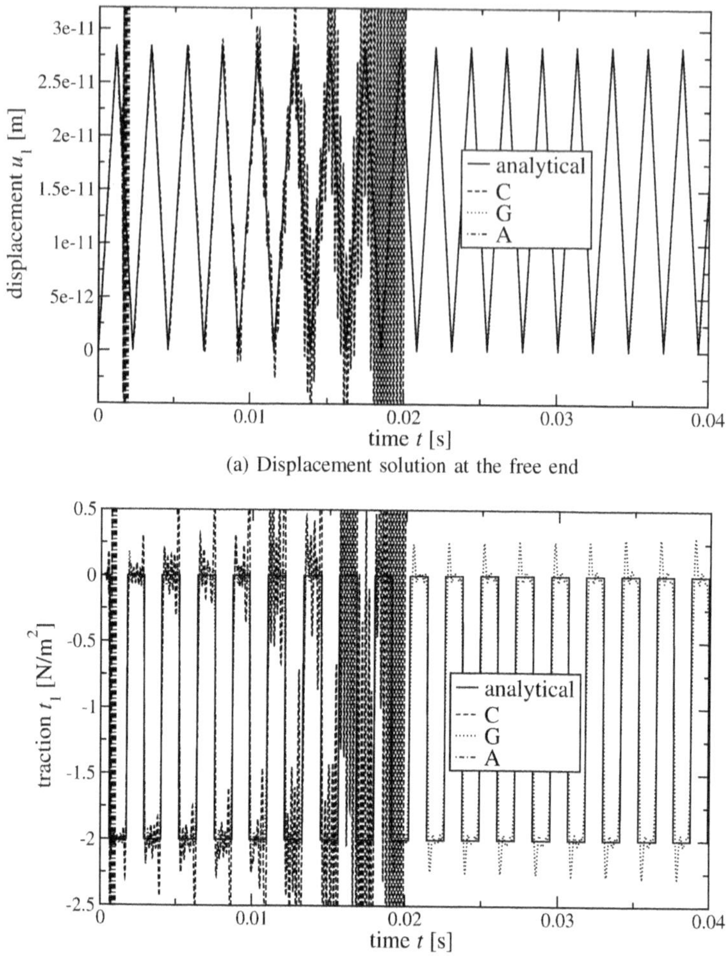

(a) Displacement solution at the free end

(b) Traction solution at the fixed end

Fig. 3 Long time behavior (Mesh 1)

Fig. 2(a) the standard collocation shows the largest numerical damping. Additionally, the phase shift for the finer time discretization is more distinctive than in the new methods. There, the damping for the Galerkin scheme is larger than those from the new collocation method.

The numerical studies close with a long time displacement and a long time traction solution for a small dimensionless value of $\beta = 0.1$. The results are depicted in Fig. 3 and, again, they base on the coarser mesh 1. There, both collocation schemes fail. While the standard collocation method fails relatively early at a time of $t \approx 0.002$ s the new collocation scheme drops out much more later at $t \approx 0.015$ s. Contrary, the symmetric Galerkin scheme is stable throughout the whole observation time.

5 Conclusions

To summarize the results, it can be concluded that due to the well structured system matrices the numerical methods presented here yield better stability properties than the standard collocation scheme does. Nevertheless, those both methods have their drawbacks. For instance, the new collocation scheme results in quite large system matrices while the symmetric Galerkin scheme requires the evaluation of the hypersingular integral kernel and involves more costly double integrations. But in the future, these additional computational costs should be reduced by, e.g., the application of some fast methods like adaptive cross approximation (Bebendorf and Rjasanow, 2003). Fast methods are known to be much more efficient if well structured systems are considered. Therefore, the new Collocation method as well as the symmetric Galerkin scheme exhibit good optimization capabilities.

References

S. Ahmad and G.D. Manolis. Dynamic analysis of 3-D structures by a transformed boundary element method. Comput. Mech., 2:185–196, 1987.

H. Antes. A boundary element procedure for transient wave propagations in two-dimensional isotropic elastic media. Finite Elem. Anal. Des., 1:313–322, 1985.

H. Antes and M. Jäger. On stability and efficiency of 3d acoustic BE procedures for moving noise sources. In S.N. Atluri, G. Yagawa, and T.A. Cruse, editors, Computational Mechanics, Theory and Applications, vol. 2, 3056–3061, Heidelberg, 1995. Springer-Verlag.

M. Bebendorf and S. Rjasanow. Adaptive low-rank approximation of collocation matrices. Computing, 70:1–24, 2003.

D.E. Beskos. Boundary element methods in dynamic analysis. AMR, 40(1):1–23, 1987.

D.E. Beskos. Boundary element methods in dynamic analysis: Part II (1986–1996). AMR, 50(3): 149–197, 1997.

M. Bonnet, G. Maier, and C. Polizzotto. Symmetric galerkin boundary element methods. AMR, 51(11):669–704, 1998.

T.A. Cruse and F.J. Rizzo. A direct formulation and numerical solution of the general transient elastodynamic problem, I. Aust. J. Math. Anal. Appl., 22(1):244–259, 1968.

J. Domínguez. Dynamic stiffness of rectangular foundations. Report no. R78-20, Department of Civil Engineering, MIT, Cambridge MA, 1978.

J. Domínguez. Boundary Elements in Dynamics. Computational Mechanics Publication, Southampton, 1993.

A. Frangi and G. Novati. Regularized symmetric Galerkin BIE formulations in the Laplace transform domain for 2D problems. Comput. Mech., 22:50–60, 1998.

K.F. Graff. Wave Motion in Elastic Solids. Oxford University Press, 1975.

E. Kausel. Fundamental Solutions in Elastodynamics. Cambridge University Press, 2006.

L. Kielhorn and M. Schanz. Convolution Quadrature Method based symmetric Galerkin Boundary Element Method for 3-d elastodynamics. Int. J. Numer. Methods. Engrg., 76(11):1724–1746, 2008.

V.D. Kupradze, T.G. Gegelia, M.O. Basheleishvili, and T.V. Burchuladze. Three-Dimensional Problems of the Mathematical Theory of Elasticity and Thermoelasticity, vol 25 of Applied Mathematics and Mechanics. North-Holland, Amsterdam, New York, Oxford, 1979.

C. Lubich. Convolution quadrature and discretized operational calculus. I/II. Numer. Math., 52: 129–145/413–425, 1988.

G.D. Manolis. A comparative study on three boundary element method approaches to problems in elastodynamics. Int. J. Numer. Methods. Engrg., 19:73–91, 1983.

W.J. Mansur. A Time-Stepping Technique to Solve Wave Propagation Problems Using the Boundary Element Method. Phd thesis, University of Southampton, 1983.

W.J. Mansur, J.A.M. Carrer, and E.F.N. Siqueira. Time discontinuous linear traction approximation in time-domain BEM scalar wave propagation. Int. J. Numer. Methods. Engrg., 42(4):667–683, 1998.

G.V. Narayanan and D.E. Beskos. Numerical operational methods for time-dependent linear problems. Int. J. Numer. Methods. Engrg., 18:1829–1854, 1982.

D. Nardini and C.A. Brebbia. A new approach to free vibration analysis using boundary elements. In C.A. Brebbia, editor, Boundary Element Methods, 312–326. Springer-Verlag, Berlin, 1982.

J.C. Nedelec. Integral equations with non integrable kernels. Integr. Equ. Oper. Theory, 5:563–672, 1982.

P.W. Partridge, C.A. Brebbia, and L.C. Wrobel. The Dual Reciprocity Boundary Element Method. Computational Mechanics Publication, Southampton, 1992.

A. Peirce and E. Siebrits. Stability analysis and design of time-stepping schemes for general elastodynamic boundary element models. Int. J. Numer. Methods. Engrg., 40(2):319–342, 1997.

D.C. Rizos and D.L. Karabalis. An advanced direct time domain BEM formulation for general 3-D elastodynamic problems. Comput. Mech., 15:249–269, 1994.

M. Schanz. Application of 3-d Boundary Element formulation to wave propagation in poroelastic solids. Eng. Anal. Bound. Elem., 25(4–5):363–376, 2001a.

M. Schanz. Wave Propagation in Viscoelastic and Poroelastic Continua: A Boundary Element Approach, vol 2 of Lecture Notes in Applied Mechanics. Springer-Verlag, Berlin, Heidelberg, New York, 2001b.

M. Schanz and H. Antes. Application of 'operational quadrature methods' in time domain boundary element methods. Meccanica, 32(3):179–186, 1997.

M. Schanz and L. Kielhorn. Dimensionless variables in a poroelastodynamic time domain boundary element formulation. Build. Res. J., 53(2–3):175–189, 2005.

O. Steinbach. Numerical Approximation Methods for Elliptic Boundary Value Problems, vol 54 of Texts in Applied Mathematics. Springer, 2008.

L.T. Wheeler and E. Sternberg. Some theorems in classical elastodynamics. Arch. Rational Mech. Anal., 31:51–90, 1968.

G. Yu, W.J. Mansur, J.A.M. Carrer, and L. Gong. Time weighting in time domain BEM. Eng. Anal. Bound. Elem., 22(3):175–181, 1998.

G. Yu, W.J. Mansur, J.A.M. Carrer, and L. Gong. Stability of Galerkin and Collocation time domain boundary element methods as applied to the scalar wave equation. Comput. Struct., 74 (4):495–506, 2000.

Some Investigations of Fast Multipole BEM in Solid Mechanics

Zhenhan Yao

Abstract Combined with the fast multipole method, the boundary element method become quite efficient, to deal with large-scale engineering and scientific problems. In this paper, some applications of the FMBEM on 2D and 3D simulation of composite materials on a PC are presented at first. A parallel algorithm of FMBEM for PC cluster is briefly introduced. And then some 3D large-scale simulation of complex fiber- and carbon nanotube-reinforced composites on PC cluster are presented. On the other hand, FMBEM are investigated to simulate 2D elastic solid containing large number of cracks and fatigue crack growth. A new approach of FMBEM for elasto-plasticity problems is also presented.

1 Introduction

Boundary element method has been developed as an efficient numerical method followed the finite element method in recent several decades, and BEM become the most important complement of the FEM in the field of solid mechanics. BEM has attractive advantages of high accuracy, dimension reduction, and it is especially suitable to deal with the problems related to the infinite or semi-infinite domain, and the problems related to singularity or high gradients. But the conventional BEM is not capable to deal with practical complex engineering problems, because the matrix of the resulted algebraic equation is dense and asymmetric, the operations increase in $O(N^3)$ for Gaussian elimination, or $O(N^2)$ for iterative solver and the memory requirement increases in $O(N^2)$, N is the number of unknowns.

In 1980s, Rokhlin and Greengard presented a multipole method with $O(N)$ operations and memory requirement, to solve the 2D potential problem and N-body simulations (Rokhlin, 1985; Greengard and Rokhlin, 1987), which has been named as one of the Top 10 algorithms in the last century by SIAM (Cipra, 2000). Later, Greengard and Rokhlin (1997) introduced the exponential expansion in multipole

Z. Yao (✉)
Department of Engineering Mechanics, Tsinghua University, Beijing, 100084, China
e-mail: demyzh@tsinghua.edu.cn

A chapter in honor of Dimitri Beskos' 65th birthday

G.D. Manolis, D. Polyzos (eds.), *Recent Advances in Boundary Element Methods*,
DOI 10.1007/978-1-4020-9710-2_28,

to local translation, which leads to further enhancement of the efficiency. Combined with the fast multipole method (FMM), the BEM become quite efficient to deal with large-scale engineering and scientific problems. Nishimura (2002) reviewed the fast multipole accelerated boundary integral equation methods.

In the field of solid mechanics, the algorithms for elasticity problems in terms of Taylor series expansion of the fundamental solution are reported in literature (Peirce and Napier, 1995). Several algorithms of new version FMM for the direct BEM of 3D elasticity problems based upon Taylor series expansion and spherical harmonic expansion are also available (Yoshida et al., 2001). In recent years, FMBEM attracted more and more researchers.

In authors' group the investigations on FMBEM started in 2000, several papers have been published in recent years: Wang and Yao (2004, 2005) and Wang et al. (2005a) investigated the FMBEM for 2D and 3D multi-domain elasticity problems, and applied to the simulation of composite material; Lei et al. (2006) presented the parallel FMBEM and applied to large-scale simulation of 3D fiber-reinforced composites; Wang et al. (2005b) presented the FMBEM for simulation of 2D solids containing large number of cracks; Wang (2006) presented the FM-dual-BEM for the analysis of 2D solids containing numerous cracks; Wang and Yao (2006) investigated the 2D FMDBEM analysis of fatigue crack growth; Wang and Yao (2007) suggested a new FMBEM for the analysis of 2D elasto-plasticity problems; and Yao et al. (2007) further applied the FMBEM to the simulation of some elastic, thermal and electrical properties of carbon nanotube (CNT) reinforced composites.

In this paper, the above-mentioned investigations of FMBEM in authors' group are briefly presented.

2 Application of FMBEM on Simulation of Composite Materials

The FMBEM has been applied in the 2D and 3D simulation of composite material and the obtained equivalent material properties have shown good agreement with the available results in micromechanics.

2.1 2D Simulation of Long-Fiber Reinforced Composite Material

Long-fiber reinforced composite material can be simulated using 2D model of plane strain. Figure 1 shows two models of square containing 1,600 randomly distributed circular inclusions discretized into 544,000 DOF. Figure 2 shows the results of equivalent volume modulus obtained using FMBEM in comparison with Mori-Tanaka method.

Both results of FMBEM and Mori-Tanaka method show good agreement. In these cases, the models containing approximately 100 inclusions are large enough.

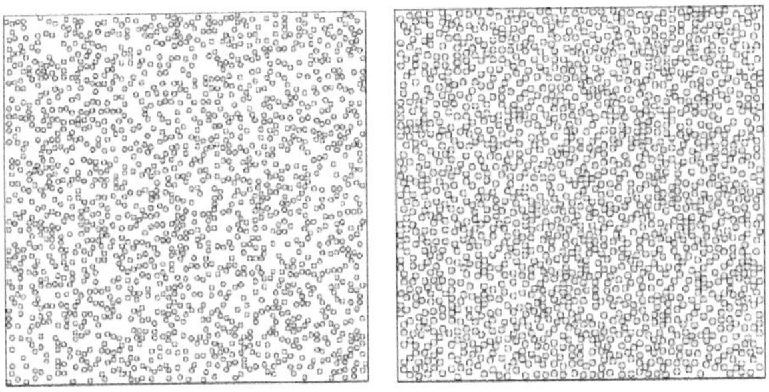

Fig. 1 Models of square containing 1,600 circular inclusions, volume fraction = 0.2 (*left*), 0.4 (*right*)

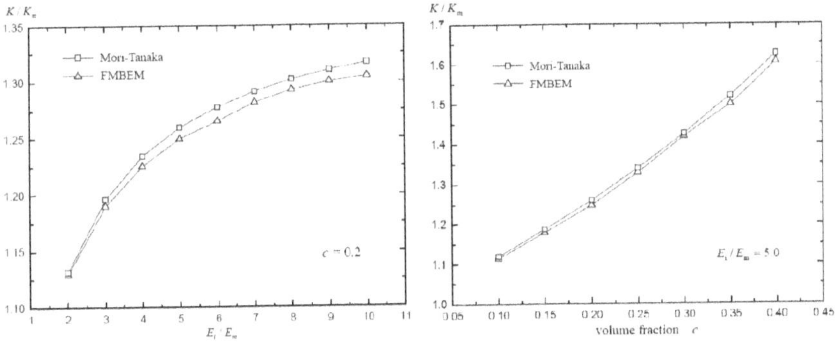

Fig. 2 The comparison of equivalent volume modulus between FMBEM and Mori-Tanaka method

2.2 3D Simulation of Particle Reinforced Composite Material

Figure 3 shows two models of cube containing 100 and 300 randomly distributed identical spherical inclusions, which are discretized into 187,000 and 372,600 DOF respectively.

Figure 4 shows the results of equivalent volume and shear moduli obtained using FMBEM in comparison with those obtained using Mori-Tanaka method. Furthermore, the computation using FMBEM can obtain the whole fields of displacement and stresses.

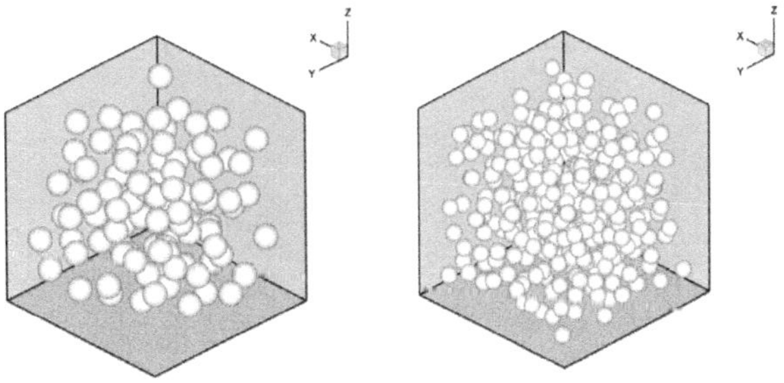

Fig. 3 Models of cube containing 100 (*left*) and 300 (*right*) spherical inclusions

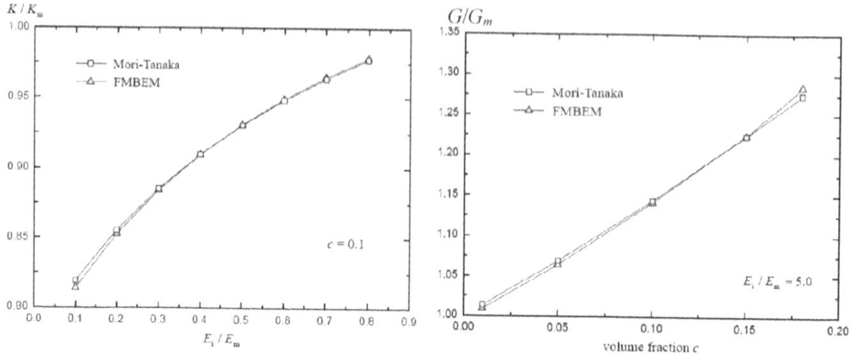

Fig. 4 The comparison of equivalent modulus between FMBEM and Mori-Tanaka method

2.3 3D Simulation of Short-Fiber Reinforced Composite Material

Figure 5 shows two models of cube containing 100 short fibers, the volume fractions of the fiber are 0.03 and 0.1 respectively, the scale of these models is 172,800 DOF.

Figure 6 shows the results of equivalent volume modulus obtained using FMBEM in comparison with the H-S bounds. Most numerical values are within the range of H-S bounds, while a few points are slightly lower than the H-S lower bound. This difference comes from the numerical errors of BEM, i.e., discretization error.

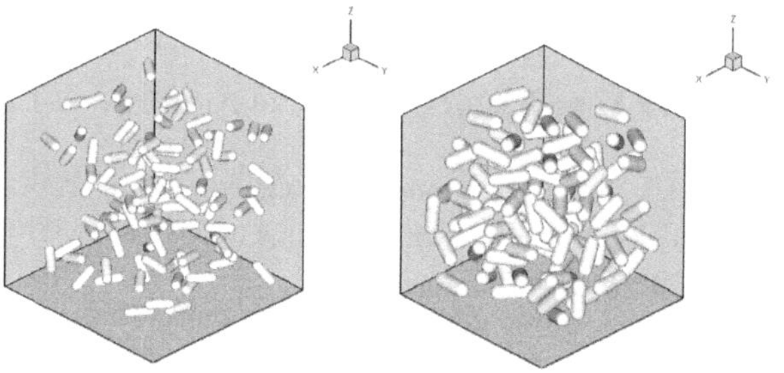

Fig. 5 Models of cube containing 100 short fibers with volume fraction $c = 0.03$ (*left*) and $c = 0.1$ (*right*)

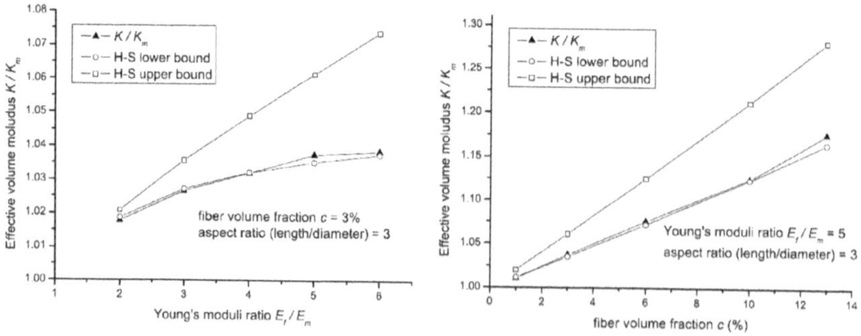

Fig. 6 The comparison of equivalent volume modulus obtained using FMBEM and the theoretical H-S bounds

Numerical examples show that as the boundary element mesh is further refined, these few points increase to approach the H-S lower bound. This means that the numerical solution obtained by fast multipole BEM does not violate the theoretical upper and lower bounds.

As the aspect ratio of fibers increases, larger-scale computation will be required, and parallel computation is necessary.

3 Large-Scale Parallel Computation of FMBEM

The FMBEM is more difficult to parallelize efficiently than the conventional BEM because of two extra key problems. Therefore some special approach has to be adopted.

3.1 Task Decomposition by Boxes

Instead of domain decomposition method (DDM) commonly used in conventional parallel BEM, box decomposition is adopted in the parallel algorithm of FMBEM. The domain is decomposed while splitting the tree by the unit of boxes. The boxes are first sorted in a 2D or 3D space order by a sorting algorithm presented by Warren and Salmon (1993). Then each task is assigned with maximal approximately even number of elements by a parameter called *dec_level*, which stands for the level the tree is decomposed at. The higher the *dec_level* is, the more balanced the decomposition will be. But it cannot be higher than the minimum level of leaves in the tree. Figure 7 depicts decomposing a model into six tasks at *dec_level=3*. The main advantage of this partition scheme is that the tasks can be decomposed regardless of different phases such as the inclusion and matrix for particle-reinforced composite.

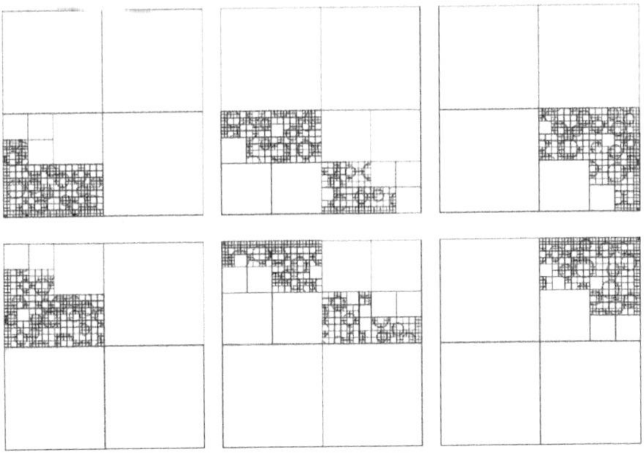

Fig. 7 Task decomposition boxes

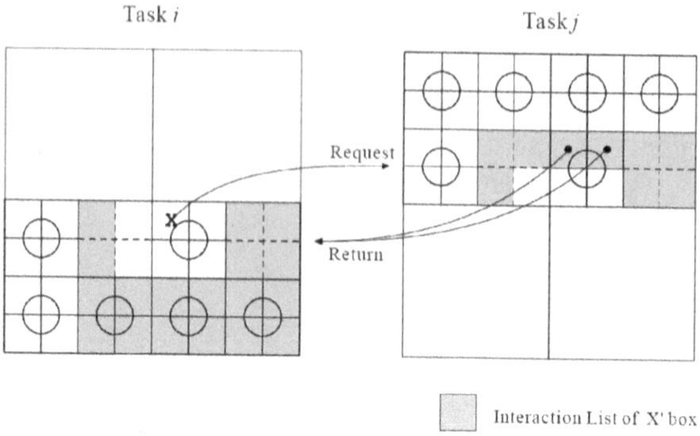

Fig. 8 Remote checks in parallel tree traversing (empty boxes are not returned)

3.2 Parallelization of the Tree Traversing

The most time-consuming step in fast multipole method is the downward stage. It is also the most complicated step in parallel formulations since interactions from interaction list and neighbor boxes are needed. And these boxes may be in the same or other tasks, and may be empty anywhere. So a remote-check procedure is used, as a simple example shown in Fig. 8 for the case of interaction list only, where box marked with X in task i need the remote boxes in task j. In that case task i sends a checking request to j and a checking result of non-empty boxes are returned. The overhead of this procedure can be neglected since it costs only a few seconds in a model with tens of thousands unknowns and also the checking result does not change from iteration to iteration for elastostatics.

3.3 Application on the Simulation of Fiber Reinforced Composites

Two fiber shapes, bone-shaped short fiber (as shown in Fig. 9) and conventional straight short fiber, are simulated and compared. Figure 10 shows a typical configuration of RVE model of well-aligned bone-shaped short-fiber reinforced composites. There are 200 fibers in the RVE, the volume fraction is 0.05. It is discretized into 2,596,800 DOF, and the computing time is approximately 38 hours. Figure 11 shows the comparison of effective tensile modulus for two types of fiber shapes, and a normalized histogram with fitted Weibull probability density functions.

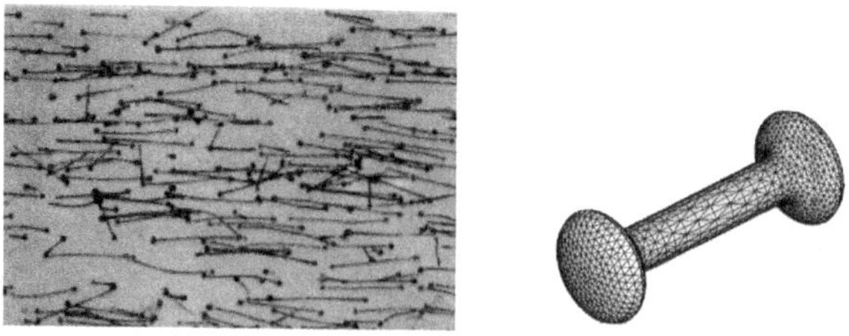

Fig. 9 Well-aligned bone-shaped Ni-fiber reinforced polyester matrix fiber composites and a fiber model

Fig. 10 The BEM model of a RVE of matrix containing 200 randomly dispersed bone-shaped short fibers

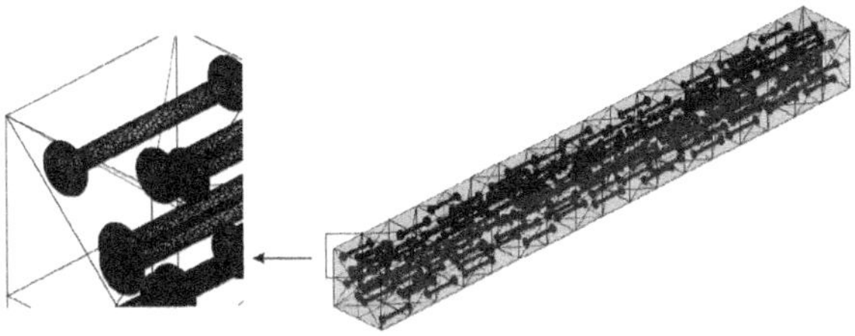

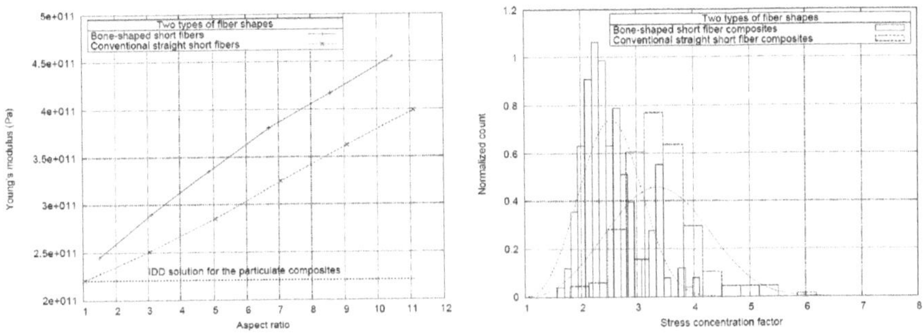

Fig. 11 Comparison of effective tensile modulus for two types of fiber shapes, and a normalized histogram of stress concentration factor with fitted Weibull probability density functions

4 Simulation of CNT Composites Using FMBEM

Based on the previous investigation on large scale simulation of particle- and fiber-reinforced composite materials using FMBEM, the CNTs are treated as effective elastic fibers for the simulation of elastic property. To enhance the efficiency, the repeated identical sub-domain approach is applied, and for large scale simulation the parallel FM-BEM. Some typical numerical examples are given to verify the applicability of the presented approach.

The effect of the CNT volume fractions on the effective elastic modulus of the CNT composites for aligned oriented case has been investigated. The results obtained have been compared with the numerical results in Liu, Nishimura and Otani (2005), and other reported data in Odegard, Gates and Wise et al. (2003), which are obtained based on MD combined with an equivalent-continuum model.

The length of the CNT fibers is fixed at 50 nm, and its radius is 0.7 nm. The volume of the CNT is considered to be 37% of the effective fiber volume. For the matrix material, the NASA LaRC-SI polymer (as adopted in above mentioned two papers) is used, with a Young's modulus of 3.8 GPa and Poisson's ratio of 0.4. For CNT, the Poisson's ratio is 0.4, and the Young's modulus is 380 GPa, which is 100 times of the matrix modulus. For saving the computational resource, up to 200 CNTs are considered. The global model of this example is shown in Fig. 12.

The estimated effective longitudinal Young's modulus of the CNT composites versus the CNT volume fractions are shown in Fig. 13 (left) in comparison with the data in literature. The satisfied agreement could be observed.

Fig. 12 A RVE containing 200 aligned CNTs

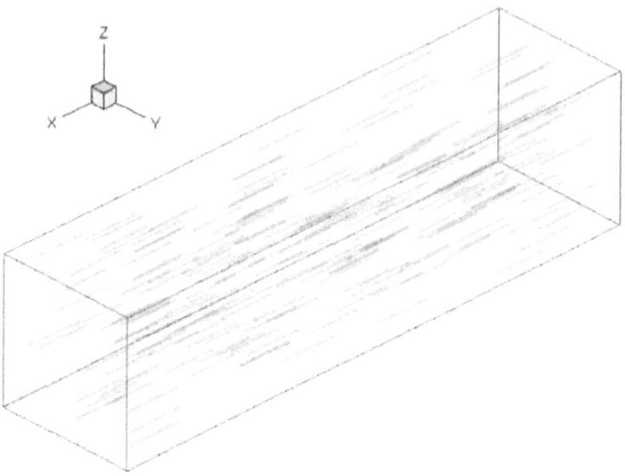

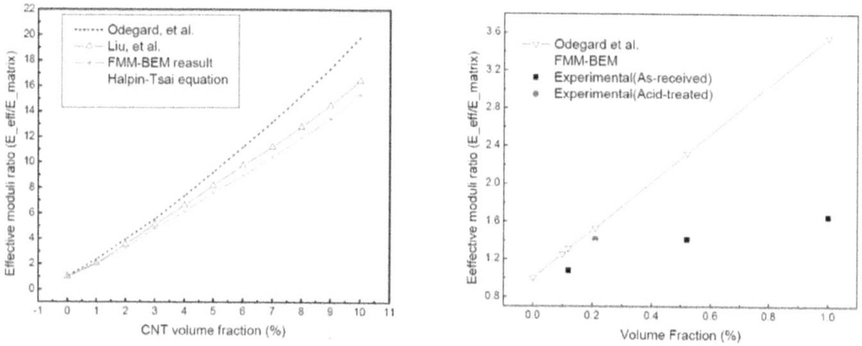

Fig. 13 Young's modulus of CNT reinforced LaRC-CP2 composite vs. CNT volume fraction

The comparison with corresponding experimental data for the composites prepared with both the as-received and acid-treated CNTs is shown in Fig. 13 (right). The properties of the composites with randomly oriented CNTs have been calculated for the aspect ratio of 60, using FM-BEM.

Although good agreement between the numerical results and the experimental data for acid-treated CNT composites has been observed, the predicted values of modulus are larger than the measured values for the as-received material, especially for values of CNT volume fraction greater than about 0.5%. The difference between the experiments and the model is most likely caused by the following factors: First, the model assumes that the effective fibers are perfectly dispersed in the polymer matrix, but a significant amount of CNTs may remain in bundles in the composite material. Second, the ideal interfacial condition is considered in the model, the stiffness may be decreased by the transfer layer between CNT and matrix. In addition, the CNT is considered as isotropic elastic fiber, but actually CNT is more close to being transversely isotropic material. Furthermore, the long CNTs will not all remain straight in the matrix. The closer agreement between the model and the acid-treated values indicates that the CNT are more dispersed in the acid-treated material than in the as-received material.

5 Simulation of Crack and Crack Growth Using FMBEM

In authors' group the FMBEM was used to solve the traction boundary integral equation for 2D crack analysis at first. And then the FMM based on complex Taylor series expansions is applied to the dual boundary element method (DBEM) for large-scale crack analysis in linear elastic fracture mechanics.

5.1 Verification of the Accuracy and Efficiency of the FMDBEM

Because the conventional boundary elements are inadequate to meet the square root displacement variation near the crack tip, a special crack tip element is employed. It is a 3-node discontinuous element, and its shape functions are shown in Fig. 14.

To verify the accuracy of the FMDBEM, a square sheet with center crack, as shown in Fig. 15, was examined, where $2w = 10\,\mathrm{mm}$, $2a$ varies form 1 to 5 mm, the crack was discretized into 16 discontinuous quadratic elements and the order of multipole and local expansion was taken as $p = 30$. Table 1 shows the normalized SIFs of the center crack with different crack sizes. Compared with the solutions by Isida (1971), the maximum error in the K_{I}^{*} values is only 0.37%. In addition, the K_{II}^{*} values are almost equal to the theoretical solution 0.

Figure 16 shows the opening displacement versus distance from the left crack tip. Figure 17 shows the CPU time versus the number of DOF. All the results have shown high accuracy and efficiency of the FMDBEM.

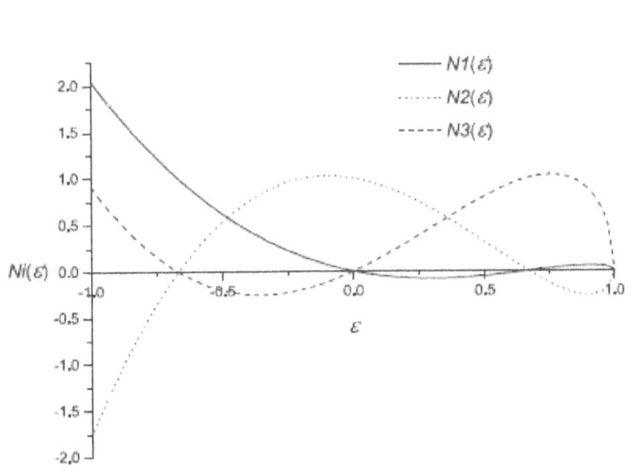

Fig. 14 Shape functions of the special crack tip element

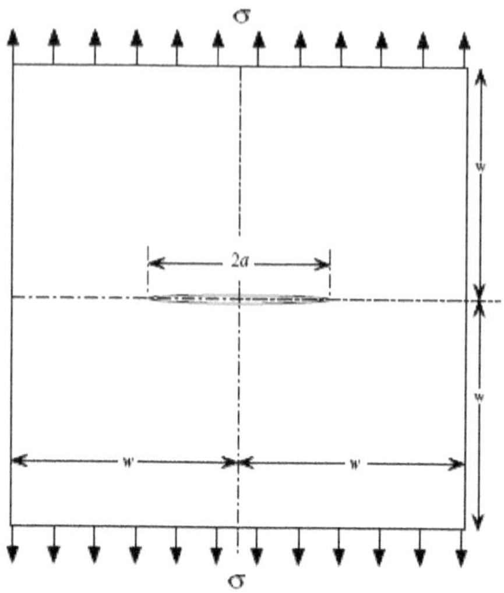

Fig. 15 A center crack in a square sheet

Table 1 Normalized SIFs for a center crack with different crack sizes

a/w	0.1	0.3	0.5
K_{I}^{*} (left tip)	1.0150	1.1248	1.3308
K_{I}^{*} (right tip)	1.0150	1.1248	1.3306
K_{I}^{*} (Isida)	1.014	1.123	1.334
K_{II}^{*} (left tip)	-6.525×10^{-8}	-2.245×10^{-7}	-1.991×10^{-7}
K_{II}^{*} (right tip)	6.419×10^{-8}	2.233×10^{-7}	2.214×10^{-7}

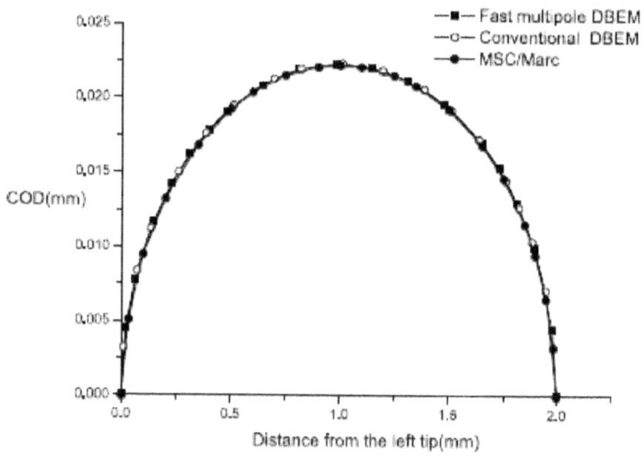

Fig. 16 Opening displacement versus distance from the left crack tip

5.2 Fatigue Crack Growth in 2D Elastic Solid Containing Numerous Cracks

An incremental crack-extension analysis is applied to simulate the fatigue crack growth. It assumes a piece-wise linear discretization of the crack-growth path. In each incremental step, the fast multipole DBEM is applied to solve the equation system. The directions and extensions of crack growth are determined in the

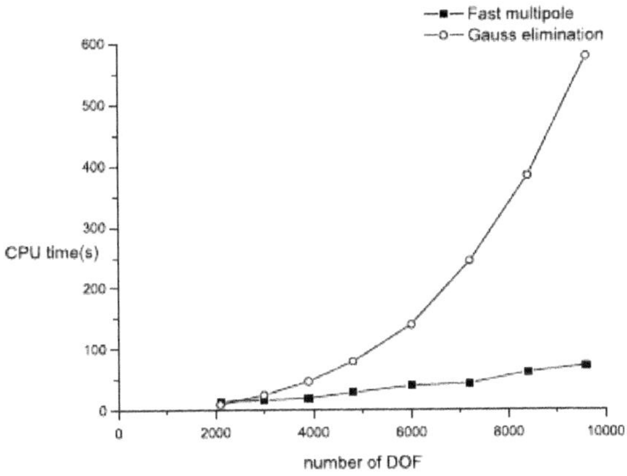

Fig. 17 CPU time versus the number of DOF

post-proceeding process. The details, including special crack tip elements, evaluation of SIF, determination of the direction and extension of fatigue crack growth and the simulation of crack coalescence, can be found in Wang and Yao (2006).

To simulate the fatigue growth of a crack under mixed-mode deformation, a cruciform plate with an initial edge crack located at one of the interior corners has been computed. Figure 18 shows a cruciform plate with an initial edge crack (left), the crack growth paths obtained by the FM DBEM analysis (center), and that presented by Portela et al. (1993) (right).

Figure 19 shows a rectangular elastic sheet containing 400 irregularly distributed cracks, and Fig. 20 shows the corresponding fatigue growth paths under horizontal tension.

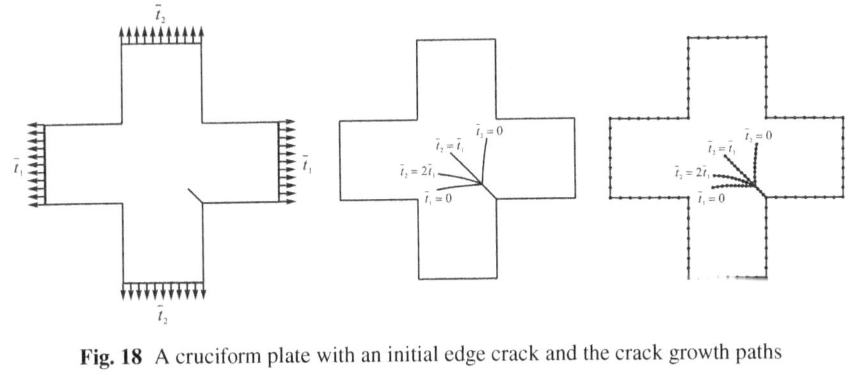

Fig. 18 A cruciform plate with an initial edge crack and the crack growth paths

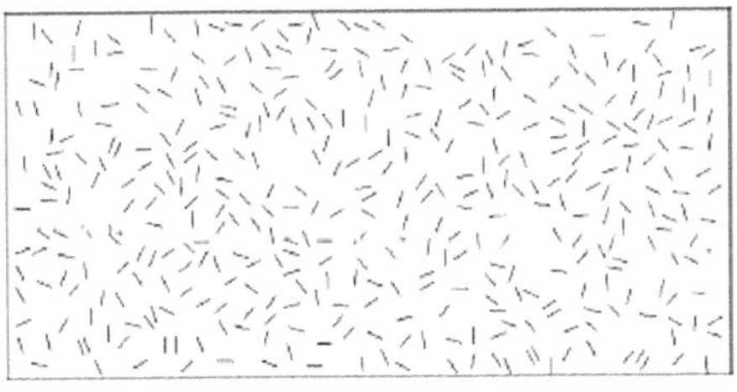

Fig. 19 400 irregularly distributed cracks in a rectangular sheet

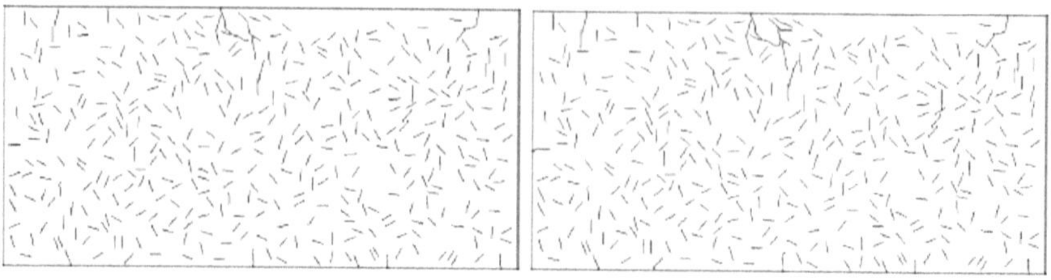

Fig. 20 Fatigue growth paths of multiple cracks after 14.85×10^8 load cycles (*left*) and 17.40×10^8 load cycles (*right*)

5.3 Computation of Effective Modulus of 2D Microcracked Solid

The FMDBEM is adopted to directly determine the effective in-plane bulk modulus of a square sheet containing 4,000 randomly oriented microcracks, as shown in Fig. 21. Figure 22 shows the numerical results of the effective in-plane bulk

Fig. 21 A sheet containing 4,000 microcracks

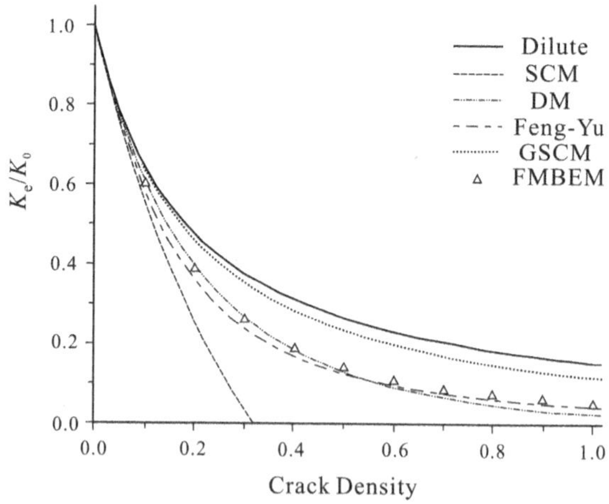

Fig. 22 Effective in-plane bulk modulus versus crack density

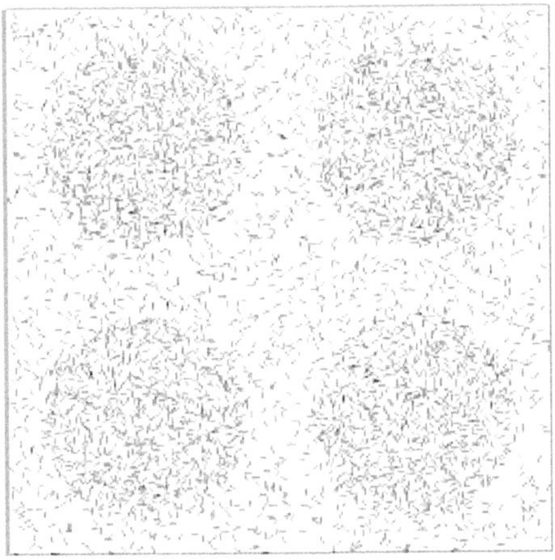

Fig. 23 A sheet containing 4 local regions

modulus by the FMDBEM in comparison with the corresponding solutions of various micromechanics methods, with the Poisson's ratio $\nu = 0.3$. The numerical results agree well with the estimation of GSCM, DM and Feng-Yu (Feng and Yu, 2000).

Furthermore, the effect of crack non-uniform distribution on effective in-plane bulk modulus is also investigated using the FMDBEM. This work assumes that the microcracked solid contains some local regions having a crack density ω_L higher or lower than the average crack density ω_0 and analyze the variation of the effective in-plane modulus with ω_L when ω_0 is fixed. Figure 23 shows a square sheet containing 4 higher crack density region, and Fig. 24 shows the results of K_e/K_0 versus ω_L/ω_0 for the case of $\omega_0 = 0.3$. The results show that the non-uniform distribution of microcracks increases the effective in-plane bulk modulus of the whole microcracked solid.

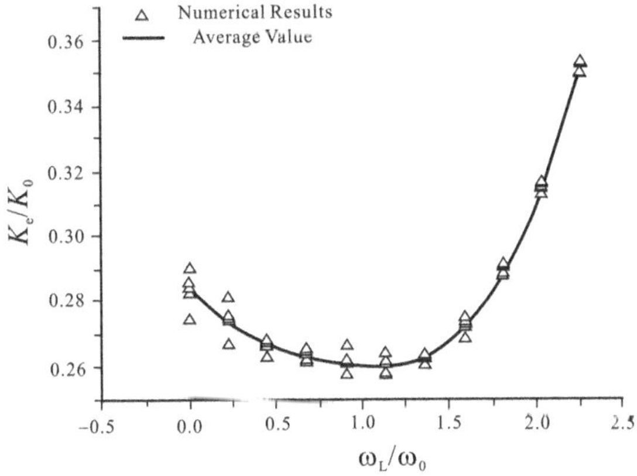

Fig. 24 K_e/K_0 versus ω_L/ω_0 ($\omega_0 = 0.3$)

6 FMBEM for 2D Elasto-Plasticity Problems

In authors' group a scheme of FMBEM for 2D elastoplasticity problems is proposed recently. Both of the boundary integrals and domain integrals are calculated by recursive operations on a quad-tree structure without explicitly forming the coefficient matrices. Some details can be found in Wang and Yao (2007). In the computation the integral equations in Telles (1983) are applied.

6.1 Implementation of the Fast Multipole Method

The fast multipole BEM uses the same discretization as the conventional BEM. The boundaries of the model are discretized using boundary elements and the internal domain where local yielding is expected to occur are discretized using internal cells. Then an adaptive quad-tree structure is constructed (Nishimura, 2002). The boundary elements and domain cells are both allocated into the leaves of the tree. An example is shown in Fig. 25, where each leaf contains at most six boundary elements or internal cells. Both of the boundary integrals and domain integrals are calculated by recursive operations on the quad-tree structure without explicitly forming the coefficient matrix. Combining multipole expansions with local expansions (Greengard and Rokhlin, 1997), computational complexity and memory requirement of the matrix-vector multiplication are both reduced to $O(N)$, where N is the number of DOFs.

In order to solve the system equations of elastoplastic problems, an incremental iterative algorithm is employed, which follows the loading history accurately. In

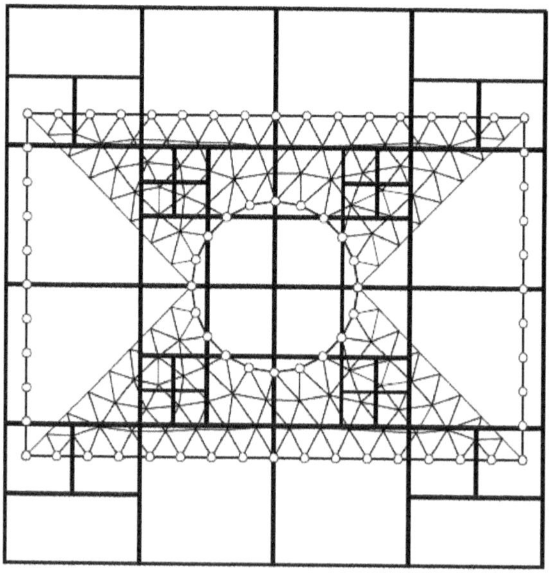

Fig. 25 A quad-tree

each loading step, an iterative procedure is executed to determine the plastic strains at the interior nodes where the material is possibly yielded. In each iteration step to obtain the new plastic strains, the equation system is solved by the generalized minimum residual method (GMRES), which uses the sparse approximate inverse type as the preconditioner.

6.2 Numerical Example

This example considered a square plate with 16×16 periodically distributed circular holes as shown in Fig. 26. The outer boundary of the plate is subjected to uniform normal displacement $\bar{u}_n = 0.04$ mm under plane strain conditions. Ideal plasticity is assumed with the following material properties:

Young's modulus $E = 42,000$ MPa, Yield stress $\sigma_0 = 105$ MPa, Poisson ratio $\nu = 0.33$.

The boundaries and expected yielding areas were discretized by quadratic elements and constant cells, respectively. The whole model has 107,520 boundary DOFs and 645,120 internal DOFs.

In the fast multipole BEM analysis, the order of the finite series was set as 25; the loading process was divided into 6 increments, and 10 iterations were employed in each increment. The analysis was done on a desktop PC and the total CPU time was 9.77 hours. Figure 27(left) shows the distribution of the equivalent von Mises stress in a unit square, where only the expected yielding area is plotted. To verify the BEM results, a unit square was isolated and analyzed using a commercial FEM software MSC/Marc, and the corresponding results are shown in Fig. 27(right).

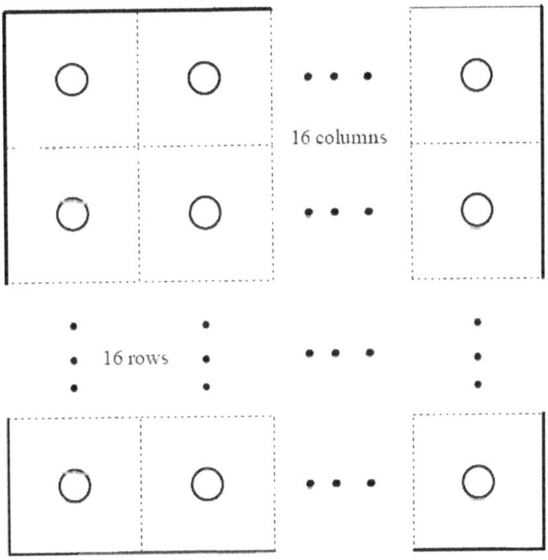

Fig. 26 A plate with 16×16 holes

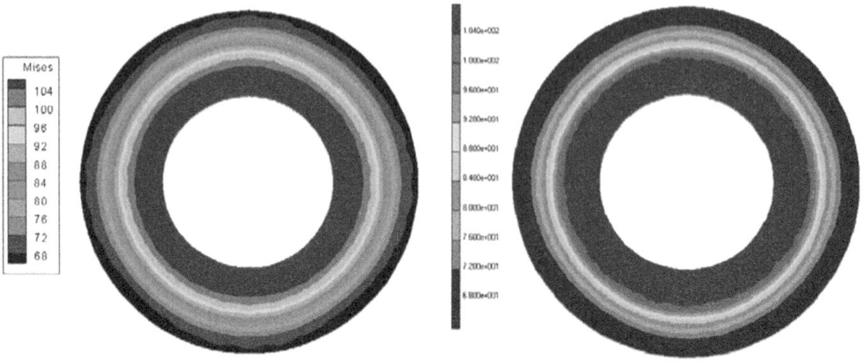

Fig. 27 Distribution of von Mises stress obtained using FMBEM (*left*) and MSC/MARC (*right*)

7 Concluding Remarks

Based on the progress of FMBEM, the authors' group carried out a series of investigation on the applications of FMBEM in solid mechanics. The investigations on large-scale FMBEM analysis in solid mechanics, including 2D and 3D elasticity and 2D fracture problems, have shown its attractive advantages, high accuracy and efficiency. A new approach of FMBEM for nonlinear problems, namely, the elasto-plasticity problems is also suggested. Combining with FMM, the boundary element method become suitable to deal with large-scale practical engineering and scientific problems.

The first author has been involved in the research on boundary element methods since 1979. The BEM is regarded as an important complement of the widely-applied FEM, but if the BEM is only capable to obtain the same results as obtained by FEM, such complement was not necessary. For the complement it is important to do something, which FEM could not do, or do something significantly better than FEM. The development of FMBEM have shown good prospects at this aspect. FMBEM have been successfully applied in the field of MEMS design and electro-magnetic field analysis. In the field of solid mechanics, the most important thing is to develop practical applications of FMBEM.

Acknowledgments Financial support for the projects from the National Natural Science Foundation of China, under grant No. 10172053, 10472051 is gratefully acknowledged.

References

Cipra BA (2000) The best of the 20th century: editors name top 10 algorithms. SIAM News 33(4).
Feng XQ, Yu SW (2000) Estimate of effective elastic moduli with microcrack interaction effects. Theor. Appl. Fract. Mech. 34: 225–233.
Greengard L, Rokhlin V (1987) A fast algorithm for particle simulations. J. Comput. Phys. 73: 325–348.

Greengard L, Rokhlin V (1997) A new version of the fast multipole method for the Laplace equation in three dimensions. Acta Numerica 6: 229–269.
Isida M (1971) Effects of width and length on stress intensity factor for the tension of internal cracked plates under various boundary conditions. Int. J. Fract. Mech. 7: 301–306.
Lei T, Yao ZH, Wang HT, Wang PB (2006) A parallel fast multipole BEM and its applications to large-scale analysis of 3-D fiber-reinforced composites. Acta Mechanica Sinica 22(3): 225–232.
Liu YJ, Nishimura N, Otani Y (2005) Large-scale modeling of carbon-nanotube composites by a fast multipole boundary element method. Comput. Mater. Sci. 34: 173–187.
Nishimura N (2002) Fast multipole accelerated boundary integral equation methods. Appl. Mech. Rev. 55: 299–324.
Odegard GM, Gates TS, Wise KE et al. (2003) Constitutive modeling of nanotube-reinforced polymer composites. Comp. Sci. Technol. 63: 1671–1687.
Peirce AP, Napier JAL (1995) A spectral multipole method for efficient solutions of large scale boundary element models in elastostatics. Int. J. Numer. Meth. Eng. 38: 4009–4034.
Portela A, Aliabadi MH, Rooke DP (1993) Dual boundary element incremental analysis of crack propagation. Comput. Struct. 46: 237–247.
Rokhlin V (1985) Rapid solution of integral equations of classical potential theory. J. Comput. Phys. 60: 187–207.
Telles JCF (1983) The boundary element method applied to inelastic problems. Springer, Berlin.
Wang HT, Yao ZH (2004) Application of a new fast multipole BEM for simulation of 2D elastic solid with large number of inclusions. Acta Mechanica Sinica 20: 613–622.
Wang HT, Yao ZH (2005) A new fast multipole boundary element method for large scale analysis of mechanical properties in 3D particle-reinforced composites. Comput. Model. Eng. Sci. 7: 85–96.
Wang HT, Yao ZH, Wang PB (2005a) On the preconditioners for fast multipole boundary element methods for 2D multi-domain elastostatics. Eng. Anal. Bound. Elem. 29: 673–688.
Wang PB, Yao ZH, Wang HT (2005b) Fast multipole BEM for simulation of 2-D solids containing large numbers of cracks. Tsinghua Sci. Technol. 10: 76–81.
Wang PB, Yao ZH (2006) Fast multipole DBEM analysis of fatigue crack growth. Comput. Mech. 38(3): 223–233.
Wang PB, Yao ZH, Lei T (2006) Analysis of solids with numerous microcracks using fast multipole DBEM. CMC Comput. Mater. Continua 3(2): 65–75.
Wang PB, Yao ZH (2007) Fast multipole boundary element analysis of two-dimensional elastoplastic problems. Commun. Numer. Meth. Eng. 23(10): 889–903.
Warren MS, Salmon JK (1993) A parallel hashed oct-trees N-body algorithm. In: Sigarch (ed) Supercomputing' 93: Proceedings Portland Oregon, US, Nov. 15–19, IEEE Computer Society Press, pp. 12–21.
Yao ZH, Xu JD, Wang HT (2007) Simulation of CNT composites using fast multipole BEM. In: Yao ZH, Yuan MW (ed) Computational Mechanics: Proceedings of the International Symposium on Computational Mechanics, Beijing, China, July 30 –August 1, 2007, Tsinghua University Press & Springer.
Yoshida K, Nishimura N, Kobayashi S (2001) Application of new fast multipole boundary integral equation method to crack problems in 3D. Engrg. Anal. Bound. Elem. 25: 239–247.

Thermomechanical Interfacial Crack Closure: A BEM Approach

Georgios I. Giannopoulos, Loukas K. Keppas and Nick K. Anifantis

Abstract In this paper a sub-regional boundary element formulation is proposed for the treatment of general two-dimensional steady-state and time-dependent thermoelastic crack closure problems considering friction and thermal resistance along the crack faces. These problems are solved by an incremental-iterative scheme since the extent and the status of the contact zone are not known in advance. The assumption of pressure-dependent thermal contact increases the degree of non-linearity and couples the thermal and mechanical fields. The present work is focused on fracture problems situated on the interface of dissimilar isotropic solids under combined mechanical and thermal loads.

1 Introduction

The study of interface cracks between materials with mismatch in their properties, under given combinations of thermal and mechanical loading has recently attracted considerable attention in the design of various composite structures. However, the research on thermoelastic fracture mechanics has been mostly focused on cracks whose faces are traction free and thermally insulated. This fact poses distinct limitations since the majority of thermal fracture problems are associated with crack heating which usually leads to the closure of the crack faces. The frictional contact phenomenon between solids due to thermomechanical loading is a non-linear problem since the extent and the nature of the contact is not known a priori. This implies that the size of the contacting area and the size of the adhesion-slip areas are load-dependent and thus the problem cannot be treated linearly. When imperfect thermal contact is assumed as well, i.e. thermal resistance between the contacting areas, the non-linearity of the problem is increased because the temperature field along the contact zone becomes pressure-dependent. Since the non-linearities are confined to the boundaries of the solids the Boundary Element Method (BEM) is specially appropriate to solve such problems.

N.K. Anifantis (✉)
Mechanical and Aeronautics Engineering Department, University of Patras, GR-26500, Greece
e-mail: nanif@mech.upatras.gr

G.D. Manolis, D. Polyzos (eds.), *Recent Advances in Boundary Element Methods*,
DOI 10.1007/978-1-4020-9710-2_29, © Springer Science+Business Media B.V. 2009

In this study a BEM formulation (Giannopoulos and Anifantis 2005b; Giannopoulos and Anifantis 2007; Keppas et al. 2008; Keppas and Anifantis 2008) is proposed to simulate thermomechanical contact between dissimilar materials by assuming frictional contact and pressure-dependent thermal resistance between the contacting areas. The formulation is based on the boundary integral equations for 2D steady-state and time-dependent thermoelasticity (Brebbia et al. 1984; Balas et al. 1989; Raveendra and Banerjee 1992) implemented in incremental form. At each stage of the incremental procedure, iterative processes are established in order to seek equilibrium state of thermal and mechanical contact conditions. The singularities in the vicinity of crack tip are approached by the use of the quarter-point elements (QPEs) (Martinez and Dominguez 1984; Gao and Tan 1992).

2 Boundary Element Analysis of Thermoelastic Crack Closure

The generalized problem considered in the present paper is depicted in Fig. 1. An interface crack lies between two dissimilar isotropic elastic media. The external boundaries of the two media are mechanically supported and thermomechanically loaded. These boundary conditions result to the partially closure of the crack as Fig. 1 illustrates. The crack faces are characterized by a coefficient of friction μ and a thermal contact resistance R which is a function the contact pressure.

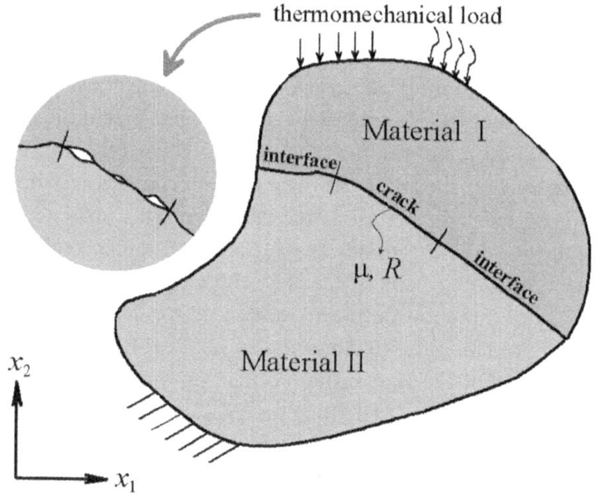

Fig. 1 The generalized problem considered

2.1 *Matrix Form of Equations*

In order to aproach the problem numerically, the boundary of each medium is discretized in a number of elements as Fig. 2 presents. The bimaterial medium is discretized using isoparametric quadratic element (IQE) anywhere except the crack

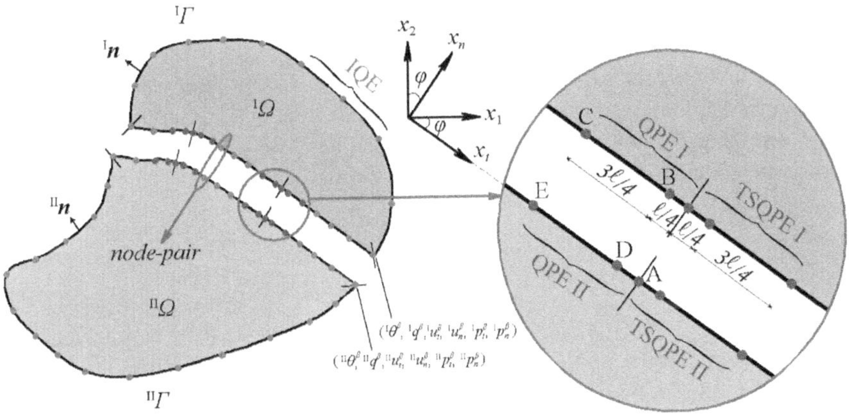

Fig. 2 Discretization of the boundaries

tips where quarter point elements (QPE) and traction singular quarter point element (TSQPE) are used. It has to be noticed here, that the boundaries of the two media which belong to the interfacial zones as well as the crack are coincidentally discretized. In this way, a number of node-pairs is created and the thermomechanical contact conditions can be expressed simply.

After the discretization, the matrix form of the boundary integral equations for each medium can be derived (Brebbia et al. 1984; Balas et al. 1989; Raveendra and Banerjee 1992):

$$\boldsymbol{Q\theta} = \boldsymbol{\Theta q} \tag{1}$$

$$\boldsymbol{Pu} = \boldsymbol{Up} + \overline{\boldsymbol{Q}}\boldsymbol{\theta} - \overline{\boldsymbol{\Theta}}\boldsymbol{q} \tag{2}$$

where the quantities $\boldsymbol{Q}$, $\boldsymbol{\Theta}$, $\boldsymbol{P}$, $\boldsymbol{U}$, $\overline{\boldsymbol{Q}}$, $\overline{\boldsymbol{\Theta}}$ are coefficient matrices of corresponding fundamental solutions while the vectors, $\boldsymbol{\theta}$, $\boldsymbol{q}$, $\boldsymbol{u}$, $\boldsymbol{p}$ represent nodal boundary values of temperatures, heat fluxes, displacements and tractions, respectively.

In order to write the matrix form of BEM equations for time-dependent thermoelasticity the time is additionally discretized in constant time steps. Then, the following expressions are derived (Brebbia et al. 1984; Balas et al. 1989; Raveendra and Banerjee 1992):

$$\boldsymbol{Q}^1\boldsymbol{\theta}^F = \boldsymbol{\Theta}^1\boldsymbol{q}^F + \sum_{f=1}^{F-1}(\boldsymbol{\Theta}^{F+1-f}\boldsymbol{q}^f - \boldsymbol{Q}^{F+1-f}\boldsymbol{\theta}^f) \tag{3}$$

$$\boldsymbol{Pu}^F - \boldsymbol{Up}^F + \sum_{f=1}^{F}(\overline{\boldsymbol{Q}}^{F+1-f}\boldsymbol{\theta}^f - \overline{\boldsymbol{\Theta}}^{F+1-f}\boldsymbol{q}^f) \tag{4}$$

where $f = 1, F$ denotes the time instant.

2.2 Incremental Procedure

To treat the inherent non-linearity of the generalized problem an incremental-iterative procedure is utilized during which the total load is applied discretely. Supposing that the load is stationary, each variable χ which describes the problem is discretized according to the following equation:

$$\chi^{m} = \chi^{m-1} + \Delta\chi^{m} \tag{5}$$

where m denotes the increment. However, if the thermal load is transient then the incrementation is implicitly achieved through the discretization of the time according to the following equation.

$$t^{f} = t^{f-1} + \Delta t^{f} \tag{6}$$

As it was previously mentioned, this discretization is accomplished through the use of constant time steps.

2.3 Assembly of Equations

The conditions due to the thermoechanical interactions between the two media for a specific node-pair and a specific load step $\beta(\beta = m$ for stationary thermal load or $\beta = f$ for transient thermal load) which may be in adhesion state (a) or slip state (s) or open state (o) or just belongs to the interface (i) are summarized in Table 1.

In order to obtain the solvable boundary element equations which describe the thermal and mechanical part of the problem for a specific load step β, Eqs. (1) and (2) or (3) and (4) are written for each medium separately and then assembled according to the themomechanical conditions included in Table 1. Then, the system of algebraic equations that arise can be solved by taking into consideration the known boundary conditions.

Table 1 Thermomechanical conditions for each node-pair

Adhesion	Slip	Open	Interface
${}^{\mathrm{I}}\theta^{\beta} = {}^{\mathrm{II}}\theta^{\beta} - R({}^{\mathrm{I}}p_n^{\beta}){}^{\mathrm{I}}q^{\beta}$	${}^{\mathrm{I}}\theta^{\beta} = {}^{\mathrm{II}}\theta^{\beta} - R({}^{\mathrm{I}}p_n^{\beta}){}^{\mathrm{I}}q^{\beta}$	${}^{\mathrm{I}}q^{\beta} = 0$	${}^{\mathrm{I}}\theta^{\beta} = {}^{\mathrm{II}}\theta^{\beta}$
${}^{\mathrm{I}}q^{\beta} = -{}^{\mathrm{II}}q^{\beta}$	${}^{\mathrm{I}}q^{\beta} = -{}^{\mathrm{II}}q^{\beta}$	${}^{\mathrm{I}}q^{\beta} = -{}^{\mathrm{II}}q^{\beta}$	${}^{\mathrm{I}}q^{\beta} = -{}^{\mathrm{II}}q^{\beta}$
${}^{\mathrm{I}}u_t^{\beta} = {}^{\mathrm{II}}u_t^{\beta}$	${}^{\mathrm{I}}u_n^{\beta} = {}^{\mathrm{II}}u_n^{\beta}$	${}^{\mathrm{I}}p_t^{\beta} = 0$	${}^{\mathrm{I}}u_t^{\beta} = {}^{\mathrm{II}}u_t^{\beta}$
${}^{\mathrm{I}}u_n^{\beta} = {}^{\mathrm{II}}u_n^{\beta}$	${}^{\mathrm{I}}p_t^{\beta} = \pm\mu\,{}^{\mathrm{I}}p_n^{\beta}$	${}^{\mathrm{I}}p_n^{\beta} = 0$	${}^{\mathrm{I}}u_n^{\beta} = {}^{\mathrm{II}}u_n^{\beta}$
${}^{\mathrm{I}}p_t^{\beta} = -{}^{\mathrm{II}}p_t^{\beta}$	${}^{\mathrm{I}}p_t^{\beta} = -{}^{\mathrm{II}}p_t^{\beta}$	${}^{\mathrm{I}}p_t^{\beta} = -{}^{\mathrm{II}}p_t^{\beta}$	${}^{\mathrm{I}}p_t^{\beta} = -{}^{\mathrm{II}}p_t^{\beta}$
${}^{\mathrm{I}}p_n^{\beta} = -{}^{\mathrm{II}}p_n^{\beta}$	${}^{\mathrm{I}}p_n^{\beta} = -{}^{\mathrm{II}}p_n^{\beta}$	${}^{\mathrm{I}}p_n^{\beta} = -{}^{\mathrm{II}}p_n^{\beta}$	${}^{\mathrm{I}}p_n^{\beta} = -{}^{\mathrm{II}}p_n^{\beta}$

2.4 Iterative Algorithm

During the incremental procedure an iterative algorithm takes place between two sequential load steps in order to attain the equilibrium of the thermomechanical contact conditions. Before the treatment of a particular load step β several assumptions are made concerning the thermal and mechanical contact state for each node-pair. After the solution is obtained the assumptions are reconsidered if necessary.

The convergence of a specific load step is controlled via two criterions. The first criterion (convergence of mechanical contact) is expressed through Table 2 while the second criterion (convergence of thermal contact) through the following equation:

$$\left| \frac{R_{last} - R_{prev}}{R_{last}} \right| \times 100 \leq \eta \tag{7}$$

The equations describing the thermal and mechanical part of the problem are solved repeatedly until the contact status does not change and the calculated thermal resistance terms of all adhesion-slip node-pairs of the last iteration R_{last} are approximately equal to the corresponding terms of the previous iteration R_{prev}. The above described iterative algorithm is depicted in Fig. 3.

Table 2 Definition of mechanical contact status

Assumption	Decision	
	Open	Adhesion or slip
Open Adhesion or slip	${}^{\mathrm{II}}u_n^{\beta} - {}^{\mathrm{I}}u_n^{\beta} < 0$ ${}^{\mathrm{II}}p_n^{\beta} \geq 0$	${}^{\mathrm{II}}u_n^{\beta} - {}^{\mathrm{I}}u_n^{\beta} \geq 0$ ${}^{\mathrm{II}}p_n^{\beta} < 0$
	Adhesion	Slip
Adhesion Slip	${}^{\mathrm{I}}p_t^{\beta} < \pm\mu^{\mathrm{I}}p_n^{\beta}$ ${}^{\mathrm{I}}p_t^{\beta}({}^{\mathrm{II}}u_t^{\beta} - {}^{\mathrm{I}}u_t^{\beta}) \geq 0$	${}^{\mathrm{I}}p_t^{\beta} \geq \pm\mu^{\mathrm{I}}p_n^{\beta}$ ${}^{\mathrm{I}}p_t^{\beta}({}^{\mathrm{II}}u_t^{\beta} - {}^{\mathrm{I}}u_t^{\beta}) < 0$

3 Fracture Characterization

The existence of an interfacial crack in a bimaterial justifies the choice of the strain energy release rate (SERR) as fracture assessment parameter. Quarter-point elements (QPEs) are used for the representation of the temperature and displacement field near the crack tip and traction singular quarter-point elements (TSQPEs) are utilized to account for the heat flux and traction field. The SERR G is related to the magnitude K_0 of complex stress intensity factor according to the equation:

$$G = \left(\frac{{}^{\mathrm{I}}\kappa + 1}{8\,{}^{\mathrm{I}}S} + \frac{{}^{\mathrm{II}}\kappa + 1}{8\,{}^{\mathrm{II}}S} \right) K_0^2 \tag{8}$$

where ${}^{\mathrm{i}}\nu$ and ${}^{\mathrm{i}}S$ present the Poisson's ratio and shear modulus, respectively. The index i = I, II denotes the medium. The magnitude K_0 is computed by the equation

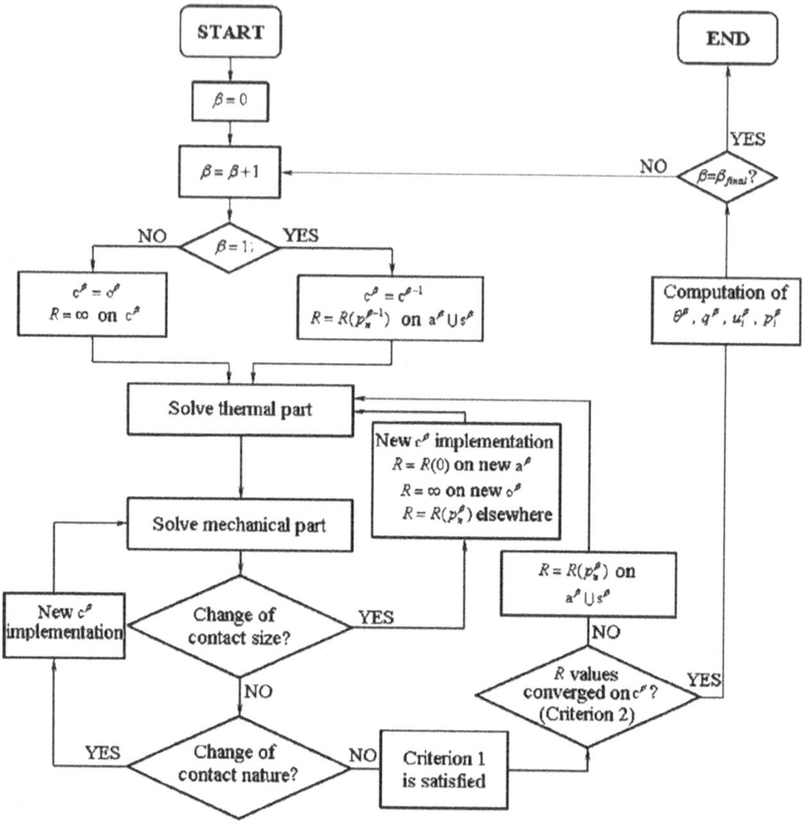

Fig. 3 Iterative algorithm

(Gao and Tan 1992):

$$K_0 = \frac{\sqrt{2\pi\ell}}{\cosh(\pi\varepsilon)}\left[(p_t^{\mathrm{A}})^2 + (p_n^{\mathrm{A}})^2\right]^{1/2} \tag{9}$$

In Eq. (9) p_t^{A} and p_n^{A} refer to the tangential and normal component of traction at the crack tip A and ℓ is the length of the crack tip element. The quantities ζ, ε are defined as $\zeta = ({}^{\mathrm{I}}S + {}^{\mathrm{I}}\kappa\,{}^{\mathrm{II}}S)/({}^{\mathrm{II}}S + {}^{\mathrm{II}}\kappa\,{}^{\mathrm{I}}S)$ and $\varepsilon = 0.5\pi \ln\zeta$ with ${}^{\mathrm{i}}\kappa = 3 - 4\,{}^{\mathrm{i}}\nu$ for plane strain or ${}^{\mathrm{i}}\kappa = (3 - {}^{\mathrm{i}}\nu)/(1 + 4\,{}^{\mathrm{i}}\nu)$ for plane stress. When the mechanical properties of the two media are the same then the fracture characteristics are computed via the equation (Martinez and Dominguez 1984):

$$\begin{Bmatrix} K_{\mathrm{I}} \\ K_{\mathrm{II}} \end{Bmatrix} = \sqrt{2\pi\ell}\begin{bmatrix} 0 & 1 \\ 1 & 0 \end{bmatrix}\begin{Bmatrix} p_t^{\mathrm{A}} \\ p_n^{\mathrm{A}} \end{Bmatrix} \tag{10}$$

where K_I and K_{II} are the opening and shearing stress intensity factors, respectively.

4 Numerical Results

4.1 Steady-State Fracture

Consider the plane strain problem of the bimaterial of Fig. 4a. The bimaterial is composed of two dissimilar isotropic semi-infinite planes I, II. The material properties are: Young's modules ${}^{I}E = {}^{II}E = 2.1 \times 10^{11}$ Pa, Poisson's ratios ${}^{I}\nu = {}^{II}\nu = 0.3$. A crack of length $2a$ is assumed on the interface. Under the influence of a uniform tension σ_0 and uniform heat flow q_0 applied to the plane at infinity and directed to the material with the lower thermal distortion d ($d = a(1+\nu)/k$ where a is the thermal expansion ratio and k is the thermal contactivity) the crack may partially close. The boundary element model utilized is depicted in Fig. 4b.

In order to approach the infinite problem, the aspect ratios $L/a = 10$ and $H/L = 1$ were selected. After convergence tests it was proved that for aspect ratios higher than those, virtually the same solutions were obtained. For all the following computations the length of the QPEs was taken equal to $\ell = a/40$. Appropriate convergence tests demonstrated that the Eqs. (9) and (10) lead to at most 2% different stress intensity factors by using $\ell = a/400$ instead of $\ell = a/40$.

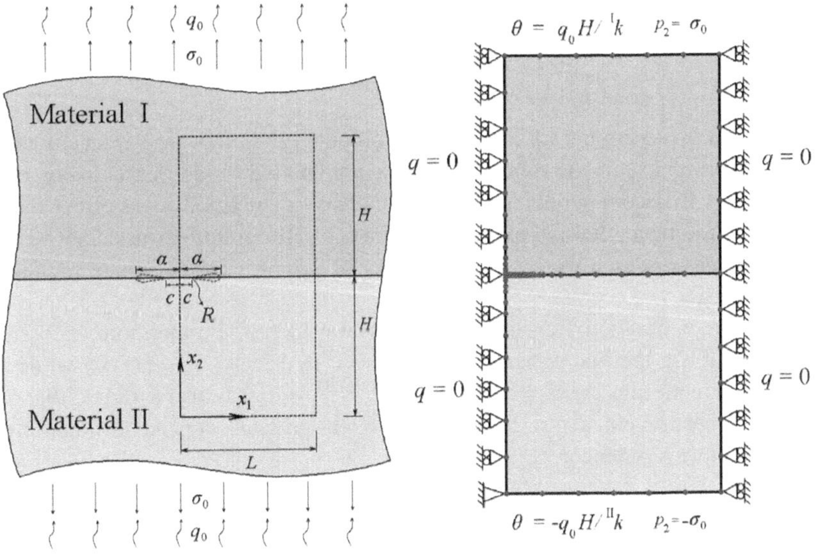

(a) (b)

Fig. 4 Problem of a bimaterial with an interfacial crack: (**a**) Geometry and (**b**) utilized boundary element model

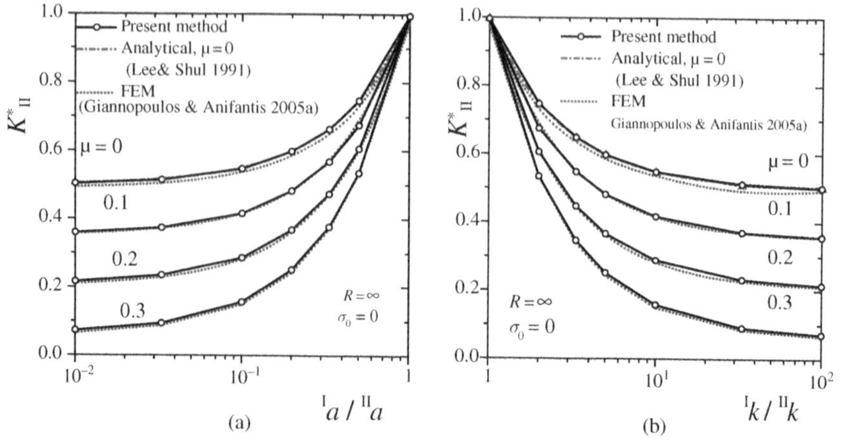

Fig. 5 Non-dimensional shearing stress intensity factor versus (**a**) thermal expansion coefficient ratio and (**b**) thermal conductivity ratio

Firstly, it is assumed that the crack faces are insulated, that the tensile stress is zero and that the thermal expansion ratio is ${}^{\mathrm{I}}a/{}^{\mathrm{II}}a < 1$ and thermal conductivity ratio ${}^{\mathrm{I}}k/{}^{\mathrm{II}}k > 1$. For the specific material ratios the crack remains fully closed (K_{I}=0). The results are presented in terms of the non-dimensional shearing stress intensity factor which is defined as:

$$K_{\mathrm{II}}^{*} = K_{\mathrm{II}}\frac{{}^{\mathrm{II}}k(1+{}^{\mathrm{II}}\kappa)}{2{}^{\mathrm{II}}a{}^{\mathrm{II}}S(1+{}^{\mathrm{II}}\nu)\sqrt{\pi}q_{0}a^{3/2}} \tag{11}$$

Figure 5a, b present the variation of the computed shearing stress intensity factor with the thermal expansion ratio and thermal conductivity ratio, respectively. From these figures it becomes obvious that the increase of μ has a great effect on K_{II} mostly for significant thermal property mismatches. The solution was achieved after several iterations which were needed in order to determine the constant regions of slip and stick.

Secondly, a non-zero tensional stress is considered. Furthermore, now it is assumed that the thermal expansions are ${}^{\mathrm{I}}a = {}^{\mathrm{II}}a/2.25 = 10^{-5}(1/{}^{\circ}\mathrm{C})$ while the thermal conductivities are ${}^{\mathrm{I}}k = {}^{\mathrm{II}}k/1.5 = 40\,(\mathrm{W/m^{\circ}C})$. It is also assumed that now the crack faces are characterized by two different pressure-dependent functions of thermal contact resistance:

$$R_a(p_n) = 0.02 \times \exp(10^{-8}\, p_n) \tag{12}$$

$$R_b(p_n) = 0.005 \times \exp(10^{-8}\, p_n) \tag{13}$$

The following dimensionless quantities are used for the results presented in the following:

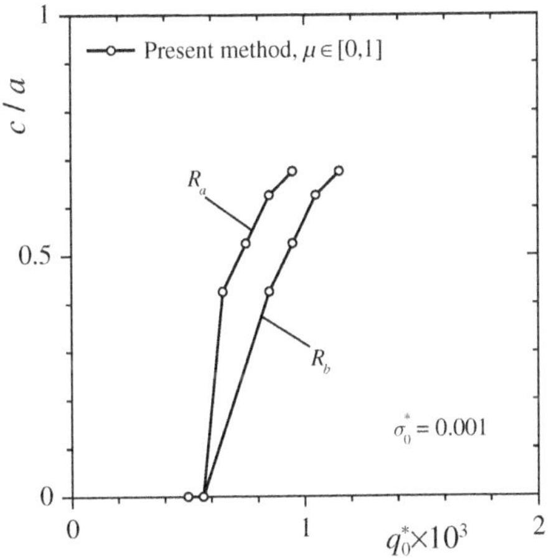

Fig. 6 Non-dimensional contact length versus non-dimensional heat flux

$$q_0^* = q_0\, a({}^{\mathrm{II}}d - {}^{\mathrm{I}}d); \quad \sigma_0^* = \frac{\sigma_0(1+{}^{\mathrm{I}}\kappa)}{{}^{\mathrm{I}}S}; \quad K_{\mathrm{I}}^* = \frac{K_{\mathrm{I}}(1+{}^{\mathrm{I}}\kappa)}{{}^{\mathrm{I}}S\sqrt{a}}; \quad K_{\mathrm{II}}^* = \frac{K_{\mathrm{II}}(1+{}^{\mathrm{I}}\kappa)}{{}^{\mathrm{I}}S\sqrt{a}} \tag{14}$$

Figures 6, 7a, b present the contact length, the opening stress intensity factor and the shearing stress intensity factor versus the uniform heat flux, respectively.

The effect of coefficient of friction on both contact length and opening stress intensity factor is negligible. Lower values of thermal resistance lead to the decrease of contact length and thus to the increase of the opening stress intensity factor. The shearing stress intensity factor becomes importantly non-linear with the assumption of pressure-dependent thermal resistance, especially when high coefficients of friction are considered. Its magnitude is slightly reduced when the coefficient of friction

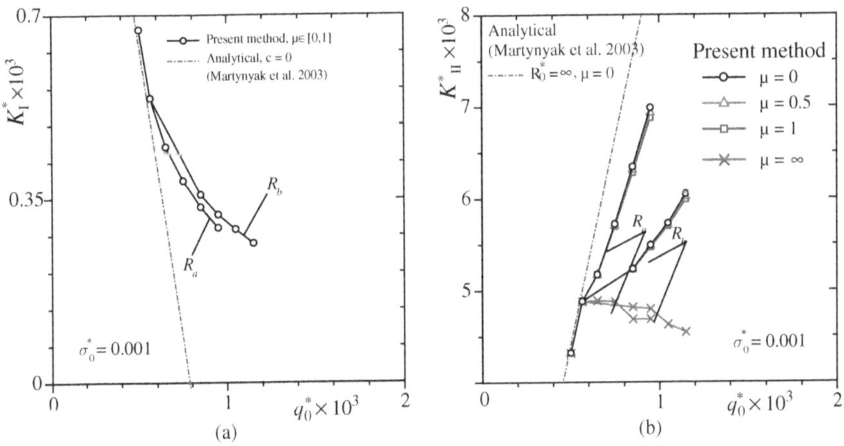

Fig. 7 Variation of (**a**) non-dimensional opening stress intensity factor and (**b**) non-dimensional shearing stress intensity factor with respect to the non-dimensional heat flux

is increased because of the previously explained reasons. The assumption of rough crack faces ($\mu = \infty$) lead to a considerable drop of shearing stress intensity factor. In all cases 10 successive increments were used for the computations even for the cases in which $\mu = 0$ due to the existence of pressure-dependent thermal resistance. Finally, the value $\eta = 0.01$ was utilized in the convergence criterion described by Eq. (7).

4.2 Time-Dependent Fracture

The developed boundary element formulation is adopted to study the problem of a fractured thermal barrier coating (TBC) system as shown in Fig. 8a. There are two subdomains, a coating layer of thickness ${}^{C}H$ and a substrate of thickness ${}^{S}H = 10{}^{C}H$ with an interfacial crack of length $2a = 2({}^{S}H/10)$ between the two layers. Nodular cast iron is used as substrate and the coating material is Zirconia with properties provided in (Kokini and Reynolds 1991).

The free surface of the coating experiences a thermal shock by the ambient temperature $\theta_\alpha \ll \theta_0$, with θ_0 being the initial uniform temperature, considering a coefficient of heat convection $h = 50\ \mathrm{W/m^{2}\,{}^\circ C}$. During the thermal shock a constant pressure p_2 acts on the surface of the coating which may be combined with a constant shear stress p_1. Plane strain conditions are assumed. Figure 8b shows the utilized boundary element model. The following functions of thermal contact resistance are assumed:

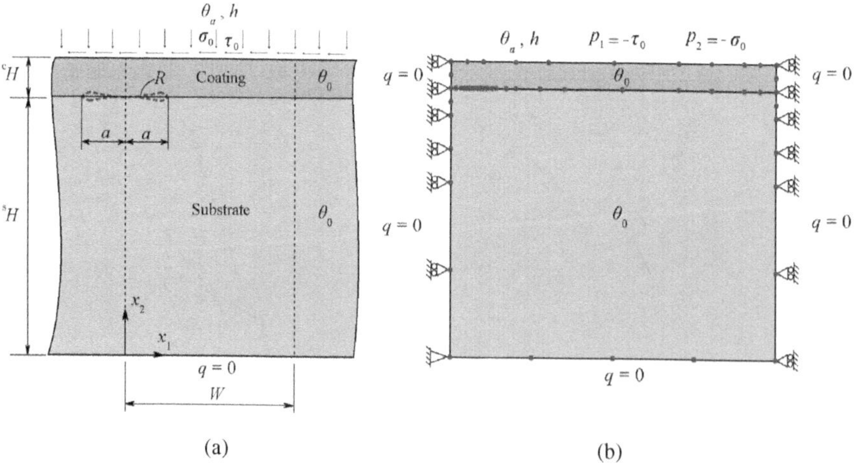

Fig. 8 Problem of a TBC system with an interfacial crack: (**a**) Geometry and (**b**) utilized boundary element model

$$R_0 = 0;\ R_1(p_n) = 0.0002 \times \exp(10^{-8}\ p_n);$$

$$R_2(p_n) = 0.002 \times \exp(10^{-8}\ p_n); \quad R_3(p_n) = 0.02 \times \exp(10^{-8}\ p_n) \tag{15}$$

The following non-dimensional SERR and time are used for the presentation of results:

$$G^* = \frac{G}{{}^{C}E({}^{C}a)^2(\theta_a - \theta_0)^{2S}H};\ t^* = \frac{{}^{C}kt}{{}^{C}\rho^{C}C({}^{S}H)^2} \tag{16}$$

where ${}^{C}\rho$ and ${}^{C}C$ is the density and the specific heat of the coating, respectively.

The evolution of non-dimensional SERR G^* is presented in Fig. 9a for several levels of thermal contact resistance. The graph refers to pure thermal shock without the presence of any mechanical load on the coating surface and frictionless contact ($\mu = 0$). The results of the present analysis are compared to FEM results of (Kokini and Reynolds 1991) which correspond to frictionless and insulated contact. Apparently, the consideration of pressure dependent thermal contact resistance has a noticeable impact on the predicted G^*. The extreme cases of thermal insulation and perfect thermal contact give a maximum difference in G^* about one order of magnitude. Low thermal contact resistance values reduce the jump in temperature at the interface and the stress distribution derives lower strain energy release rates.

In Fig. 9b, G^* is plotted against the time for several combinations of coefficient of friction and thermal contact resistance. The friction coefficient has not a remarkable influence on G^* because the normal contact stresses has relatively low values resulting to low tangential frictional stresses (Coulomb's law).

However, when the thermal shock is combined with constant pressure $p_2 = \sigma_0$ the crack closure is better pronounced. On the contrary to the previous case the effect

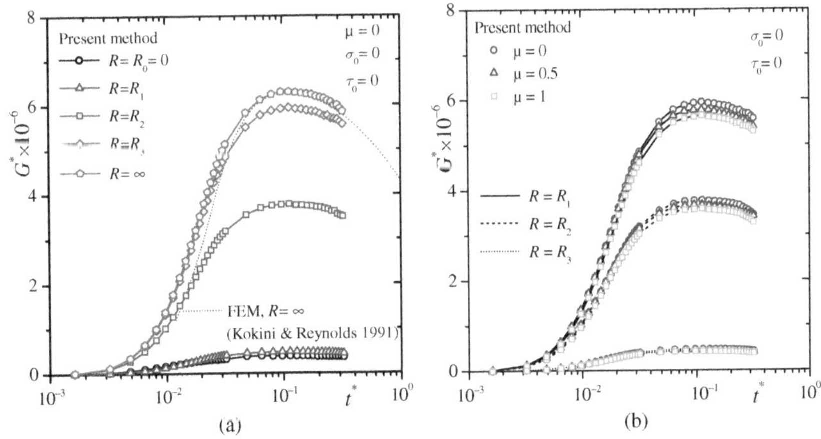

Fig. 9 Non-dimensional SERR versus non-dimensional time for several (**a**) functions of thermal contact resistance and (**b**) coefficients of friction

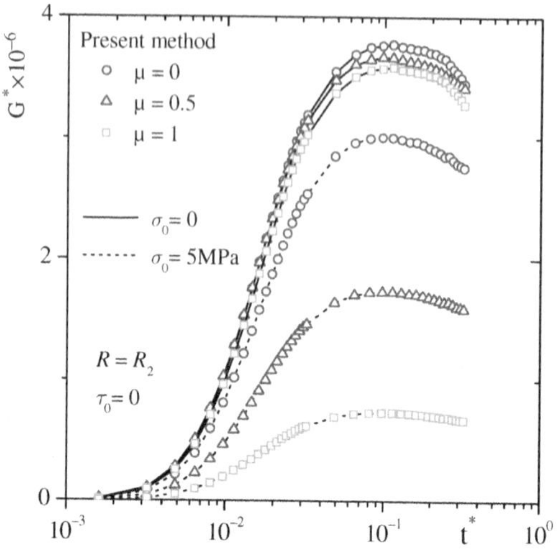

Fig. 10 Non-dimensional SERR versus non-dimensional time for combined thermal and mechanical load

of coefficient of friction is not negligible here. The results are plotted in Fig. 10 for an intermediate resistance $R = R_2$. The diagram indicates that the assumption for frictionless contact can be justified only for pure thermal loading. Higher values of μ result to lower G^* when a pressure load is applied during a thermal shock.

The fracture mode II is predominant for the cases under consideration where partial or approximately full crack closure occurs. Therefore, a reduction of the peak value of SERR G_p^* is expected when the load σ_0 and the coefficient of friction are increased causing restriction of sliding between crack faces. This is evident in

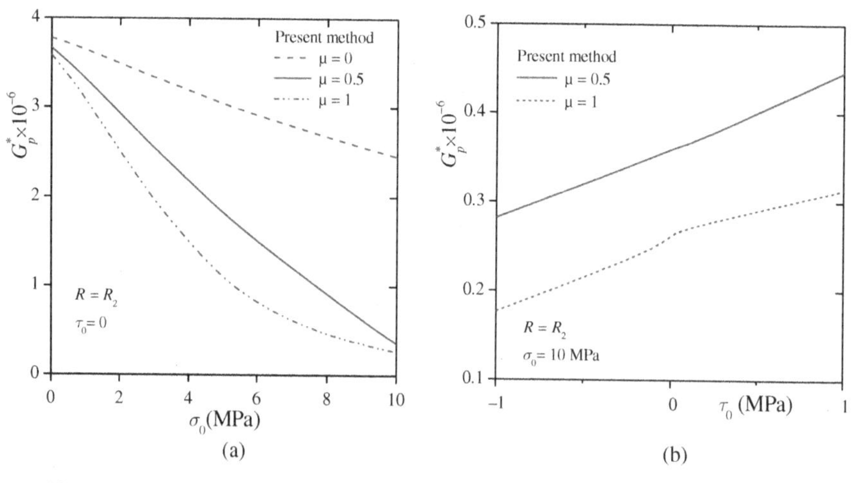

Fig. 11 Non-dimensional peak values of SERR versus (**a**) pressure and (**b**) shear stress

Fig. 11a for $R = R_2$. The relation between G_p^* and σ_0 is linear for frictionless contact and exhibits non-linear behavior mostly when $\mu = 1$. In addition, the slope of $G_p^* = G_p^*(\sigma_0)$ for $\mu = 1$ is very intense initially and it is reduced at $\sigma_0 = 5\,\mathrm{MPa}$. In some cases the components protected by thermal barrier coatings may withstand a combined action of thermal shock, pressure and shear stress. This case was examined in the present analysis where frictional contact and $\sigma_0 = 10\,\mathrm{MPa}$ was assumed and the magnitude of shear stress τ_0 takes values between $-1/10$ and 1/10 of σ_0. The results are provided in Fig. 11b. For both coefficients of friction G_p^* exhibits a linear variation with the magnitude of σ_0 and it is reduced considerably when σ_0 directs towards the fractured edge of TBC (see Fig. 8).

5 Conclusions

A boundary element formulation capable of treating steady-state and time-dependent, coupled thermoelastic crack closure problems, where pressure-dependent thermal resistance between contacting faces is involved, was presented. The approach is based on sub-regional technique and an incremental-iterative procedure. Apparently, it can be concluded that (i) the increase of the contact area along the crack leads to a significant decrease of opening stress intensity factor, (ii) the increase of the coefficient of friction leads to the decrease of shearing stress intensity factor and (iii) the increase of the thermal contact resistance leads to a considerable increase of the peak values of SERR. Therefore, the numerical implementation of crack closure phenomenon is of high importance. The present boundary element formulation is proved to be efficient to treat this class of problems

References

Balas J, Sladek J, Sladek V (1989) Stress Analysis by Boundary Element Methods. Elsevier, Amsterdam

Brebbia CA, Telles JCF, Wrobel LC (1984) Boundary Element Techniques. Springer, Berlin

Gao YL, Tan CL (1992) Determination of characterizing parameters for bimaterial interface cracks using the boundary element method. Eng. Fract. Mech. 41: 779–784

Giannopoulos GI, Anifantis NK (2005a) Finite element analysis of crack closure in two-dimensional bodies subjected to heating. Comput. Struct. 83: 303–314

Giannopoulos GI, Anifantis NK (2005b) Interfacial steady-state and transient thermal fracture of dissimilar media using the boundary element contact analysis. Int. J. Num. Meth. Eng. 62: 1399–1420

Giannopoulos GI, Anifantis NK (2007) A BEM analysis for thermomechanical closure of interfacial cracks incorporating friction and thermal resistance. Comput. Meth. Appl. Mech. Eng. 196: 1018–1029

Keppas LK, Anifantis NK (2008) BEM prediction of TBC fracture resistance. Proceedings of the 1th International Conference of Engineering Against Fracture, 28–30 May, University of Patras, Greece

Keppas LK, Giannopoulos GI, Anifantis NK (2008) Transient coupled thermoelastic contact problems incorporating thermal resistance: A BEM approach. CMES–Comput. Model. Eng. Sci. 25: 181–196

Kokini K, Reynolds RR (1991) Transient heating vs cooling of interfacial cracks in ceramic-to-metal bonds. Eng. Fract. Mech. 38: 371–383
Lee KY, Shul CW (1991) Determination of thermal stress intensity factors for an interface crack under vertical uniform heat flow. Eng. Fract. Mech.40: 1067–1074
Martinez J, Dominguez J (1984) On the use of quarter-point boundary elements for stress intensity factor computations. Int. J. Num. Meth. Eng. 20: 1941–1950
Martynyak RM, Honchar KhI, Nahalka SP (2003) Simulation of thermomechanical closure of an initially open interface crack with heat resistance. Mater. Sci. 39: 672–681
Raveendra ST, Banerjee PK (1992) Boundary element analysis of cracks in thermally stressed planar structures. Int. J. Solids Struct. 29: 2301–2317

Author Index

Subject Index

GPSR Compliance

The European Union's (EU) General Product Safety Regulation (GPSR) is a set of rules that requires consumer products to be safe and our obligations to ensure this.

If you have any concerns about our products, you can contact us on ProductSafety@springernature.com

In case Publisher is established outside the EU, the EU authorized representative is:

Springer Nature Customer Service Center GmbH
Europaplatz 3
69115 Heidelberg, Germany

Batch number: 10761599

Printed by Printforce, the Netherlands